CAP-NSERC Summer Institute in Theoretical Physics, Edmonton, Alberta, 10-25 July 1987

CAP-NSERC SUMMER INSTITUTE IN
THEORETICAL PHYSICS
EDMONTON, ALBERTA,
10–25 JULY, 1987

CAP-NSERC SUMMER INSTITUTE IN THEORETICAL PHYSICS EDMONTON, ALBERTA, 10–25 JULY, 1987

Vol. II

Field Theory in Two Dimensions and Related Topics.

Editors

G. Kunstatter

H.C. Lee

F.C. Khanna

H. Umezawa

Published by

World Scientific Publishing Co. Pte. Ltd.
P.O. Box 128, Farrer Road, Singapore 9128

U. S. A. office: World Scientific Publishing Co., Inc.
687 Hartwell Street, Teaneck NJ 07666, USA

Library of Congress Cataloging-in-Publication Data

CAP-NSERC Summer Institute in Theoretical Physics
(1987 : Edmonton, Alta)
CAP-NSERC Summer Institute in Theoretical Physics,
Edmonton, Alberta, 10–25 July 1987.

1. Quantum field theory –– Congresses. 2. Mathematical
physics –– Congresses. I. Title.
QC174.45.A1C36 1987 530.1'43 88–113
ISBN 9971-50-433-2

Printed in Singapore by General Printing Services Pte. Ltd.

Preface

In June, 1985 The CAP/NSERC Advisory Council for Summer Institutes received two different proposals for workshops in Quantum Field Theory. It was soon realized that an unique opportunity existed to bring together many young physicists from all over Canada, with expertise in a variety of disciplines, to discuss recent developments in Quantum Field Theory. The result was a joint Summer Institute on Quantum Field Theory that was held at Edmonton in July, 1987. The Institute was composed of two workshops that were run concurrently. These two volumes assemble reports of participants of the Institute. The reports of the workshop on "Quantum Field Theory as an Interdisciplinary Basis" are presented in Vol. I while Vol. II contains contributions to the workshop on "Field Theory in Two Dimensions".

The Institute started with a two day common symposium on a general survey of the subjects of both workshops. This was followed by twelve days of relaxed, intimate but intense communications among the participants. Owing to the high degree of overlap between the topics, the decision as to which workshop was best suited to individual contributions was often made somewhat arbitrarily. We feel that the Institute fulfilled its purpose and we owe this to all the participants who contributed in formal and informal discussions.

It is frequently stated that advancement in modern physics enhanced the specialization of physicists up to the point that scientific communication and collaboration among physicists in different disciplines was extremely difficult. However, in the past two decades a new situation developed that broke this unhappy pattern. Nowadays, physicists from many different disciplines (such as high energy particle physics, cosmology, condensed matter physics, non-linear mathematics) share many common intuitions and common methods. Some examples are: symmetry breakdown and ordered states, long range correlations, topological objects, conformal invariance, critical phenomena and renormalization groups, condensation and coherent representation.

Our strong desire to take advantage of the new situation and further enhance the interaction among physicists in different disciplines motivated us to organize a workshop on interdisciplinary subjects in physics. Since all common notions mentioned above are associated with many body effects, their theoretical basis may be well-supplied by quantum field theory. Therefore, we chose "Quantum Field Theory as an Interdisciplinary Basis" for the subject of one of the workshop.

The subject of the second workshop "Field Theory in Two Dimensions" was chosen because it has recently revolutionized our understanding of two dimensional systems at critical point on the one hand, and on the other it is the basis for string and superstring theories, which at this moment is the only viable candidate for a consistent unified quantum description of all the interactions, including gravity. In selecting talks for the workshop, we have also kept in mind the overall goal for interdisciplinary communication, and believe the resulting diversity in topics serves that goal well.

We would like to express our gratitude to the international advisors who expressed their thoughts to us so that we could pick the right combination of topics. The initial financial support in 1985 from Gordin Kaplan, Vice-President (Research), was greatly instrumental in our pursuing vigorously to hold this workshop in Edmonton. We wish to express our sincere appreciation to Natural Sciences and Engineering Research Council (NSERC) for a generous support. Additional support from Central Research Funds and the Conference Fund at University of Alberta, from Institute of Particle Physics, University of Toronto, University of Winnipeg, Atomic Energy of Canada Ltd. and from TRIUMF is gratefully acknowledged. A workshop of this size cannot succeed without a generous support of numerous local organisers. The chairman of the Physics Department, Dr. G.L. Cumming, provided a great deal of the infrastructure support and a free access to office space and facilities in the Avadh Bhatia Physics Laboratory. The local organising committee, A.Z. Capri, H. Schiff, Y. Takahashi, and S.B. Woods provided valuable support. The local graduate students gave assistance whenever asked for. Finally a task of this magnitude requires some real dedicated workers. Our sincere thanks go to Mrs. M. Yiu, Mrs. P. Anderson and Mrs. L. Chandler who handled all the details admirably and put in an effort well-beyond the call of duty.

F.C. Khanna

G. Kunstatter

H.C. Lee

H. Umezawa

December 15, 1987

TABLE OF CONTENTS

CAP-NSERC SUMMER INSTITUTE IN THEORETICAL PHYSICS EDMONTON, ALBERTA, 10–25 JULY, 1987

Recent Developments in the Light-Cone Gauge Approach to Superstrings

J.G. Taylor *
International Centre for Theoretical Physics
Miramare, Trieste, 34100 Italy.

ABSTRACT

The question of stabilising the vacuum and closing the supersymmetry algebra is answered by presenting an analysis of the closure of the algebra using equal-time commutation relations for the second-quantized closed superstring field operators. A new term, quartic in the closed fields, is presented both for the super-charges and the Hamiltonian. The effect of the latter term on multi-loop string amplitudes is briefly analysed for type II and heterotic superstrings.

1. INTRODUCTION

String field theory has recently become of considerable interest due to the need to distinguish between the plethora of different theories [1] constructed in the last year or so. It is necessary to use both experimental data and theoretical ingenuity to distinguish between the host of possible models of quantum gravity now available. Part of the requisite ingenuity may be by way of non-perturbative analyses (tunneling effects through instantons, non-trivial vacua, etc.) In order to see if such non-perturbative effects are helpful a non-perturbative approach to strings is essential. Various avenues for this have been put forward, in particular the elegant formulation in terms of projective vector bundles over universal moduli space [2] and its related approach through Grassmannian [3] or loop space [4] techniques.

---

* On leave of absence from Department of Mathematics, King's College, Strand, London WC2R 2LS, United Kingdom.

The above developments have not yet led to any clear formulation of a covariant field theory with interacting strings for the closed string case; even the beautiful approach of Witten [5] for open strings has proved difficult to extend to closed strings [6]. In order to make progress in this quest for a covariant closed string field theory it may be helpful to return to the lightcone (L.C.) gauge approach. There is already a complete L.C. field theory of closed bosonic strings [7] which was depended upon as a guide in the construction of a covariant string field theory [8]. Moreover this LC field theory was used recently to give a proof [9] of the unitarity of the covariant first quantised bosonic string theory. It would therefore seem useful to extend the LC bosonic string field theory to the superstring case.

Such an extension has already been given by Green and Schwarz [10] where a purely cubic interaction term was proposed, the total LC Hamiltonian for the closed superstring case having the schematic form

$$H(\Psi) = \int [\Psi K \Psi + g V \Psi^3]$$

where K is the kinetic term, V the interaction δ-function sewing the three strings together, g the string coupling constant and Ψ the closed superstring LC field (depending on transverse bosonic and fermionic coordinates). But SUSY implies that H is positive definite, which cannot be valid for $H(\Psi)$ above due to the cubic term. This was pointed out by Greensite and Klinkhamer, both for the open and closed superstring [11]. The latter remarked on the need for at least a quartic addition to $H(\psi)$ of contact type (in addition to the zipper diagram in the open case). They have suggested a form for this quartic term, but not discussed the closure of the super-Poincaré algebra beyond hinting at possible non-polynomiality in the complete $H(\Psi)$. After the present work [12] was reported on at the meeting, a further paper [13] was received by the authors giving more details of their proposed quartic interaction term; this will be discussed later.

The present account is to present preliminary results of our analysis [12] of the above situation, both for the open and closed cases. In particular, we give a suggested form of new quartic terms in both the Hamiltonian and supersymmetry charges for the closed superstring, and also discuss the possibility as to whether or not there are further terms in the closed case. This latter is important for the non-perturbative approach to compactification since it would seem possible that closure of the SUSY algebra could lead to a non-polynomial theory in the string fields and so make non-perturbative analysis very difficult. We also present an analysis of the manner in which these new terms may modify multi-loop amplitudes [14].

2. SUPERSTRINGS AT CUBIC ORDER

We use the notation of Ref. 10, so the $SO(8) \rightarrow SU(4) \times U(1)$ description of the Grassmann valued string variables S_α is (for type II) $S_\alpha \rightarrow (\lambda^A, \theta_A), (\tilde{\lambda}^A, \tilde{\theta}_A)$, where $\lambda, \tilde{\lambda}$ are left and right moving momentum-type and coordinate type variables with $A = 1$ to 4, λ^A transforming as $\bar{4}$, θ_A as 4 of $SU(4)$. In the Wick-rotated space-time with $\rho = \tau + i\sigma$, $t = -i\tau$ denoting the real time, the free first quantized lagrangian is

$$L_0 = \lambda \partial_{\bar{\rho}} \theta + \tilde{\lambda} \partial_\rho \tilde{\theta}$$

The generators $(J_{\mu\nu}, Q_i^A, Q_i^{\bar{A}}, q_i^{-A}, q_i^{-\bar{A}}, i = 1,2)$ may be written down explicitly [10], such as

$$J_{\mu\nu} = M_{\mu\nu} + S_{\mu\nu}$$

$$S_{ij} = -\frac{1}{2} i \int (\lambda \rho_{ij} \theta + \tilde{\lambda} \rho_{ij} \tilde{\theta}) d\sigma$$

$$q_1^{-A} = \int (\sqrt{2}\, \rho^i \partial_\rho X^i \theta^A + 2\pi \partial_\rho X^L \lambda^A) d\sigma$$

and the other quantities are constructed similarly. Then the second quantized generators of the super-Poincaré algebra are written as

$$(\hat{J}_2, \hat{Q}_2) = \int \Psi^+ (J_{\mu\nu}, Q^{A,\bar{A}}, q^{-A,-\bar{A}}) \Psi(\underline{X}, \theta)$$

where $\underline{X}$ denotes the transverse bosonic modes and the suffix 2 denotes the number of string fields involved (where the type I superstring has supercharges generated from $q_1 + q_2$, $Q_1 + Q_2$).

Interaction is introduced as is usual for LC bosonic string field theory [7] by a cubic interaction which has a δ-function along the three fusing strings 1,2,3 which fuses, say 1 and 2 with 3. This latter, in the closed case, must therefore be self-interacting at the assumed common junction or interaction point P of 1 and 2. In order to preserve the super-Poincaré group (SP) it is necessary to include insertion factors. For generators of SP outside the light-cone set (the only ones non-linearly realized) the cubic terms, for the open string denoted by the field Φ, are

$$\hat{Q}_3^{-\bar{A}} = c\int \theta^{\bar{A}}(P')\ \Delta(1,2,3)(\Phi^+\Phi\Phi + \text{h.conj})$$

with $\Delta(1,2,3)$ being the three-string functional δ-function for the string fusion. For closed superstrings the above is modified by the Hamiltonian insertion factor considered below for the right moving mode (and $L \leftrightarrow R$ for $\hat{\tilde{Q}}^{-\bar{A}}$). The constant c is rather important due to the singular nature of the operator $\theta^{\bar{A}}$ at the interaction point P; P' is chosen to be close to P on one or other of the strings with $c \propto (P-P')^{1/2}$ (since θ transforms with conformal weight 1). A similar expression for $\hat{Q}_3^{-A}$ is available, with $c\theta$ replaced by $(c\theta)^3$, so leading [10], through

$$[\hat{Q}_2^{-(A},\ \hat{Q}_3^{-\bar{B})}]_+ = \delta^{A\bar{B}}\hat{H}_3$$

to an expression for $\hat{H}_3$ for open strings similar to that for $\hat{Q}_3^{-\bar{A},A}$ but now with the insertion factor

$$V = c[2^{-1/2}\partial_{\bar{\rho}}X^L - \partial_{\bar{\rho}}X^i c^2\theta\rho^i\theta + (\sqrt{2}/3)\partial_{\bar{\rho}}X^R c^4\theta^4]$$

where $X^{L,R} = X^{7\pm i8}$ and $1 \leq i \leq 6$. For closed superstrings there is such an insertion factor for both L and R modes. These results were proved by using a detailed mode analysis in Ref. 10, but can also be shown valid by conformal operator techniques with

$$\overline{\lambda^{A}(z)\theta^{\bar{B}}}(w) \sim \delta^{A\bar{B}}(z-w)^{-1}, \quad \overline{\partial_z X^I \partial_w X^J} \sim \delta^{IJ}(z-w)^{-2}$$

In order to justify that the SUSY algebra is closed it is necessary to show, for example, that

$$[\hat{Q}^{-\bar{A}}, \hat{Q}^{-\bar{B}}]_+ = 0, \; [\hat{Q}^{-(A}, \hat{Q}^{-\bar{B})}] \propto \delta^{A\bar{B}}$$

If

$$\hat{Q}^{-\bar{A}} = \hat{Q}_2^{-\bar{A}} + \hat{Q}_3^{-\bar{A}}$$

this requires both

$$[\hat{Q}_2^{-(\bar{A}}, \hat{Q}_3^{-\bar{B})}]_+ = 0$$

and

$$[\hat{Q}_3^{-(\bar{A}}, \hat{Q}_3^{-\bar{B})}]_+ = 0 \tag{1}$$

The former of these equations was proven true in Ref. 10. We turn to consider the validity of the latter in the next section.

3. SUPERSTRINGS AT QUARTIC ORDER

It is possible to calculate the last expression $[\hat{Q}_3^{-(\bar{A}}, \hat{Q}_3^{-\bar{B})}]_+$, and similar expressions involving pairs of non-linearly realised generators of SP at cubic order, by making direct use of the equal time CCR's for the fields Φ or Ψ:

$$[\Psi[1],\Psi[2]] = \frac{1}{p_2^+}\,\delta(p_1^+ + p_2^+)\int_0^{\pi p_2^+} \frac{d\sigma_0}{\pi |p_2^+|}\,\Delta^{16}[z_1(\sigma) - z_2(\sigma+\sigma_0)] \tag{2}$$

where z denotes the $\underline{X}$ and $\theta, \tilde{\theta}$ variables of a particular string, and the integral removes the origin dependence. The results of these calculations [12] show that, in particular, $[\hat{Q}_3^{-(\bar{A}}, \hat{Q}_3^{-\bar{B})}]_+$ and $[\hat{Q}_3^{-(A}, \hat{Q}_3^{-\bar{B})}]_+$ have no contribution from string configurations other than for coinciding interaction points. This may be seen by noting that the integration over the position σ_0 of the origin in Eq.(2) corresponds to spinning the intermediate string in the anticommutator of Eq.(1). This leads to cancellation due to an anticommutator of the insertion factors

$$[\theta^{\bar{A}}(P_1), \theta^{\bar{B}}(P_2)]_+ = 0 \tag{3}$$

at the two interaction points, following the arguments of the fourth paper in Ref. 8. If matrix elements of the l.h.s. of (1) are taken, this cancellation corresponds to integration over the equal but opposite contours denoted by P,Q,R, in that reference, in one of the external Koba-Nielsen source variables, z. Such integration is the generalisation to the closed string of the well-known z-plane discussion [7),15)] for the open bosonic string case.

The cancellation associated with the anticommutator (3) must be handled with care when P_1 approaches P_2, since the constant c entering the insertion factor becomes infinite c behaves like $(F'')^{-1/2}$, where F is the Mandelstam mapping from the z-plane to the LC diagram; as $P_1 \sim P_2$ then $F'' \sim 0$. Thus it is necessary to regularise the anticommutator with great care. This can be done by one of two methods: (a) by considering (1) only at equal times for the two $\hat{Q}_3$'s (which will be denoted by equal time regularisation, or ETR for short) or (b) by taking one of the $\hat{Q}_3$'s at a slightly different time T from the other in (1) and letting $T \to 0$ at the end of the calculation (which will be denoted by constant time regularisation, or CTR for short). There is a third method of regularisation, used by Greensite and Klinkhamer [13)]. This corresponds to removing a small disc in the z-plane around the value corresponding to $P_1 = P_2$ (coincidence of the interacting points). This appears to overcount, by a factor of 2, the results obtained from CTR as can be seen in specific cases; since this third method appears to require contributions in (1) at differing times it is difficult to judge its validity. Since, as we have noted, it would appear that this third method may give results differing only by a factor from CTR, only ETR or CTR will be considered from now on. The two regularisation schemes, ETR and CTR, seem to lead to very different forms for $\hat{Q}_4^{-\bar{A},A}$ and higher order terms $\hat{Q}_n^{-\bar{A},A}$ $(n > 4)$. We will consider in the rest of the section only the $n=4$ case, however, and delay till the next two sections any discussion of terms with $n > 4$.

In ETR, integration over the variable σ_0 in (2) corresponds to moving the intersection point P_2 round the intermediate string. This leads to the anticommutator (3), which is zero except for $P_1 = P_2$, as was remarked on earlier. In ETR a small segment of angular length ε is removed from the integration range about the position $P_1 = P_2$, and then $\varepsilon \to 0$ at the end of the calculation. Thus annihilation of the anticommutator (1) by (3) occurs for each $\varepsilon \neq 0$, and hence leads to zero. Such cancellation of (1) in ETR does not arise, however, in the special configuration when there is no intermediate string at all, and $P_1 \equiv P_2$. This corresponds to identification of the configurations $z_1 = z_1'$, $z_2 = z_2'$ in the initial and final states (z denoting $\underline{X}$, θ, $\tilde{\theta}$ along a string) in $\hat{Q}_3^{-\bar{A}}$, $\hat{Q}_3^{-\bar{B}}$ respectively. Such a lack of cancellation corresponds, in the language of the fourth paper in Ref. 8, to the disappearance of the region Q corresponding to a 3-string intermediate state. It is then necessary to regularise by preserving the absence of the Q region. This can be done by taking a small contour in the P region near the common interaction point x_-, without the corresponding contour in the Q region to cause cancellation. A non-zero value of the l.h.s. of (1) then ensues.

When CTR is used, integration along the common-time contours leads to a non-zero value of (1) even when there is a non-zero intermediate string, but P_1 approaches P_2. Such a value is of $O(1)$, as in the case of a vanishing intermediate string, and in either case must be cancelled by a non-zero counter-term from a proposed $\hat{Q}_4^{-\bar{A}}$ in the expression

$$[\hat{Q}_3^{-\bar{A}}, Q_3^{-\bar{B}}]_+ + [\hat{Q}_2^{-(\bar{A}}, \hat{Q}_4^{-B)}]_+ = 0 \tag{4}$$

The expression for $\hat{Q}_4^{-\bar{A}}$ which appears to do this is

$$\hat{Q}_4^{\bar{A}} = \lim_{\rho_{21}} \int \prod_{i=1}^{4} \mathcal{D} z_i \; \Psi(z_i) \; \Delta(\Sigma z_i)(c\theta_1^{\bar{A}} V_1 + c'\theta_2^{\bar{A}} V_2 + c''\theta_1^{\bar{A}} V_2 + c'''\theta_2^{\bar{A}} V_1)\tilde{V}_1 \tilde{V}_2 \tag{5}$$

In (5), c, c', c'', c''' are suitable numerical constants (independent of the separation ρ_{21} of the two interaction points 1 and 2) and the limit is as $P_1 \to P_2$. If CTR is used, integration in (5) is over all configurations with the common interaction point $P_1 = P_2$;

if ETR is used only those configurations with $z_1 = z_4$, $z_2 = z_3$ are to be taken.

A similar situation to the above occurs for the open superstring, as was originally pointed out in Ref. 10. However the situation is considerably complicated due to the presence of the Φ^4 "zipper" diagram. This gives additional terms in the analysis of (4), both from a field-theoretic or amplitude approach, and we will leave the details of such terms to elsewhere [12].

It is also necessary to check that with (5) the Hamiltonian part of the SUSY algebra is correctly given. This seems to also be the case [12], for both the closed and open superstring situations. There is a general argument for the closed case which rapidly indicates the form which H_4 must take. $\hat{Q}_3^{-\bar{A}}$, in the closed case, must have an insertion factor proportional to $\tilde{V}(P)$ at the interaction point P, so that any non-zero contribution to the l.h.s. of (1) must contain the insertion factor $\tilde{V}(P_1)\tilde{V}(P_2)$. This factor must therefore be present in any possible choice of $\hat{Q}_4^{-\bar{A},A}$ for the L-moving modes, and hence be present in $\hat{H}_4$ resulting from computation of $[\hat{Q}^{-A},\hat{Q}^{-\bar{B}}]_+$ at fourth order in the fields. By symmetry between L and R, the net insertion factor must therefore be proportional to $V_1V_2\tilde{V}_1\tilde{V}_2$, with constant of proportionality given by ρ_{21} (to within a numerical constant). Thus

$$\hat{H}_4 = \lim c_1 \int \prod_{i=1}^{4} \mathcal{D}^4 z_i \; \Psi(z_i) \; \Delta(\Sigma z_i) \rho_{21} V_1 V_2 \tilde{V}_1 \tilde{V}_2 \tag{6}$$

where the configurations z_i to be integrated over in ETR or CTR are those discussed after Eq.(5), above, so determining $\hat{Q}_4^{-\bar{A}}$ in these two regularisations; c_1 is a numerical constant.

We finally note that if $\hat{Q}_3$ in (1) was defined, as it should be, in terms of normal ordering of the field operators Ψ, then an additional term $\hat{Q}_2$ will arise which will be only quadratic in the field operators but quadratic in the string interaction strength G, as is $\hat{Q}_4$. This will have the same insertion factors as $\hat{Q}_4$ in (5), but only contains the configuration $z = z'$ for the two string fields. This term appears to play the role of a self-energy counter-term [15].

Thus we have the additional term

$$\hat{H}_2 = \lim c_2 \int \mathfrak{D} z \, \Psi^+(z)\Psi(z)\rho_{21}V_1V_2\tilde{V}_1\tilde{V}_2 \tag{7}$$

to be added to the Hamiltonian, for some further numerical constant c.

4. SUPERSTRINGS AT HIGHER ORDER

It is important to answer the question as to the necessity or otherwise of higher order terms in $\hat{Q}$ or $\hat{H}$. Thus we write

$$\hat{Q}^{-\bar{A}} = \hat{Q}_2^{-\bar{A}} + \hat{Q}_3^{-\bar{A}} + \hat{Q}_4^{-\bar{A}} + \ldots$$

and consider the condition

$$[\hat{Q}^{-\bar{A}}, \hat{Q}^{-\bar{B}}]_+ = [\hat{Q}_2^{-(\bar{A}}, \hat{Q}_3^{-\bar{B})}]_+ + \left\{[\hat{Q}_3^{-\bar{A}}, \hat{Q}_3^{-\bar{B}}]_+ + [\hat{Q}_2^{-(\bar{A}}, \hat{Q}_4^{-\bar{B}}]_+\right\}$$
$$+ \left\{[\hat{Q}_3^{-(\bar{A}}, \hat{Q}_4^{-\bar{B})}]_+ + [\hat{Q}_2^{-(\bar{A}}, \hat{Q}_5^{-\bar{B})}]_+\right\} + \ldots \tag{8}$$

Then we note from (8) that only if

$$[\hat{Q}_3^{-(\bar{A}}, \hat{Q}_4^{-\bar{B})}]_+ = 0 \tag{9}$$

can we hope that the process of adding counter terms can terminate at the quartic term. Moreover, the further condition

$$[\hat{Q}_4^{-\bar{A}}, \hat{Q}_4^{-\bar{B}}]_+ = 0 \tag{10}$$

must also be shown to be valid. We will not consider (9) and (10) in the open superstring case, where in any case there are a number of further three-string interactions besides that of only open superstrings, as well as the quartic zipper interaction. The situation is simpler when only closed strings are present.

In that case the results appear to depend crucially on which of the regularisations ETR or CTR, described in the previous section, are being used. In the former, a second quantised analysis leads to the conclusion that both (9) and (10) are completely cancelled. This can be seen directly as due to the fact that there is no situation in which there is no internal string configuration which contributes to the anticommutators in (9) or (10). Thus the lack of cancellation arising from such a configuration in the anticommutator of (1) does not occur in (9) or (10) and they are both identically zero.

Such cancellation does not appear to occur for CTR, and we expect an infinite number of non-zero terms $\hat{Q}_n^{-A}$ in the expansion of $\hat{Q}^{-A}$, for $n > 4$. These will be considered elsewhere, but can be analysed in terms of counter-terms to the amplitudes to ensure Lorentz invariance and supersymmetry, along the lines of Ref. 13. We will turn to consider that after some brief remarks about the construction of multi-loop amplitudes from $\hat{H}_3$.

5. MULTILOOP AMPLITUDES

A construction of multi-loop amplitudes for superstrings in the LC gauge has been given in Ref. 16 and an analysis of the associated divergences can be found in Ref. 17. However this latter analysis was based on that part of the construction of Ref. 16 which used a short-string limit (SSL); this required that all but two of the external strings have values $P_r^+ \sim 0$. Such a limit allowed [18] the calculation of the positions of the interacting vertices which are not on closed slits in the LC diagram (called external interacting vertices in Ref.16) explicitly in terms of the nearby Koba-Nielsen sources z_r. Such a procedure led to a removal of many contributions to the multi-loop amplitudes, and was useful in the resulting divergence analysis of Ref. 17. It has more recently been noted that the SSL cannot be justified if there is a Lorentz anomaly present in the amplitudes [19]. Thus in order to determine the finiteness, or otherwise, of multiloop amplitudes we must abandon the SSL and consider the complete multi-loop expressions in the LC gauge. This we can now do using a compact notation [14].

The procedure of calculation developed in Ref. 16 was to use the Kaku-Kikkawa [7] reduction of the second to first quantised perturbation amplitudes, but now including the insertion factors described in Sec. 2 as part of the amplitudes. The resulting functional integrals

$$\int \mathcal{D} X \mathcal{D}\lambda\, \mathcal{D}\theta\, \exp[+\iint \underline{X}\, \Delta\, \underline{X} + \lambda \partial_{\bar{\rho}} \theta + JX + K\theta] \times \prod_P V(P) \tag{11}$$

(where P denotes the interacting positions, J and K the external bosonic and Grassmann-valued sources, and only L-fermions are considered in (11)) may now be performed. For λ this leads to the constraint $\partial_{\bar{\rho}}\theta = 0$, whose solution in terms of the external sources θ_r and the g internal θ_i's, with one θ_i, associated with each loop [16], is

$$\theta(\tilde{z}) = \Theta = [F''(\tilde{z})]^{-1/2}[\sum_{r=1}^{N} p_r^+ G'(\tilde{z},z_r)\theta_r + \sum_{i=1}^{g} \theta_i u_i'(\tilde{z})] \tag{12}$$

where $\tilde{z}$ is the interaction position in the z-plane related to the LC variable ρ by the Mandelstam map [18]

$$\rho = \sum_{r=1}^{N} p_r^+ G(z,z_r) \tag{13}$$

and G is the Green function of the associated Riemann surface of genus g. We note also that the interaction factors do not depend on λ but only the value Θ at the associated point. The final $\underline{X}$-integration may be reduced to a total Gaussian by introducing Grassmann variables c_α, c'_α for each interaction point P_α, and letting V(P) be exponentiated as

$$V(P_\alpha) = \int dc_\alpha dc'_\alpha \exp[\iint \underline{X}\, \underline{N}_\alpha\, c_\alpha c'_\alpha] \tag{14}$$

where N_α has L, i and R components respectively,

$$F''^{-1/2}\Big\{2^{-1/2}\partial_{\bar{\tilde{z}}_\alpha}\delta^2(z-\tilde{z}_\alpha), -\partial_{\bar{\tilde{z}}_\alpha}(\Theta^{2i}(\tilde{z}_\alpha)\delta^2(z-\tilde{z}_\alpha)),$$
$$(\sqrt{2}/3)\,\partial_{\bar{\tilde{z}}_\alpha}(\Theta^4(\tilde{z}_\alpha)\delta^2(z-\tilde{z}_\alpha))\Big\}$$

The Gaussian integration over $\underline{X}$ may then be performed with the inclusion of suitable factors of det Δ, etc. arising from the remaining θ-integration [16],[20]. This leads to the final expression for the multi-loop amplitude (again dropping R-moving factors in integration, etc.) for N external strings with zero-mode superfields $\Phi_r(u_r,\zeta_r,\theta_r)$,

$$\int \prod_{\alpha=1}^{2g+N-3} d^2\tilde{\rho}_\alpha \prod_{i=1}^{g} d\alpha_i d\beta_i (\det \mathrm{Im}\pi)^{-4} \prod_{r=1}^{N} d^4\theta_r \Phi_r(u_r, \zeta_r, \theta_r) \prod_{i=1}^{g} d^4\theta_i$$

$$\times\ \delta^4(\Sigma p_r^+ \theta_r)\ \exp[p_r p_s G(z_r, z_s)]. \int \exp M \exp N \prod_\alpha dc_\alpha dc'_\alpha \qquad (15)$$

where in (15) J has been represented as given only by the zero mode momenta of the external strings,

$$M = \sum_\alpha \iint \underline{J}\, G\, \underline{N}_\alpha\, c_\alpha\, c'_\alpha \qquad (16a)$$

$$N = \sum_{\alpha,\beta} \iint \underline{N}_\alpha\, G(\tilde{z}_\alpha, \tilde{z}_\beta) \underline{N}_\beta\, c_\alpha c'_\alpha c_\beta c'_\beta \qquad (16b)$$

The expressions (16a) & (16b) are a compact way of expressing the many contractions of the insertion factors which we attempted to reduce by the SSL in Refs. 16 and 18. In particular we note that evaluation of the Berezhin integrations over the c_α, c'_α leads to G" and G' factors evaluated at interaction vertices. In particular, it is the former of these which lead to ∞^2-divergences considered in Ref. 17 for the type II superstring.

6. DIVERGENCE ANALYSIS

There are two different types of divergences in (15) (and the companion heterotic amplitude which has been constructed carefully in Refs. 21). These arise either from (a) the degeneration of the Riemann surface Σ or (b) the coincidence of interaction points $\tilde{\rho}_\alpha \sim \tilde{\rho}_\beta$. These were both analysed in Ref. 17, though the distinction between the two was not then made clear. It is now apparent to us[14] that these occur independently, the former involving degeneration of the period matrix π of Σ, the latter not. This is a little puzzling since in the LC strip diagram when $\tilde{\rho}_\alpha \sim \tilde{\rho}_\beta$ it appears that there is a change of genus, and so a degeneration of Σ. However this does not happen, as can be seen for the tree level amplitude, when Σ is a sphere, which has no moduli in terms of which such degeneration could arise. At the multi-loop level this is also clear from our specific use, in Ref. 17, of the degeneration analysis of Lebowitz [22]. It was shown in Ref. 17 that the interaction points $\tilde{\rho}_\alpha, \tilde{\rho}_\beta$ did not converge either under a handle or a dividing geodesic degeneration of Σ. It was necessary to use further moduli on Σ to achieve convergence of $\tilde{\rho}_\alpha, \tilde{\rho}_\beta$ which does not correspond, therefore, to a degeneration of Σ in the fundamental domain being considered for the definition of the amplitude, Eq.(15).

The divergence analysis of Ref. 17 showed that the amplitude (15) was finite under the surface degenerations of type (a) of Σ. However when the further analysis of type (b) coincidences was made for type II superstrings $a|\delta|^{-4}$ divergence was noted from the factor $|G''|^2$, where $\delta = \tilde{z}_\alpha - \tilde{z}_\beta$. An attempt was made to remove this by partial integration in δ, and although the resulting integral in $d^2\delta$ may then be made convergent the resulting boundary term was not fully discussed. We may now complete that discussion by addition of the term $\hat{H}_4$ of Eq.(6) above. The divergent boundary term arising from $|G''|^2$ has the form

$$\int_C [|\delta|^2 \bar{\delta}]^{-1} d\bar{\delta} \tag{17}$$

where the above expression is only local. The contour C is defined in terms of the type of integration discussed in the fourth paper of Ref. 8, involving equal-time contours transversed in opposite directions. If we define (17) by (ETR) (as defined earlier) then exactly the corresponding term in $\hat{H}_4$ of (6) will be expected to cancel [17]. If we define (17) by (CTR) we expect $\hat{H}_4$ of (6) defined by (CTR) to achieve such cancellation (the details of such cancellations will be reported in detail elsewhere [14]).

We now consider the possible higher order terms $\hat{H}_n$ $(n > 4)$ that may be required by (CTR). If three or more $\tilde{z}_\alpha$'s coincide there are new singularities which arise which are of form (17) but cannot be cancelled by any contribution from $\hat{H}_4$'s at pairs of converging interaction points. Since the singularities like (17) are expected to be non-Lorentz covariant, then closure of the SP algebra will force the introduction of $\hat{H}_n$'s for arbitrary high n. This is one way of seeing why (CTR) is expected to lead to a non-polynomial LC superstring field theory.

The result of the above divergence analysis is that type II superstrings have finite multi-loop amplitudes. The earlier result of finiteness in Ref. 17 for the heterotic case remains unaltered, since the equivalent to the boundary terms in this case arises only from $\bar{G}''$ singularities (only R-moving 10-dimensional supersymmetry is required) with the following form

$$\int_C \bar{\delta}^{-2} d^2 \delta$$

This vanishes, by phase integration, either for ETR or CTR.

Finally, we note that since type II or heterotic strings appear finite, they are not expected to have anomalies. In particular, the SSL may be taken and the amplitudes of Ref. 16 recovered.

7. CONCLUSIONS

Closure of the SP algebra has not yet been completely achieved, and it is necessary to investigate the other generators following the analysis of Linden [23]. The present results lead to optimism that the full SP algebra can be closed, and leads to a quartic interaction theory when using (ETR). Closure may be more problematic if (CTR) is used, since the generators of the algebra appear to be non-polynomial in the string field operators. Thus in spite of obtaining apparently identical S-matrix elements for multi-string states in perturbation theory by (ETR) or (CTR) techniques the underlying physics is expected to be very different. Indeed, it may be very hard to consider non-trivial vacua by (CTR), whilst this need not be so for (ETR).

Consider the type II superstring action

$$\hat{L}_2 + \hat{L}_3 + \hat{L}_4$$

constructed from the usual kinetic terms in $\hat{L}_2$ and the three string interactions $\hat{L}_3$, as well as the new quadratic and quartic terms in $\hat{L}_2$ and $\hat{L}_4$ given by the terms (6) and (7) in the Hamiltonian. Then one may suggest the non-trivial solution of the field equations

$$\Psi = \Psi_0 \propto (G\rho_{21} V\hat{V})^{-1} \tag{18}$$

on configurations with a self-interaction. If it is possible to make sense of this solution (18) its total energy and super-charge is of interest. If the first is negative and the second non-zero this may give the first example of a non-trivial vacuum which may be preferred to the Fock vacuum. Its physics may therefore be decidedly non-trivial.

ACKNOWLEDGMENTS

I would like to thank Professors G. Kunstatter and H.C. Lee for their kind invitation to give these talks in the stimulating atmosphere of the Edmonton Summer Institute for Theoretical Physics, 1987, and Professor Abdus Salam for his very kind hospitality at the International Centre for Theoretical Physics, Trieste, where further developments on the subject of the talks were achieved. My thanks also go to Noah Linden for a collaborative endeavour in closing the light cone superstring algebra and finally by Alvaro Restuccia for his deep and persistent thinking without which many of the developments reported on here could not have been achieved.

REFERENCES

1) Schellekens A.N., "Four-Dimensional Strings", CERN preprint, TH. 4807/87, July 1987;
Schwarz J.H., "Recent Developments in Superstrings", in Superstrings, Eds. K.T. Mahanthappa and P.G.O. Freund (Plenum Pub. Co., New York, 1988).
2) Friedan D. and Shenker S., Phys. Lett. 175, 287 (1986); ibid. Nucl. Phys. B281, 509 (1987).
3) Alvarez-Gaume L., Gomez C. and Reina C., Phys. Lett. 190,55(1987); ibid. "New Methods in String Theory", CERN preprint, TH. 4775/87; Witten. E., "Quantum Field Theory, Grassmannians and Algebraic Curves", Princeton Preprint PUPT-1057.
4) Bowick M. and Rajeev S., Phys. Lett. 58, 535 (1987); ibid. "The Holomorphic Geometry of Closed Bosonic String Theory and Diff S^1/S^1", MIT preprint, CTP #1450 (1987).
5) Witten E., Nucl. Phys. B268, 253 (1986); also ibid. B276,2(1987).
6) For a discussion of these and other approaches see Nouri-Moghadam M. and Taylor J.G., "A Review of String Field Theory", to appear in Proc. A.M.S. Workshop on Theta Functions.
7) Kaku M. and Kikkawa K., Phys. Rev. D10, 1110, 1832 (1974).

8) Hata H., Itoh K., Kugo T., Kunitomo H. and Ooguri H., Phys. Lett. 172B, 186, 195 (1986); ibid. Phys. Rev. D34, 2360 (1986); ibid. Phys. Rev. D35, 1318 (1987);
Neveu A. and West P.C., Phys. Lett. 168B, 192 (1986); ibid. Nucl. Phys. B278, 601 (1986).
9) Giddings S.B. and D'Hoker E., Nucl. Phys. B291, 90 (1987).
10) Green M.B. and Schwarz J.H., Nucl. Phys. B243, 475 (1984).
11) Greensite J. and Klinkhamer F.R., Nucl. Phys. B281, 269 (1987); ibid. B291, 557 (1987).
12) Linden N., Restuccia A. and Taylor J.G., in preparation.
13) Greensite J. and Klinkhamer F.R., "Superstring Amplitudes and Contact Interactions", Niels Bohr preprint NBI-HE-87-58, August 1987.
14) Restuccia A. and Taylor J.G., in preparation.
15) Cremmer E. and Gervais J.L., Nucl. Phys. B90, 410 (1975).
16) Restuccia A. and Taylor J.G., Phys. Rev. D, July 15, 1987.
17) Restuccia A. and Taylor J.G., Comm. Math. Phys. 112, 447 (1987).
18) Mandelstam S., Phys. Rep. 13C, 260 (1974).
19) We would like to thank M.B. Green for this useful comment.
20) Restuccia A. and Taylor J.G., Phys. Lett. 192B, 89 (1987); 174B, 56 (1986); 177B, 39 (1986); 186B, 57 (1987).
21) Bressloff P.C., Restuccia A. and Taylor J.G., Phys. Lett. 185B, 99 (1987) and Int. J. Mod. Phys. (to appear).
22) Lebowitz A., in "Advances in the Theory of Riemann Surfaces", Annals of Math. Studies, Eds. L.V. Ahlfors et al., Princeton University Press, 1971.
23) Linden N., Nucl. Phys. B286, 429 (1987).

ZETA FUNCTION REGULARIZATION OF QUANTUM GRAVITY

R.B. Mann
Guelph-Waterloo Program for Graduate Work in Physics
Department of Physics
University of Waterloo
Waterloo, Ontario
CANADA
N2L 5V1

Abstract

In regularization of quantum gravity and of quantum field theory in curved spacetime, the zeta-function regularization technique has proven to be most useful to one-loop order. This paper reviews this technique, and its applications in the aforementioned areas. A new technique called Operator Regularization (recently introduced by McKeon and Sherry) is then presented. Being in certain respects an extension of zeta-function regularization techniques, Operator Regularization has the following noteworthy features. (A) No infinities appear in any stage of the calculation. (B) All symmetries are manifestly preserved, apart from anomalies (which may also be evaluated using this scheme). (C) The method may be extended to higher loops. Applications of operator regularization to quantum gravity and quantum field theory in curved spacetime are discussed.

1. Introduction

Computations of radiative processes in quantum field theory are typically accompanied by the presence of infinite quantities. In the context of perturbative quantum field theory, the objects one calculates are given in terms of a power series in small powers of a dimensionless coupling parameter, the coefficients of which are typically infinite. The renormalization program was developed so that physically meaningful results may be extracted from such apparently meaningless quantities. In general this involves altering the original (unquantized) action so that the infinities which appear may be parametrized in some manner that preserves the symmetries present before quantization. Although quantities evaluated in the regulated theory are finite, as the regulating parameter that characterizes the altered theory approaches its physical value divergences occur. These divergences are removed by absorbing them into the parameters that define the initial theory. Of course any parameters which absorb such infinites cannot be calculated; rather they must be input by experiment. The renormalization program will be successful only if a finite number of parameters are needed to absorb such infinite quantities to all orders in perturbation theory. Consequently only a finite number of experiments are needed to define such parameters; upon doing so, the theory has predictive power and so may meaningfully be compared to all remaining relevant experiments.

The renormalization program has enjoyed a large measure of success in its application to gauge theories(1). Renormalizable gauge theories form the foundation of our present day understanding of the strong, weak and electromagnetic interactions. Unfortunately in the case of gravity the renormalization program is not viable(2). For gravity coupled to matter in the context of general relativity, it is not possible to absorb divergences occuring in the one loop effective Lagrangian into the coupling constants appearing in the original (unquantized) Lagrangian.

This non-renormalizability problem is one of many that one faces in constructing a quantum theory of gravitation. Indeed, quantizing gravity involves rather subtle conceptual issues concerning causality, separability, and locality[3]. In contrast to these, the non-renormalizability problem is technical in nature; one might imagine that an appropriate symmetry principle or reformulation of the theory could ameliorate the ubiquitous infinities which occur. In this sense, the 'right' theory of gravity coupled to matter would solve the non-renormalizability problem by banishing all infinities: all quantities calculated (*ie.*, all Green's functions) would be finite, thus obviating the need to remove infinites by redefining wavefunctions and coupling constants as in standard gauge theories. Such is the hope of superstring theories: by considering the fundamental objects of nature to be extended (strings) rather than pointlike, the short-distance (high-momentum) divergences which occur in point-particle field theory are 'smeared out' . Although finiteness for superstrings has not been completely proven, there are hopeful indications that for at least some such theories this is the case[4].

Even in the case of quantum field theory in curved spacetime, implementation of the renormalization program involves subtle and difficult issues that are absent in the flat-space case[5]. Quantum field theory in curved spacetime, apart from being a useful theoretical laboratory for quantum gravity, is interesting in its own right for studying quantum systems in a background gravitational field. In this manner the effects of classical curvature on quantum systems may be explored.

A variety of regulating schemes have been employed to study the renormalization program in quantum gravity and quantum field theory in curved spacetime.[5] In this regard, zeta-function regularization has been quite popular[5]. Unfortunately, it has the limitation of being applicable to one-loop processes.[5,6] Recently a new regularization technique called Operator Regularization was introduced by McKeon and Sherry[7]. Being in some sense an extension of the zeta-function technique, Operator Regularization[7] is a means of computing radiative corrections in a given quantum field theory to any given loop order (*ie.*, order in $\hbar$). In this procedure, the background field quantization formalism is used in conjuction with the Feynman path integral[8]. The path integral is evaluated prior to an expansion in powers of the background field, in contradistinction to the usual Feynman diagrammatic approach. The resulting expression for the generating functional is expressed in terms of determinants and inverses of operators. Rather than regulating the theory by the insertion of a regulating parameter into the initial Lagrangian, in operator regularization the functions of operators occuring in the expression for the generating functional are regulated. This ensures that symmetries present in the original theory are not explicitly broken. Remarkably, no explicit divergences arise in this procedure, even as the regulating parameter approaches its physical value.

This paper will review the application of zeta-function regularization schemes to quantum field theory in curved spacetime and to quantum gravity. Particular attention will be given to the aforementioned operator regularization technique, both in comparing it to other implementations of zeta-function techniques in the renormalization program and in demonstrating its applicability in several examples. In the next section an introduction to the background field method in the context of path-integral quantization will be given. In section three, the zeta-function formalism will be introduced and implemented both in the context of heat-kernel methods and in the context of Operator Regularization. In section four, applications will be discussed in four areas: (a) flat space quantum field theory, (b) curved space quantum field theory, (c) quantum gravity, and (d) the trace anomaly in the sigma model. A summary and some brief conclusions are presented in section five.

2. Review of Background Field methods in the Path Integral Formalism

The generic form of the Feynman path integral in Euclidian space is

$$Z[J] = \int d\Phi \exp\{-\frac{1}{\hbar}\int dx(\mathcal{L}[\Phi] + J\cdot\Phi)\} \tag{2.1}$$

Here Φ is some set (perhaps a multiplet or supermultiplet) of quantum fields, fermionic and/or bosonic. J is an external source, and $\int dx$ is a Euclidean space measure, denoting either $\int d^n x$ or $\int d^n x\sqrt{g}$ depending on whether or not the space is curved. In this section indices will be suppressed on all fields and sources.

In the background field method[8], the field Φ is assumed to be separable into a classical part ϕ and a quantum part φ:

$$\Phi = \phi + \hbar\varphi \tag{2.2}$$

The field φ commutes or anticommutes depending on whether it is bosonic or fermionic. Consequently (2.1) may be written as:

$$Z[\phi, J] = \int d\varphi \exp\{-\frac{1}{\hbar}\int dx(\mathcal{L}[\phi + \hbar\varphi] + \hbar J\cdot\varphi)\} \tag{2.3}$$

which is the generating functional for Green's functions in the presence of a source functional J.

Expanding in $\mathcal{L}$ in powers of $\hbar$ gives

$$\begin{aligned}\frac{1}{\hbar}\mathcal{L}[\phi + \hbar\varphi] =& \frac{1}{\hbar}\mathcal{L}[\phi] + \left.\frac{\delta\mathcal{L}}{\delta\Phi}\right|_{\phi}\varphi \\ &+ \frac{\hbar}{2!}\varphi M[\phi]\varphi + \frac{\hbar^2}{3!}a[\phi]\varphi^3 + \frac{\hbar^3}{4!}b[\phi]\varphi^4 + \cdots\end{aligned} \tag{2.4}$$

In the first line, the term $\left.\frac{\delta\mathcal{L}}{\delta\Phi}\right|_{\phi}\varphi$ vanishes if ϕ obeys the classical equations of motion; this will be assumed henceforth. The second line in the above expression contains all possible terms which would appear in a renormalizable theory in four dimensions. The last line is non-vanishing for non-renormalizable theories such as general relativity (it is generically non-vanishing for two dimensional renormalizable theories as well). Using (2.4) in (2.3) yields

$$\begin{aligned} Z[\phi] &= e^{-\frac{1}{\hbar}S[\phi]}\int d\varphi \exp\{-\int dx\Big(\frac{\hbar}{2!}\varphi M\varphi + \frac{\hbar^2}{3!}a\varphi^3 + \frac{\hbar^3}{4!}b\varphi^4 + \cdots + J\cdot\varphi\Big)\} \\ &= e^{-\frac{1}{\hbar}S[\phi]}\Bigg\{\sum_N^\infty \frac{1}{N!}\Bigg[\int dx\Bigg(\frac{\hbar^2}{3!}a_{ijk}(\phi)\frac{\delta^3}{\delta J_i \delta J_j \delta J_k} \\ &\quad + \frac{\hbar^3}{4!}b_{ijk\ell}(\phi)\frac{\delta^4}{\delta J_i\delta J_j\delta J_k\delta J_\ell} + \cdots\Bigg)\Bigg]^N \int d\varphi e^{-\int dx\left(\frac{\hbar}{2!}\varphi M[\phi]\varphi + J\cdot\varphi\right)}\Bigg\}\Bigg|_{J=0} \end{aligned} \tag{2.5}$$

which is an expression for $Z[\phi, J]$ in terms of a power series in $\hbar$; here indices have been explicitly included. The $e^{-\frac{1}{\hbar}S[\phi]} = e^{-\frac{1}{\hbar}\int dx\mathcal{L}[\phi]}$ term is of course what would arise if there were no quantization; it is the zeroth order contribution to the effective action Γ. In what follows it shall be denoted by Z_0.

Relative to the remaining functional integral in the last line of (2.5), the $N = 0$ term in the above expression is of lower order than all the other terms in the sum. In this sense, the 'lowest order' term in (2.5) is given by the functional integral over the term bilinear in φ; this is referred to as the 'one-loop' generating functional:

$$Z_{(1)}[\phi, J] = Z_0 \int d\varphi e^{-\int dx\left(\frac{\hbar}{2!}\varphi M[\phi]\varphi + J\cdot\varphi\right)} \tag{2.7}$$

This expression may be formally evaluated by decomposing φ into its bosonic and fermionic parts, $\varphi = (b, f)$, so that $M(\phi)\varphi$ is given by

$$M\varphi = \begin{pmatrix} M_{bb}(\phi) & M_{bf}(\phi) \\ M_{fb}(\phi) & M_{ff}(\phi) \end{pmatrix}\begin{pmatrix} b \\ f \end{pmatrix} \tag{2.8}$$

and so

$$\varphi M\varphi = \sum_i \left[b_i M^i_{bb} b_i + f_i M^i_{fb} b_i + b_i M^i_{bf} f_i + f_i M^i_{ff} f_i\right] \tag{2.9}$$

where (b_i, f_i) form a complete set of states for (b, f), and where M^i_{bb}, M^i_{fb}, etc., are the eigenvalues of $M[\phi]$; they are functionals of the classical background field

ϕ. Using (2.8) and (2.9) in (2.7), completing the square and performing a shift of the functional integration variables gives[7,9]:

$$Z_{(1)}[\phi, J] = Z_0 s\det^{-\frac{1}{2}}\left[\hbar M(\phi)\right] \exp\left[-\tfrac{1}{2}\hbar^{-1}\int dx J M^{-1}(\phi) J\right] \tag{2.10}$$

where

$$s\det\begin{pmatrix} M_{bb} & M_{bf} \\ M_{fb} & M_{ff}\end{pmatrix} = \det(M_{bb})\det{}^{-1}\left(M_{ff} - M_{fb}M_{bb}^{-1}M_{bf}\right) \tag{2.11a}$$

$$= \det(M_{bb} - M_{bf}M_{ff}M_{fb})\det{}^{-1}(M_{ff}) \tag{2.11b}$$

and

$$\begin{pmatrix} M_{bb} & M_{bf} \\ M_{fb} & M_{ff}\end{pmatrix}^{-1}$$

$$= \begin{pmatrix} 1 & -M_{bb}^{-1}M_{bf}(M_{ff} - M_{fb}M_{bb}^{-1}M_{bf})^{-1} \\ 0 & (M_{ff} - M_{fb}M_{bb}^{-1}M_{bf})^{-1}\end{pmatrix}\begin{pmatrix} M_{bb}^{-1} & 0 \\ -M_{fb}M_{bb}^{-1} & 1\end{pmatrix} \tag{2.12a}$$

$$= \begin{pmatrix} (M_{bb} - M_{bf}M_{ff}^{-1}M_{fb})^{-1} & 0 \\ -M_{ff}^{-1}M_{fb}(M_{bb} - M_{bf}M_{ff}^{-1}M_{fb})^{-1} & 1\end{pmatrix}\begin{pmatrix} 1 & -M_{bf}M_{ff}^{-1} \\ 0 & M_{ff}^{-1}\end{pmatrix} \tag{2.12b}$$

If complex fields are employed, then one obtains $s\det^{-1}\left[\hbar M(\phi)\right]$ in (2.10) instead of $s\det^{-\frac{1}{2}}\left[\hbar M(\phi)\right]$. If no fermionic fields are present, then the superdeterminant is replaced by the usual determinant.

The 'all-loop' generating functional (2.5) is therefore given by

$$Z[\phi] = Z_0 s\det^{-\frac{1}{2}}\left[\hbar M(\phi)\right]\left\{\sum_N^\infty \frac{1}{N!}\left[\int dx\left(\frac{\hbar^2}{3!}a_{ijk}(\phi)\frac{\delta^3}{\delta J_i \delta J_j \delta J_k} + \frac{\hbar^3}{4!}b_{ijk\ell}(\phi)\frac{\delta^4}{\delta J_i\delta J_j\delta J_k\delta J_\ell}\cdots\right)\right]^N \exp\left[-\tfrac{1}{2\hbar}\int dx J M^{-1}(\phi) J\right]\right\}\Bigg|_{J=0} \tag{2.13}$$

In general, the matrix $M[\phi]$ will be a differential operator acting on bosonic and fermionic fields. In the case of a charged scalar field φ in a curved background

$$M[A, g]\varphi \equiv \Box\varphi = \frac{1}{\sqrt{g}}D_\mu(A)(\sqrt{g}g^{\mu\nu}D_\nu(A)\varphi) \tag{2.14}$$

whereas for charged fermions ψ one has

$$M[A,g]\psi = \gamma^a e^\mu_a (D_\mu(A) - \frac{i}{4}\sigma^{ab}\omega_{\mu ab})\psi \tag{2.15}$$

where the conventions

$$\delta^{ab} e^\mu_a e^\nu_b = g^{\mu\nu} \qquad D_\mu(A) = \partial_\mu - ieA_\mu \tag{2.16a, b}$$

$$\omega^\mu_{ab} = [e^\nu_{[a}(\partial^\mu e_{b]\nu} - \partial_\nu e_{b]\mu}) + \tfrac{1}{2} e^\rho_{[a} e^\sigma_{b]} e^{c\mu}(\partial_\sigma e_{c\rho} - \partial_\rho e_{c\sigma})] \tag{2.16c}$$

$$\{\gamma^a, \gamma^b\} = 2\delta^{ab} \qquad \sigma^{ab} = \frac{i}{2}[\gamma^a, \gamma^b] \tag{2.16d, e}$$

have been used, ω being the spin connection. In the case of supersymmetry for, say, the Wess-Zumino model with chiral and antichiral superfields $\Phi(x,\theta)$, $\overline{\Phi}(x,\theta)$

$$M[\phi_B, \overline{\phi}_B] = \begin{pmatrix} \frac{1}{4}(m+\lambda\phi_B)\overline{D}^2 & \frac{\overline{D}^2 D^2}{16} \\ \frac{D^2\overline{D}^2}{16} & \frac{1}{4}(m+\lambda\overline{\phi}_B)\overline{D}^2 \end{pmatrix} \tag{2.17}$$

where ϕ_B and $\overline{\phi}_B$ are the chiral and antichiral background fields, and where the quantities D^2 and $\overline{D}^2$ are the squares of the superderivatives

$$D_a = \frac{\partial}{\partial\theta^a} - i\theta^{\dot{a}}(\sigma^\mu_{a\dot{a}})\partial_\mu \tag{2.18a}$$

and

$$\overline{D}_{\dot{a}} = \frac{\partial}{\partial\theta^{\dot{a}}} - i\theta^a(\sigma^\mu_{a\dot{a}})\partial_\mu \tag{2.18b}$$

respectively.

The formal expression

$$\det[M] = \prod_n \lambda_n \tag{2.19}$$

appearing in (2.10) and (2.13) is problematical to evaluate since the eigenvalues λ_n in general increase without bound. This implies that a regulator of some sort is needed. One particularly attractive solution to this problem is to employ zeta-function techniques, which will be the focus of the remainder of this paper.

3. Zeta-function Formalism and Operator Regularization

The functional determinants encountered at the end of the previous section may be evaluated according to a technique introduced by Salam and Strathdee[(6)], Dowker and Critchley[(6)] and Hawking[(6)]. Consider the following function of s :

$$\varsigma(s) = \sum_n \lambda_n^{-s} \tag{3.1}$$

Of course $\varsigma(s)$ is a functional of ϕ whenever the λ_n are; however this notation will be suppressed. Now formally,

$$\begin{aligned} \det[M(\phi)] &= \exp\Big[Tr(\ln(M(\phi))\Big] \\ &= \exp\Big[\sum_n \ln(\lambda_n)\Big] \\ &= \exp\Big[-\frac{d}{ds}\varsigma(s)\Big|_{s=0}\Big] \end{aligned} \tag{3.2}$$

or

$$= \exp\Big[-\sum_n \lim_{s\to 0}(s^{-1}\lambda_n^{-s} - s^{-1})\Big]$$

where $M(\phi)\hat{\phi}_n = \lambda_n\hat{\phi}_n$ and the $\hat{\phi}_n$ form a complete set of states for the quantum field φ in the presence of the background field ϕ. The former expression (3.2) is that used by Hawking[(6)] , whereas the latter is that used by Dowker and Critchley[(6)].

The zeta function $\varsigma(s)$ derives its name from the Riemann zeta function $\varsigma_R(s)$; indeed for $\lambda_n = n$ the two are identical. The Riemann zeta function may be used to 'regulate' certain sums; for example

$$\sum_{n=1}^{\infty}(1) = \sum_{n=1}^{\infty}(\lim_{s\to 0} n^{-s}) = \lim_{s\to 0}(\varsigma_R(s)) = \frac{1}{2}. \tag{3.3}$$

Here the regularization procedure using $\varsigma_R(s)$ involves the interchanging of the summation and limit operations. $\varsigma_R(s)$ is a meromorphic function of s with a pole at $s = 1$. It and its derivatives are regular at $s = 0$. These properties carry through for $\varsigma(s)$[10] (provided M is an elliptic operator); consequently $\varsigma(s)$ is a powerful regulator.

Three distinct uses of $\varsigma(s)$ have been employed in the literature, each of which will be discussed in turn.

The simplest manner to employ $\varsigma(s)$ as a regulator is when the eigenvalues λ_n are known explicitly. A simple example (due to Hawking[6]) should suffice to illustrate. Consider a free massless scalar field φ at finite temperature β^{-1} in a box of volume V. Finite temperature means periodicity in Euclidean time τ, so

$$\varphi(\tau, x) = \varphi(\tau + \frac{2\pi}{\beta}, x). \tag{3.4}$$

The operator M is merely $M = \partial_\tau^2 + \nabla^2$, so

$$\lambda_n = (\frac{2\pi n}{\beta})^2 + k^2 \tag{3.5}$$

and consequently

$$\sum_n \lambda_n^{-s} = \frac{8\pi V}{(2\pi)^3} \int d^3k \left[\frac{1}{2}k^{-2s} + \sum_{n=1}^{\infty} (\frac{4\pi^2}{\beta^2}n^2 + k^2)^{-s}\right]$$

yielding

$$\varsigma(s) = \frac{8\pi V}{(2\pi)^3} \left(\frac{2\pi}{\beta}\right)^{3-2s} \varsigma_R(3-2s) \frac{\sqrt{\pi}}{4} \frac{\Gamma(s-\frac{3}{2})}{\Gamma(s)} \tag{3.6}$$

Note that this is regular (in fact vanishing) at $s = 0$. The partition function Z is[6]

$$Z = \frac{1}{2}\ln(\det M) = \frac{1}{2}\varsigma'(0) = \frac{\pi^2}{90}VT^3 \tag{3.7}$$

in agreement with earlier results derived by different methods[11].

If the eigenvalues of M are not known explicitly (as is usually the case) then other means must be employed to evalute the zeta function. Consider the equation

$$\frac{\partial}{\partial\tau}F(x, y, \tau) + M_x F(x, y, \tau) = 0 \tag{3.8a}$$

where

$$F(x,y,0) = \delta(x-y) \tag{3.8b}$$

Equation (3.8a) is the heat equation for the function F, with (3.8b) being the boundary condition[8,12]. Formally, this equation has the solution

$$F(x,y,\tau) = \sum_n e^{-\lambda_n \tau} \hat{\phi}_n(x)\hat{\phi}_n(y) \tag{3.9}$$

where the $\hat{\phi}$'s are defined above. The completeness relations $(\sum_n \hat{\phi}_n(x)\hat{\phi}_n(y) = \delta(x-y))$ for the $\hat{\phi}$'s imply that this solution for F obeys the boundary condition (3.8b) whereas the orthogonality relations for the $\hat{\phi}$'s $(\int dx\sqrt{g}\hat{\phi}_n(x)\hat{\phi}_m(x) = \delta_{nm})$ imply

$$\int dx\sqrt{g}F(x,x,\tau) = \sum_n e^{-\lambda_n \tau} \tag{3.10}$$

so formally

$$\begin{aligned} \varsigma(s) &= \sum_n \lambda_n^{-s} = \frac{1}{\Gamma(s)} \int_0^\infty dt t^{s-1} \sum_n e^{-\lambda_n t} \\ &= \int dx \sqrt{g} \frac{1}{\Gamma(s)} \int_0^\infty dt t^{s-1} F(x,x,t). \end{aligned} \tag{3.11}$$

So if the heat equation can be solved for $F(x,y,\tau)$ (at least in the limit $x \to y$), then it is possible to evaluate $\varsigma(s)$ using (3.11). Usually the heat equation is solved using a series expansion of some sort for $F(x,y,\tau)$[12], with some sort of regularization required to evalute (3.10).

Recently McKeon and Sherry[7] have proposed a new regularization technique which makes use of the aforementioned attractive features of the zeta-function scheme in a new way. Recall from (2.10) that the one-loop generating functional is given by

$$Z_{(1)}[\phi,J] = Z_0 \mathrm{sdet}^{-\frac{1}{2}} \left[\hbar M(\phi)/\mu^2\right] \tag{3.12}$$

where the μ^2 appears because of the arbitrariness in the normalization of Z. Equation (3.12) is an operator equation. The basic idea of operator regularization is to regulate these formal expressions for the operators themselves directly, rather than the eigenvalues of these expressions. For example, for an operator H, formally

$$\begin{aligned} \ln(H) &= -\frac{d}{ds} H^{-s}\Big|_{s=0} \\ &\left(= -\frac{d^m}{ds^m}\left(\frac{s^m}{m!} H^{-s}\right)\Big|_{s=0}\right) \qquad (m = 1, 2, 3, \ldots) \\ &= -\frac{d}{ds}\left[\frac{1}{\Gamma(s)} \int_0^\infty dt t^{s-1} e^{-Ht}\right]\Big|_{s=0} \end{aligned} \tag{3.13}$$

and so

$$\det(H) = \exp\Big[Tr(\ln(H))\Big] = \exp\Big[-\varsigma'(0)\Big] \tag{3.14}$$

where now the zeta function is given by

$$\varsigma(s) = \frac{1}{\Gamma(s)} \int_0^\infty dt t^{s-1} Tr(e^{-Ht}). \tag{3.15}$$

So to one loop order operator regularization is very similar to the usual zeta function methods[6,12] with the subtle but important difference that the operator H is regulated rather than its eigenvalues. Of course if a complete set of states is inserted in (3.15), then eq. (3.1) is recovered.

In order to evaluate (3.15) it is necessary to either (a) be able to diagonalize H over some space of states or (b) decompose H into a part that is diagonalizable over some space of states and a part that is perturbatively small. The first case has been discussed above. Suppose then that the latter possibility holds, *ie.*, suppose

$$H = H_0 + H_I \tag{3.16}$$

where H_0 is diagonalizable on some space of states (typically it will be independent of the background, but not necessarily) and H_I. Now using an expansion introduced by Schwinger[13]

$$Tr(e^{-Ht}) = Tr\Big[e^{-H_0 t} + (-t)e^{-H_0 t}H_I + \frac{(-t)^2}{2}\int_0^1 due^{-(1-u)H_0 t}H_I e^{-uH_0 t}H_I$$
$$+ \frac{(-t)^3}{3}\int_0^1 udu\int_0^1 dve^{-(1-u)H_0 t}H_I e^{-u(1-v)H_0 t}H_I e^{-uvH_0 t}H_I$$
$$+\cdots\Big] \tag{3.17}$$

eq. (3.14) becomes

$$\det(H) = \exp\Big[-\frac{d}{ds}\Big\{\frac{1}{\Gamma(s)}\int_0^\infty dt t^{s-1}Tr\Big[e^{-H_0 t} - te^{-H_0 t}H_I$$
$$+ \frac{(-t)^2}{2}\int_0^1 due^{-(1-u)H_0 t}H_I e^{-uH_0 t}H_I + \cdots\Big]\Big\}\Big]\Big|_{s=0}$$
$$= \exp\Big[-\sum_n \varsigma_n'(s)\Big|_{s=0}\Big] \tag{3.18}$$

where $\varsigma_n(s)$ is the n-th term in the series in the middle line of (3.18). The derivative of the n-th term at $s = 0$, $\varsigma_n'(s)\Big|_{s=0}$ is the n-point one particle irreducible (1PI) Green's function to 'one-loop' order when H is identified with $M(\phi)$ and (3.18) is used to evalute the generating functional (3.12). Note that (3.18) makes sense only when H_0 has positive eigenvalues.

Before describing the general procedure for operator regularization it will be instructive to first consider a simple example to illustrate the above technique[7]. Consider Φ^3 theory in 6 dimensions; the Lagrangian is

$$\mathcal{L} = \frac{-1}{2}(\partial_\mu\Phi)^2 - \frac{\lambda}{3!}\Phi^3. \tag{3.19}$$

Expanding Φ into a background plus quantum part as in the previous section gives

$$\mathcal{L} = \mathcal{L}[\phi] + \frac{1}{2}\varphi(\partial^2 - \lambda\phi)\varphi + \cdots \tag{3.20}$$

implying

$$M(\phi) = -(\partial^2 - \lambda\phi) = (\hat{p}^2 + \lambda\phi) \tag{3.21}$$

where $\hat{p}_\mu = -i\partial_\mu$. The two point 1PI Green's function is given by the derivative at $s = 0$ of

$$\varsigma_2(s) = \frac{\mu^{-2s}}{\Gamma(s)} \int_0^\infty dt \frac{t^{s+1}}{2} Tr\{\int_0^1 e^{-(1-u)\hat{p}^2 t} \lambda\phi e^{u\hat{p}^2 t} \lambda\phi\}. \tag{3.22}$$

At this point a complete set of states which diagonalizes $\hat{p}$ may be inserted to evaluate the trace

$$\int d^6 p |p><p| = 1 \tag{3.23a}$$

$$<p|\phi|q> = \frac{1}{(2\pi)^3}\phi(p-q) \tag{3.23b}$$

yielding

$$\begin{aligned}\varsigma_2(s) &= \frac{\lambda^2\mu^{-2s}}{2\Gamma(s)} \int_0^\infty dt t^{s+1} \int \frac{d^6 p d^6 q}{(2\pi)^6} \int_0^1 du e^{-[q^2+u(1-u)p^2]t} \phi(p)\phi(-p) \\ &= \frac{\lambda^2\mu^{-2s}}{2} \frac{\Gamma(s-1)\Gamma^2(2-s)}{\Gamma(s)\Gamma(4-2s)} \int \frac{d^6 p}{(4\pi)^3} \phi(p)(p^2)^{1-s}\phi(-p).\end{aligned} \tag{3.24}$$

The derivative of this at $s = 0$ is

$$\varsigma_2'(0) = \frac{\lambda^2}{24} \int \frac{d^6 p}{(4\pi)^3} \phi(p) p^2 \left[\ln(p^2/\mu^2) - \frac{8}{3}\right] \phi(-p) \tag{3.25}$$

which is the renormalized 1PI two point function in Φ^3 theory in 6 dimensions. The constant $-\frac{8}{3}$ may be absorbed by redefining the arbitrary parameter μ.

The general procedure for operator regularization is as follows.

(1) Use the background field formalism ($\Phi = \phi + \hbar\varphi$) to identify the relevant terms in the action $S[\Phi] = \int dx \mathcal{L}[\Phi]$ that are bi-linear ($M(\phi)$), tri-linear ($a(\phi)$), etc., in the quantum field φ, and compute $Z[\phi, J]$ as in the previous section:

$$\begin{aligned}Z[\phi] = Z_O \text{sdet}^{-\frac{1}{2}} \left[\hbar M(\phi)/\mu^2\right] &\Bigg\{ \sum_N^\infty \frac{1}{N!} \Bigg[\int dx \Bigg(\frac{\hbar^2}{3!} a_{ijk}(\phi) \frac{\delta^3}{\delta J_i \delta J_j \delta J_k} \\ &+ \frac{\hbar^3}{4!} b_{ijk\ell}(\phi) \frac{\delta^4}{\delta J_i \delta J_j \delta J_k \delta J_\ell} \cdots \Bigg) \Bigg]^N \exp\left[-\tfrac{1}{2\hbar} \int dx J M^{-1}(\phi) J\right] \Bigg\} \Bigg|_{J=0} \\ &= Z_O \text{sdet}^{-\frac{1}{2}} \left[\hbar M(\phi)/\mu^2\right]. \qquad \text{(to one loop)}\end{aligned} \tag{3.26}$$

(2) Regulate the (super)-determinant in this expression using the formula

$$\det(H) = \exp\left[-\varsigma'(0)\right] \tag{3.27a}$$

where

$$\varsigma(s) = \frac{1}{\Gamma(s)} \int_0^\infty dt t^{s-1} Tr(e^{-Ht}). \tag{3.27b}$$

To n-loop order, products of inverses of operators appear; represent these using

$$H^{-N} = \frac{1}{\Gamma(N)} \lim_{s\to 0} \frac{d^m}{ds^m}\left[\frac{s^{m-1}}{m!}\frac{\Gamma(s+N)}{\Gamma(s)} H^{-s-N}\right] \tag{3.28}$$

and

$$H_1^{-1}H_2^{-1}...H_p^{-1} = \lim_{s\to 0}\frac{d^m}{ds^m}\left[\frac{s^m}{m!} H_1^{-s-1}H_2^{-s-1}...H_p^{-s-1}\right]. \tag{3.29}$$

(3) Decompose the operator(s) of interest into diagonalizable and (perturbatively) interacting parts ($H = H_0 + H_I$), using the expansions[13]

$$\begin{aligned} Tr(e^{-Ht}) = Tr\Big[& e^{-H_0 t} + (-t)e^{-H_0 t}H_I + \frac{(-t)^2}{2}\int_0^1 du e^{-(1-u)H_0 t}H_I e^{-uH_0 t}H_I \\ & + \frac{(-t)^3}{3}\int_0^1 u du \int_0^1 dv e^{-(1-u)H_0 t}H_I e^{-u(1-v)H_0 t}H_I e^{-uvH_0 t}H_I \\ & + \cdots\Big] \end{aligned} \tag{3.17}$$

and

$$\begin{aligned} e^{-(H_0+H_1)t} = & e^{-H_0 t} + (-t)\int_0^1 du e^{-(1-u)H_0 t}H_I e^{-uH_0 t} \\ & + (-t)^2\int_0^1 du\, u \int_0^1 dv\, e^{-(1-u)H_0 t}H_I e^{-u(1-v)H_0 t}H_I e^{-uvH_0 t} + ...] \end{aligned} \tag{3.30}$$

to evalute the $\varsigma(s)$ and expression which occur at higher loops.

(4) When the above expansions are inserted into the right hand side of (3.26) the $n-th$ term in H_I gives, upon evaluation, the amplitude corresponding to the n-point 1PI Green's function.

Operator Regularization has several distinctive features which will now be outlined.

As previously mentioned, the full one-loop generating functional as computed using Operator Regularization is given by

$$Z[\phi] = Z_0 \exp\left[-\frac{1}{2}\sum_n \varsigma_n'(0)\right] \tag{3.31}$$

and so the effective action to one-loop is given by

$$\Gamma = S[\phi] + \frac{1}{2}\sum_n \varsigma_n'(0). \tag{3.32}$$

It is important to note that the decomposition $M = M_0 + M_I$ is quite general; one need not require that M_I be proportional to some small dimensionless coupling constant. For example, in the case of quantum field theory in curved spacetime it is possible to require M_I to be small in powers of derivatives of the metric[14] (the adiabatic approximation[15]).

A second noteworthy feature of Operator Regularization is that there are no Feynman diagrams. Indeed, it is not possible to obtain the series (3.32) by judiciously inserting s in some systematic manner in the individual terms in the usual diagrammatic series [7]. Hence this method differs from that of Speer[16] and the analytic regularization technique of Lee and Milgram[17]. This is easily understood by explicitly comparing the two approaches. From (2.5) one has in the usual diagrammatic approach

$$Z[\phi] = Z_0\left\{\sum_N^{\infty}\frac{1}{N!}\left[\int dx\left(\frac{\hbar}{2!}(M_{ij})_I\frac{\delta^2}{\delta J_i \delta J_j} + \frac{\hbar^2}{3!}a_{ijk}(\phi)\frac{\delta^3}{\delta J_i \delta J_j \delta J_k} + \cdots\right)\right]^N \times \int d\varphi e^{-\int dx\left(\frac{\hbar}{2!}\varphi M_0 \varphi + J\cdot\varphi\right)}\right\}\Bigg|_{J=0} \tag{3.33}$$

because the decomposition $M = M_0 + M_I$ is made *before* evaluating the functional integral[18]. Here M_0 is independent of the background field. However in Operator regularization, this decomposition is made *after* evaluation of the functional integral as in (2.13):

$$Z[\phi] = Z_0\left\{\sum_N^{\infty}\frac{1}{N!}\left[\int dx\left(\frac{\hbar^2}{3!}a_{ijk}(\phi)\frac{\delta^3}{\delta J_i \delta J_j \delta J_k} + \cdots\right)\right]^N\right.$$

$$\times \int d\varphi e^{-\int dx\left(\frac{\hbar}{2!}\varphi M(\phi)\varphi + J\cdot\varphi\right)}\Big\}\Big|_{J=0}$$

$$= Z_0 s\det{}^{-\frac{1}{2}}\left[\hbar M(\phi)/\mu^2\right]\left\{\sum_N^{\infty}\frac{1}{N!}\left[\int dx\left(\frac{\hbar^2}{3!}a_{ijk}(\phi)\frac{\delta^3}{\delta J_i\delta J_j\delta J_k}+\cdots\right)\right]^N\right.$$

$$\times \exp\left[-\tfrac{1}{2}\hbar^{-1}\int dx J M^{-1}(\phi)J\right]\Big\}\Big|_{J=0} \tag{3.34}$$

Only after this stage is the decompostion $M = M_0 + M_I$ made, (3.34) being further evaluated using the expansions (3.17) and (3.30). It is this particular order of operations that distinguishes Operator Regularization from the usual graphical approach.

A third feature of Operator Regularization is the absence of any infinite quantities at all stages of the calculation, even as the regulating parameter s approaches its limiting value of zero. This is a consequence of the regularizations employed in step (2) above. Again, it is useful to compare to other methods. Consider, for example the regularization of det$[M]$. One has

$$\begin{aligned}\det[M] &= \exp\left[Tr(\ln(M))\right]\\ &= \exp\left[\int_0^{\infty}\frac{dt}{t}Tr\left(e^{-Mt}\right)\right]\end{aligned} \tag{3.35}$$

in the usual proper-time regularization scheme. This representation of the determinant is clearly divergent; each term in the series expansion associated with the decomposition $M = M_0 + M_I$ will therefore need to be regulated in some manner. A variety of prescriptions exist for doing this and are given by Lee, Pac and Rim[9,19], Fradkin and Tseytlin[20], Ford[21], Guven[22], Parker, Jack[23] and Toms[24] and Zuk[25]. In contrast to these approaches, in Operator Regularization (3.35) becomes

$$\det[M] = \exp\left\{-\frac{d}{ds}\left[\frac{1}{\Gamma(s)}\int_0^{\infty} dt t^{s-1} Tr\left(e^{-Mt}\right)\right]\Big|_{s=0}\right\} \tag{3.36}$$

which is the same as (3.35) if the derivative operation is carried out before the trace and t-integration operations. Ultraviolet divergences, then, are removed in Operator Regularization by employing the zeta function in this particular way. When the expansion (3.17) is applied to (3.36) every term in the series is manifestly ultraviolet finite, even as the parameter s approaches its limiting value.

Infrared divergences will be present, as in the usual diagrammatic approaches, when massless particles mediate interactions; they typically take the form $\int d^n k(k^2)^{-s}$. These divergences may be removed either by adding an infrared regulator mass or by analytically continuing them to zero in the following way:

$$\begin{aligned}
\int d^n k(k^2)^{-s} &= \int d^n k \frac{(k+p)^2}{k^{2s}(k+p)^2} \\
&= s\int d^n k \int_0^1 dx(1-x)^{s-1}\frac{(k+p)^2}{[(k+px)^2+p^2x(1-x)]^{s+1}} \\
&= \frac{1}{\Gamma(s)}\int d^n k \int_0^1 dx(1-x)^{s-1}\frac{k^2+(1-x)^2p^2}{[k^2+p^2x(1-x)]^{s+1}} \\
&= \pi^{\frac{n}{2}}\frac{\Gamma(s-\frac{n}{2})}{\Gamma(s+1)}(p^2)^{\frac{n}{2}-s}\int_0^1\left[\frac{n}{2}x^{\frac{n}{2}-s}(1-x)^{\frac{n}{2}-1}+(s-\frac{n}{2})x^{\frac{n}{2}-s-1}(1-x)^{\frac{n}{2}}\right] \\
&= \pi^{\frac{n}{2}}\frac{\Gamma(s-\frac{n}{2})}{\Gamma(s+1)}(p^2)^{\frac{n}{2}-s}\left[\frac{n}{2}\frac{\Gamma(\frac{n}{2}-s+1)\Gamma(\frac{n}{2})}{\Gamma(n-s+1)}+(s-\frac{n}{2})\frac{\Gamma(\frac{n}{2}-s)\Gamma(\frac{n}{2}+1)}{\Gamma(n-s+1)}\right] \\
&= 0.
\end{aligned} \tag{3.37}$$

Note that this result holds even if n is integer-valued from the outset.

Green's functions calculated in Operator Regularization differ by a finite renormalization from results obtained using standard schemes such as MS, $\overline{\text{MS}}$, and MOM[26]. In this sense renormalization is automatic in Operator Regularization (the renormalization parameter being the aforementioned μ^2), enhancing the efficiency of the procedure.

A fourth feature, related to the previous one, is that Operator Regularization is symmetry preserving. Because it is the *operators* which occur in the generating functional which are regulated, there is no need to alter the original Lagrangian (as in the Pauli-Villars or Dimensional Regularization schemes).

Furthermore, the absence of infinities at all stages of the calculation (including the last) means that no additional cutoffs or continuations need be introduced. The procedure will respect all of the symmetries that the original Lagrangian does (except in the case of anomalies), including evidently supersymmetry, as the number of Bose and Fermi degrees of freedom are preserved at each stage in the calculation. Indeed, it is possible to use operator regularization entirely within the context of superfield formalism[27]. Anomalies manifest themselves in a breakdown of the expansion of $\varsigma(s)$ into $\sum_n \varsigma_n(s)$. For example, the usual chiral anomaly arises because Hermiticity of the Euclidean effective action (which is crucial for a meaningful definition of all terms in the sum $\sum_n \varsigma_n$) can be preserved only at the expense of chiral invariance[28].

Finally it should be pointed out that operator regularization is designed to be an all-loop procedure (unlike other zeta-function schemes[6]). Provided that m in (3.28) and (3.29) is chosen to be greater than or equal to the loop order, the above features of operator regularization will hold in higher loops[7] (although complete self-consistency to higher loops has not yet been fully demonstrated).

4. Applications

The formalism developed in the preceding sections will now be applied in several specific cases. Particular attention will be given to the Operator Regularization technique.

(a) Flat Space

Before addressing problems in quantum field theory in curved spacetime and quantum gravity, it will be instructive to look at a few examples of the application of Operator Regularization in flat space.

Consider first the 2-point function in Yang-Mills theory[7]. The Lagrangian is

$$\mathcal{L} = -\frac{1}{4} tr\big(F_{\mu\nu}(W) F^{\mu\nu}(W)\big) \tag{4.1}$$

Decomposing the gauge field W_μ^a into its background and quantum parts $W_\mu^a = V_\mu^a + Q_\mu^a$ yields for the quadratic part of the gauge-fixed Lagrangian:

$$\mathcal{L} = -\frac{1}{2}Q_\mu^a\Big[D^2(V)^{ab}\delta_{\mu\nu} - (1-\frac{1}{\alpha})D_\mu^{ac}(V)D_\nu^{cb}(V) + 2gf^{acb}F_{\mu\nu}^c(V)\Big]Q_\nu^b - \bar{c}^a D^2(V)^{ab}c^b \tag{4.2}$$

where α is the gauge parameter associated with the gauge-fixing $D_\mu^{ab}(V)Q_\mu^b = 0$ and c, $\bar{c}$ are the ghost and antighost fields respectively. Here

$$D_\mu^{ab}(V) \equiv \delta^{ab}\partial_\mu + gf^{acb}V_\mu^c. \tag{4.3}$$

The matrix M of the previous section may be read off directly from (4.2) and so the one-loop generating functional is

$$Z_{(1)}[V] = \det\big[-D^2(V)^{ab}\big]\det^{-\frac{1}{2}}\Big[-D^2(V)^{ab}\delta_{\mu\nu} + (1-\frac{1}{\alpha})D_\mu^{ac}(V)D_\nu^{cb}(V) - 2gf^{acb}F_{\mu\nu}^c(V)\Big]. \tag{4.4}$$

Choosing $M_0 = p^2\delta^{ab}$ and $\alpha = 1$, performing the expansion (3.17) for both of the above determinants, and computing the trace for the piece quadratic in V_μ gives[7]

$$\begin{aligned}\varsigma_2(s) &= \frac{1}{2}\varsigma_2^{(Q)}(s) - \varsigma_2^{(ghost)} \\ &= \frac{g^2C_2\delta^{ab}}{(4\pi)^2}\int d^4p(\frac{p^2}{\mu^2})^{-s}V_\mu^a(p)\big[-\delta_{\mu\nu}p^2 + p_\mu p_\nu\big]V_\nu^b(-p) \\ &\quad\times\Big[\frac{\Gamma(2-s)\Gamma(1-s)}{\Gamma(4-2s)} - 2\frac{\Gamma^2(1-s)}{\Gamma(2-2s)}\Big]\end{aligned} \tag{4.5}$$

where C_2 is the usual Casimir quantity. Note that the above expression is completely finite at $s = 0$. The 2-point function is given by the derivative of this term at $s = 0$:

$$\varsigma_2'(0) = \frac{g^2C_2\delta^{ab}}{(4\pi)^2}\int d^4p(\frac{64}{9} - \frac{11}{3}\ln(\frac{p^2}{\mu^2}))V_\mu^a(p)\big[-\delta_{\mu\nu}p^2 + p_\mu p_\nu\big]V_\nu^b(-p) \tag{4.6}$$

which, up to a finite renormalization, is the standard result for the renormalized vacuum polarization tensor in Yang-Mills theory.

As mentioned in the previous section, Operator Regularization is applicable to superfield formalism as well. Consider the Wess-Zumino model with chiral and antichiral superfields:

$$S = \int d^4x \Big\{ \int d^4\theta \overline{\phi}\phi - \int d^2\theta \Big(\frac{m}{2}\phi^2 + \frac{\lambda}{3!}\phi^3 \Big) - \int d^2\overline{\theta} \Big(\frac{m}{2}\overline{\phi}^2 + \frac{\lambda}{3!}\overline{\phi}^3 \Big) \Big\} \tag{4.7}$$

By decomposing ϕ and $\overline{\phi}$ into background plus quantum parts $(\phi_B + \varphi)$, it is straightforward to show that the matrix M is given by (2.17):

$$M[\phi_B, \overline{\phi}_B] = \begin{pmatrix} \frac{1}{4}(m + \lambda\phi_B)\overline{D}^2 & \frac{\overline{D}^2 D^2}{16} \\ \frac{D^2 \overline{D}^2}{16} & \frac{1}{4}(m + \lambda\phi_B)\overline{D}^2 \end{pmatrix} \tag{2.17}$$

Choosing M_0 to be the λ-independent part of this expression, M_I to be the remaining part, and applying Operator Regularization yields[27] (for $m = 0$):

$$\zeta_2(s) = \frac{\lambda^2}{16} \int \frac{d^4p d^4\theta}{(4\pi)^2} (\frac{p^2}{\mu^2})^{-s} \varphi(p,\theta)\overline{\varphi}(-p,\theta) \frac{\Gamma^2(1-s)}{\Gamma(2-2s)} \tag{4.8}$$

and so the 2-point function is

$$\zeta_2'(0) = \frac{\lambda^2}{16} \int \frac{d^4p d^4\theta}{(4\pi)^2} \varphi(p,\theta) \left[2 - \ln(\frac{p^2}{\mu^2})\right] \overline{\varphi}(-p,\theta) \tag{4.9}$$

This result is the same as that obtained by Grisaru, Siegel and Rocek[29]. In ref. (29), however, it was necessary to employ a complicated regularization scheme (dimensional reduction) to preserve supersymmetry and then specify an appropriate subtraction procedure (the $\overline{\mathrm{MS}}$ scheme). In Operator regularization, supersymmetry is manifestly preserved[27].

(b)Curved Space

Consider next the evaluation of the one-loop generating functional associated with the action

$$S = \frac{1}{2} \int dx \sqrt{g} \Big[(D_\mu\Phi)^\dagger (D^\mu\Phi) - (m^2 + \xi R)|\Phi|^2 \Big] \tag{4.10}$$

using both heat-kernel techniques and Operator Regularization

Consider first the heat kernel approach[15]. The Euler-Lagrange equations imply that the associated Green's function obeys

$$(-\Box + m^2 + \xi R)G(x,y) = -\frac{1}{\sqrt{g}}\delta(x-y) \tag{4.11}$$

where

$$\Box\Phi = \frac{1}{\sqrt{g}}D_\mu(\sqrt{g}g^{\mu\nu}D_\nu\Phi) \tag{4.12}$$

and where D_μ is defined in (2.16). In operator form (4.11) may be rewritten as

$$(H + m^2 + \xi R)\mathcal{G} = 1 \tag{4.13a}$$

$$< x|G|y > \equiv G(x,y) \tag{4.13b}$$

$$\mathcal{G} \equiv g^{\frac{1}{4}} G g^{\frac{1}{4}} \tag{4.13c}$$

$$H \equiv g^{\frac{1}{4}}\pi_\mu\sqrt{g}g^{\mu\nu}\pi_\nu g^{-\frac{1}{4}} \tag{4.13d}$$

$$\pi_\mu \equiv p_\mu - eA_\mu \tag{4.13e}$$

and so the heat kernel equation is

$$\frac{\partial}{\partial\tau}F(x,y,\tau) + HF(x,y,\tau) = 0. \tag{4.14}$$

A method of solving this equation was derived by DeWitt[30]. If x and y are sufficiently close so that there is only one geodesic connecting them then the following ansatz for F

$$F(x,x',\tau) = \frac{(\det(\partial_\mu\partial'_\nu\sigma(x,x'))^{\frac{1}{2}}}{(4\pi\tau)^2}e^{-\sigma/2\tau}\sum_n^\infty a_n\tau^n \tag{4.15}$$

allows one to iteratively solve for the a_n in (4.14). The a_n are functionals of the background fields $g_{\mu\nu}$ and A_μ, and $\sigma(x,x')$. (Alternatively one could expand (4.11) in normal coordinates and iteratively invert the result in fourier space[31]. The a_n's obey the recursion relations

$$\partial^\mu\sigma\partial_\mu a_0 = 0 \tag{4.16a}$$

$$[\partial^\mu\sigma\partial_\mu + (n+1)]a_{n+1} = \Delta^{-\frac{1}{2}}\Box(\Delta^{-\frac{1}{2}}a_n) \tag{4.16b}$$

where $\triangle$ is the van-Vleck determinant[5]

$$\triangle = g^{\frac{1}{2}}(x)\det(\partial_\mu \partial'_\nu \sigma(x,x'))g^{\frac{1}{2}}(x') \tag{4.17}$$

In the coincidence limit $x = x'$, $\sigma = 0$ and

$$\begin{aligned}
a_0 &= 1 \\
a_1 &= (\frac{1}{6} - \xi)R(x) \\
a_2 &= \frac{1}{2}(\frac{1}{6} - \xi)^2 R^2(x) + \frac{1}{6}(\frac{1}{6} - \xi)\Box R(x) + \frac{1}{180}\Box R(x) \\
&\quad - \frac{1}{180}R^{\alpha\beta}(x)R_{\alpha\beta}(x) + \frac{1}{180}R^{\alpha\beta\mu\nu}(x)R_{\alpha\beta\mu\nu}(x) + \frac{1}{12}F_{\mu\nu}F^{\mu\nu}
\end{aligned} \tag{4.18}$$

(In fact it is possible to show[23] that the series $\sum_n^\infty a_n\tau^n$ can be partially summed to factor out $e^{-(\frac{1}{6}-\xi)R\tau}$). Hence, from (3.11), the heat kernel solution is

$$G(x,x') = \frac{\triangle^{\frac{1}{2}}(x,x')}{(4\pi)^{n/2}} \int_0^\infty dt t^{-\frac{n}{2}} e^{-[m^2 t - \frac{\sigma}{2t}]} F(x,x',t) \tag{4.19}$$

in n dimensions.

Consider now the same problem in the context of Operator Regularization[14]. The generating functional is

$$\begin{aligned}
Z[g,A] &= \int d\varphi e^{-\frac{1}{2}\int dx\sqrt{g}(\varphi(-\Box + m^2 + \xi R)\varphi)} \\
&= \det{}^{-1}(-\Box + m^2 + \xi R) \\
&= \exp\big(tr[\ln((-\Box + m^2 + \xi R)]\big) \\
&= \frac{d}{ds}\varsigma(s)\Big|_{s=0}
\end{aligned} \tag{4.20}$$

where now

$$\varsigma(s) = \frac{1}{\Gamma(s)}\int_0^\infty dt t^{s-1} \int dx < x|e^{-(-\Box + m^2 + \xi R)t}|x > . \tag{4.21}$$

In order to proceed further, it is necessary to make use of some approximation scheme for $(-\Box + m^2 + \xi R)$ as the eigenvalues of this operator are not known exactly for a general background metric g. Two useful such approximations are the normal coordinate expansion and the weak field expansion.

Consider first the normal coordinate expansion. Locally, for any operator $\mathcal{O}(x,x')$

$$
\begin{aligned}
< x|\mathcal{O}(x,x')|x' > &= < y|\mathcal{O}(y,x')|0 > \\
&= \int dpdq < y|p >< p|\mathcal{O}(i\frac{\partial}{\partial p}, x')|q >< q|0 >
\end{aligned}
\tag{4.22}
$$

where $y = x - x'$. By carrying out an expansion of $\mathcal{O}$ in powers of y about the point x' one can then apply Operator Regularization. In the normal coordinate expansion, the coordinates x' are chosen so that the connection $\Gamma^{\lambda}_{\mu\nu}$ vanishes at that point. The series expansion of $\mathcal{O}$ is then an expansion in powers of metric derivatives (*ie.*, curvatures), and is fully covariant. The expansion is valid provided the curvature tensor and its derivatives are are sufficiently small so that there is one unique geodesic connecting the points x and x'. This is sometimes referred to as an adiabatic expansion[5,30].

For $(-\Box + m^2 + \xi R)$ this expansion is, to adiabatic order 4

$$
\begin{aligned}
(-\Box + m^2 + \xi R) &= \frac{-1}{\sqrt{g}}(-i\hat{\pi}_\mu)(\sqrt{g}g^{\mu\nu})(-i\hat{\pi}_\nu) + m^2 + \xi R \\
&= \hat{p}^2 - \hat{p}_\mu B^{\mu\nu}_{2\alpha\beta}\hat{\partial}^\alpha\hat{\partial}^\beta\hat{p}_\nu - A_{2\alpha\beta}\hat{\partial}^\alpha\hat{\partial}^\beta\hat{p}^2 \\
&\quad - i\hat{p}_\mu B^{\mu\nu}_{3\alpha\beta\gamma}\hat{\partial}^\alpha\hat{\partial}^\beta\hat{\partial}^\gamma\hat{p}_\nu - iA_{3\alpha\beta\gamma}\hat{\partial}^\alpha\hat{\partial}^\beta\hat{\partial}^\gamma\hat{p}^2 \\
&\quad + \hat{p}_\mu B^{\mu\nu}_{4\alpha\beta\gamma\delta}\hat{\partial}^\alpha\hat{\partial}^\beta\hat{\partial}^\gamma\hat{\partial}^\delta\hat{p}_\nu + A_{4\alpha\beta\gamma\delta}\hat{\partial}^\alpha\hat{\partial}^\beta\hat{\partial}^\gamma\hat{\partial}^\delta\hat{p}^2 \\
&\quad + A_{2\alpha\beta}\hat{\partial}^\alpha\hat{\partial}^\beta\hat{p}_\mu B^{\mu\nu}_{2\gamma\delta}\hat{\partial}^\gamma\hat{\partial}^\delta\hat{p}_\nu \\
&\quad + m^2 + \xi R + i\xi R_{;\alpha}\hat{\partial}^\alpha - \xi R_{;\alpha\beta}\hat{\partial}^\alpha\hat{\partial}^\beta \\
&\quad + \frac{1}{2}e(\hat{p}_\mu[F^\mu_\nu\hat{\partial}^\nu + \frac{i}{3}F^\mu_{\nu;\beta}\hat{\partial}^\nu\hat{\partial}^\beta - \frac{1}{8}F^\mu_{\nu;\beta\gamma}\hat{\partial}^\nu\hat{\partial}^\beta\hat{\partial}^\gamma] \\
&\quad + [F^\mu_\nu\hat{\partial}^\nu + \frac{i}{3}F^\mu_{\nu;\beta}\hat{\partial}^\nu\hat{\partial}^\beta - \frac{1}{8}F^\mu_{\nu;\beta\gamma}\hat{\partial}^\nu\hat{\partial}^\beta\hat{\partial}^\gamma]\hat{p}_\mu) \\
&\quad - \frac{e^2}{4}F^\mu_\nu\hat{\partial}^\nu F_{\mu_\alpha}\hat{\partial}^\alpha + \cdots
\end{aligned}
\tag{4.23}
$$

where $\hat{p}$ and $\hat{\partial}^\mu \equiv \frac{\partial}{\partial p^\mu}$ are to be considered as operators on the space of states $|p>$ and where

$$A_{2\alpha\beta} = \frac{1}{6}R_{\alpha\beta} \tag{4.24a}$$

$$A_{4\alpha\beta\gamma} = \frac{1}{12}R_{\alpha\beta;\gamma} \tag{4.24b}$$

$$A_{4\alpha\beta\gamma\delta} = \frac{1}{2}(\frac{1}{36}R_{\alpha\beta}R_{\gamma\delta} + \frac{1}{90}R^{\kappa}{}_{\lambda\alpha\beta}R^{\lambda}{}_{\gamma\delta\kappa} + \frac{1}{20}R_{\alpha\beta;\gamma\delta}) \tag{4.24c}$$

$$B^{\mu\nu}_{2\alpha\beta} = \frac{1}{3}(R^{\mu}{}_{\alpha}{}^{\nu}{}_{\beta} - \frac{1}{2}\eta^{\mu\nu}R_{\alpha\beta}) \tag{4.24d}$$

$$B^{\mu\nu}_{3\alpha\beta\gamma} = \frac{1}{6}(R^{\mu}{}_{\alpha}{}^{\nu}{}_{\beta;\gamma} - \frac{1}{2}\eta^{\mu\nu}R_{\alpha\beta;\gamma}) \tag{4.24e}$$

$$\begin{aligned} B^{\mu\nu}_{4\alpha\beta\gamma\delta} &= \frac{1}{2}\eta^{\mu\nu}(\frac{1}{36}R_{\alpha\beta}R_{\gamma\delta} - \frac{1}{90}R^{\kappa}_{\lambda\alpha\beta}R^{\lambda}_{\gamma\delta\kappa} - \frac{1}{20}R_{\alpha\beta;\gamma\delta}) \\ &+ \frac{1}{20}R^{\mu}{}_{\alpha}{}^{\nu}{}_{\beta;\gamma\delta} + \frac{1}{15}R^{\mu}{}_{\alpha}{}^{\sigma}{}_{\beta}R^{\sigma}{}_{\gamma}{}^{\nu}{}_{\delta} - \frac{1}{18}R^{\mu}{}_{\alpha}{}^{\sigma}{}_{\beta}R_{\gamma\delta} \end{aligned} \tag{4.24f}$$

with the semicolon denoting the covariant derivative and where all quantities are evaluated at x'.

Applying operator regularization to (4.21) yields[14]

$$\begin{aligned} \Gamma &= \sum_n \zeta_n'(0) \\ &= \frac{d}{ds}\Big[\frac{\Gamma(s+2-\frac{n}{2})}{\Gamma(s)(4\pi)^{\frac{n}{2}}}(\frac{\mu^2}{m^2})^2 \int dx\sqrt{g(x)}\{\frac{(m^2)^{\frac{n}{2}}}{(s-\frac{n}{2})(s+1-\frac{n}{2})} + \frac{(m^2)^{\frac{n}{2}-1}}{(s-\frac{n}{2})}(\frac{1}{6}-\xi)R(x) \\ &+ (m^2)^{\frac{n}{2}-2}[\frac{1}{30}\Box R + \frac{1}{180}R^{\mu\nu\alpha\beta}R_{\mu\nu\alpha\beta} - \frac{1}{180}R^{\mu\nu}R_{\mu\nu} - \frac{1}{6}\xi\Box R \\ &+ \frac{1}{2}(\xi-\frac{1}{6})^2R^2 + \frac{1}{12}tr(F_{\mu\nu}F^{\mu\nu})]\}\Big]\Big|_{s=0} + \zeta_5'(0) + \cdots \end{aligned} \tag{4.25}$$

for the effective action to adiabatic order 4. This result is in agreement with that obtained by DeWitt[30] (equation (4.18)). In contrast to the manner in which (4.18) was obtained (in which it was necessary to solve the recursion relations (4.16) which arise from the ansatz (4.15) inserted into the Green's function equation), in Operator Regularization this result follows from a straightforward evaluation of (4.21) in normal coordinates[14].

It is also possible to make use of the weak field expansion to solve for the effective action from (4.21). In this case one writes

$$g_{\mu\nu} = \delta_{\mu\nu} + h_{\mu\nu} \tag{4.26}$$

yielding

$$\Box = \partial^2 - \partial_\mu (h^{\mu\nu} - \frac{1}{2}\delta^{\mu\nu} h) - h^{\mu\nu}\partial_\mu \partial_\nu + \cdots \tag{4.27}$$

Inserting these expansions in (4.21) yields for the h–h 2 point function

$$\begin{aligned} \zeta_{hh}(s) = \frac{1}{8\Gamma(s)} \frac{\Gamma(s-\frac{n}{2})(\Gamma(\frac{n}{2}-s+1))^2}{\Gamma(n-2s+2)} \int \frac{d^n p}{(4\pi)^{\frac{n}{2}}} (p^2)^{n/2-(s+2)} (\mu^2)^s \\ \Big\{ 2p^2 h^{\mu\nu}(p) p^2 h_{\mu\nu}(-p) - 4p^\alpha p_\mu h^{\mu\nu}(p) p_\alpha p_\beta h^\beta_\nu(-p) \\ +(n-2s)(n-2s-2) p_\mu p_\nu h^{\mu\nu}(p) p_\alpha p_\beta h^{\alpha\beta}(-p) \\ +(p^2 h(p) p^2 h(-p) - 2p^2 h(p) p_\mu p_\nu h^{\mu\nu}(-p))[(n-2s)^2 - 2(n-2s) - 2] \Big\} \end{aligned} \tag{4.28}$$

which is leading term in $h^{\mu\nu}$ in the expansion of

$$\begin{aligned} \zeta_{hh}(s) = \frac{\pi^{\frac{n}{2}} \Gamma(s-n/2)(\Gamma(\frac{n}{2}-s+1))^2}{\Gamma(s)\Gamma(n-2s+2)} \\ \times \int d^n x (-\Box)^{\frac{n}{2}-2+s} (\mu^2)^s \sqrt{g(x)} \Big\{ R_{\alpha\beta} R^{\alpha\beta} - \frac{n-2s}{4(n-2s-1)} R^2 \\ + \frac{(n-2s-2)^2 (n-2s+1)}{8(n-2s-1)} R^2 \Big\}. \end{aligned} \tag{4.29}$$

in n dimensions. Taking account of the gauge fields and the $m^2 + \xi R$ term gives

$$\begin{aligned} \zeta_2(s;n) = \frac{\pi^{\frac{n}{2}} \Gamma(s-\frac{n}{2})(\Gamma(\frac{n}{2})-s+1))^2}{\Gamma(s)\Gamma(n-2s+2)} \\ \times \int d^n x \sqrt{g} (-\Box)^{\frac{n}{2}-2+s} (\mu^2)^s \sqrt{g(p)} Tr \Big\{ (n-2s+1) F_{\mu\nu} F^{\mu\nu} \\ +2((n-2s)^2-1)[m^2 + (\xi - \frac{n-2s-2}{4(n-2s-1)})R][m^2 + (\xi - \frac{(n-2s-2)}{4(n-2s-1)})R] \end{aligned}$$

$$+R_{\alpha\beta}R^{\alpha\beta} - \frac{n-2s}{4(n-2s-1)}R^2\Big\} \qquad (4.30)$$

Again, note that for any integer n, $\frac{d\varsigma}{ds}|_{s=0}$ is finite. Furthermore, if one now considers n to be a continuous parameter $n = 4 - \epsilon$, then

$$\frac{d\varsigma}{ds}(0; 4-\epsilon) = \frac{1}{8\pi^2\epsilon}\int d^4x Tr\Big\{\frac{1}{12}F^{\mu\nu}F_{\mu\nu} + \frac{1}{2}(m^2 + (\xi - \frac{1}{6})R)^2$$
$$+ \frac{1}{60}(R_{\mu\nu}R^{\mu\nu} - \frac{1}{3}R^2)\Big\} \qquad (4.31)$$

which is the result of 't Hooft and Veltman[2]. Both (4.30) (and consequently (4.31)) are obtained in operator regularization not only without resorting to Feynman rules or diagrams, but also without going through the intermediate step of considering the conformal case $g^{\mu\nu} = \frac{1}{F}\delta^{\mu\nu}$ as was done in ref. (2). This result is also in agreement with the weak field limit of (4.25) in 4 dimensions provided the Gauss bonnet identity $R^{\mu\nu\tau\sigma}R_{\mu\nu\tau\sigma} = 4R^{\mu\nu}R_{\mu\nu} - R^2$ is employed. In general, the weak field expansion cannot pick terms of the form $R^{\mu\nu\tau\sigma}R_{\mu\nu\tau\sigma}$ because in *any* number of dimensions $R^{\mu\nu\tau\sigma}R_{\mu\nu\tau\sigma} = 4R^{\mu\nu}R_{\mu\nu} - R^2$ up to a total derivative to $\mathcal{O}(h^3)$. Similarly, the weak field expansion does not pick the $\Box R$ terms to this order.

(c) Quantum Gravity to one-loop

As a third example, consider the case of quantum gravity. In diagrammatic language, one is considering the case where internal gravitons circulate through the loops[2]; in Operator Regularization, one functionally integrates out the graviton field h in the presence of a classical background metric.

The generating functional is now taken to be

$$Z[g_{\mu\nu}, \tilde{\varphi}] = \int [dh][d\varphi]exp\Big[\int d^nx\sqrt{\bar{g}}(\bar{R} + \frac{1}{2}\partial_\mu\bar{\varphi}\bar{g}^{\mu\nu}\partial_\nu\bar{\varphi})\Big] \qquad (4.32)$$

where $\bar{R}$ is the curvature scalar formed out of the full metric $\bar{g}_{\mu\nu}$ with

$$\bar{g}_{\mu\nu} = g_{\mu\nu} + h_{\mu\nu} \qquad (4.33a)$$

and

$$\bar{\varphi} = \tilde{\varphi} + \varphi \tag{4.33b}$$

so that $g_{\mu\nu}$ and $\tilde{\varphi}$ are the classical background fields.

In order to apply Operator Regularization to one-loop order, it is necessary to extract the part of the action that is quadratic in the quantum fields h and φ. This was previously carried out in ref. (2); upon supplementing the Lagrangian with a gauge fixing term and a ghost term as a consequence of general co-ordinate invariance, the result is[(2)]

$$\begin{aligned}\mathcal{L}_Q = \mathcal{L}_{(2)} + \mathcal{L}_{G.F.} + \mathcal{L}_{GHOST} = \sqrt{g}[h^*_{\alpha\beta}\{P^{\alpha\beta\mu\nu} D_\gamma D^\gamma + \frac{1}{2}(X^{\alpha\beta\mu\nu} + X^{\mu\nu\alpha\beta})\}h_{\mu\nu} \\ + \varphi^* D_\mu D^\mu \varphi + \varphi^* Y^{\alpha\beta} h_{\alpha\beta} + h^*_{\alpha\beta} Y^{\alpha\beta}\varphi + \varphi^* Z\varphi \\ + \eta^*_\mu \left\{g^{\mu\nu} D_\alpha D^\alpha - R^{\mu\nu} - \partial^\mu\tilde{\varphi}\partial^\nu\tilde{\varphi}\right\}\eta_\nu\Big]\end{aligned} \tag{4.34}$$

where the doubling trick[(2)] has been employed. Here the doubled fields $h_{\mu\nu}, h'_{\mu\nu}$ (φ, φ') have been combined into one complex field $h_{\mu\nu}(\varphi)$, η_μ is the anticommuting Fadeev-Popov ghost, and

$$P^{\alpha\beta\mu\nu} = \frac{1}{2} g^{\alpha\mu} g^{\beta\nu} - \frac{1}{4} g^{\alpha\beta} g^{\mu\nu} \tag{4.35a}$$

$$X^{\alpha\beta\mu\nu} = g^{\alpha\gamma} g^{\nu\pi}(-\delta^\beta_\pi Q^\mu_\gamma + \delta^\beta_\gamma Q^\mu_\pi + \frac{1}{4}(\delta^\beta_\pi \delta^\mu_\gamma - \delta^\beta_\gamma \delta^\mu_\pi)Q - Q^{\beta\mu}{}_{\pi\gamma}) \tag{4.35b}$$

$$Y^{\alpha\beta} = \frac{1}{2} g^{\alpha\beta} D_\nu D^\nu \tilde{\varphi} - D^\alpha D^\beta \tilde{\varphi} \tag{4.35c}$$

$$Z = -D_\mu \tilde{\varphi} D^\mu \tilde{\varphi} \tag{4.35d}$$

with

$$Q^{\beta\mu}{}_{\pi\gamma} \equiv R^{\beta\mu}{}_{\pi\gamma} + \frac{1}{2}\delta^\beta_\pi D^\mu\tilde{\varphi} D_\gamma \tilde{\varphi} \tag{4.36a}$$

$$Q^\mu_\nu = Q^{\beta\mu}{}_{\nu\beta} = R^\mu_\nu + \frac{1}{2} D^\mu \tilde{\varphi} D_\nu \tilde{\varphi} \tag{4.36b}$$

$$Q \equiv Q^\mu_\mu = R + \frac{1}{2} D^\tau \tilde{\varphi} D_\tau \tilde{\varphi} \tag{4.36c}$$

All indices are raised/lowered with the background metric $g_{\mu\nu}$, D_μ is the covariant derivative with respect to $g_{\mu\nu}$.

By considering $h_{\mu\nu}$ and φ as the components of an $\frac{n^2+n+2}{2}$ vector $\hat{\varphi}_I$ it is possible to rewrite $\mathcal{L}_Q$ in the form[2,32]

$$\mathcal{L}_Q = \sqrt{g}\Big\{\partial_\mu\hat{\varphi}_I^* g^{\mu\nu}\partial_\nu\hat{\varphi}_I + 2\hat{\varphi}_I^*\hat{\mathcal{N}}_{IJ}^\mu\partial_\mu\hat{\varphi}_J + \varphi_I^*\hat{\mathcal{M}}_{IJ}\varphi_J\Big\} \tag{4.37}$$

This form can be identified with the $\mathcal{L}$ for a set of 'charged' scalars in a curved background, which corresponds to the previous case (b).

The result[14] is that $\varsigma_2(s)$ is completely finite; consequently it is possible to evaluate $\frac{d\varsigma_2}{ds}|_{s=0}$ for any n. μ^2 dependence occurs only for n even, consistent with the result of Duff and Toms in dimensional regularization[32]. In four dimensions, when the equations of motion are applied to the background fields, the result is[14]

$$\varsigma_2'(0) = \int \frac{d^4x}{16\pi^2}\sqrt{g}\Big(-\frac{203}{80}\ln\Big(-\Box/\mu^2\Big) + \frac{5893}{1200}\Big)R^2 \tag{4.38}$$

The first term is the manifestation of the non-renormalizability of quantum gravity to one loop in Operator Regularization. Although no divergences ever appear, the final answer contains a term in the action proportional to logarithm of the arbitrary parameter μ which is not present in the original action. The coefficient of R^2 is completely arbitrary; the coefficient of this arbitrary term $-\frac{203}{80}$ agrees with that obtained by 't Hooft and Veltman[2] for the divergent part of the one-loop effective action.

(d) Non-Linear Sigma Model in 2 Dimensions

Consider, finally, the non-linear sigma model in 2 Dimensions[33]. The action is

$$S = \frac{1}{2}\int d^2x\sqrt{\gamma}\partial_\mu\Phi^i\gamma^{\mu\nu}\partial_\nu\Phi^j g_{ij}(\Phi) \tag{4.39}$$

The Φ^i map the two dimensional manifold (with metric $\gamma_{\mu\nu}$) into some target space manifold with metric g_{ij} and coordinates Φ^i. In string theory, the two dimensional manifold is interpreted as the world sheet of the string and the target space manifold is interpreted as spacetime[34].

Classically, the action is conformally invariant, so

$$\frac{2}{\sqrt{\gamma}}\gamma^{\mu\nu}\frac{\delta S}{\delta\gamma^{\mu\nu}}=0 \tag{4.40}$$

The generating functional in the background field formalism is

$$Z[\gamma,\phi]=\int d\xi e^{-S[\gamma,\phi+\pi(\xi)]}. \tag{4.41}$$

The background fields are γ and ϕ where

$$\frac{d^2}{d\tau^2}\phi^i(\tau)+\Gamma^i_{jk}\frac{d\phi^j}{d\tau}\frac{d\phi^k}{d\tau}=0 \tag{4.42}$$

is a geodesic in field space connecting ϕ to a nearby point in field space $\phi+\pi(\xi)$; hence $\phi(\tau=0)\equiv\phi$ and $\phi(\tau=1)=\phi+\pi(\xi)$, where $\xi\equiv\frac{d\phi}{d\tau}|_{\tau=0}$ is the tangent vector along the target space geodesic at $\tau=0$.

Although it is technically possible to carry out an expansion in powers of π, this is impractical because π does not transform as a target space vector; it is much simpler to consider ξ as the quantum field, and carry out a normal coordinate expansion of π in terms of ξ[33]:

$$\pi^i(\xi)=\xi^i-\frac{1}{2}\Gamma^i_{jk}(\tau=0)\xi^j\xi^k+\cdots \tag{4.43}$$

where target space coordinates may be chosen so that $\Gamma^i_{jk}(\tau=0)=0$. The quadratic part of the action is

$$\begin{aligned}S_2&=\frac{1}{2}\int d^2x\sqrt{\gamma}\gamma^{\mu\nu}\left\{\nabla_\mu\xi^i\nabla_\nu\xi^j g_{ij}(\phi)+R_{ijkl}\partial_\mu\phi^i\partial_\nu\phi^k\xi^j\xi^l\right\}\\&=\frac{1}{2}\int d^2x\sqrt{\gamma}\gamma^{\mu\nu}\left\{D_\mu\xi^a D_\nu\xi^a+R_{iajb}\partial_\mu\phi^i\partial_\nu\phi^j\xi^a\xi^b\right\}\end{aligned} \tag{4.44}$$

where the last line has been rewritten in terms of tangent space quantities in the target space: $\xi^a=e^a_i\xi^i$, $g_{ij}=e^a_ie^b_j\delta_{ab}$ and $D_\mu\xi^a=\partial_\mu\xi^a+\omega^a_{\mu b}\xi^b$.

The action (4.44) is like that of (4.10). Working in the weak-field limit of the 2 dimensional metric (*ie.*,$\gamma_{\mu\nu} = \delta_{\mu\nu} + h_{\mu\nu}$), the result (4.29) applies, and so the h–h 2 point function associated with the 2 dimensional box operator $\Box$ is

$$\varsigma_{hh}(s) = \frac{d}{32\pi}\int d^2p\left(\frac{p^2}{\mu^2}\right)^{-s}\frac{\Gamma(s-1)\Gamma(2-s)}{\Gamma(s)\Gamma(4-s)}\{\frac{(p^\mu p^\nu h_{\mu\nu} - p^2h)^2}{p^2}(2s)(2s-2)\} \tag{4.45}$$

Rewriting this is terms of generally covariant quantities gives, upon taking the derivative with respect to s at $s = 0$:

$$\varsigma'(0) = \frac{d}{48\pi}\int d^2x\sqrt{\gamma}R\Box^{-1}R(\gamma) \tag{4.46}$$

and so the trace anomaly is[35]

$$\frac{2}{\sqrt{\gamma}}\gamma^{\mu\nu}\frac{\delta S_{eff}}{\delta\gamma^{\mu\nu}} = \frac{d}{24\pi}R(\gamma) - \frac{1}{4\pi}R_{ij}\partial_\mu\phi^i\partial_\nu\phi^j \tag{4.47}$$

where the last term arises from the second term in the action (4.44).

The effective action as calculated in Operator regularization is finite and unambiguously leads to the trace anomaly (4.47). This is in contrast to continuous dimension schemes[36] such as dimensional regularization which do not unambiguously lead to the trace anomaly. In dimensional regularization, the one loop diagrams associated with the h–h part yield[35,36]

$$\begin{aligned}\frac{d}{16\pi}\left(\frac{p^2}{\mu^2}\right)^{\frac{\epsilon}{2}}\Gamma(1-\frac{\epsilon}{2})B(2+\frac{\epsilon}{2},2+\frac{\epsilon}{2})\Big\{&\frac{(p^\mu p^\nu\bar{h}_{\mu\nu} - p^2h)^2}{p^2}\\ &-\frac{2}{\epsilon(1+\frac{\epsilon}{2})}\left[-(p^\mu p^\nu\bar{h}_{\mu\nu})^2 - \frac{1}{2}p^2(\bar{h}_{\mu\nu})^2 + \frac{\epsilon}{8}p^2h^2\right]\Big\}\end{aligned} \tag{4.48}$$

for the one loop part of the effective action. Here $\bar{h}$ is defined to be $\bar{h}_{\mu\nu} = h_{\mu\nu} - \frac{1}{2}\delta_{\mu\nu}h$ in $n = 2+\epsilon$ dimensions. The second term in the above expression appears to have a pole at $\epsilon = 0$; this is not the case, however, since the term it multiplies identically vanishes in 2 dimensions and is therefore of $\mathcal{O}(\epsilon)$. Hence the last term is finite. This term makes a non-vanishing contribution to the trace anomaly. Furthermore, the contribution of this term is a non-trivial function of ϵ; its value at $\epsilon = 0$ depends upon how the continuation away from 2 dimensions is performed[35]. Consequently it is not possible to evaluate the trace anomaly unambiguously in dimensional regularization. Of course this is to be expected. The conformal group is very special in 2 dimensions, having an infinite number of generators. Hence a continuation away from 2 dimensions is a continuation away from this very special case and there is no unambiguous way to carry out such a continuation, (in contrast to gauge symmetries, whose associated Ward identities are insensitive to the number of dimensions). In contrast to this, in Operator Regularization all steps of the calculation are performed in *exactly* 2 dimensions, leading to a unique result for the trace anomaly[35].

5. Summary

Zeta function regularization techniques have the characteristic feature of removing infinities that appear in other schemes. All terms in the effective action are either finite or proportional to $\ln(\mu^2)$; these latter terms in other regulating schemes have infinite coefficients and so must be subtracted out by an appropriate choice of counterterms. In this sense zeta function methods 'automatically' renormalize the theory in question at some scale μ^2. As such they have been very useful in evaluating quantities associated in quantum gravity and quantum field theory in curved spacetime.

Traditionally, zeta function techniques have been implemented either by explicit knowledge of the eigenvalues of the Hamiltonian of the system[6] or by heat kernel techniques. A new method has recently been introduced by McKeon and Sherry[7], called Operator Regularization, which exploits zeta function technology in a manner that does not entail the modification of the original action that other regulating schemes do. The (one-loop) examples presented here illustrate the utility and versatility of Operator Regularization in both flat and curved space as a finite, symmetry-preserving method for computing perturbative Green's functions.

Some open questions concerning Operator Regularization are outstanding. The method appears to be self-consistent to all-loop order, although this has not been fully demonstrated. An understanding of the renormalization group equations to all-loop order in the context of Operator Regularization has yet to be worked out. Despite this, the examples presented here provide encouraging evidence that Operator Regularization may be a generic finite symmetry-preserving regulator for quantum field theory.

Acknowledgements

I would like to thank Gerry McKeon for introducing me to Operator Regularization and to the participants of the CAP-NSERC Summer Institute on Field Theory for interesting discussions on many of the above points. This work was supported by the Natural Sciences and Engineering Research Council of Canada.

References

1. G. 't Hooft, Nucl. Phys. **B33**, 173 (1971); **B35**, 167 (1971); E.S. Abers and B.W. Lee, Phys. Rep. **C9**, 1 (1973).
2. G. 't Hooft and M. Veltman, Ann. Inst. Henri Poincaré A XX, **69** (1974); S. Deser and P. van Nieuwenhuizen, Phys. Rev. Lett. **32**, 245 (1974); Phys. Rev. **D10**, 401 (1974); **D10**, 411 (1974).
3. For a collection of review articles, see *Quantum Gravity* vols. I and II, eds. C. Isham, R. Penrose and D. Sciama (Oxford University Press, 1982).
4. J.G. Taylor, *Proceedings of the NSERC-CAP Summer Institute on Theoretical Physics*, eds. F.C. Khanna, G. Kunstatter, H.C. Lee and H. Umezawa, Edmonton, Alberta (to be published 1987).
5. N. Birrell and P.C.W. Davies, *Quantum Field Theory in Curved Spacetime*, Cambridge University Press, Cambridge, UK (1982).
6. J. Dowker and R. Critchley, Phys. Rev. **D13**, 3224 (1976); S.W. Hawking, Comm. Math. Phys. **55**, 133 (1977); Salam and J. Strathdee, Nucl. Phys. **B90**, 203 (1975).
7. D.G.C. McKeon and T.N. Sherry, Phys. Rev. Lett. **59**, 532 (1987); Phys. Rev. **D35**, 3854 (1987); Can. J. Phys. (in press).
8. B. DeWitt, Phys. Rev. **162**, 1195 (1967); L. Abbott, Nucl. Phys. **B185**, 189 (1981).
9. C.Lee and C.Rim, Nucl. Phys. **B255**, 439 (1985); P. van Nieuwenhuizen, Phys. Rep.**68**, 189 (1981).
10. Heuristically speaking, we might expect this for free field theories for which

$M \sim \partial^2$ for bosons or $M \sim \not\partial$ for fermions, (whose eigenvalues are $\lambda \sim n^2$, $\sim n$ in the approximation of discreted spacetime). Interactions, when considered perturbatively, presumably do not change this behaviour significantly, so it is reasonable to expect $\varsigma(s)$ to enjoy the same properties as $\varsigma_R(s)$.

11. C. Bernard, Phys. Rev. **D9**, 3312 (1974); L. Dolan and R. Jackiw, Phys. Rev. **D9**, 3320 (1974).
12. J. Honerkamp, Nucl. Phys. **B48**, 269 (1972); N.K. Nielsen, Nordita report 78/24 (1978) unpublished; P.B. Gilkey, *The Index Theorem and the Heat Equation*, (Boston: Publish or Perish, 1974).
13. J. Schwinger, Phys. Rev. **82**, 664 (1951).
14. R.B. Mann, D.G.C. McKeon, L. Tarasov and T. Steele, University of Toronto preprint (1987).
15. B. DeWitt in *Relativity, Groups and Topology* II, eds. B. DeWitt and R. Stora (North Holland, 1983).
16. E.R. Speer, J. Math. Phys. **9**, 1404 (1968).
17. H.C. Lee and M. Milgram, Phys. Lett. **B133**, 320 (1983).
18. P. Ramond, *Field Theory: A Modern Primer* (Benjamin Cummings, 1981); C. Itzykson and B. Zuber, *Quantum Field Theory* (McGraw Hill, 1980).
19. C. Lee and M. Pac, Nucl. Phys. **B201**, 429 (1982); **B202**, 336 (1982); **B207** 157 (1982).
20. E.S. Fradkin and A.A. Tseytlin, Nucl. Phys. **B201**, 469 (1982); **B203**, 157 (1982); **B227**, 252 (1983); **B234**, 472 (1982); **B234**, 509 (1982).
21. L.H. Ford, Nucl. Phys. **B204**, 35 (1982); Phys. Rev. **D31**, 704 (1985); Phys. Rev. **D31**, 710 (1985).
22. J. Guven, Phys. Rev. **D35**, 2378 (1987).
23. L.Parker and I. Jack, Phys. Rev. **D31**, 2349 (1985).
24. L. Parker and D. Toms, Phys. Rev. **D31**, 953 (1985).

25. M. Zuk, Phys. Rev. **D33**, 3645 (1986).

26. H. Georgi and H.D. Politzer, Phys. Rev. **D14**, 1829 (1976); G. 't Hooft, Nucl. Phys. **B61**, 455 (1973); S. Weinberg, Phys. Rev. **D8**, 3497 (1973).

27. D.G.C. McKeon, T. Sherry and S. Rajpoot, Phys. Rev. **D35**, 3873 (1987).

28. D.G.C. McKeon and T. Sherry, University of Western Ontario preprint (1987).

28. M. Grisaru, W. Siegel and M. Rocek, Nucl. Phys. **B159**, 429 (1979).

30. B. DeWitt in *Relativity, Groups and Topology* I, eds. B. DeWitt and R. Stora (North Holland, 1963).

31. T. Bunch and L. Parker, Phys. Rev. **D20**, 2499 (1979).

32. M. Duff and D. Toms, *Unification of Fundamental Interactions II*, eds. S. Ferrara and J. Ellis (Plenum Press 1982).

33. L. Alvarez-Gaume, D.Z. Freedman and S. Mukhi, Ann. Phys. **134**, 85 (1981).

34. M. Green, J. Schwarz and E. Witten, *Superstring Theory*, Cambridge University Press (1987).

35. M. LeBlanc, B. Shadwick and R. Mann, University of Waterloo preprint (1987).

36. C. Hull and P. Townsend, Nucl. Phys. **B274**, 349 (1986).

APPLICATIONS OF THE SCHRÖDINGER PICTURE IN QUANTUM FIELD THEORY*

Roberto Floreanini

Center for Theoretical Physics
Laboratory for Nuclear Science
and Department of Physics
Massachusetts Institute of Technology
Cambridge, Massachusetts 02139 U.S.A.

and
Luc Vinet

Laboratoire de Physique Nucléaire
Université de Montréal
Montréal, Québec, H3C 3J7 CANADA

TABLE OF CONTENTS

* This work is supported in part by funds provided by the U. S. Department of Energy (D.O.E.) under contract #DE-AC02-76ER03069, the Natural Science and Engineering Research Council (NSERC) of Canada, the Québec Ministry of Education, and the Istituto Nazionale di Fisica Nucleare, Rome, Italy.

I. INTRODUCTION

It often proves useful to adopt the Schrödinger picture in quantum field theory. Explicit wavefunctionals can be written to describe effects of interest, and intuition with ordinary quantum mechanics may be used to advance understanding. One of the main advantages of this picture is the possibility of discussing kinematical problems [*i.e.* representation theory] without reference to dynamics. In the usual procedure, the need to regularize and renormalize kinematical objects, like generators of transformations, brings in dynamics through the normal ordering algorithm, which is defined with reference to a Fock vacuum, the lowest eigenstate of some quadratic Hamiltonian. Since there are infinitely many different Fock vacua, all connected by unitarily inequivalent Bogolubov transformations, one can obtain an infinity of inequivalent representations for the same operators.

On the contrary, within the Schrödinger picture one can regularize and renormalize intrinsically, without reference to any particular vacuum state. In this way one can study quantum representations in field theory independently of dynamics, exactly as in ordinary quantum mechanics. Moreover, when there is no well-defined notion of Fock vacuum, as in de Sitter space, the intrinsic method is the only one available.

We shall discuss the Schrödinger picture both for bosonic and fermionic quantum field theory.[1] We shall confine the discussion to two-dimensional models, which are relatively simple. The formalism may be used in higher dimensions, but regularization procedures may differ owing to differences in singularities.[2] It can be applied to linear and non-linear theories, although in the latter case only approximate functionals can be constructed, using for example variational methods.[3] For boson fields the technique is well-known, although not widely used. We shall review the representation theory for the two-dimensional conformal transformation group and the corresponding Lie algebra. The theory of complex scalar fields in de Sitter space interacting with an $SO(2,1)$-invariant Abelian gauge potential will also be considered in detail. The intrinsic functional method of renormalization will be used to well-define the $SO(2,1)$ symmetry charges of the theory and a solution to the problem of finding a vacuum state for the model will be given.

The Schrödinger picture for fermion fields has been introduced only recently and will be reviewed in detail. A functional representation for the two-dimensional conformal group in terms of anticommuting variables in analogy to that in terms of Bose fields will be presented. Fermionic dynamics governed by a quadratic Hamiltonian will also be discussed.

II. BOSONIC QUANTUM FIELD THEORIES IN MINKOWSKI SPACE

A. The Schrödinger Picture

For a scalar field theory the canonical variables $\Phi(x)$ and $\Pi(x)$ are Hermitian operators which satisfy the Heisenberg commutation relations.

$$\begin{aligned} [\Phi(x),\, \Phi(y)] &= [\Pi(x),\, \Pi(y)] = 0 \\ [\Phi(x),\, \Pi(y)] &= i\delta(x-y) \end{aligned} \tag{2.1}$$

This algebra may be realized on the space of functionals $\Psi(\varphi)$ of a classical field $\varphi(x)$. These functionals are viewed as kets, $|\Psi\rangle \leftrightarrow \Psi(\varphi)$. An inner product in the space is defined by functional integration,

$$\langle\Psi_1|\Psi_2\rangle = \int \mathcal{D}\varphi \Psi_1^*(\varphi)\Psi_2(\varphi) \tag{2.2}$$

so that the dual space of bras consists of complex conjugated functionals: $\langle\Psi| \leftrightarrow \Psi^*(\varphi)$. Operators are represented by functional kernels.

$$\mathcal{O}|\Psi\rangle \leftrightarrow \int \mathcal{D}\bar{\varphi}\mathcal{O}(\varphi,\bar{\varphi})\Psi(\bar{\varphi}) \tag{2.3}$$

In particular, the field operator $\Phi(x)$ is represented by a diagonal kernel $\varphi(x)\delta(\varphi-\bar{\varphi})$; the canonical momentum operator $\Pi(x)$ by $-i\frac{\delta}{\delta\varphi(x)}\delta(\varphi-\bar{\varphi})$. Note that the previous δ functions are functional and defined by

$$\delta(\varphi-\bar{\varphi}) = \int \mathcal{D}\alpha\, e^{i\int dx\,\alpha(x)(\varphi(x)-\bar{\varphi}(x))} \quad . \tag{2.4}$$

The algebra (2.1) is thus realized with Φ and Π Hermitian operators with respect to the scalar product (2.2).

In this functional space one can introduce a *Fock basis*. The Fock vacuum $|\Omega\rangle$ is represented by a Gaussian functional with specific covariance Ω, which is symmetric, possible complex, but with a positive definite real part Ω_R.

$$\begin{gathered}|\Omega\rangle \leftrightarrow \det{}^{1/4}\left(\frac{\Omega_R}{\pi}\right)\exp -\frac{1}{2}\int \varphi\Omega\varphi \\ \Omega = \Omega_R + i\Omega_I, \qquad \Omega(x,y)=\Omega(y,x)\end{gathered} \tag{2.5a}$$

$$\langle\Omega| \leftrightarrow \det{}^{1/4}\left(\frac{\Omega_R}{\pi}\right)\exp -\frac{1}{2}\int \varphi\Omega^*\varphi \tag{2.5b}$$

[An obvious functional notation is used throughout, $\int\varphi\Omega\varphi \equiv \int dx\,dy\,\varphi(x)\Omega(x,y)\varphi(y)$, and the determinant is functional.] Higher basis states are polynomials in φ multiplying $|\Omega\rangle$, and they are orthonormalized if linear combinations corresponding to "functional Hermite polynomials" are taken. This defines a *Fock space* within our functional space.

The reason for the nomenclature is that the above Fock vacuum is annihilated by an operator A that is linear in Φ and Π.

$$A = \frac{1}{\sqrt{2}}\int \Omega_R^{-1/2}\left\{\Omega\varphi + \frac{\delta}{\delta\varphi}\right\} \tag{2.6}$$

$$A|\Omega\rangle = 0 \tag{2.7}$$

Moreover, together with its Hermitian conjugate, A satisfies the standard relations of creation and annihilation operators.

$$\begin{gathered}[A(x),\, A(y)] = [A^\dagger(x),\, A^\dagger(y)] = 0 \\ [A(x),\, A^\dagger(y)] = \delta(x-y)\end{gathered} \tag{2.8}$$

Finally, for a quadratic Hamiltonian,

$$H = \frac{1}{2} \int \{\Pi^2 + \Phi\, h\, \Phi\} \tag{2.9}$$

the energy eigenstates are precisely those elements of the Fock space that are constructed by repeated applications of the creation operator $A^\dagger$ on the vacuum with covariance $\Omega = h^{1/2}$.

In our function space, Fock bases corresponding to vacua with different covariances can be inequivalent. This happens because field theory possesses an infinite number of degrees of freedom. Indeed, consider two Fock vacua with covariances Ω_1 and Ω_2. Their overlap is

$$e^{-N} \equiv \langle \Omega_1 | \Omega_2 \rangle \tag{2.10a}$$

$$N = \frac{1}{2} \operatorname{tr} \ln \frac{\Omega_1 + \Omega_2^*}{2(\Omega_{1R}\Omega_{2R})^{1/2}} \quad . \tag{2.10b}$$

In the case where the covariances are real and translation invariant [*i.e.* diagonal in Fourier space] N is given by

$$N = \frac{1}{2} V \int_k \ln \frac{1}{2} \left(\sqrt{\frac{\Omega_1(k)}{\Omega_2(k)}} + \sqrt{\frac{\Omega_2(k)}{\Omega_1(k)}} \right) \quad . \tag{2.11}$$

Here $\Omega_i(k)$ are the Fourier transformed kernels, and V is the volume of space. Even ignoring the [infrared] infinity associated in the spatial volume, N/V will still diverge in the ultraviolet unless Ω_1 and Ω_2 approach each other rapidly at large k. Since the integrand is positive, the divergence sets e^{-N} to zero, and the overlap (2.10a) vanishes. It also follows that all higher states built on $|\Omega_1\rangle$ and $|\Omega_2\rangle$, respectively, are mutually orthogonal. We conclude that our functional space contains all the inequivalent Fock spaces.

On the other hand, calculations can also be performed directly on the functional space, by introducing a *field basis* instead of a Fock basis. It consists of field eigenstates $|\varphi\rangle$, $\Phi(x)|\varphi\rangle = \varphi(x)|\varphi\rangle$, with functional δ function normalization.

$$\langle \varphi_1 | \varphi_2 \rangle = \delta(\varphi_1 - \varphi_2) \tag{2.12}$$

By definition this basis is complete in the functional space. The functionals $\Psi(\varphi)$ may be viewed as overlaps,

$$\langle\varphi|\Psi\rangle = \Psi(\varphi) \qquad \langle\Psi|\varphi\rangle = \Psi^*(\varphi) \tag{2.13}$$

and the functional kernels $\mathcal{O}(\varphi,\bar{\varphi})$ as matrix elements.

$$\langle\varphi|\mathcal{O}|\bar{\varphi}\rangle = \mathcal{O}(\varphi,\bar{\varphi}) \tag{2.14}$$

It follows that

$$\langle\varphi|\Phi(x)|\Psi\rangle = \varphi(x)\Psi(\varphi) \qquad \langle\varphi|\Pi(x)|\Psi\rangle = -i\frac{\delta}{\delta\varphi(x)}\Psi(\varphi) \ . \tag{2.15}$$

In this way, provided the functional kernels can be well-defined, one gets results that make no reference to any pre-selected Fock basis, *i.e.* they are independent of Ω.

When specific dynamics is in mind, and a specific Hamiltonian given, it may be that a unique Fock vacuum is determined by the quadratic part of the Hamiltonian. In this case a "natural" choice for the covariance is at hand, one of the inequivalent Fock spaces is selected and normal ordering can be defined. However, there certainly exists Hamiltonians for which the concept of ground state is inapplicable, as in the case of field theory in de Sitter space discussed in Section III. It is therefore preferable to use a framework which gives well-defined and unique results without the necessity of pre-selecting dynamics or particular states.

B. Representing Transformation Groups in Bosonic Quantum Field Theory

In a canonical, fixed time framework for field theory, infinitesimal transformations of the variables Φ and Π are generated via Poisson brackets with charges Q which are typically polynomials in the canonical variables. After quantization the charges are promoted to operators that generally seem to obey the commutation relations of the Lie algebra of the transformation group.

$$[Q_1, Q_2] = iQ_{(1,2)} \tag{2.16}$$

This result is however often misleading. The source of the problem resides in the fact that the charges involve products of operators at the same point and are consequently ill-defined.

To arrive at well-defined generators, a three-step procedure is adopted. First, the formal expression for Q is regulated in some fashion so that no ill-defined products occur: $Q \to Q^R$. Second, the singular portions of Q^R, which are ill-defined in the absence of regulators are isolated and removed. For the simple models that we shall consider, a c-number subtraction q^R will suffice. Third, the regulators are removed from the subtracted expression, leaving well-defined generators, which we denote by $:Q:$, even though the colons do not necessarily signify normal ordering.

$$:Q: \equiv \lim_R \left(Q^R - q^R\right) \tag{2.17}$$

The generators, $:Q:$, well-defined in the above manner, continue to generate the infinitesimal transformations on the canonical variables. However, non-linear relations like (2.16) can be modified. From (2.17) one gets

$$[:Q_1:, :Q_2:] = i:Q_{(1,2)}: + i \lim_R q^R_{(1,2)} \ . \tag{2.18}$$

If the limit of $q^R_{(1,2)}$ is non-zero, the quantum field theoretic realization of the Lie algebra acquires an extension, not seen in the classical theory — we get anomalous commutators.

Still one must provide a prescription for determining the renormalizing subtraction q^R. In the conventional approach, a Fock vacuum is chosen, q^R is defined as the expectation value of Q^R in that state and $:Q:$ is the standard normal ordered operator with respect to the selected vacuum. While this procedure may produce well-defined results, they depend on the covariance of the vacuum; in particular, the extension of the Lie algebra may depend on the vacuum.

In the Schrödinger picture the subtraction can be carried out intrinsically in terms of field states. Because the representation of the regulated generator

$$\left\langle \varphi_1 \left| Q^R(\Phi, \Pi) \right| \varphi_2 \right\rangle = Q^R\left(\varphi_1, \frac{\delta}{i\delta\varphi_1}\right) \delta(\varphi_1 - \varphi_2) \tag{2.19}$$

is a functional distribution involving a functional delta function, expression (2.19) does not yield useful information about the singularities of Q^R when the regulators are removed. However, one may also consider the functional representation kernel for the finite transformation,

$$U^R(\varphi_1, \varphi_2) = \left\langle \varphi_1 \left| e^{-i\tau Q^R} \right| \varphi_2 \right\rangle \tag{2.20}$$

which implements the [regulated] transformation on states.

$$e^{-i\tau Q^R} |\Psi\rangle \leftrightarrow \int \mathcal{D}\tilde{\varphi} U^R(\varphi, \tilde{\varphi}) \Psi(\tilde{\varphi}) \tag{2.21}$$

Clearly, U^R satisfies a functional Schrödinger-like equation,

$$i \frac{\partial}{\partial \tau} U^R(\varphi_1, \varphi_2) = Q^R \left(\varphi_1, \frac{\delta}{i\delta\varphi_1} \right) U^R(\varphi_1, \varphi_2) \tag{2.22}$$

with boundary condition

$$U^R(\varphi_1, \varphi_2)\big|_{\tau=0} = \delta(\varphi_1 - \varphi_2) \ .$$

Unlike $\langle \varphi_1 | Q^R | \varphi_2 \rangle$, $U^R(\varphi_1, \varphi_2)$ is a specific functional of φ_1 and φ_2 [rather than a functional distribution] and its behavior when the regulators are removed may be explicitly studied.

Generically, U^R becomes singular in the absence of regulators, but in the simple models that we consider, the infinities are confined to a φ-independent phase, $e^{-i\tau q^R}$. Thus $e^{i\tau q^R} U^R(\varphi_1, \varphi_2)$ possesses a well-defined limit and provides a representation for the finite transformation. Since

$$e^{i\tau q^R} U^R(\varphi_1, \varphi_2) = \left\langle \varphi_1 \left| e^{-i\tau (Q^R - q^R)} \right| \varphi_2 \right\rangle \ , \tag{2.23}$$

the regularizing subtraction for the generator is q^R. In this approach the renormalizing subtraction q_R is determined intrinsically, without referring to any Fock space or pre-selecting any vacuum. Note finally that when $q^R_{(1,2)}$ survives in the limit where the regulators are removed, the above construction yields a projective representation for the transformation group, with an explicitly determined 2-cocycle whose infinitesimal form is $\lim_R q^R_{(1,2)}$.

C. Two-Dimensional Conformal Transformations

Quantum field theoretic representation for conformal transformations on two-dimensional space-time illustrates well our program. Conformal transformations in two dimensions form a doubly infinite transformation group, whereby $x \pm t$ are taken into arbitrary, and in general, different functions of $x \pm t$. For simplicity we consider the special case when $x - t$ is unaffected and only $x + t$ transforms. At fixed time, the infinitesimal transformation law for the coordinate x, $\delta_f x = -f(x)$, obeys a Lie algebra given by the Lie bracket.

$$[\delta_f, \delta_g]x = -\delta_{(f,g)}x \tag{2.24a}$$

$$(f,g) = fg' - gf' \tag{2.24b}$$

[The dash always signifies differentiation with respect to the argument.]

To obtain a representation in terms of quantized boson fields, we consider the simple conformally invariant theory describing a self-dual field, with Lagrangian[4]

$$L = \frac{1}{4}\int dx\, dy\, \chi(t,x)\epsilon(x-y)\dot{\chi}(t,y) - \frac{1}{2}\int dx\, \chi^2(t,x) \quad . \tag{2.25}$$

[The dot means time differentiation.] The Euler–Lagrange equations satisfied by the field χ imply

$$\dot{\chi} = \chi' \quad . \tag{2.26}$$

Canonical quantization of this first order theory produces the commutator

$$[\chi(x), \chi(y)] = i\delta'(x-y) \equiv k(x,y) = \int \frac{dp}{2\pi}\, e^{-ip(x-y)} p \tag{2.27}$$

and the Hamiltonian density is simply $\frac{1}{2}\chi^2$. One can check that the system is Poincaré invariant; the Poincaré algebra is now contracted since the momentum is equal to the negative energy. The Lagrangian (2.25) actually possesses a larger symmetry; it is in fact invariant under two-dimensional conformal transformations. At fixed time $[t = 0]$, the formal generator of the transformation

$$Q_F = \frac{1}{2}\int dx\, \chi(x) f(x) \chi(x) \tag{2.28}$$

transforms the field operator χ as

$$\delta_f \chi = i[Q_f, \chi] = (f\chi)' \ . \tag{2.29}$$

The generators follow the Lie algebra (2.24),

$$[Q_f, Q_g] = iQ_{(f,g)} \tag{2.30}$$

when the commutator is evaluated formally, without care about the product of χ with itself at the same point.

To regulate the generator, we promote f to a bilocal function $F(x,y)$ and define,

$$Q^R \equiv Q_F = \frac{1}{2}\int \chi F \chi \tag{2.31}$$

while removing the regulator consists of passing to the local limit.

$$F(x,y) \to \frac{1}{2}\left(f(x) + f(y)\right)\delta(x-y) \tag{2.32}$$

$F(x,y)$ is taken to be real and symmetric in (x,y) and sufficiently well-behaved near $x \sim y$ to permit all formal manipulations. Of course, Q_F no longer generates conformal transformations, it gives rise instead to general linear canonical transformations.

In the field representation for the above quantities, we represent χ by

$$\langle \varphi_1 | \chi(x) | \varphi_2 \rangle = \frac{1}{\sqrt{2}}\left(\frac{1}{i}\frac{\delta}{\delta\varphi_1(x)} + \varphi_1'(x)\right)\delta(\varphi_1 - \varphi_2) \ . \tag{2.33}$$

The regulated transformation kernel $U^R(\varphi_1, \varphi_2) \equiv U(\varphi_1, \varphi_2; \tau F) = \langle \varphi_1 | e^{-i\tau Q_F} | \varphi_2 \rangle$ satisfies the equation

$$\begin{aligned} & i\frac{\partial}{\partial\tau} U(\varphi_1, \varphi_2; \tau F) \\ & \quad = \frac{1}{4}\int \left(\varphi_1' - i\frac{\delta}{\delta\varphi_1}\right) F \left(\varphi_1' - i\frac{\delta}{\delta\varphi_1}\right) U(\varphi_1, \varphi_2; \tau F) \end{aligned} \tag{2.34}$$

with a boundary condition at $\tau = 0$

$$U(\varphi_1, \varphi_2; 0) = \delta(\varphi_1 - \varphi_2) \ . \tag{2.35}$$

The solution is a Gaussian times a normalization factor N_F.[1a,b]

$$U(\varphi_1,\varphi_2;F) = N_F \exp - \int \varphi_1 k \varphi_2 \exp \frac{i}{2} \int (\varphi_1 - \varphi_2) K_F (\varphi_1 - \varphi_2) \tag{2.36}$$

$$N_F = \det{}^{-1/2} F^{1/2} \left(\frac{2\pi i}{k_F} \sin \frac{k_F}{2} \right) F^{1/2} \tag{2.37}$$

$$K_F = F^{-1/2} \left(k_F \operatorname{ctn} \frac{k_F}{2} \right) F^{-1/2} \tag{2.38}$$

$$k_F \equiv F^{1/2} k F^{1/2} \tag{2.39}$$

The representation is clearly unitary. The composition law

$$\begin{aligned} &\int \mathcal{D}\varphi U(\varphi_1,\varphi;F)\, U(\varphi,\varphi_2;G) = U(\varphi_1,\varphi_2;F\circ G) \\ &F \circ G = F + G - \frac{i}{2}(FkG - GkF) + \cdots \end{aligned} \tag{2.40}$$

may be verified with the help of trigonometric identities when G is proportional to F; the general case is checked by expanding $F \circ G$.

In the local limit, K_F attains a well-defined expression.[1a,b]

$$\begin{aligned} K_F(x,y) \to K_f(x,y) &= \frac{1}{f(x)} \left\{ \int \frac{d\lambda}{2\pi} \left(\lambda \operatorname{ctn} \frac{1}{2}\lambda \right) \exp \left(-i\lambda \int_y^x \frac{dz}{f(z)} \right) \right\} \frac{1}{f(y)} \\ &= -i\pi \frac{1}{f(x)} \left\{ P \csc^2 \pi \int_y^x \frac{dz}{f(z)} \right\} \frac{1}{f(y)} \end{aligned} \tag{2.41}$$

[P means principal value.] The normalization constant, N_F, however, diverges. The divergence resides in an unimportant constant factor Z [which may be removed by redefining the measure of functional integration] and in a phase e^{-iq_F}, which is determined for imaginary $\tau[F \to -iF]$; this continuation, rather than $F \to iF$, is appropriate when the energy spectrum is bounded below.

$$q_F = \frac{1}{4} \operatorname{tr} F\omega \tag{2.42}$$

$$\omega(x,y) \equiv |k|(x,y) = \int \frac{dp}{2\pi} e^{-ip(x-y)} |p| = -P \frac{1}{\pi (x-y)^2} \tag{2.43}$$

It follows that*

$$Z^{-1}e^{iq_F}U(\varphi_1,\varphi_2;F) = Z^{-1/2}\left\langle\varphi_1\left|e^{-i(Q_F-q_F)}\right|\varphi_2\right\rangle Z^{-1/2} \tag{2.44}$$

possesses a well-defined local limit, and we are led to define a renormalized generator as

$$:Q_f: \equiv \lim_{F\to f}\left(Q_F - \frac{1}{4}\operatorname{tr} F\omega\right) \quad . \tag{2.45}$$

Notice that the renormalizing subtraction $q^R \equiv q_F$ has been determined without choosing any vacuum state. Since q_F is a numerical quantity, independent of φ_1 and φ_2, the subtraction is a c-number which does not change the transformation law of the field χ, see (2.29). However, the non-linear commutator (2.30) is modified as in (2.18), and the Lie algebra of the renormalized generators acquires a central extension.**

$$\begin{aligned}[:Q_f::Q_g:] &= i{:}Q_{(f,g)}{:} + \frac{c}{24\pi}\int f'\,k\,g' \\ &= i{:}Q_{(f,g)}{:} - \frac{i}{48\pi}c\int(fg''' - gf''') \end{aligned} \tag{2.46}$$

$$c = 1$$

Of course, the subtraction is ambiguous up to terms that are finite in the local limit: these are obviously "trivial" in the sense that they may be adjusted at will by a finite redefinition of the generators. Nevertheless, the result for the non-trivial part of the extension — not removable by redefining generators — is unique and it is not specific to the representation (2.33) for χ, which may be generalized to

$$\langle\varphi_1|\chi|\varphi_2\rangle = -\frac{i}{\sqrt{2}}\left(\alpha\frac{\delta}{\delta\varphi_1} + \beta k\varphi_1\right)\delta(\varphi_1 - \varphi_2) \tag{2.47a}$$

---

* Notice that in the local limit and for $f = 1$ the expression (2.44) gives the propagation kernel for the time evolution of the self-dual field χ; in fact, Q_f is then the Hamiltonian.

** Representations with $c > 1$ can be obtained using an inhomogeneous transformation law for the field χ, $\delta_f\chi = (\chi f)' + \frac{\lambda}{\sqrt{2\pi}}f''$. This is still a symmetry of the Lagrangian (2.25) and the corresponding renormalized generators $:Q: + \frac{\lambda}{\sqrt{2\pi}}\int f'\chi$ satisfy (2.46) with $c = 1 + 12\lambda^2 > 1$. For details, see Ref. [1b].

provided

$$\frac{1}{2}(\alpha k\beta^T + \beta k\alpha^T) = k \tag{2.47b}$$

as required by (2.27). One may verify that the representation kernel which arises from the more general formula (2.47) possess a different Gaussian φ_1, φ_2 dependence and the infinite constant Z is modified. But the F dependent infinity (2.42) is unaffected by the generalization (2.47) and the center in (2.46) remains the same.

Finally, note that our functional transformation kernel allows one to compute how states transform under conformal transformations.

$$\Psi(\varphi) \underset{F}{\longrightarrow} \Psi_F(\varphi) = e^{iq_F} \int \mathcal{D}\tilde{\varphi} U(\varphi, \tilde{\varphi}; F)\Psi(\tilde{\varphi}) \tag{2.48}$$

In particular, for a Gaussian with covariance Ω, the transformed state is again a Gaussian with transformed covariance.

$$\Omega_F = \Omega - (\Omega + k)(\Omega - iK_F)^{-1}(\Omega - k) \tag{2.49}$$

In addition, the transformed state acquires an extra phase θ_F,

$$\theta_F = \operatorname{Im}\left\{\gamma_F - \frac{1}{2}\operatorname{tr}\ln(\Omega - iK_F)\right\} \tag{2.50}$$

where $\gamma_F \equiv \ln N_F + iq_F$. Notice that the transformation (2.49) provides a representation for the conformal algebra without center; the center resides in the corresponding representation of the phase (2.50).[1b]

D. Discussion

The above approach is to be contrasted with the conventional one, wherein the subtraction is given by the expectation of Q_F in a Fock vacuum of the form (2.5). [For simplicity, here we take Ω to be real.] The problem is that without specifying a dynamical Hamiltonian, which determines a unique ground state, the covariance is undetermined. The expectation value of Q_F in the state $|\Omega\rangle$ is

$$q_F^{\Omega} \equiv \langle\Omega|Q_F|\Omega\rangle = \frac{1}{4}\operatorname{tr} F\Big(\Omega - (\Omega - k)\rho(\Omega + k)\Big) \tag{2.51}$$

where ρ is the two-point function for the field.

$$\rho(x,y) \equiv \langle \Omega | \Phi(x)\Phi(y) | \Omega \rangle = \frac{1}{2}\Omega^{-1}(x,y) \tag{2.52}$$

The conventional subtraction therefore depends on Ω,

$$q_F^{\Omega} = \frac{1}{8}\,\mathrm{tr}\, F(\Omega + k\Omega^{-1}k) \tag{2.53}$$

as do the conventionally renormalized generators.

$$:Q_F^{\Omega}: \equiv \lim_{F \to f} \left(Q_F - q_F^{\Omega}\right) \tag{2.54}$$

The Ω dependence survives in the center of the algebra (2.30). For example, for translation invariant vacua,

$$\Omega(x,y) = \int \frac{dp}{2\pi} e^{-ip(x-y)}\Omega(p) \tag{2.55}$$

the last term in (2.46) is replaced by,

$$-\frac{i}{48\pi}\int (fg''' - gf''') \to \frac{1}{4}\int dx\, dy f(x) g(y) \\ \times \int \frac{dp}{2\pi} e^{-ip(x-y)} \int \frac{dq}{2\pi}\left[\left(q+\frac{p}{2}\right) C\left(q-\frac{p}{2}\right) - \left(q-\frac{p}{2}\right) C\left(q+\frac{p}{2}\right)\right] \tag{2.56}$$

where C is constructed from Ω.

$$C(p) \equiv \frac{|p|}{2}\left[\frac{\Omega(p)}{|p|} + \frac{|p|}{\Omega(p)}\right] \tag{2.57}$$

For centers that are non-trivially different, the representations of the algebra are inequivalent;[5] in other words inequivalent vacua have been used in the renormalization of the generators.

On the contrary, the Schrödinger picture result yields a unique [up to finite terms] covariance $\Omega = \omega$, and $C(p) = |p|$. The reason for this more specific result is that in the Schrödinger approach we impose the additional requirement that the representation of the infinitesimal generators be exponentiable.

In the conventional vacuum subtraction method for renormalizing the conformal algebra, we may place additional regularity requirements, which limit the allowed vacua — the allowed covariances Ω — and within this limited class, the center is essentially independent of Ω.[1d] However, this cannot be done in all circumstances. For example, in theories with a time-dependent Hamiltonian stationary energy eigenstates do not exist; a Fock space cannot be defined and transformation generators cannot be renormalized using conventional normal ordering methods. In these cases, the intrinsic renormalization procedure based on the Schrödinger picture is the only one available.

III. FIELD THEORY IN DE SITTER SPACE

A. The Classical Theory

In the previous section we have discussed how in the Schrödinger picture symmetry generators in a field theory can be regulated and renormalized intrinsically, without reference to any particular Fock vacuum. An interesting application of this formalism is the study of two-dimensional field theory in de Sitter space.[1c,e] The system is governed by a time-dependent Hamiltonian so that a vacuum state cannot be defined as the lowest energy eigenstate. Nevertheless, since non-trivial isometries are present, the vacua can be defined as the states which are invariant under the corresponding transformations. To make operational use of this definition it is clearly necessary to understand how the symmetry generators can be constructed without *a priori* knowledge of the vacuum states.

The classical theory of complex scalar fields of mass m in de Sitter space with metric $g_{\mu\nu}$, minimally coupled to $SO(2,1)$-invariant Maxwell potential $\mathcal{A}_\mu$, is described by the following Lagrangian density:

$$\mathcal{L} = \sqrt{-g}\left\{g^{\mu\nu}\left(D_\mu\phi\right)^* D_\nu\phi - m^2\phi^*\phi\right\} \tag{3.1}$$

with $D_\mu = \partial_\mu + i A_\mu$. In conformal coordinates* [with which we shall work] the de Sitter metric is proportional to the Minkowski metric $\eta_{\mu\nu}$.[6]

$$g_{\mu\nu} = \frac{1}{h^2 t^2} \eta_{\mu\nu} \tag{3.2}$$

It possesses three isometries since the Killing equation

$$f_{\mu;\nu} + f_{\nu;\mu} = 0 \tag{3.3}$$

admits three linearly independent solutions.

$$f_P^\mu = (0, 1) \qquad \text{translations} \tag{3.4a}$$

$$f_D^\mu = (t, x) \qquad \text{dilatations} \tag{3.4b}$$

$$f_K^\mu = \left(tx, \frac{1}{2}(t^2 + x^2) \right) \qquad \text{spacial conformal transformations} \tag{3.4c}$$

The three infinitesimal isometries (3.4) close on the $SO(2,1)$ de Sitter Lie algebra.

$$\begin{gathered} [f_P, f_D]^\mu = f_P^\mu \qquad [f_P, f_K]^\mu = f_D^\mu \qquad [f_D, f_K]^\mu = f_K^\mu \\ [f_i, f_j]^\mu \equiv f_i^\alpha \partial_\alpha f_j^\mu - f_j^\alpha \partial_\alpha f_i^\mu \end{gathered} \tag{3.5}$$

The most general $SO(2,1)$-invariant Abelian gauge potential is given [in the Weyl gauge][1e] by

$$A_0 = 0 \quad A_1 = \frac{\lambda}{t} \qquad \lambda \in \mathbf{R} \ . \tag{3.6}$$

One may check that A is invariant under the isometries of the de Sitter space by computing its Lie derivative with respect to the three Killing vectors (3.4) and observing that a pure gauge results in all cases.[7]

$$\begin{gathered} \mathcal{L}_f A_\mu \equiv (\partial_\mu f^\alpha) A_\alpha + f^\alpha \partial_\alpha A_\mu = \partial_\mu \rho_f \\ \rho_P = \rho_D = 0 \qquad \rho_K = \lambda t \end{gathered} \tag{3.7}$$

The Lagrangian (3.1) thus describes the most general linear theory of scalar fields in two-dimensions which is invariant under the group $SO(2,1)$.[8]

---

* These coordinates cover only an open slicing of the de Sitter space, which is defined by the surface of a hyperboloid of revolution in $1+2$-dimensions. The space possesses constant curvature $2h^2$.

The symmetries of (3.1) lead to various conservation laws. Invariance under global gauge transformations entails the covariant conservation of the electromagnetic current.

$$
\begin{aligned}
J^{\mu} &= i\left(\phi^{*} D^{\mu}\phi - \left(D^{\mu}\phi\right)^{*}\phi\right) \\
J^{\mu}{}_{;\,\mu} &= 0
\end{aligned}
\tag{3.8}
$$

Moreover, since the background gauge potential and metric satisfy (3.6) and (3.3), the variation of the Lagrangian under

$$\delta_f\phi = f^{\alpha}\partial_{\alpha}\phi + i\rho_f\phi \tag{3.9}$$

is a covariant divergence

$$\delta\mathcal{L} = \left(f^{\mu}\mathcal{L}\right)_{;\mu} \quad . \tag{3.10}$$

From Noether's theorem, the following currents are seen to be covariantly conserved

$$J^{\mu}_{f} = f_{\alpha}\left(T^{\mu\alpha} + A^{\alpha}J^{\nu}\right) - \rho_f J^{\mu} \quad ; \tag{3.11}$$

$T^{\mu\nu}$ stands for the energy-momentum tensor

$$T^{\mu\nu} = \left(D^{\mu}\phi\right)^{*} D^{\nu}\phi + \left(D^{\nu}\phi\right)^{*} D^{\mu}\phi - g^{\mu\nu}\left(\left(D_{\alpha}\phi\right)^{*}\left(D^{\alpha}\phi\right) - m^2\phi^{*}\phi\right) \tag{3.12}$$

and satisfies

$$T^{\mu\nu}{}_{;\nu} = F^{\mu\nu}J_{\nu} \quad , \qquad F^{\mu\nu} = \partial^{\mu}A^{\nu} - \partial^{\nu}A^{\mu} \quad . \tag{3.13}$$

The corresponding conserved charges (one for each Killing vector of Eq. (3.4)) read

$$
\begin{aligned}
Q_f \equiv \int dx\,\sqrt{-g}\,J^{0}_{f} = \int dx \Big\{ & f^{0}\Big[\pi\pi^{*} + \phi^{*\prime}\phi' - i\frac{\lambda}{t}\left(\phi^{*}\phi' - \phi^{*\prime}\phi\right) \\
&+ \frac{1}{t^2}\left(\lambda^2 + \frac{m^2}{h^2}\right)\phi^{*}\phi\Big] + f^{1}\left(\pi\phi' + \phi^{*\prime}\pi^{*}\right) - i\rho_f\left(\phi^{*}\pi^{*} - \phi\pi\right)\Big\} \quad .
\end{aligned}
\tag{3.14}
$$

The canonical momenta π and π^{*} are defined by

$$\pi = \frac{\delta\mathcal{L}}{\delta\dot{\phi}} = \dot{\phi}^{*} \qquad \pi^{*} = \frac{\delta\mathcal{L}}{\delta\dot{\phi}^{*}} = \dot{\phi} \quad , \tag{3.15}$$

and differentiation with respect to x is denoted by a dash. Using Poisson brackets one can verify that the charges (3.14) generate the transformation (3.9) and that they satisfy the structure relations (3.5). Finally, using the explicit form of the Hamiltonian

$$H = \int dx \left\{ \pi\pi^* + \phi^{*\prime}\phi' - i\frac{\lambda}{t}\left(\phi^*\phi' - \phi^{*\prime}\phi\right) + \frac{1}{t^2}\left(\lambda^2 + \frac{m^2}{h^2}\right)\phi^*\phi \right\} \tag{3.16}$$

one can also check that ($\{\ ,\ \}$ denotes Poisson brackets):

$$\frac{d}{dt}Q_f \equiv \frac{\partial}{\partial t}Q_f + \{Q_f, H\} = 0 \ . \tag{3.17}$$

B. Intrinsic Renormalization of the Symmetry Generators

To implement the symmetries at the quantum level we shall follow Section II.B. [The use of complex fields will require a few obvious modifications.] Therefore, quantum states are represented by functionals $\Psi(\phi, \phi^*) \equiv \langle\phi|\Psi\rangle$ and field states $|\phi\rangle$ normalized to a δ function.

$$\begin{aligned} \langle\phi_1|\phi_2\rangle &= \delta\left(\phi_1 - \phi_2\right) \\ &= \int \mathcal{D}\alpha\mathcal{D}\alpha^* \exp\left\{ i\int dx \left[\alpha(x)\left(\phi_1(x) - \phi_2(x)\right) + \alpha^*(x)\left(\phi_1^*(x) - \phi_2^*(x)\right)\right]\right\} \end{aligned} \tag{3.18}$$

The field operators Φ, $\Phi^\dagger$ and their conjugate momenta Π, $\Pi^\dagger$ act on the states in the standard way.

$$\begin{aligned} \langle\phi|\Phi(x)|\Psi\rangle &= \phi(x)\Psi(\phi,\phi^*) & \langle\phi|\Phi^\dagger(x)|\Psi\rangle &= \phi^*(x)\Psi(\phi,\phi^*) \\ \langle\phi|\Pi(x)|\Psi\rangle &= -i\frac{\delta}{\delta\phi(x)}\Psi(\phi,\phi^*) & \langle\phi|\Pi^\dagger(x)|\Psi\rangle &= -i\frac{\delta}{\delta\phi^*(x)}\Psi(\phi,\phi^*) \end{aligned} \tag{3.19}$$

At the quantum level the symmetry generators (3.14) are beset by singularities owing to the ultraviolet divergences when field operators are multiplied at the same point. Well-defined renormalized generators can be obtained, once a regularization prescription is adopted, by introducing suitable subtractions to the formal expressions (3.14). As explained in Section II.B, these renormalizing subtractions are uniquely fixed by the requirement that the representation kernels for the finite symmetry transformations be well-defined as the regulators are removed.

Let us now apply this method to the charges at hand. The electric charge

$$Q_e = i \int dx \left\{ \Phi^\dagger(x)\Pi^\dagger(x) - \Phi(x)\Pi(x) \right\} \tag{3.20}$$

is a perfectly well-defined operator and no subtraction is needed. Moreover, since Q_e is linear in the momenta, the matrix elements for the finite transformations, are given by δ-functions*

$$U_e(\phi_1, \phi_2; \tau) = \delta\left(e^{i\tau}\phi_1 - \phi_2\right) \quad . \tag{3.21}$$

We now turn to the de Sitter group generators. The charges (3.14) can be rewritten as

$$\begin{aligned} Q_f = \int \Big\{ & \Pi f^0 \Pi^\dagger + \Phi^\dagger \left[k f^0 k - \frac{\lambda}{t}\left(k f^0 + f^0 k\right) + \frac{1}{t^2}\left(\lambda^2 + \frac{m^2}{h^2}\right) f^0 \right] \Phi \\ & + i\left(\Phi^\dagger k f^1 \Pi^\dagger - \Pi f^1 k \Phi\right) - i\rho_f\left(\Phi^\dagger \Pi^\dagger - \Phi\Pi\right) \Big\} \end{aligned} \tag{3.22}$$

with $k(x,y) \doteq i\delta'(x-y)$. We suppress the integration variables and adopt an obvious functional matrix notation. A regularization of the previous charges can be obtained by replacing k in (3.22) by

$$k_\Delta \equiv \Delta k \Delta \tag{3.23}$$

with $\Delta(x,y)$ a real, symmetric, well-behaved kernel, approaching the δ-function in the local limit. As can be shown by making the canonical transformation

$$\begin{aligned} \Pi &= \Delta\tilde{\Pi} \qquad & \Pi^\dagger &= \tilde{\Pi}^\dagger\Delta \\ \Phi &= \Delta^{-1}\tilde{\Phi} \qquad & \Phi^\dagger &= \tilde{\Phi}^\dagger\Delta^{-1} \quad , \end{aligned} \tag{3.24}$$

* Had we taken a different ordering in the definition of Q_e, an infinite phase would have occurred in U_e; in this case the prescribed subtraction would give back (3.20) and (3.21).

the resulting expression for the charges is equivalent to one where all ambiguous products are point-split. Our next task is to determine the renormalizing subtraction.

We first observe that the momentum whose regularized form is

$$P^{\Delta} = i \int \left[\Phi^{\dagger} k_{\Delta} \pi^{\dagger} - \Pi k_{\Delta} \Phi \right] \tag{3.25}$$

requires no subtraction, as is familiar. The same is true also for the conformal generator, since special conformal transformations correspond to translations in the inverted coordinate system.

According to our renormalization scheme, the subtraction for the dilatation generator will be obtained by examining the representation functional $U_D^{\Delta}(\phi_1, \phi_2; \tau) = \left\langle \phi_1 \left| e^{-i\tau Q_D^{\Delta}} \right| \phi_2 \right\rangle$. Since Q_D^{Δ} is quadratic in the canonical variables, U_D^{Δ} must be a Gaussian in ϕ_1, ϕ_1^*, ϕ_2, ϕ_2^*. Taking into account the fact that Q_D^{Δ} is invariant under global gauge transformations, an appropriate *Ansatz* for U_D^{Δ} is thus

$$U_D^{\Delta}(\phi_1, \phi_2; \tau) = N \exp\left\{ - \int \left[\phi_1^* A \phi_1 - \phi_1^* B \phi_2 - \phi_2^* C \phi_1 + \phi_2^* D \phi_2 \right] \right\} \quad . \tag{3.26}$$

Note that unitarity, *i.e.* $U_D^{\Delta}(\phi_1, \phi_2; \tau) = \left[U_D^{\Delta}(\phi_2, \phi_1; -\tau) \right]^*$, implies the following constraints:

$$N(\tau) = N^*(-\tau) \tag{3.27a}$$

$$B(x, y; \tau) = B^*(y, x; -\tau) \tag{2.37b}$$

$$C(x, y; \tau) = C^*(y, x; -\tau) \tag{3.27c}$$

$$D(x, y; \tau) = A^*(y, x; -\tau) \quad . \tag{3.27d}$$

The Schrödinger-like equation which U_D^{Δ} obeys produces the following set of differential equations for the unknown kernels in (3.26) $(X(x, y) \equiv x\delta(x - y))$:

$$i \frac{\partial}{\partial \tau} \ln N = t \operatorname{tr} A \tag{3.28a}$$

$$i \frac{\partial}{\partial \tau} A = t A^2 - t k_{\Delta}^2 + 2\lambda k_{\Delta} - \frac{1}{t} \left(\lambda^2 + \frac{m^2}{h^2} \right) - A X k_{\Delta} + k_{\Delta} X A \tag{3.28b}$$

$$i \frac{\partial}{\partial \tau} B = t A B + k_{\Delta} X B \tag{3.28c}$$

$$i \frac{\partial}{\partial \tau} C = t C A - C X k_{\Delta} \tag{3.28d}$$

$$i \frac{\partial}{\partial \tau} D - = t C B \quad . \tag{3.28e}$$

We have not succeeded in solving explicitly these equations. However, in order to determine the renormalizing subtraction for Q_D, we only need to identify the infinities that arise in U_D^Δ when $\Delta \to \delta$. These infinities are ϕ-independent and confined to the normalization factor N in the form of a phase $e^{-i\tau q_D}$. Now notice that Eq. (3.28a) relates N to A. In the local limit, the divergent part of $\ln N$, linear in τ, can thus be extracted from the τ-independent divergent part of $\operatorname{tr} A$ as $\Delta \to \delta$. Our procedure for arriving at an expression for q_D will be to remove the regulator first, to determine $A(x, y; \tau)$ and to isolate its infinite part as $x \to y$ after τ has been continued to imaginary values $\tau \to -i\tau$.[1c]

The Riccati-like equation (3.28b), when $\Delta = \delta$, can be solved in Fourier space. One gets[1e]

$$A(x, y; \tau) = i \int \frac{dp}{2\pi} e^{-ip(x-y)}$$
$$\times p \left\{ \frac{\Theta''_{\lambda,\mu}(tp)}{\Theta'_{\lambda,\mu}(tp)} - \Theta'_{\lambda,\mu}(tp) \cot\left(\Theta_{\lambda,\mu}(tp) - \Theta_{\lambda,\mu}\left(tpe^{-\tau}\right)\right) \right\} \tag{3.29}$$

where $\Theta_{\lambda,\mu}(tp)$ is the phase of the Whittaker function $W_{-i\lambda,\mu}(-2itp)$ and $\mu = \left[\frac{1}{4} - \left(\lambda^2 + \frac{m^2}{h^2}\right)\right]^{1/2}$. We can now obtain the renormalizing subtraction for Q_D using Eq. (3.28a), *i.e.*, by computing the ultraviolet divergence that occurs in the symmetric part $A_s(x, y; \tau)$ of $A(x, y; \tau)$ as $x \sim y$. We find that A_s behaves like[1e]

$$A_s(x, y; \tau) \sim \int \frac{dp}{2\pi} e^{-ip(x-y)} \left(|p| + \frac{m^2}{2|p|t^2h^2}\right) \tag{3.30}$$

for $x \sim y$ and this identifies the renormalizing subtraction. The renormalized generators are then written as

$$\tilde{Q}_f = \lim_{\Delta \to \delta} \left(Q_f^\Delta - q_f^\Delta\right) = \lim_{\Delta \to \delta} \left(Q_f^\Delta - \operatorname{tr} f_\Delta^0 \omega^m\right) \tag{3.31}$$

with $f_\Delta^0 = \Delta f^0 \Delta$ and

$$\omega^m(x, y) = \int \frac{dp}{2\pi} e^{-ip(x-y)} \sqrt{p^2 + \frac{m^2}{t^2h^2}} \ . \tag{3.32}$$

Observe that ω^m differs only by finite terms from the expression on the right-hand side of (3.30). At this point, the finite part of the subtraction is arbitrary, it will be determined in the next section by conservation requirements. Finally, note that (3.31) can be used for the translation and conformal generators as well since the subtraction actually vanishes in these cases.

C. The Vacuum State

Owing to the time-dependence of the Hamiltonian (3.16), the Schrödinger equation does not separate in time and it makes no sense to define the vacuum as the lowest energy eigenfunction. It shall rather be defined as a Gaussian solution to the time-dependent Schrödinger equation, constrained to be invariant under the transformations corresponding to the $SO(2,1)$ symmetry group of the theory. This is a plausible strategy; in flat space and without gauge potentials, one may solve the Schrödinger equation with a Gaussian *Ansatz*, and upon imposing Poincaré invariance recover the unique Fock vacuum of the theory is obtained.

In keeping with the regularization of the symmetry generators, the regularized Schrödinger equation reads

$$\begin{aligned} i\frac{\partial}{\partial t}\Psi(\phi,\phi^*;t) = \int\Big\{ & -\frac{\delta}{\delta\phi}\frac{\delta}{\delta\phi^*} \\ & + \phi^*\left[k_\Delta^2 - \frac{2\lambda}{t}k_\Delta + \frac{1}{t^2}\left(\lambda^2 + \frac{m^2}{h^2}\right)\right]\phi\Big\}\Psi(\phi,\phi^*;t) \ . \end{aligned} \tag{3.33}$$

We demand the vacuum to be charge neutral, thus a suitable *Ansatz* for its wave functional is

$$\Psi(\phi,\phi^*;t) = N\, e^{-\int \phi^*\Omega\phi} \ . \tag{3.34}$$

Substituting in (3.34), we find that the covariance $\Omega(x,y;t)$ solves

$$i\frac{\partial}{\partial t}\Omega = \Omega^2 - k_\Delta^2 + \frac{2\lambda}{t}k_\Delta - \frac{1}{t^2}\left(\lambda^2 + \frac{m^2}{h^2}\right) \tag{3.35}$$

and that the normalization factor $N(t)$ satisfies

$$i\frac{\partial}{\partial t}\ln N = \mathrm{tr}\,\Omega \ . \tag{3.36}$$

To obtain solutions to (3.35) and (3.36) we impose the requirement that Ψ be "translation" invariant even in the presence of the regulator

$$P^\Delta\Psi = 0 \ . \tag{3.37}$$

This is possible because the regularized Hamiltonian commutes with P^{Δ}. Condition (3.37) reduces to

$$[k_{\Delta}, \Omega] = 0 \tag{3.38}$$

implying that Ω is a function of the kernel k_{Δ}.

The solutions to (3.35) and (3.36) can now be expressed in terms of the phase $\Theta_{\lambda,\mu}(tk_{\Delta})$ of the Whittaker function $W_{-i\lambda,\mu}(-2itk_{\Delta})$, with $\mu = \left[\frac{1}{4} - \left(\lambda^2 + \frac{m^2}{h^2}\right)\right]^{1/2}$. For the covariance $\Omega = \Omega_R + i\Omega_I$, $\Omega^{\dagger}_{R,I} = \Omega_{R,I}$, one finds[1e]

$$\Omega_R = k_{\Delta} \frac{\Theta'_{\lambda,\mu}(tk_{\Delta})}{\cos^2\left(\Theta_{\lambda,\mu}(tk_{\Delta}) - \chi\right) + r^2 \sin^2\left(\Theta_{\lambda,\mu}(tk_{\Delta}) - \chi\right)} \tag{3.39a}$$

$$\Omega_I = k_{\Delta} \left\{ \frac{1}{2} \frac{\Theta''_{\lambda,\mu}(tk_{\Delta})}{\Theta'_{\lambda,\mu}(tk_{\Delta})} + \Theta'_{\lambda,\mu}(tk_{\Delta}) \frac{(1 - r^2)\tan\left(\Theta_{\lambda,\mu}(tk_{\Delta}) - \chi\right)}{1 + r^2 \tan^2\left(\Theta_{\lambda,\mu}(tk_{\Delta}) - \chi\right)} \right\} \tag{3.39b}$$

while the normalization factor is given by[1e]

$$N = \det{}^{-1} \left\{ \left(\frac{1 - r^2}{|r|}\right)^{1/2} \left(\frac{\pi}{k_{\Delta}\Theta'_{\lambda,\mu}(tk_{\Delta})}\right)^{1/2} \cos\left(\Theta_{\lambda,\mu}(tk_{\Delta}) - \alpha(k_{\Delta})\right) \right\} . \tag{3.40}$$

The integration constant α is complex and depends on k_{Δ}; χ stands for its real part, while r is given in terms of its imaginary part, $r = \tanh(\operatorname{Im}\alpha)$. Various multiplicative constants have been adjusted so that $\int \mathcal{D}\phi\mathcal{D}\phi^*\Psi^*\Psi = 1$.

To restrict Ω further we demand that the vacuum functional be also invariant under dilatations and special conformal transformations. The action of the two generators on Ψ takes the form

$$\left(Q_f^{\Delta} - q_f^{\Delta}\right)\Psi = \operatorname{tr}\left(f^0\Omega - f^0_{\Delta}\omega^m\right)\Psi + \int dx\, dy\, \phi^*(x)\delta_f\Omega(x, y; t)\phi(y)\Psi \tag{3.41}$$

where

$$\begin{aligned} \delta_f\Omega = \int \Big\{ & -\Omega f^0\Omega + k_{\Delta} f^0 k_{\Delta} - \frac{\lambda}{t}\left(k_{\Delta} f^0 + f^0 k_{\Delta}\right) \\ & + \frac{1}{t^2}\left(\lambda^2 + \frac{m^2}{h^2}\right) f^0 + \Omega f^1 k_{\Delta} - k_{\Delta} f^1 \Omega \Big\} . \end{aligned} \tag{3.42}$$

To control the singularities as the regulators are removed we consider the overlap of (3.41) with the various states of the Fock space built on Ψ. This can be done systematically be examining the convolution of $\left(Q_f^\Delta - q_f^\Delta\right)\Psi$ with $e^{\int (J^*\phi+\phi^* J)}\Psi^*(\phi,\phi^*;t)$, as repeated functional differentiation with respect to the sources J and J^* will yield the required matrix elements. One obtains

$$\langle J\,|Q_f^\Delta - q_f^\Delta|\,\Psi\rangle = e^{1/2\int J^*\Omega_R^{-1}J} \times\left\{\mathrm{tr}\left(f^0\Omega - f_\Delta^0\omega^m + \frac{1}{2}\delta_f\Omega\Omega_R^{-1}\right) + \frac{1}{4}\int J^*\Omega_R^{-1}\delta_f\Omega\Omega_R^{-1}J\right\}. \tag{3.43}$$

The first term in brackets, involving the trace, is the diagonal matrix element $\left\langle\Psi\left|Q_f^\Delta - q_f^\Delta\right|\Psi\right\rangle$. The remaining term generates the connected off-diagonal matrix elements.

A priori we would like to set the above expression equal to zero, for all times. However, this is not possible in the presence of the regulator, since the regulated charges are not conserved relative to the regulated Hamiltonian H^Δ: $\frac{\partial}{\partial t}Q_f^\Delta \neq i\left[Q_f^\Delta, H^\Delta\right]$. We shall therefore only demand that (3.43) be zero in the limit $\Delta\to\delta$. In this limit and with

$$\Omega(x,y;t) = \int\frac{dp}{2\pi}\,e^{-ip(x-y)}\Omega(p;t)\ , \tag{3.44}$$

the conditions $\delta_{f_D}\Omega = 0$ and $\delta_{f_K}\Omega = 0$, that guarantee the vanishing of the off-diagonal matrix elements, imply

$$t\frac{\partial}{\partial t}\frac{\Omega(p;t)}{p} = p\frac{\partial}{\partial p}\frac{\Omega(p;t)}{p}\ . \tag{3.45}$$

The above equation requires α to be of the form

$$\alpha(k) = \alpha_+\theta(k) + \alpha_-\theta(-k)\ ; \tag{3.46}$$

with θ the standard step function and α_+, α_- complex constants. This leaves the vacuum state depending on two complex parameters. Let us now look at the diagonal part of (3.43). From the cyclicity of the trace and the fact that k_Δ, Ω_R and Ω_I all commute, it follows that

$$\operatorname{tr}\left(f^0\Omega_I + \frac{1}{2}\delta_f\Omega_I\Omega_R^{-1}\right) = 0 \ .$$

Hence,

$$\langle\Psi|Q_f^\Delta - q_f^\Delta|\Psi\rangle = \operatorname{tr}\left(f^0\Omega_R - f_\Delta^0\omega^m + \frac{1}{2}\delta_f\Omega_R\Omega_R^{-1}\right) \ . \tag{3.47}$$

When the regulator is removed $\delta_f\Omega_R$ vanishes and the above equation becomes

$$\left\langle\Psi\left|\tilde{Q}_f\right|\Psi\right\rangle = \int dx\, f^0(x)\int\frac{dp}{2\pi}\left\{\Omega_R(p;t) - \omega^m(|p|)\right\} \ . \tag{3.48}$$

In general, the p integration diverges. However, for precisely one set of values for the parameters, that is

$$r(k) = \epsilon(k) \ , \tag{3.49}$$

it converges ($\epsilon(k) = \operatorname{sign} k$), giving a finite time-dependent value for $\left\langle\Psi\left|\tilde{Q}_f\right|\Psi\right\rangle$. At this point we realize that our renormalization did not preserve the conservation of the generators. Indeed, matrix elements of conserved charges on states that solve the Schrödinger equation should be time-independent. This problem can easily be fixed by effecting a complete subtraction through the definition

$$:Q_f: = \lim_{\Delta\to\delta}\left\{Q_f^\Delta - \operatorname{tr} f_\Delta^0\,|k_\Delta|\,\Theta'_{\lambda,\mu}(tk_\Delta)\right\} \tag{3.50}$$

which differs from $\tilde{Q}_f$ of (3.31) only by finite terms.

In conclusion, we have shown that the state with $r = \epsilon(k)$ is completely invariant under $SO(2,1)$ transformations

$$:Q_f:|\Psi\rangle\Big|_{r=\epsilon(k)} = 0 \ . \tag{3.51}$$

The other states with $r \neq \epsilon(k)$ satisfy

$$\begin{aligned} :Q_f:|\Psi\rangle = &\int dx\, f^0(x)\int\frac{dp}{2\pi}|p|\Theta'_{\lambda,\mu}(pt) \\ &\times\left\{\frac{(\epsilon(p)r-1) - (r^2-1)\sin^2\left(\Theta_{\lambda,\mu}(tp)-\chi\right)}{1+(r^2-1)\sin^2\left(\Theta_{\lambda,\mu}(tp)-\chi\right)}\right\}|\Psi\rangle \ . \end{aligned} \tag{3.52}$$

As already mentioned, the eigenvalue is non-zero only in the case of dilatations; it vanishes for translations because $f^0 = 0$ and for conformal transformations because f^0 is odd in x. When non-zero, however, the eigenvalue is not only infrared divergent, but also ultraviolet divergent owing to the p-integral; moreover, it is time-dependent. Thus these vacua are only phase-invariant, and an infinite one-cocycle occurs. We should finally point out that it is not possible to redefine the generators so that some other vacuum with $r \neq \epsilon(k)$ becomes invariant. The reason is that only finite redefinitions are permitted at this stage and since the phase in (3.52) is infinite it cannot be removed.

IV. FERMIONIC QUANTUM FIELD THEORIES

A. The Schrödinger Picture

A development for fermion theories, analogous to the one described in Section II for boson theories, is presented here.[1d] We shall discuss two-dimensional charge-neutral Majorana–Weyl fermions, which are described by one component Hermitian field, $\psi = \psi(x)$. [Charged fermions of the Weyl variety can be described by a pair of neutral ones: $\psi = \frac{1}{\sqrt{2}}(\psi_1 + i\psi_2)$, $\psi^\dagger = \frac{1}{\sqrt{2}}(\psi_1 - i\psi_2)$].

At fixed time, the Hermitian operator ψ,

$$\psi^\dagger = \psi \tag{4.1}$$

satisfies anticommutation relations.

$$\{\psi(x), \psi(y)\} = \delta(x-y) \equiv I(x,y) \tag{4.2}$$

In analogy with the boson case, this algebra may be realized in the space of functionals $\Psi(u)$ of the field $u(x)$, which is now Grassmannian.

$$\{u(x), u(y)\} = 0 \tag{4.3}$$

We associate each functional with the ket $|\Psi\rangle \leftrightarrow \Psi(u)$, and represent the action of $\psi(x)$ on the state $|\Psi\rangle$ by

$$\psi(x)|\Psi\rangle \leftrightarrow \frac{1}{\sqrt{2}}\left[u(x) + \frac{\delta}{\delta u(x)}\right]\Psi(u) \ . \tag{4.4}$$

The relation (4.2) is then satisfied. To verify (4.1) we must define an inner product on the functional space with respect to which $\psi(x)$ as defined in (4.4) is Hermitian.

The inner product involves a Grassmann integration over u of an element in the functional space, composed with an element in the dual functional space. For the bosonic case, the dual is constructed by complex conjugation. Here this will not suffice, as the following example shows. It is possible for a functional to be u-independent. If the dual functional were its complex conjugate, the inner product would vanish, since a Grassmann u integral over an u-independent quantity is zero. The "state" would have zero norm, and this is undesirable.

To understand how the dual must be constructed, we analyze first the problem on a space consisting of two points with two fermion operators $\psi(i)$, $i = 1, 2$, satisfying a Clifford algebra.

$$\{\psi(i), \psi(j)\} = \delta_{ij} \tag{4.5}$$

[When we model generic continuum field theories with a discrete example, we must use an even number of discrete points. This is appropriate to the charge conjugation invariant situation with no unpaired, charge self-conjugate states.] A specific state $|\Psi_f\rangle$ is represented by a function of $u(i)$ that can be expanded in a four-dimensional basis.

$$\begin{aligned} |\Psi_f\rangle \leftrightarrow \Psi_f(u) &= f_0 + \sum_i f_1(i)u(i) + \frac{1}{2}\sum_{i,j} f_2(i,j)u(i)u(j) \\ &= f_0 + f_1(1)u(1) + f_1(2)u(2) + f_2(1,2)u(1)u(2) \end{aligned} \tag{4.6}$$

The f_i's are ordinary numbers with $f_2(1,2) = -f_2(2,1)$. The inner product with a second state $|\Psi_g\rangle$ is defined in the natural way.

$$\begin{aligned} \langle\Psi_g|\Psi_f\rangle &= g_0^* f_0 + \sum_i g_1^*(i) f_1(i) + \frac{1}{2}\sum_{i,j} g_2^*(i,j) f_2(i,j) \\ &= \langle\Psi_f|\Psi_g\rangle^* \end{aligned} \tag{4.7}$$

This can be expressed as*

$$\langle\Psi_g|\Psi_f\rangle = \int d^2u\, \Psi_g^*(u)\Psi_f(u) \tag{4.8}$$

* This definition of inner product can also appear in ordinary quantum mechanics. See the Appendix in Ref. 1d.

provided the dual of $|\Psi_g\rangle$ is represented by

$$\langle\Psi_g| \leftrightarrow \Psi_g^*(u) = g_2^*(1,2) + g_1^*(2)u(1) - g_1^*(1)u(2) + g_0^* u(1)u(2) \ . \tag{4.9}$$

Equation (4.7) follows from (4.6), (4.8) and (4.9) since only one Grassmann integral is non-vanishing.

$$\int d^2u\, u(1)u(2) = 1 \tag{4.10}$$

Thus the star operation signifies complex conjugation on numbers like f_n and g_n, but it dualizes Grassmann variables in a way similar to differential forms. Since the two-dimensional Grassmann δ-function is also given by a product,

$$\delta^2(u - \tilde{u}) = \Big(u(1) - \tilde{u}(1)\Big)\Big(u(2) - \tilde{u}(2)\Big) \tag{4.11a}$$

$$\int d^2\tilde{u}\, \delta^2(u - \tilde{u})\Psi(\tilde{u}) = \Psi(u) \tag{4.11b}$$

the dual (4.9) may be written as

$$\Psi_g^*(u) = \left(g_0^* - \sum_i g_1^*(i)\frac{\delta}{\delta u(i)} + \frac{1}{2}\sum_{i,j} g_2^*(i,j)\frac{\delta^2}{\delta u(j)\delta u(i)}\right)\delta^2(u) \ . \tag{4.12}$$

Of particular interest are the states $|\Omega\rangle$ which are represented by Gaussian functions.

$$|\Omega\rangle \leftrightarrow \Psi_\Omega(u) = \det{}^{-1/4}\Omega \exp\frac{1}{2}(u\Omega u) \tag{4.13}$$

[Here Ω is an anti-symmetric 2×2 matrix.] Using the definition (4.9), one easily finds that the dual state is represented by

$$\langle\Omega| \leftrightarrow \Psi_\Omega^*(u) = \det{}^{-1/4}\left(\Omega^{\dagger -1}\right) \exp\frac{1}{2}\left(u\Omega^{\dagger -1}u\right) \ . \tag{4.14}$$

With these definitions, the Hermitian conjugate of $u(i)$ is $\frac{\delta}{\delta u(i)}$ and ψ is given by

$$\psi(i) = \frac{1}{\sqrt{2}}\left(u(i) + \frac{\delta}{\delta u(i)}\right) \tag{4.15}$$

is Hermitian.

The representation (4.15) of the Clifford algebra (4.5) is four-dimensional, as is seen from (4.6). Consequently, it is reducible, since a two-dimensional irreducible representation is given in terms of Pauli matrices: $\psi(i) = \sigma^i/\sqrt{2}$. More generally, a $2n$-dimensional Clifford algebra possess an irreducible matrix representation with dimensionality 2^n, while our formalism gives a reducible, 2^{2n}-dimensional representation.

It is possible to give an irreducible representation in terms of Grassmann variables by splitting the Clifford elements in two, representing half of them as in (4.5) and the other half by $\frac{1}{\sqrt{2}\,i}\left(u(i) - \frac{\delta}{\delta u(i)}\right)$, which is also hermitian, satisfies (4.5) and anti-commutes with (4.15). However, for the continuum field theory we do not adopt this approach for the following two reasons. First, there is no *a priori* natural choice for the splitting. [When a Hamiltonian is posited, one could effect a splitting by reference to the positive and negative frequencies — this is essentially the holomorphic representation.[9] Our whole purpose however is to develop representation theory without reference to dynamics; moreover, the division into positive and negative frequencies can change if the Hamiltonian is time-dependent, or depends on other varying parameters.] Second, as will be discussed below, the reducibility seems desirable, since it allows making inequivalent choices for filling the Dirac sea when defining the vacuum of a dynamical model.

For the continuum field theory, we use the above results extended to a continuous infinity of points. A member of the functional space

$$|\Psi\rangle \leftrightarrow \Psi_f(u) = f_0 + \int f_1(x)u(x) + \frac{1}{2}\int f_2(x_1, x_2)u(x_1)u(x_2) + \cdots \tag{4.16}$$

possesses the dual

$$\langle\Psi_f| \leftrightarrow \Psi_f^*(u) = \left\{f_0^* - \int f_1^*(x)\frac{\delta}{\delta u(x)} + \frac{1}{2}\int f_2^*(x_1, x_2)\frac{\delta}{\delta u(x_2)\delta u(x_1)} + \cdots\right\}\delta(u) \tag{4.17}$$

where the functions f_n are totally antisymmetric in their arguments. The inner product is defined by functional Grassmann integration.*

$$\begin{aligned} \langle \Psi_g | \Psi_f \rangle &= \int \mathcal{D}u \Psi^*_g(u) \Psi_f(u) \\ &= \langle \Psi_f | \Psi_g \rangle^* \end{aligned} \tag{4.18}$$

For a Gaussian state, which we also call a Fock vacuum, in analogy with the bosonic case,

$$|\Omega\rangle \leftrightarrow \Psi_\Omega(u) = \det{}^{-1/4}\Omega \exp \frac{1}{2} \int u\Omega u \tag{4.19}$$

the dual is

$$\langle \Omega | \leftrightarrow \Psi^*_\Omega(u) = \det{}^{-1/4} \left(\Omega^{\dagger -1}\right) \exp \frac{1}{2} \int u \Omega^{\dagger -1} u \ . \tag{4.20}$$

Here Ω is an antisymmetric kernel. Note that these states are not normalized to unity, $\langle\Omega|\Omega\rangle = \det{}^{1/2}\left(\Omega^{1/2}\Omega^{\dagger 1/2} + \Omega^{-1/2}\Omega^{\dagger -1/2}\right)$.

The above concerns charge-neutral Majorana fermions. Because charged fermions are described as a pair of Majorana fields,

$$\begin{aligned} \psi &= \frac{1}{\sqrt{2}}(\psi_1 + i\psi_2) \\ \psi^\dagger &= \frac{1}{\sqrt{2}}(\psi_1 - i\psi_2) \end{aligned} \tag{4.21}$$

their representation is constructed accordingly: functionals depend on $u = \frac{1}{\sqrt{2}}(u_1 + iu_2)$ and $u^\dagger = \frac{1}{\sqrt{2}}(u_1 - iu_2)$, while operators are realized by

$$\psi = \frac{1}{\sqrt{2}}\left(u + \frac{\delta}{\delta u^\dagger}\right) \qquad \psi^\dagger = \frac{1}{\sqrt{2}}\left(u^\dagger + \frac{\delta}{\delta u}\right) \tag{4.22}$$

et cetera.

---

* Note that unlike the bosonic case functionals of u are not overlaps with field states and operator kernels are not matrix elements.

B. Two-Dimensional Conformal Transformations

Two-dimensional conformal transformations of the type discussed in Section II.C also act on fermion Majorana fields. The formal generator

$$Q_f = \frac{i}{4}\int dx\Big(\psi(x)f(x)\psi'(x) - \psi'(x)f(x)\psi(x)\Big) \tag{4.23}$$

gives the field transformation law,

$$\delta_f\psi = i\,[Q_f,\psi] = (f\psi)' - \frac{1}{2}f'\psi \ . \tag{4.24}$$

Q_f formally satisfies the algebra (2.30), but suffers from singularities owing to the coincident-point operator product. For a well-defined regularized generator we take

$$Q_F = \frac{1}{2}\int \psi F\psi \ . \tag{4.25}$$

Equation (4.23) is regained when the antisymmetric Hermitian kernel $F(x,y)$ tends to

$$F(x,y) \to \frac{i}{2}\left(f(x)+f(y)\right)\delta'(x-y) = \frac{1}{2}\left(f(x)+f(y)\right)k(x,y) \ . \tag{4.26}$$

[Notice that this is like (2.32) with the interchange $k \leftrightarrow \delta$; see also below.] Q_F satisfies

$$\begin{aligned}[] [Q_F,Q_G] &= iQ_{(F,G)} \\ (F,G) &= -i[F,G] \ . \end{aligned} \tag{4.27}$$

The representation and intrinsic renormalization of these quantities is an important application of the Schrödinger picture for fermions.

The kernel $U(u_1,u_2;F)$ that represents the finite transformation satisfies the differential equation

$$\begin{aligned} i\frac{\partial}{\partial\tau}&U(u_1,u_2;\tau F) \\ &= \frac{1}{4}\int dx\,dy\left(u_1(x)+\frac{\delta}{\delta u_1(x)}\right)F(x,y)\left(u_1(y)+\frac{\delta}{\delta u_1(y)}\right)U(u_1,u_2;\tau F) \end{aligned} \tag{4.28}$$

with a boundary condition at $\tau = 0$.

$$U(u_1, u_2; 0) = \delta(u_1 - u_2) \tag{4.29}$$

The solution is Gaussian.[1d]

$$U(u_1, u_2; F) = N_F \exp - \int u_1 u_2 \exp \frac{i}{2} \int (u_1 - u_2) K_F (u_1 - u_2) \tag{4.30}$$

$$N_F = \det{}^{1/2} i \sin \frac{F}{2} \tag{4.31}$$

$$K_F = \operatorname{ctn} \frac{F}{2} \tag{4.32}$$

The composition law

$$\begin{aligned} \int \mathcal{D}u U(u_1, u; F) U(u, u_2; G) &= U(u_1, u_2; F \circ G) \\ F \circ G = F + G + \frac{1}{2}(F, G) + \cdots & \end{aligned} \tag{4.33}$$

can be verified explicitly when G is proportional to F and the general case can be checked by expanding $F \circ G$.

The definition of the inner product on our space determines the form of the adjoint kernel.

$$U^\dagger(u_1, u_2; F) = N_F^* \exp - \int u_1 u_2 \exp - \frac{i}{2} \int (u_1 - u_2) K_F^\dagger (u_1 - u_2) \tag{4.34}$$

Since F and K_F are Hermitian, the above is just $U(u_1, u_2; -F)$ and the representation is unitary.

The kernel (4.30) should be compared to the corresponding bosonic one (2.36). The first exponential is similar with the commutator of the bosonic field, k, replaced by the anticommutator of the fermionic fields, $I \equiv \delta$. The analogy of the remaining formulas is brought out, if we similarly replace k in k_F by I. Of course, differences in the Jacobian factor between fermions and bosons have to be taken into account.

The local limit when the regulator is removed can be evaluated as in in the bosonic case. K_F attains a well-defined expression.

$$K_F(x,y) \to K_f(x,y) = \frac{1}{\sqrt{f(x)}} \left\{ \int \frac{d\lambda}{2\pi} \left(\operatorname{ctn} \frac{\lambda}{2} \right) \exp\left(-i\lambda \int_y^x \frac{dz}{f(z)} \right) \right\} \frac{1}{\sqrt{f(y)}}$$
$$= \frac{1}{\sqrt{f(x)}} \left\{ P \operatorname{ctn} \int_y^x \frac{dz}{f(z)} \right\} \frac{1}{\sqrt{f(y)}} \tag{4.35}$$

The normalization factor diverges. The divergence resides, apart from an infinite constant Z, in a phase e^{-iq_F}, which as in the bosonic case is determined for imaginary $\tau[F \to -iF]$. One finds

$$q_F = -\frac{1}{4} \operatorname{tr} \left\{ F \frac{k}{|k|} \right\} \quad . \tag{4.36}$$

To renormalize, we absorb the infinite constant Z in the definition of the functional integration measure, and remove the divergent phase. Thus $Z^{-1} e^{iq_F} U(u_1, u_2; F)$ possesses a finite limit but the composition law (4.33) acquires a trivial cocycle,

$$\omega_2(F,G) = \frac{1}{4} \operatorname{tr} \left(F \circ G - F - G \right) \frac{k}{|k|} \quad . \tag{4.37}$$

In the local limit $\omega_2(F,G)$ becomes non-trivial and its infinitesimal form reads

$$\delta\omega_2(f,g) = \lim \left(-\frac{i}{8} \operatorname{tr} [F,G] \frac{k}{|k|} \right) \quad . \tag{4.38}$$

This implies that the renormalized charges

$$:Q_f: = \lim \left(Q_F + \frac{1}{4} \operatorname{tr} F \frac{k}{|k|} \right) \tag{4.39}$$

satisfy (2.46) with $c = 1/2$.

As in the bosonic case this central extension is not sensitive to the way in which the field operator is represented. One can verify that the divergent phase in (4.36) is unaffected by the following generalization of (4.4),

$$\psi = \frac{1}{\sqrt{2}} \left(\alpha u + \alpha^* \frac{\delta}{\delta u} \right) \tag{4.40a}$$

$$\frac{1}{2}\left(\alpha^*\alpha^T + \alpha\alpha^\dagger\right) = I \tag{4.40b}$$

the last condition being required by (4.2).

The action of the transformation kernel on a generic state $\Psi(u)$ is given by

$$\Psi(u) \leftrightarrow \Psi_F(u) = e^{iq_F} \int \mathcal{D}\tilde{u} U(u, \tilde{u}; F)\Psi(\tilde{u}) \ . \tag{4.41}$$

In particular, the transform of a normalized Gaussian with covariance Ω is again a Gaussian with transformed covariance

$$\Omega_F = \Omega + (I - \Omega)\left(\Omega + iK_F\right)^{-1}(I + \Omega) \tag{4.42}$$

with an additional phase θ_F given by

$$\theta_F = \mathrm{Im}\left\{\gamma_F + \frac{1}{2}\,\mathrm{tr}\,\ln\left(\Omega + iK_F\right)\right\} \tag{4.43}$$

where $\gamma_F \equiv \ln N_F + iq_F$. As in the bosonic case, (4.42) gives a representation for the conformal algebra without center; the center resides in the representation provided by the phase (4.43).

Charged fermions give similar results. The formal generator is

$$Q_f = \frac{i}{2}\int dx \Big(\psi^\dagger(x) f(x)\psi'(x) - \psi'^\dagger(x) f(x)\psi(x)\Big) \tag{4.44}$$

and the transformation kernel is modified from (4.30) in an obvious way, due to the doubling in degrees of freedom

$$\begin{aligned} &U\left(u_1^\dagger, u_1, u_2^\dagger, u_2; F\right) \\ &\quad = N_F^2 \exp - \int \left(u_1^\dagger u_2 - u_2^\dagger u_1\right) \exp i \int \left(u_1^\dagger - u_2^\dagger\right) K_F \left(u_1 - u_2\right) \end{aligned} \tag{4.45}$$

As a consequence, the center on the algebra is twice the Majorana value, *i.e.* $c = 1$ in (2.46).*

---

* For charged fermions representations with $c > 1$ can be obtained by adding an extra $U(1)$ gauge transformation to the conformal transformation law for the field ψ: $\delta_f\psi = (f\psi)' - \left(\frac{1}{2} + i\lambda\right) f'\psi$. The new renormalized generators, $:Q_f: + \frac{\lambda}{2}\int f' \left[\psi^\dagger, \psi\right]$, satisfy (2.46) with $c = 1 + 12\lambda^2 > 1$. Note that this method of increasing the center does not work for Majorana fermions, since there the extra pieces in the generators vanish.

C. Fock Space Dynamics

Using the Schrödinger picture, we have obtained in the previous section a fermionic representation for the two-dimensional conformal group, without reference to any specific Hamiltonian. However, the formalism presented in Section IV.A can be equally well-applied to discuss fermionic dynamics.[1d]

As an example, let us consider the case of a free Majorana–Weyl fermion, with Hamiltonian

$$H = \frac{i}{2}\int \psi\psi' \equiv \frac{1}{2}\int \psi h \psi \tag{4.46}$$

$$h(x,y) = i\delta'(x-y) \quad . \tag{4.47}$$

The "first quantized" Hamiltonian h is anti-symmetric and imaginary, and possesses a complete, orthonormal set of "first quantized" eigenmodes.

$$h v_k = k v_k \tag{4.48a}$$

$$v_k(x) = \frac{1}{\sqrt{2\pi}} e^{-ikx} \tag{4.48b}$$

The field operator may be expanded in these modes.

$$\psi(x) = \frac{1}{\sqrt{2\pi}} \int dk\, e^{-ikx} a_k \tag{4.49}$$

$$a_k = a^\dagger_{-k} = \frac{1}{\sqrt{2\pi}} \int dx\, e^{ikx} \psi(x) \tag{4.50}$$

$$\left\{ a_k, a^\dagger_{k'} \right\} = \delta(k-k') \tag{4.51}$$

The operator a_k is a shift operator for the second quantized Hamiltonian.

$$[a_k, H] = k\, a_k \tag{4.52}$$

Hence, the spectrum of H is unbounded from above and below, unless a_k annihilates states. All this is of course familiar.

We now seek eigenstates of H within our Grassmann functional space. For the Fock vacuum we chose a Gaussian,

$$|\Omega\rangle = \det{}^{-1/4}\Omega \exp \frac{1}{2} \int u \Omega u \tag{4.53}$$

where Ω is anti-symmetric. The eigenvalue equation

$$H|\Omega\rangle = \frac{1}{4}\int \left(u + \frac{\delta}{\delta u}\right) h \left(u + \frac{\delta}{\delta u}\right)|\Omega\rangle = E_V|\Omega\rangle \tag{4.54}$$

requires that

$$(I - \Omega)\, h\, (I + \Omega) = 0 \tag{4.55}$$

and the vacuum energy is

$$E_V = \frac{1}{4}\operatorname{tr} h\,\Omega \ . \tag{4.56}$$

Excited states are polynomials in ψ operating on $|\Omega\rangle$; in our formalism they become polynomials in $\frac{1}{2}(I+\Omega)u \equiv u_+$ multiplying the Fock vacuum.

We now show that Ω is not determined uniquely by (4.55). Take Ω to be simultaneously diagonalized with h; (4.55) then requires that $\Omega^2 = I$. In momentum space we have

$$\Omega(k,k') = \Omega(k)\delta(k-k') \tag{4.57a}$$

$$\Omega(k) = -\Omega(-k) \tag{4.57b}$$

$$\Omega(k) = \pm 1 \ . \tag{4.57c}$$

The sign in (4.57c) can be chosen independently for any k. In other words, there is an infinity of solutions for Ω depending on the different ways one assigns signature.

To understand further the form of Ω, and to select a single covariance out of this infinite class, we compute the effect of a_k on $|\Omega\rangle$.

$$\begin{aligned} a_k|\Omega\rangle &= \frac{1}{\sqrt{2\pi}}\int dx\, e^{ikx}\psi(x)|\Omega\rangle = \frac{1}{\sqrt{4\pi}}\int dx\, e^{ikx}\left(u(x) + \frac{\delta}{\delta u(x)}\right)|\Omega\rangle \\ &= \frac{1}{\sqrt{4\pi}}\left(1+\Omega(k)\right)\int dx\, e^{ikx}u(x)|\Omega\rangle \end{aligned} \tag{4.58}$$

From (4.58) it is seen that a_k annihilates $|\Omega\rangle$ whenever $\Omega(k)$ is -1. Thus choosing Ω is equivalent to choosing a prescription for filling the Dirac sea in order to define a field theoretic vacuum. When $\Omega(k) = -1$ for positive k and $+1$ for negative k, *i.e.*

$$\Omega(k) = \epsilon(-k) \tag{4.59}$$

a_k annihilates $|\Omega\rangle$ for $k > 0$ but not for $k < 0$. This is the conventional choice, it corresponds to a filled negative energy sea, with

$$E_V = -\frac{V}{4\pi}\int_0^\infty dk\, k \tag{4.60}$$

where V is the volume of the space.

More generally, $\frac{1}{2}(1 + \Omega(k))$ is the filling factor, vanishing for empty states. Choices other than (4.59) for Ω are also possible; they correspond to other unconventional filling prescriptions and in general define inequivalent theories. Note that the overlap between two vacua is proportional to

$$\langle\Omega_1|\Omega_2\rangle \propto \det{}^{1/2}(\Omega_1 + \Omega_2) \quad . \tag{4.61}$$

Since Ω_1 and Ω_2 can differ only in the sign of one or more eigenvalues, $\Omega_1 + \Omega_2$ has a zero eigenvalue. If a sufficiently infinite number of modes are differently filled between Ω_1 and Ω_2, then the determinant vanishes. When the vacuum overlap is zero, so will the overlap between corresponding excited states, and the Fock spaces — the different theories built with different Ω's — are inequivalent.

Let us further observe that the energy of states built by multiplication of $|\Omega\rangle$ by ψ will in general depend on Ω. But repeated application of the shift operator a_k will result in the same energy spectrum being attained regardless of the choice of Ω. Hence, there is a large degeneracy in our formalism — this comes from the reducibility of our representation. However, we see no difficulty with this; on the contrary, the degeneracy reflects the true circumstance that in a fermionic quantum field theory, with a given Hamiltonian, a prescription for defining a vacuum must also be chosen.

Therefore, in our Schrödinger picture the first step in defining a theory, even with specific dynamics at hand, is to choose Ω. Then there yet remains another subtlety in the form of an additional degeneracy. As mentioned before, higher Fock states are polynomials in $\frac{1}{2}(I + \Omega)u \equiv u_+$ multiplying the vacuum Gaussian. However, we may also consider polynomials in $\frac{1}{2}(I - \Omega)u \equiv u_-$. Because $\frac{1}{2}(I \pm \Omega)$

are projection operators, polynomials in u_+ are orthogonal to those with u_-. Also, multiplying by u_- does not affect the energy of a state. But there is no operator constructed from ψ which can product factors of u_-. Hence, we may safely ignore these states, provided we remember to consider only operations with the Fermi fields. [Note, however, that the Gaussian *does* contain u_-, since $\frac{1}{2}u\Omega u = u_-u_+$.]

As a final check of our formalism, we compute the equal-time correlation function. Following the rules that we have put forward, it is easy to show that

$$\rho(x,y) \equiv \langle\Omega|\psi(x)\psi(y)|\Omega\rangle = \frac{1}{2}(I-\Omega)(x,y) = \int \frac{dp}{2\pi} e^{-ip(x-y)}\theta(p) \quad , \tag{4.62}$$

which is the conventional result.

In summary, let us contrast our fermionic Schrödinger picture with the familiar bosonic one. In both cases the functional space contains inequivalent Fock spaces. Choosing a specific quadratic Hamiltonian can select a specific Fock space for bosons, but not for fermions. In the former case, there is no sign ambiguity for the Gaussian covariance Ω because we require convergence of a Gaussian normalization integral, hence $\mathrm{Re}\,\Omega > 0$; in the latter, the integral is Grassmannian, all integrals converge, and the sign of Ω is not fixed. Stated differently, a particle state is localized in the bosonic functional space, while there is no concept of localization in the Grassmann space. A unique fermionic Fock space requires prescribing a filling factor and restricting to properly projected polynomials in $\frac{1}{2}(I+\Omega)u$.

To conclude, let us mention that other effects in fermionic field theory can be investigated using the Schrödinger picture. For example in a theory of Weyl fermions coupled with an external gauge field, one can explicitly compute the chiral anomaly and the Berry's phase.[1d] Furthermore, supersymmetric theories can now be treated within the Schrödinger picture, and in particular a functional representation for the two-dimensional superconformal group can be constructed.[10]

REFERENCES

1. The material presented in Sections II and IV is based on the following research papers: (a) R. Floreanini and R. Jackiw, *Phys. Lett.* **B175** (1986) 428; (b) R. Floreanini, *Ann. of Phys.* (1987), in press; (c) R. Floreanini, C. Hill and R. Jackiw, *Ann. of Phys.* **175** (1987) 354; (d) R. Floreanini and R. Jackiw, MIT preprint, CTP#1468, 1987. Section III is drawn from: (e) R. Floreanini and L. Vinet, *Phys. Rev.* **D36** (1987) 1731. Two review articles on the subject are: R. Jackiw in *Superfields*, Vancouver 1986 (Plenum, New York, 1987), and in the *Proceedings of the First Asia Pacific Workshop on High Energy Physics*, Singapore, 1987 (World Scientific, Singapore, 1987).

2. In this matter, see J. Mickelsson and S. Rajeev, MIT preprint CTP#1482, 1987.

3. O. Éboli, R. Jackiw and S.-Y. Pi, MIT preprint CTP#1519, 1987.

4. R. Floreanini and R. Jackiw, MIT preprint CTP#1512, 1987.

5. A. Niemi and G. Semenoff, *Phys. Lett.* **B176** (1986) 108; H. Neuberger, A. Niemi and G. Semenoff, *Phys. Lett.* **B181** (1986) 244.

6. N. Birrell and P. Davis, *Quantum Fields in Curved Space* (Cambridge University Press, Cambridge, UK, 1982).

7. The problem of determining the gauge potentials that are invariant under a given group action is discussed in: J. Harnad, S. Shnider and L. Vinet, *J. Math. Phys.* **21** (1980) 2719. For a review, see R. Jackiw, *Acta Phys. Austriaca*, Suppl. XXII (1980) 383.

8. L. Vinet, in *Lecture Notes in Physics*, Vol. 135, p. 191 (Springer-Verlag, New York, 1980).

9. See for instance, L. Faddeev and A. Slavnov, *Gauge Fields* (Benjamin/Cummings, Menlo Park, CA, 1980).

10. R. Floreanini and L. Vinet, in preparation.

Infinite-Component Field Theory

J. W. Moffat*
W. W. Hansen Laboratories of Physics
Stanford University
Stanford, California 94305-4085, USA

ABSTRACT

The properties of infinite-component field theories and their wave equations are studied using the nonsymmetric gravitation theory (NGT) as a basis. The local fiber bundle gauge group of NGT is $U(3,1,\Omega)$ based on hyperbolic complex numbers, isomorphic to the local gauge group $GL(4,R)$ that generalizes the group of Lorentz transformations $SO(3,1)$. The spinor fields of $GL(4,R)$ are infinite-dimensional and this forces us to consider infinite-dimensional field theories. The irreducible representations of $GL(4,R)$ are decomposed into infinite sums of finite-dimensional representations of the Lorentz group and this leads to a field theory free of causality problems. The problem of gauging away all unphysical modes in infinite-component field theories is discussed. The idea of infinite-parameter gauge fields plays an important role in removing all unphysical modes, independently of the number of space-time dimensions. A model of an infinite-component quantum field theory is formulated using perturbation theory. There are no ultraviolet divergences and the S-matrix is causal and unitary. Einstein's theory of gravity and NGT are formulated within this scheme and an expansion about flat-space leads to a finite theory of gravity.

*Permanent address: Department of Physics, University of Toronto, Toronto, Ontario M5S 1A7, Canada.

Contents

I. *Introduction*

In Einstein's General Relativity (GR), the group of diffeomorphisms $\mathcal{C}$ corresponds to the group of general coordinate transformations in the four-dimensional manifold M_4. The group $\mathcal{C}$ contains the sub-group $GL(4,R)$ which is the most general group of transformations among the linear frames associated with M_4. The invariance group in the fiber bundle (tangent space) associated with a pseudo-Riemannian metric tensor $g_{(\mu\nu)}$ is $SO(3,1)$, the homogeneous group of Lorentz transformations.

We observe that whereas GR contains the general group of transformations of coordinate frames in the real four-dimensional manifold, corresponding to the group of transformations $\mathcal{C}$, this is not true for the fiber bundle (tangent space) of GR. The symmetries of the Lorentz group $SO(3,1)$ are those that are apparently observed in Nature. But we have learned from quantum field theory and particle physics that there may be a larger (local) gauge group of transformations that contains the Lorentz group. This situation can already be seen in relation to the Schrödinger and Dirac equations. The Schrödinger equation is invariant only under the Newtonian Galilean invariance group, whereas the Dirac equation was discovered by searching for a wave equation that possesses invariance under the larger group of Lorentz transformations. Moreover, symmetries such as the isotopic spin group $SU(2)$ are considered to be contained in larger compact groups that are broken by spontaneous symmetry breaking or some dynamical symmetry breaking mechanism.

The most general real group of transformations associated with the inertial frames or linear frame bundle of a four-dimensional fiber bundle is $GL(4,R)$. Since GR purports to be the *general theory of relativity*, why does it not obey the most general gauge invariance permissible in the four-dimensional fiber bundle? It is, of course, possible that Nature simply deems that GR is the most general structure allowed i.e., it is the only mathematical structure that leads to a consistent, dynamically complete description of space-time. Then we would be forced to conclude that GR is the most general description of space-time consistent with experiment.

However, our experience with the development of Special Relativity and GR would lead us to suspect otherwise. This logical conundrum does appear to have worried Einstein, in the years following his publication of the final form of GR, in 1916. The standard interpretation of his work, after 1916, is that he sought to unify gravitation with electromagnetism, i.e. to construct a unified field theory. However, there is another deeper reason perhaps for his apparent unhappiness with GR, which surfaces in his work with the nonsymmetric field structure(1). He seemed to be searching for the most general description of space-time compatible with a complete dynamical scheme much like GR itself. Such a description of Nature would have completed his life's goal: the discovery of the most general consistent description by a mathematical theory of four-dimensional space-time. The so-called "unification" of gravitation and electromagnetism was merely a vehicle used by him to achieve this goal.

These considerations lead one inevitably to ask the question: Does there exist a complete dynamical scheme that possesses the most general fiber bundle gauge group of transformations $GL(4,R)$, as well as the diffeomorphism group C in the manifold M_4? The answer to this question resides in the Nonsymmetric Gravitation Theory (NGT)(2)–(7).

It is important to realize that simply adding local fields to GR with its pseudo-Riemannian metric, such as in Brans-Dicke[8] theories where a scalar field is included in the gravitational scheme, does not fundamentally alter the local gauge structure or group-theoretical properties of the manifold. Only by extending the symmetries of the fundamental tensor $g_{\mu\nu}$ do we succeed in obtaining a radical departure from GR.

One of the important problems to be resolved in modern physics is the split that exists between gravitation theory and quantum mechanics. There have been several serious attempts to construct a consistent quantum theory of gravity in recent years. The most recent attempts are supergravity [9] and superstring theory[10][11]. These theories were formulated in higher dimensions so as to attempt simultaneously a unification of the four known forces of Nature. If we cannot solve the problem of quantum gravity, then we are faced with an "ultraviolet" catastrophe comparable to that which lead to the discovery of quantum mechanics. Even though there obviously is no foreseeable practical application of quantum gravity, since it only becomes important at the Planck energy $\simeq 10^{19}$ GeV, and we shall never build accelerators at these energies, the solution of the quantum gravity problem will lead to fundamental changes in all other quantum field theories including probably quantum electrodynamics, in spite of the considerable success of this theory.

Supergravity did not lead to a renormalizable theory of gravity as was initially hoped. This theory was based on the notion of a point particle and it was thought that supersymmetry was a strong enough symmetry to lead to the necessary cancellations of ultraviolet infinities to all orders. Out of a possible sense of frustration, caused by the failure of these attempts to obtain a finite theory of gravity, the idea of an extended one-dimensional object that would replace the zero-dimensional point particle was resurrected using the concept of a string as the fundamental object of matter. The strings would only interact locally, permitting the construction of a string field theory in two-dimensions. The supersymmetric string (or superstring) was found to be free of ghost poles and tachyons and seemed to describe a consistent theory including gravity in ten dimensions. Both supergravity and superstrings are formulated using supersymmetry that requires the existence of a superfermion partner for all known bosons including the graviton.

In spite of many efforts by experimentalists there has not been a single discovery of a super partner i. e. no trace of supersymmetry has been observed in Nature.

One problem with superstring theory is that it requires compactification to four dimensions to produce observable predictions. Such a compactification scheme is not unique. Indeed, it is possible to produce a very large number of *bona fide* superstring theories in four dimensions that appear to be consistent i.e. free of ghost poles, tachyons and anomalies(12)(13). Moreover, there is increasing evidence that the compactification of the superstring cannot be carried out using perturbation theory. Nothing less than an exact solution would solve the problem of the superstring. Apart from mathematical niceties such as conformal invariance, the superstring theories do not seem to possess a fundamental physical justification for their existence. Why a string and not a higher-dimensional membrane? If a string really is the fundamental object of the universe, would Nature not provide a strong physical reason and motivation for this fact? As yet, there is no agreement among theorists that superstrings provide a finite consistent theory of gravity, although there appear to be strong reasons for believing this to be true(14). Indeed, since strings are non-local extended objects that interact locally, we would expect the theory to be ultraviolet finite to all orders of perturbation theory.

On the phenomenological side, it has recently been shown that on the basis of perturbation theory the compactification of superstring models must lead to a massless dilaton field, and the consequent gravity theory in four dimensions has the form of a Brans-Dicke theory with a coupling constant of order unity(15). Such a theory disagrees strongly with solar system tests e.g. the time-delay of radar signals passing the limb of the Sun(16) and the anomalous precession of Mercury (even allowing for a large solar quadrupole moment such as is observed by Hill(17) and Dicke). Morever, Brans-Dicke theory possesses the unfortunate feature that it violates the equivalence of inertial and gravitational mass at all orders, in disagreement with accurate Nordtvedt effect observations using Lure data for the Earth and moon(18). At a more fundamental level, the test particle in Brans-Dicke theory is not "self-effacing" i.e. we have to know all the self-accelerations and self-energy properties of the test particle to be able to make a physical prediction. To overcome this problem, it would appear necessary to appeal to strongly coupled superstrings or to find an exact solution to superstring theories. Neither of these possibilities has any foreseeable solution.

As their is no clear physical reason for choosing strings as the extended fundamental object of Nature, we could envisage using n-dimensional membranes embedded in an (n+p)-dimensional space-time. The three-dimensional membrane is the simplest extension of the string but it runs afoul of Poincaré's conjecture (or question): Let Σ be a closed three-manifold with $\Pi_1(\Sigma) = 0$. Is $\Sigma \approx S^3$? Since this question has not yet been answered by mathematicians, it means that there is no classification of three-manifolds and this would cause serious difficulties in the construction of a three-dimensional membrane field theory. The situation is much better for higher-dimensional membranes with $n \geq 4$, since n-dimensional manifolds with $n \geq 4$ have been classified by mathematicians. Thus, it is conceivable that you could construct a four-dimensional membrane field theory. Another problem is that for $n > 2$, one loses the infinite- parameter conformal invariance of the string theory. This is equivalent to saying that one is dealing with an n-dimensional pseudo-Riemannian gravity theory for $n > 2$ with all its technical encumbrances, not least of which is that the equations of motion are highly non-linear.

It would appear that trying to specify the exact topological properties of the fundamental extended object leads to serious problems. In what follows, we shall attempt to construct an infinite-component field theory that can be solved without having to specify the precise nature of the nonlocal extended object.

We shall pursue the idea that perhaps a fundamental alteration of quantum field theory is required to construct a finite consistent theory of quantum gravity. In NGT, the local gauge group of the theory is $GL(4,R) \supset SL(4,R)$[19]. The non-compact group $SL(4,R)$ *only possesses inifinite-dimensional spinor fields*[20], so that NGT must be formulated as an infinite-component field theory of gravity in four dimensions. Such theories share with string theory the feature that they contain an infinite number of particle states and yield a description of the Regge-like particle spectrum observed in Nature. Thus, NGT forces upon us a radical departure from GR in that it demands a field theoretic description of Nature, based on the idea of an extended particle that interacts locally with other such particles. This kind of theory would be expected to lead to a finite theory of gravity, in the same way that string theory does, without the new theory being necessarily tied directly to the concept of a string as the fundamental object of matter. Moreover, the theory could be formulated in four dimensions, thereby avoiding the difficulties of compactification.

II. *The Nonsymmetric Gravitation Theory*

In this theory, whose initial formulation was first published in 1979[(2)], the fundamental tensor $g_{\mu\nu}$ is chosen to be nonsymmetric

$$g_{\mu\nu} = g_{(\mu\nu)} + g_{[\mu\nu]}, \tag{2.1}$$

where

$$g_{(\mu\nu)} = \frac{1}{2}(g_{\mu\nu} + g_{\nu\mu}), \quad g_{[\mu\nu]} = \frac{1}{2}(g_{\mu\nu} - g_{\nu\mu}). \tag{2.2}$$

A contravariant tensor $g^{\mu\nu}$ satisfies the relation

$$g^{\mu\nu} g_{\sigma\nu} = g^{\nu\mu} g_{\nu\sigma} = \delta^{\mu}_{\sigma}. \tag{2.3}$$

The fundamental geometrical object in NGT is the connection $W^{\lambda}_{\mu\nu}$ which is also nonsymmetric

$$W^{\lambda}_{\mu\nu} = W^{\lambda}_{(\mu\nu)} + W^{\lambda}_{[\mu\nu]}. \tag{2.4}$$

This connection can be related to another nonsymmetric connection in the theory denoted by $\Gamma^{\lambda}_{\mu\nu}$, which is related to $W^{\lambda}_{\mu\nu}$ by a projective transformation

$$W^{\lambda}_{\mu\nu} = \Gamma^{\lambda}_{\mu\nu} - \frac{2}{3}\delta^{\lambda}_{\mu} W_{\nu}, \tag{2.5}$$

where

$$W_{\nu} \equiv W^{\alpha}_{[\nu\alpha]} = \frac{1}{2}(W^{\alpha}_{\nu\alpha} - W^{\alpha}_{\alpha\nu}). \tag{2.6}$$

It can be shown by using the last equation that

$$\Gamma_{\mu} = \Gamma^{\alpha}_{[\mu\alpha]} = 0. \tag{2.7}$$

We see that the geometrical notion of torsion as described by Cartan[21], has entered the scheme of NGT in a natural way. This is to be expected in a theory that describes space-time in the most general manner possible in a classical theory. The restriction in GR that the torsion should be identically zero cannot be justified either on theoretical or experimental grounds.

The primary concept in NGT is the displacement field $\Gamma^\lambda_{\mu\nu}$, which is defined by the parallel displacement law for a vector A^μ:

$$\delta A^\lambda = -\Gamma^\lambda_{\mu\nu}\, dx^\mu\, A^\nu. \tag{2.8}$$

It is clear already from (2.8) that there is no justification for taking $\Gamma^\lambda_{\mu\nu}$ to be symmetric in the subscripts μ and ν. We learn from this that GR is a specialization of a more general theory of space-time in which both $g_{\mu\nu}$ and $\Gamma^\lambda_{\mu\nu}$ are taken to be nonsymmetric quantities.

A curvature tensor can be formed from the connection $W^\lambda_{\mu\nu}$:

$$B^\sigma_{\mu\nu\rho} = W^\sigma_{\mu\nu,\rho} - W^\sigma_{\mu\rho,\nu} - W^\sigma_{\alpha\nu} W^\alpha_{\mu\rho} + W^\sigma_{\alpha\rho} W^\alpha_{\mu\nu} \tag{2.9}$$

and a contracted curvature tensor

$$B_{\mu\nu} = W^\beta_{\mu\nu,\beta} - W^\beta_{\mu\beta,\nu} - W^\beta_{\alpha\nu} W^\alpha_{\mu\beta} + W^\beta_{\alpha\beta} W^\alpha_{\mu\nu}, \tag{2.10}$$

where we have used the notation $X_{,\nu} = \partial/\partial x^\nu$. By symmetrizing $B_{\mu\nu}$ in the second term, we get

$$R_{\mu\nu}(W) = W^\beta_{\mu\nu,\beta} - \frac{1}{2}(W^\beta_{\mu\beta,\nu} + W^\beta_{\nu\beta,\mu}) - W^\beta_{\alpha\nu} W^\alpha_{\mu\beta} + W^\beta_{\alpha\beta} W^\alpha_{\mu\nu}. \tag{2.11}$$

Substituting (2.5) into (2.11) gives

$$R_{\mu\nu}(W) = R_{\mu\nu}(\Gamma) + \frac{2}{3} W_{[\mu,\nu]}, \tag{2.12}$$

where

$$R_{\mu\nu}(\Gamma) = \Gamma^\beta_{\mu\nu,\beta} - \frac{1}{2}(\Gamma^\beta_{(\mu\beta),\nu} + \Gamma^\beta_{(\nu\beta),\mu}) - \Gamma^\beta_{\alpha\nu} \Gamma^\alpha_{\mu\beta} + \Gamma^\beta_{(\alpha\beta)} \Gamma^\alpha_{\mu\nu}. \tag{2.13}$$

The property of transposition symmetry - or Hermitian symmetry - of fundamental tensors and invariants in NGT, plays as fundamental a role in restricting the arbitrariness of NGT, as the symmetry of tensor quantities in Riemannian geometry restricts the choice of physical laws in GR. A quantity $A_{\mu\nu}(\Gamma)$ is called transposition symmetric or Hermitian, if $A_{\mu\nu}(\tilde{\Gamma}) = A_{\nu\mu}(\Gamma)$ where $\tilde{\Gamma}^{\lambda}_{\mu\nu} = \Gamma^{\lambda}_{\nu\mu}$. If a tensor $A_{\mu\nu}(\tilde{\Gamma})$ is transposition symmetric, then a field equation of the form $A_{\mu\nu}(\Gamma) = 0$ implies that the field equation $A_{\mu\nu}(\tilde{\Gamma}) = 0$ and we say that the system of equations is transposition or Hermitian invariant. The operation of raising the suffixes α and β for a tensor $A_{\alpha\beta}$, while simultaneously retaining transposition or Hermitian symmetry , can be implemented by using the expression $A^{\mu\nu} = g^{\mu\alpha} g^{\beta\nu} A_{\alpha\beta}$, where the tensor $A_{\alpha\beta}$ satisfies $\tilde{A}_{\alpha\beta} = A_{\beta\alpha}$ and we recall that by definition $\tilde{g}_{\mu\nu} = g_{\nu\mu}$.

The Lagrangian density has the form

$$\mathcal{L} = \mathbf{g}^{\mu\nu} R_{\mu\nu}(W) + \mathcal{L}_{\text{matter}}, \tag{2.14}$$

where for any tensor quantity, we define $\mathbf{X}^{\mu\nu} = (-g)^{\frac{1}{2}} X^{\mu\nu}$ and $\mathcal{L}_{\text{matter}}$ is given by

$$\mathcal{L}_{\text{matter}} = -8\pi g^{\mu\nu} \mathbf{T}_{\mu\nu} + \frac{8\pi}{3} W_{\mu} \mathbf{S}^{\mu}. \tag{2.15}$$

$\mathbf{T}^{\mu\nu}$ is the energy-density tensor for matter and $\mathbf{S}^{\mu}$ is a conserved current density in NGT:

$$\mathbf{S}^{\mu}{}_{,\mu} = 0. \tag{2.16}$$

This conservation law is a consequence of the invariance of the Lagrangian density with respect to the Abelian transformation

$$W'_{\mu} = W_{\mu} + \lambda_{,\mu}, \tag{2.17}$$

where λ is an arbitrary scalar field. Thus the group of general coordinate transformations C in the real manifold M_4 is extended, in NGT, to include the Abelian $U(1)$ (or R_{+}) group of transformations (2.17).

From the variational principle[3]

$$\delta \int \mathcal{L} d^4 x = 0 \tag{2.18}$$

we obtain the field equations

$$G_{\mu\nu}(W) = 8\pi T_{\mu\nu} \tag{2.19}$$

$$\mathbf{g}^{[\mu\nu]}{}_{,\nu} = 4\pi \mathbf{S}^{\mu}, \tag{2.20}$$

where $G_{\mu\nu}(W)$ is the NGT Einstein tensor

$$G_{\mu\nu}(W) = R_{\mu\nu}(W) - \frac{1}{2}g_{\mu\nu}R(W). \tag{2.21}$$

The energy-momentum tensor for a perfect fluid can be derived from a Lagrangian density[22], and has the form

$$T^{\mu\nu} = (\rho + p)u^{\mu}u^{\nu} - pg^{\mu\nu}, \tag{2.22}$$

where $u^{\mu} = dx^{\mu}/ds$ is the four-velocity of a fluid element and s is the proper time. The NGT charge of a body is given by the equation

$$\ell^2 = \int \mathbf{S}^0 \, d^3x, \tag{2.23}$$

where ℓ has the dimensions of a length.

The problem of the motion of test particles and of massive, extended bodies has been solved in NGT[5][7] using the generalized Bianchi identities and the resulting conservation laws for matter that follow from the variational principle. A test particle is defined by $m_p \to 0, \ell_p^2 \to 0$ and $\ell_p^2/m_p \neq 0$ and the unique equation of motion for the test particle is

$$Du^{\mu}/D\tau = \kappa H^{\mu}{}_{\nu}u^{\nu}, \tag{2.24}$$

where $\kappa = \ell_p^2/m_p$ and

$$Du^{\mu}/D\tau = du^{\mu}/d\tau + \left\{ {\mu \atop \alpha\beta} \right\} u^{\alpha}u^{\beta} \tag{2.25}$$

$$H^{\mu}{}_{\nu} = \frac{1}{2}\gamma^{(\mu\alpha)}R_{[\alpha\nu]}(\Gamma). \tag{2.26}$$

In (2.26), the NGT Christoffel symbols are defined by

$$\left\{ \begin{matrix} \lambda \\ \mu\nu \end{matrix} \right\} = \frac{1}{2}\gamma^{(\lambda\alpha)}(g_{(\alpha\nu),\mu} + g_{(\alpha\mu),\nu} - g_{(\mu\nu),\alpha}) \tag{2.27}$$

and $\gamma^{(\mu\nu)}$ is defined by

$$\gamma^{(\mu\nu)}\gamma_{(\nu\lambda)} = \delta^{\mu}_{\lambda}. \tag{2.28}$$

In the weak field approximation to NGT, the $g_{\mu\nu}$ are expanded as

$$g_{\mu\nu} = \eta_{\mu\nu} + h_{\mu\nu}, \tag{2.29}$$

where $|h_{\mu\nu}| \ll 1$ and $\eta_{\mu\nu} = \mathrm{diag}(+1,-1,-1,-1)$ is the flat-space Minkowski metric. The quadratic part of the Lagrangian for the $g_{[\mu\nu]}$ sector, in the weak field approximation, does not obey manifest Kalb-Ramond gauge invariance associated with the effective field theory of closed-strings[23][24]. This gauge invariance guarantees the absence of ghost poles. In the case of NGT, this gauge invariance is broken by the vector field W_μ and its derivative[25], but since W_μ does not have a kinetic energy piece in the Lagrangian, ghost poles are canceled when only causal solutions to the linearized field equations are chosen[26]-[30]. Therefore, NGT represents a viable theory of gravitation.

The macroscopic predictions of NGT have been extensively investigated[5][7]. It has been shown that NGT is consistent with all solar system tests of gravity, with the data for the binary pulsar PSR 1913-16, and with the anomalously low periastron shift data for the non-degenerate eclipsing binary systems DI Herculis and AS Cam[31][32].

III. *The Fiber Bundle Gauge Group of Transformations* $GL(4,R)$

The fiber bundle group of transformations, in NGT, is extended to the local gauge group of transformations $U(3,1,\Omega)$, which is the unitary group of transformations over hyperbolic complex numbers Ω that form a *ring* with the pure imaginary number $\omega^2 = +1$[4][19][35]-[37]. The transformations of the group $U(3,1,\Omega)$ preserve the fundamental nonsymmetric tensor $g_{\mu\nu} = g_{(\mu\nu)} + g_{[\mu\nu]}$, where $g_{[\mu\nu]} = \omega a_{[\mu\nu]}$ with $a_{[\mu\nu]}$ a real skew symmetric tensor. When we choose

$i^2 = -1$ for the *field* of ordinary complex numbers, then the analogous group of transformations would be $U(3,1,C)$ that, according to the Cartan classification, preserves the invariance of the Hermitian nonsymmetric tensor $g_{\mu\nu} = s_{(\mu\nu)} + ia_{[\mu\nu]}$. It can be proved[37] that $U(3,1,\Omega)$ *is isomorphic to* $GL(4,R)$. This amounts to something of a miracle. Such isomophisms do not exist for the group $U(3,1,C)$, based on the ordinary complex numbers (with $i^2 = -1$). Thus the local gauge group in the tangent space or linear frame bundle space is $GL(4,R)$.

The reason that we use the hyperbolic complex group $U(3,1,\Omega)$ and its real isomorphism $GL(4,R)$ as the gauge group of NGT, is that this guarantees the absence of ghost poles in the theory. Thus, if we used $U(3,1,C)$ as the gauge group based on ordinary complex numbers, then we would have unphysical ghost poles in the spectrum of the theory[27]–[30].

Let us consider the hyperbolic complex ring of numbers Ω and its associated functions in more detail. The first known use of hyperbolic complex numbers was made by the mathematician Clifford[38], in 1873, who called them 'motors'. In 1878[39], he published work in which he used the hyperbolic complex numbers to form the biquaternion algebra Ω_2. Consider $z = a + \omega b$ (a and b are real numbers) and $\tilde{z} = a - \omega b$. Then we have that $|z|^2 = z\tilde{z} = a^2 - b^2$. z does not vanish in Ω but $|z|$ may vanish (e.g. consider $z = 1 + \omega$). z^{-1} exists, if and only if $|z|^2 \neq 0$. We have

$$z^{-1} = \tilde{z}/|z|^2 = (a - \omega b)/(a^2 - b^2). \tag{3.1}$$

There is a line of singularities at the points where $a = \pm b$, and the ring Ω has more than one divisor of zero. This feature does not pose any difficulties for physics, where singularities of this type are commonly associated with fields in Minkowski space. We therefore employ the elements in Ω as we do complex numbers in C, except along the singular lines. Since Ω is an Abelian associative division ring with zero divisors, we can refer to Ω as a singular field. The ring Ω bypasses Frobenius' theorem[40] on the construction of associative division algebras over the real numbers R, which does not allow singular points other than the point corresponding to the zero element. In the Frobenius construction of division algebras, it is taken that when a product of terms is zero at least one of the terms is zero. This is not the case with Ω, which explains why Ω is not well-known as a division algebra in the usual sense. Fields are usually constructed using the fundamental theorem of algebra, but the equation $x^2 - 1 = 0$ has a solution in both R and Ω i.e. $x = \pm 1$ in R and $x = \pm\omega$ in Ω.

There are two special elements in Ω, i. e. $y = \frac{1}{2}(1+\omega)$ and $\tilde{y} = \frac{1}{2}(1-\omega)$. It follows that

$$y^2 = y, \quad \tilde{y}^2 = \tilde{y}, \quad |y|^2 = 0. \tag{3.2}$$

The numbers y and $\tilde{y}$ play the roles of 1 and 0 in a hyperbolic complex matrix.

The sixteen generators Σ_{ab} of the non-compact group $GL(4, R)$ can be split into the one-parameter group of dilations and the $SL(4, R)$ group, the latter being the group of volume preserving transformations in a non-simply connected parameter space. We picture the group elements of $GL(4, R)$ as being described by 4×4 matrices. The sub-group of dilations consists of constant, diagonal matrices that commute with those of the semisimple noncompact Lie group $SL(4, R)$. The maximal compact sub-group of $SL(4, R)$ is $SO(4)$ and the universal covering group of $SL(4, R)$ is $\mathcal{SL}(4, R)$ which has the same Lie algebra as $SL(4, R)$. $\mathcal{SL}(4, R)$ is simply connected and contains the maximal subgroup $\mathcal{SO}(4)$ which is isomorphic to $SU(2) \times SU(2)$. $\mathcal{SL}(4, R)$ and $\mathcal{SO}(4)$ are the double covering groups of $SL(4, R)$ and $SO(4)$, respectively. The complete center Z_4 of $\mathcal{SL}(4, R)/Z_4 \simeq SO(3,3)$.

We identify $\Sigma_{[ab]} = M_{[ab]}$ as the six generators $M_{[ab]}$ of the homogeneous Lorentz group, formed from the angular momentum and boost operators J_i and K_i (i = 1,2,3). The remaining nine generators of $SL(4, R)$ are the shear tensor $S_{(ab)}$ with $\mathrm{Tr}(S_{(ab)}) = 0$. The dilation generator S and the nine generators $S_{(ab)}$ together determine the ten generators $\Sigma_{(ab)}$. The shear tensor describes the "stretching" of the space and time coordinates of the linear inertial frames. The commutation relations of the algebra $SL(4, R)$ are[41][42]

$$[K_{ab}, K_{cd}] = i\eta_{bc} K_{ad} - i\eta_{ad} K_{cb}, \tag{3.3}$$

where $K_{(ab)} = S_{(ab)}$ and $K_{[ab]} = \Sigma_{[ab]} = M_{[ab]}$. The generators Σ^a_b of the algebra $GL(4, R)$ are given by

$$\Sigma^a_b = \frac{1}{2}(M^a_b + S^a_b + \frac{1}{2}\delta^a_b S), \tag{3.4}$$

and they satisfy the commutation relations

$$[\Sigma_{ab}, \Sigma_{cd}] = i\eta_{bc}\Sigma_{ad} - i\eta_{ad}\Sigma_{cb}. \tag{3.5}$$

The sub-group $SL(3, R)$ of $SL(4, R)$ exists in the spatial subspace of Minkowski space. The group $SL(3, R)$ has the eight generators formed from the angular momentum operators $M_{[ij]}$(i,j = 1,2,3) and the five shear operators $S_{(ij)}$, which transform under the sub-group $SO(3)$ of $SL(3, R)$ as a quadrupole operator. The non-compact sub-group $SO(3, 3)$ possesses the compact sub-group $SO(3) \times SO(3)$ with the double covering $SU(2) \times SU(2)$. The Lie algebra $SL(4, R)$ is isomorphic to the algebra $SO(3, 3)$. The generators of the group $SU(2) \times SU(2)$ are

$$J_i^{(1)} = \frac{1}{4}\epsilon_{ijk}M_{[jk]} + \frac{1}{2}S_{(0i)}$$

$$J_i^{(2)} = \frac{1}{4}\epsilon_{ijk}M_{[jk]} - \frac{1}{2}S_{(0i)}. \tag{3.6}$$

The Poincaré algebra $\mathcal{P}$ is replaced by the algebra $\mathcal{A} = T_4 \times GL(4, R)$ where T_4 are the translations in spacetime. The algebra $\mathcal{A}$ cannot be gauged directly by the fiber bundle of NGT. This situation is similar to GR in which the Poincaré algebra $\mathcal{P}$ cannot be derived as a local gauge structure.

The physical spinor representations, in NGT, are the *infinite- dimensional* irreducible representations, which are double-valued unirreps of $SL(4, R)$. The existence of infinite-dimensional, multivalued spinor representations of $SL(4, R)$ was proved by Ne'eman[20].The homogeneous Lorentz group $SO(3, 1)$ is a sub-group of $SL(4, R)$ and, consequently, the Lorentz double covering group $\mathcal{SO}(3, 1) \simeq SL(2, C)$ is a sub-group of SL(4,R). In contrast to $SL(n, R)$, the group $SL(n, C)$ does possess finite-dimensional, multivalued spinor representations.

The fact that only infinite-dimensional spinor fields exist within the local gauge group $GL(4, R)$ represents a fundamental physical departure of NGT from GR[43][44]. It forces us to consider an infinite-dimensional field theory when we couple spinor fields to gravity as described by NGT, and this means that particles in NGT are non-localizable extended objects that interact *locally* and could lead to a finite quantum theory of gravity.

The result that $GL(4,R)$ is the local gauge group of NGT, has also been proved by treating the fiber bundle as a real eight-dimensional space, which has a hyperbolic complex structure imposed upon it that is generated by a complex operator E satisfying $E^2 = +1$[19]. The sub-group of $GL(8,R)$ of the eight-dimensional fiber bundle space that preserves E is isomorphic to $GL(4,R) \times GL(4,R)$ and this group reduces to $GL(4,R)$ when a symmetric metric $g_{AB} = g_{BA}$ $(A, B = 1, 2, ..., 8)$ is introduced. More specifically, the group $GL(8,R)$ reduces to $GL(4,R) \times GL(4,R)$ when we require that $\nabla E = 0$, where ∇ denotes the covariant derivative with respect to the connection in the eight-dimensional fiber bundle. The introduction of the metric g_{AB} in the fiber bundle and the requirement that $\nabla g_{AB} = 0$, then reduces $GL(4,R) \times GL(4,R)$ to $GL(4,R)$. The latter group preserves the nonsymmetric fundamental tensor $g_{\mu\nu} (\mu, \nu = 0, 1, 2, 3)$ under linear frame transformations in the fiber bundle.

$GL(4,R)$ is the general group of (real) transformations in space-time and $GL(4,R) \supset SO(3,1)$. The group $GL(4,R)$ is the most general linear group of transformations among linear inertial frames of reference. In addition to the Lorentz frame rotations, there are space and time stretching (shear) and dilations (scale transformations).

We have thus arrived at the important result that NGT is the most general desciption of space-time in the associated fiber bundle or tangent space as well as in the manifold M_4, although the group of diffeomorphism transformations C in the manifold is extended to include a $U(1)$ (or R_+) Abelian group of transformations. Because $GL(4,R) \supset SO(3,1)$, it follows that NGT automatically contains GR as the minimal description of space-time.

IV. *The Vierbein Structure of NGT*

We can define hyperbolic complex *vierbeins* by means of the equation[43][44]

$$e^a_\mu = \mathrm{Re}(e^a_\mu) + \omega \mathrm{Im}(e^a_\mu), \tag{4.1}$$

where $a, b = 0, 1, 2, 3$. The sesquilinear form of $g_{\mu\nu}$ is given by

$$g_{\mu\nu} = e^a_\mu \tilde{e}^b_\nu \eta_{ab}. \tag{4.2}$$

Here $\eta_{ab} = \text{diag}(1, -1, -1, -1)$ is the Minkowski flat-space metric tensor associated with the tangent space (anholonomic coordinates) and $\tilde{e}^a_\mu$ is the complex conjugate of the hyperbolic complex *vierbein* e^a_μ. This formalism has been extended to an n-dimensional space elsewhere[33][34].

The *vierbeins* e^a_μ obey

$$e^\mu_c e^b_\mu = \delta^b_c, \quad e^a_\sigma e^\rho_a = \delta^\rho_\sigma \tag{4.3}$$

and they obey the equation of compatibility

$$e^a_{\mu,\sigma} + (\omega_\sigma)^a_c e^c_\mu - W^\rho_{\sigma\mu} e^a_\rho = 0, \tag{4.4}$$

where ω_σ is the spin connection in NGT and $W^\lambda_{\mu\nu}$ is the nonsymetric affine connection, defined by Eq.(2.4); the skew part $W^\lambda_{[\mu\nu]}$ is pure imaginary in the hyperbolic complex sense: $W^\lambda_{[\mu\nu]} = \omega L^\lambda_{[\mu\nu]} (L^\lambda_{[\mu\nu]}$ is a real skew-symmetric tensor).

A solution of W in terms of e and ω is given by

$$W_{\sigma\lambda\rho} = g_{\delta\rho} W^\delta_{\sigma\lambda} = \eta_{ab}(D_\sigma e^a_\lambda)\tilde{e}^b_\rho, \tag{4.5}$$

where D_σ is the covariant derivative operator

$$D_\sigma e^a_\mu = e^a_{\mu,\sigma} + (\omega_\sigma)^a_c e^c_\mu. \tag{4.6}$$

Differentiating (4.2) yields

$$g_{\mu\nu,\sigma} - g_{\rho\nu} W^\rho_{\mu\sigma} - g_{\mu\rho} \tilde{W}^\rho_{\nu\sigma} = 0, \tag{4.7}$$

where we have required that

$$(\omega_\sigma)_{ca} = -(\tilde{\omega}_\sigma)_{ac} \tag{4.8}$$

This shows that the spin connection $(\omega_\sigma)_{ab}$ is skew- Hermitian in the indices a and b. For a Hermitian connection $\tilde{W}^\lambda_{\mu\nu} = W^\lambda_{\nu\mu}$, we obtain the compatibility condition

$$g_{\mu\nu,\sigma} - g_{\rho\nu} W^\rho_{\mu\sigma} - g_{\mu\rho} W^\rho_{\sigma\nu} = 0. \tag{4.9}$$

We can define a group of isometries by the equation

$$e_\sigma^a = e_\sigma'^b (U)_b^a, \tag{4.10}$$

where U is an element of $GL(4, R)$ that leaves the fundamental form $g_{\mu\nu}$ invariant. Moreover, the connection W will remain invariant under the transformation (4.10) for the non-Abelian transformation

$$(\omega_\sigma)_b^a \rightarrow [U\omega_\sigma U^{-1} - (\partial_\sigma U)U^{-1}]_b^a. \tag{4.11}$$

The curvature tensor in this formalism is defined by

$$([D_\mu, D_\nu])_b^a = (R_{\mu\nu})_b^a, \tag{4.12}$$

where

$$(R_{\mu\nu})_b^a = (\omega_\nu)_{b,\mu}^a - (\omega_\mu)_{b,\nu}^a + ([\omega_\mu, \omega_\nu])_b^a. \tag{4.13}$$

For the non-Abelian transformation (4.11) with

$$(U^{-1})_b^a = \eta_{bd}(\tilde{U})_e^d \eta^{ea}, \tag{4.14}$$

it follows that

$$(R_{\mu\nu})_b^a \rightarrow U_c^a (R_{\mu\nu})_d^c (U^{-1})_b^d. \tag{4.15}$$

We can express the curvature tensor in holonomic coordinates

$$R^\lambda{}_{\sigma\mu\nu} = (R_{\mu\nu})_b^a e_a^\lambda e_\sigma^b. \tag{4.16}$$

The scalar curvature is given by

$$R = e^{\mu a} \tilde{e}^{\nu b} (R_{\mu\nu})_{ab}. \tag{4.17}$$

For the case of the vacuum, the action takes the form

$$S = -\frac{1}{16\pi} \int d^4x |e(x)| R(x), \tag{4.18}$$

where $|e| = (e\tilde{e})^{\frac{1}{2}}$ with $e = \det(e^a_\mu)$. A variation of the action S with respect to ω and e leads to the field equations

$$[|e|(e^{\mu a}\tilde{e}^{\nu b} - e^{\nu a}\tilde{e}^{\mu b})]_{,\nu} + |e|[(\omega_\mu)^b_c(e^{\mu a}\tilde{e}^{\nu c} - e^{\nu a}\tilde{e}^{\mu c})$$

$$+(\omega_\mu)^a_c(e^{\nu c}\tilde{e}^{\mu b} - e^{\mu c}\tilde{e}^{\nu b})] = 0, \tag{4.19}$$

$$R_{\mu a} = 0, \tag{4.20}$$

where

$$R_{\mu a} = \tilde{e}^{\nu b}(R_{\mu\nu})_{ab}. \tag{4.21}$$

Contracting (4.19) over the suffixes a and b gives

$$[|e|(e^{\mu a}\tilde{e}^{\nu}_a - e^{\nu a}\tilde{e}^{\mu}_a)]_{,\nu} = 0. \tag{4.22}$$

This equation can be reduced to the form

$$[(-g)^{\frac{1}{2}}g^{[\mu\nu]}]_{,\nu} = 0. \tag{4.23}$$

The metric tensor in NGT is $g_{(\mu\nu)}$ and this takes its Minkowski values $\eta_{\mu\nu}$ at a point $x = x'$ in a local inertial frame. The metric tensor can be defined in terms of the real *vierbeins* E^a_μ according to

$$g_{(\mu\nu)} = E^a_\mu E^b_\nu \eta_{ab}, \tag{4.24}$$

and we have $E^a_\mu = \delta^a_\mu$ at the point $x = x'$. The equivalence of inertial and gravitational masses still holds in NGT[(7)], although a new 'fifth' tensor force that behaves like $1/r^5$ causes test particles to fall at different rates in a gravitational field.

V. *Spinors in GL(4,R), Infinite-Component Wave Equations and the Mass Spectrum*

The particles in standard relativistic quantum field theory span unitary representations of the Poincaré group P and its double-cover $\mathcal{P}$. The fields transform as finite and non-unitary representations of $GL(4, R)$ when tensorial and of

$SL(2,C)$ when spinorial, where $SL(2,C)$ is the double covering of the Lorentz group $SO(3,1)$. The overall unitarity of a quantum theory is physically guaranteed, because the Hermiticity of the Lagrangian requires the addition of complex conjugate terms that cancel the non-Hermitian parts of the densities determined by Noether's theorem.

Majorana[45][46] first studied infinite component fields by using irreducible, infinite-dimensional unitary representations of $SL(2,C)$ to introduce an invariant, linear wave equation

$$(iY^\mu \partial_\mu - \kappa)\psi(x) = 0, \tag{5.1}$$

where the operators Y^μ close on the algebra $\mathcal{SP}(4,R) \simeq \mathcal{SO}(3,2)$. The ladder representation of $Sp(4,R)$ is unitary and separates into the direct sum of two Majorana representations. Majorana's results were unphysical owing to his using *unitary* infinite-dimensional representations of $SL(2,C)$. He used these representations to solve the problem of the Dirac negative energy states, which at the time, in 1932, were creating difficulties. Soon after the publication of Majorana's paper, the positron was discovered by Anderson, and there was no need at the time to pursue further the properties of equation (5.1). The mass spectrum predicted by Majorana's equation decreases in mass with increasing spin contrary to our experience. Barut, Fronsdal and Nambu[47]–[50] have studied more general infinite-component wave equations in the context of the hydrogen atom by using representations of $SO(4,2)$, and by suitably choosing certain parameters the correct hydrogen spectrum was obtained in the non-relativistic limit.

Various potential difficulties inherent in certain kinds of infinite-component field theories have been discovered by several authors:[48][51][52] i) problems with spin and statistics and the associated PCT theorem, and ii) space-like solutions with $P_\mu P^\mu \leq 0$. However, all these studies were based on the use of infinite-dimensional unitary representations of the Lorentz group. It can be proved that a Lorentz invariant infinite-component field theory with a spinor field that is an infinite irreducible representation of $SL(2,C)$ has a degenerate mass spectrum[53]–[56] Put in other words, a field theory with an infinite-component field that is invariant under the transformations of the Lorentz group, *violates locality unless the towers of particles are all degenerate.* This theorem will also hold true, if we extend the transformation group $SO(3,1)$ to that of $GL(4,R)$. But if we break the symmetry group $GL(4,R)$ down to that of the Lorentz group, then we can retain locality and realize simultaneously a non-degenerate infinite mass spectrum, if for each mass level there is associated a *finite*, non-unitary representation of the Lorentz group $SO(3,1)$. In string theories, each mass level of the infinite mass spectrum has associated with it a finite representation of the Lorentz group. In this way, we can avoid the common diseases of infinite-component field theories discussed above. Because we do not use infinite-dimensional unitary representations of the Lorentz group, *we have no trouble with the causality of the theory.*

In studies of the non-relativistic hydrogen atom[50], the symmetry of the Hamiltonian is $O(4)$ instead of $SO(3,1)$, and one considers the spectrum- generating group $SO(4,1)$, the irreducible representations of which contains all the bound states of hydrogen. In analogy with the problem of the hydrogen atom, we could have used instead of $GL(4,R)$ the group $SO(3,2)$ which is the group of linear homogeneous transformations that leaves the quadratic form

$$Q = x_0^2 - x_1^2 - x_2^2 - x_3^2 + x_5^2 \tag{5.2}$$

invariant. $SO(3,2)$ has ten generators $M_{\mu\nu} = -M_{\nu\mu}(\mu,\nu = 0,1,2,3,5)$ that satisfy the commutation relations

$$[M_{\mu\nu}, M_{\rho\sigma}] = i(M_{\mu\rho}g_{\nu\sigma} + M_{\nu\sigma}g_{\mu\rho} - M_{\nu\rho}g_{\mu\sigma} - M_{\mu\sigma}g_{\nu\rho}), \tag{5.3}$$

where we take $g_{00} = g_{55} = +1, g_{11} = g_{22} = g_{33} = -1$, and $g_{\mu\nu} = 0$ for $\mu \neq \nu$. Then the infinite towers of particles in the irreducible representations of $SO(3,1)$ decompose into infinite sums of finite-dimensional representations of the Lorentz sub-group $SO(3,1)$[56]. As before, because we do not introduce any infinite-dimensional unitary representations of the Lorentz group, we can avoid the common diseases of infinite-component field theories. We could, in fact, formulate an infinite-component field theory based on the five-dimensional group $SO(3,2)$. However, we prefer to work with NGT and its $GL(4,R)$ local gauge structure, since the latter contains a complete dynamical description of four-dimensional space-time.

Ne'eman and Šijački[(57)–(61)] have constructed a classification of the multiplici free irreducible unitary representations of $SL(4,R)$ by using the maximal compact sub-group $SU(2) \times SU(2)$. Work on this problem has also been carried out by Speh[(59)], and Friedman and Sorkin[(60)]. No complete classification of all the unirreps of $SL(4,R)$ exists in the literature. It has been noted by Ne'eman and Šijački[(62)] that the $\mathcal{SL}(4,R)$ algebra commutation relations (3.3) are invariant under the automorphism:

$$J_i' = J_i, \quad K_i' = iN_i, \quad N_i' = iK_i, \quad (i = 1,2,3) \tag{5.4}$$

where $N_i = S_{(0i)}$ and the (J_i, iK_i) form a new algebra $SO(4)'$ and the (J_i, iN_i) form the algebra $SL(2,C)'$. We can use this automorphism to convert the irreducible representations of $\mathcal{SO}(4) \simeq SU(2) \times SU(2)$ into infinite sums of *non-unitary finite* representations of the Lorentz group $SO(3,1)$. The boost operators K_i become anti-Hermitian operators under the deunitarizing automorphism. In $SO(3,1)$ only non-unitary spinor representations have a physical meaning, because the scalar density $\psi^\dagger\psi = E/m$, where E is the particle energy has the correct boosting property; the anti-Hermitian boosting operator cancels the intrinsic pieces e.g. $\frac{1}{2}[\gamma^0, \gamma^i] + h.c. = 0$, as in the case of standard finite-component spinors. In this way, we avoid the unphysical excitation of a given spin state to other spin states and masses that follows from using unitary spinor representations that satisfy $\psi^\dagger\psi = 1$. The physical spin -1/2 nucleon is observed to *accelerate* under a boost rather than be excited to a spin -3/2 N^* state.

To obtain a physical picture, we are required to break $SL(4,R)$ down to $\mathcal{SO}(4)$ and at the same time invoke the automorphism, so that we obtain an infinite sum of *physical* finite, non-unitary spinor representations of the Lorentz group. The automorphism transformation corresponds to a transition from an Euclidean space to the Minkowski space-time with the signature $(1,-1,-1,-1)$. Alternatively, we could have started with *non-unitary* infinite-dimensional representations of $SL(4,R)$ and proceeded directly to a symmetry breaking pattern consisting of an infinite sum of non-unitary, finite representations of the Lorentz group. The deunitarizing automorphism (DUA) is a generalization of the 'Weyl unitary trick' and accomplishes the same results. More importantly, only the unitary infinite-dimensional representations of $SL(4,R)$ have a (incomplete) classification. To summarize these results, the infinite unirreps of $\mathcal{SL}(4,R) \Rightarrow$ infinite non-unitary irreps of $\mathcal{SL}(4,R)$, where $\Rightarrow$ denotes the action of the DUA. The infinite unirreps of $\mathcal{SL}(4,R)$ decompose into an infinite sum of finite unirreps of $\mathcal{SO}(4) \Rightarrow$ an infinite sum of finite non-unitary irreps of $\mathcal{SO}(3,1) \simeq SL(2,C)$.

For the standard Dirac equation, each particle is labelled with a mass m and a spin j in terms of the four-component Dirac spinor. In the $GL(4,R)$ scheme each particle is labelled by the two spin variables j_1, j_2 associated with the compact subgroup $\mathcal{SO}(4) \simeq SU(2) \times SU(2)$. The wave equation in NGT will be required to be an infinite-component equation invariant under $SL(2,C)$ transformations with a spinor that is a unitary irreducible representation of $\mathcal{SL}(4,R)$ given by $\mathcal{D}^{\text{disc}}(j_1, j_2,$ for $j_1 = p_1+1, j_2 = 0$ or $j_1 = 0, j_2 = p_1+1$, where $p_1 = -1/2, 1/2, 3/2, 5/2, ..., p_2 = 0$, and $|j_1 - j_2| \geq p_1 + 1$. The physical spinor particles will be described by the DUA of the representation $\psi(x) \simeq \mathcal{D}^{\text{disc}}(1/2,0) \oplus \mathcal{D}^{\text{disc}}(0,1/2)$.

The Y^μ operators that take their values in the Hilbert space $\mathcal{H}$ of the representations of $SL(4,R)$ transform as an $SL(2,C)$ vector, i.e. Y^μ behaves like a vector under Lorentz transformations. With this requirement the $\mathcal{SL}(4,R)$ towers of particles that make up the representations of $\psi(x)$ are not mass degenerate and under the action of the DUA the formalism will display causality.

The spin content described by (j_1, j_2) in the $SL(4,R)$ unirreps has been shown by Ne'eman and Šijački(63) to provide a shell-model like description of the baryon and meson mass spectrum for each flavor. The infinite-dimensional representations of the sub-group $SL(3,R)$ have been shown to possess a Regge behavior. This scheme gives a phenomenological picture of the towers of particle states at the lowest mass level.

The $GL(4,R)$ invariant wave equation takes the form(43)(44)

$$(iY^\mu \partial_\mu + K(x) - \kappa)\psi(x) = 0, \tag{5.5}$$

where the operators $\psi(x)$ and Y^μ take their values in the Hilbert space $\mathcal{H}$ of the representations of $\mathcal{SL}(4,R)$, and the operator-valued function $K(x)$ is

$$K(x) = Y^\mu(\omega_\mu(x))_{ab}\Sigma^{ab}. \tag{5.6}$$

The $K(x)$ is obtained from a solution of the NGT spin connection $(\omega(x)_\mu)_{ab}$ which is determined by the field equations. This is analogous to the situation in electrodynamics in which the minimal coupling assumption

$$\partial_\mu \to \partial_\mu + ieA_\mu(x) \tag{5.7}$$

is substituted into the Dirac equation and the potential A_μ is found from Maxwell's equations.

There will be three mass scales in NGT: (1) the zero masses associated with the vacuum state, (2) the Planck mass scale: $G^{-1/2} = 1.2 \times 10^{19}$ Gev, determined by the skew spin connection $(\omega_\mu(x))_{[ab]}$ governing the gravitational coupling, and a third mass scale,(3) $m \geq 10^6$ GeV determined by the coupling constant f associated with the symmetric part of the spin connection $(\omega_\mu(x))_{(ab)}$, that measures the strength of the coupling of $g_{[\mu\nu]}$ to matter. The coupling constant f has the dimensions of a length so that $f^{-1} \geq 10^6$ GeV.

VI. *Infinite-Component Field Theory*

A field theory formalism for fields of arbitrary spin was developed by Weinberg and we shall use his notation in the following. We take it as given that we can construct an infinite-component field theory that does not have any unphysical modes and satisfies the required gauge invariances. This issue will be addressed in the next Section. We adopt the point of view that we can solve the theory using perturbation theory, and we assume forthwith that the S-matix can be calculated from Dyson's formula:

$$S = \sum_{n=0}^{\infty} \frac{(-i)^n}{n!} \int_{-\infty}^{\infty} dt_1 ... dt_n T\{L_I(t_1)...L_I(t_n)\}, \tag{6.1}$$

where we have split the Lagrangian operator into a free part L_0 and an interaction part L_I, where L_I is defined in the interaction picture by

$$L_I(t) = \exp(iL_0 t) L_I \exp(-iL_0 t). \tag{6.2}$$

We also require that S is invariant under orthochronous Lorentz transformations and that to every inhomogenous Lorentz transformation $x_\mu \to \Lambda^\mu{}_\nu x^\nu + a^\mu$ there corresponds a unitary operator $U[\Lambda, a]$ such that

$$U[\Lambda, a]\mathcal{L}(x)U^{-1}[\Lambda, a] = \mathcal{L}(\Lambda x + a), \tag{6.3}$$

where $\mathcal{L}$ is a scalar density and

$$L(t) = L_0(t) + L_I(t) = \int d^3x \mathcal{L}(\mathbf{x}, t). \tag{6.4}$$

For space-like $(x - y)$ separations

$$[\mathcal{L}_\mathcal{I}(x), \mathcal{L}_\mathcal{I}(y)] = 0. \tag{6.5}$$

We shall form $\mathcal{L}_\mathcal{I}$ from fields $\phi_n(x)$ which are linear combinations of creation and annihilation operators. The fields transform as

$$U[\Lambda, a]\phi_n(x)U^{-1}[\Lambda, a] = \sum_m \mathcal{D}_{nm}[\Lambda^{-1}]\phi_m(\Lambda x + a), \tag{6.6}$$

where $\mathcal{D}_{nm}[\Lambda]$ is some representation of the Lorentz group. For space-like separations

$$[\phi_n(x), \phi_m(y)]_\pm = 0, \tag{6.7}$$

where $[\ ]_\pm$ denotes either a commutator (+) or an anti-commutator (−).

We shall not encounter any problems with spin and statistics or causality, if we keep to the infinite sums of finite, non-unitary representations of the Lorentz group. Let us consider the finite representations. Any representation is specified by a representation of the infinitesimal Lorentz transformations. They are defined by

$$\Lambda^\mu{}_\nu = \delta^\mu{}_\nu + \omega^\mu{}_\nu, \tag{6.8}$$

where the ω's form the infinitesimal coefficients

$$\omega_{\mu\nu} = -\omega_{\nu\mu}. \tag{6.9}$$

The unitary operators are

$$U[1+\omega] = 1 + (1/2)M_{[\mu\nu]}\omega^{\mu\nu}. \tag{6.10}$$

From the six operators $M_{[\mu\nu]}$, we form the two Hermitian vectors

$$J_i = \frac{1}{2}\epsilon_{ijk}M_{jk}, \tag{6.11}$$

$$K_i = M_{[i0]}. \tag{6.12}$$

These vectors satisfy the algebra

$$[J_i, J_j] = i\epsilon_{ijk}J_k, \tag{6.13}$$

$$[J_i, K_j] = i\epsilon_{ijk}K_k, \tag{6.14}$$

$$[K_i, K_j] = -i\epsilon_{ijk}J_k. \tag{6.15}$$

The J_i and the K_i correspond to the generators of rotation and boosts, respectively. The unitary operator for the finite Lorentz boost is

$$U[(p)] = \exp(-ip\cdot K\theta). \tag{6.16}$$

It is convenient to define a pair of anti-Hermitian operators

$$\mathbf{A} = \frac{1}{2}[\mathbf{J} + i\mathbf{K}], \tag{6.17}$$

$$\mathbf{B} = \frac{1}{2}[\mathbf{J} - i\mathbf{K}], \tag{6.18}$$

with the commutation rules

$$\mathbf{A}\times\mathbf{B} = i\mathbf{A}, \tag{6.19}$$

$$\mathbf{B}\times\mathbf{B} = i\mathbf{B}, \tag{6.20}$$

$$[A_i, B_j] = 0. \tag{6.21}$$

The $(2A+1)(2B+1)$-dimensional irreducible representations (A, B) are defined by

$$< a, b|\mathbf{A}|a', b' >= \delta_{bb'}\mathbf{J}^{(A)}_{aa'}, \tag{6.22}$$

$$< a, b|\mathbf{B}|a', b' >= \delta_{aa'} \mathbf{J}^{(B)}_{bb'}, \tag{6.23}$$

where a and b run over the values $-A$ to $+A$ and $-B$ to $+B$ in units steps, respectively, and $\mathbf{J}^{(i)}$ is the $2j+1$-dimensional representation of the rotation group:

$$(J^{(j)}_x \pm i J^{(j)}_y)_{\sigma'\sigma} = \delta_{\sigma',\sigma\pm 1}[(j \mp \sigma)(j \pm \sigma + 1)]^{1/2},$$

$$(J^{(j)}_z)_{\sigma'\sigma} = \delta_{\sigma'\sigma}\sigma. \tag{6.24}$$

All the finite non-unitary irreducible representations of the homogeneous Lorentz group are given by the representations (A, B). We are most interested in the irreducible representations $(j, 0)$ and $(0, j)$ described by

$$\mathbf{J} \to \mathbf{J}^{(j)}, \quad \mathbf{K} \to -i\mathbf{J}^{(j)}, \quad \text{for} \quad (j, 0) \tag{6.25}$$

$$\mathbf{J} \to \mathbf{J}^{(j)}, \quad \mathbf{K} \to +i\mathbf{J}^{(j)}, \quad \text{for} \quad (0, j), \tag{6.26}$$

where the $\mathbf{J}^{(j)}$ are given by (6.24). The $2j + 1$-dimensional matrix representing a finite Lorentz transformation Λ is denoted by $\mathcal{D}^{(j)}[\Lambda]$ and $\bar{\mathcal{D}}^{(j)}[\Lambda]$ for the $(j, 0$ and $(0, j)$ representations, respectively. They are related by

$$\mathcal{D}^{(j)}[\Lambda] = \bar{\mathcal{D}}^{(j)}[\Lambda^{-1}]^{\dagger}. \tag{6.27}$$

The field operator for arbitrary spin takes the form

$$\phi^j_\sigma(x) = (2\pi)^{-3/2} \int \frac{d^3\mathbf{p}}{[2\omega(\mathbf{p})]^{1/2}} \sum_{\sigma'} [\xi \mathcal{D}^{(j)}_{\sigma\sigma'} a(\mathbf{p}, \sigma') e^{-ip\cdot x}$$

$$+ \eta \{\mathcal{D}^{(j)} C^{-1}\}_{\sigma\sigma'} c^*(\mathbf{p}, \sigma') e^{ip\cdot x}], \tag{6.28}$$

where C is a $(2j + 1) \times (2j + 1)$ matrix with

$$C^* C = (-)^{2j}; \quad C^\dagger C = 1, \tag{6.29}$$

and C is used to define the ordinary complex conjugate of the representation $\mathcal{D}$:

$$\mathcal{D}^{(j)}[R]^* = C \mathcal{D}^{(j)}[R] C^{-1}. \tag{6.30}$$

According to our particle interpretation, the a's and c's are creation and annihilation operators that satisfy Bose or Fermi commutation rules

$$[a(\mathbf{p},\sigma), a^*(\mathbf{p}',\sigma')]_\pm = \delta(\mathbf{p}-\mathbf{p}')\delta_{\sigma\sigma'},$$

$$[c(\mathbf{p},\sigma), c^*(\mathbf{p}',\sigma')]_\pm = \delta(\mathbf{p}-\mathbf{p}')\delta_{\sigma\sigma'}, \tag{6.31}$$

and all the other commutation relations vanish.

The fields ϕ^j_σ satisfy the commutation or anticommutation relations

$$[\phi^j_\sigma(x), \phi^j_{\sigma'}{}^\dagger(y)]_\pm$$

$$= \frac{m^{-2j}}{(2\pi)^3}\int \frac{d^3\mathbf{p}}{2\omega(\mathbf{p})}\Pi^j_{\sigma\sigma'}(\mathbf{p},\omega(\mathbf{p}))\{|\xi|^2\exp[-ip\cdot(x-y)]\pm|\eta|^2\exp[ip\cdot(x-y)]\}, \tag{6.32}$$

where $\Pi(p)$ is the matrix

$$m^{-2j}\Pi(\mathbf{p},\omega) = \mathcal{D}^{(j)}[L(\mathbf{p})]\mathcal{D}^{(j)}[L(\mathbf{p})]^\dagger = \exp(-2\hat{p}\cdot\mathbf{J}\theta) \tag{6.33}$$

with $\cosh\theta = p^0/m = \omega(\mathbf{p})/m$. From crossing symmetry and Lorentz invariance, we must have $\xi = \eta = 1$ and $\mp(-)^{2j} = 1$, and the commutation or anticommutation relation becomes

$$[\phi^j_\sigma(x), \phi^j_{\sigma'}{}^\dagger(y)]_\pm = i(-im)^{-2j} t^{\mu_1\mu_2\ldots\mu_{2j}}_{\sigma\sigma'}\,\partial_{\mu_1}\partial_{\mu_2}\ldots\partial_{\mu_{2j}}\Delta(x-y), \tag{6.34}$$

where Δ is the causal Green's function

$$\Delta(x-y) = \frac{-i}{(2\pi)^3}\int\frac{d^3\mathbf{p}}{2\omega(\mathbf{p})}[e^{-ip\cdot(x-y)} - e^{ip\cdot(x-y)}] \tag{6.35}$$

and the t is a constant, traceless tensor symmetric in $\mu_1\mu_2\ldots\mu_{2j}$. We have

$$\Pi^j_{\sigma\sigma'}(p) = (-)^{2j} t^{\mu_1\mu_2\ldots\mu_{2j}}_{\sigma\sigma'}\, p_{\mu_1}p_{\mu_2}\ldots p_{\mu_{2j}}\,. \tag{6.36}$$

The commutator (anticommutator) (6.34) vanishes outside the light-cone, thus guaranteeing causality and the correct statistics and crossing symmetry. The covariant propagator is

$$D^j_{\sigma\sigma'}(x-y) = < T\{\phi^j_\sigma(x)\phi^j{}^\dagger_{\sigma'}(y)\} >_0$$

$$= -i(-im)^{-2j} t_{\sigma\sigma'}^{\mu_1\mu_2\ldots\mu_{2j}} \partial_{\mu_1}\partial_{\mu_2}\ldots\partial_{\mu_{2j}} \Delta^C(x-y), \tag{6.37}$$

where Δ^C is the spin-zero propagator

$$-i\Delta^C(x) = i\theta(x)\Delta_+(x) + i\theta(-x)\Delta_+(-x)$$

$$= \frac{1}{2}[\Delta_1(x) + i\epsilon(x)\Delta(-x)]. \tag{6.38}$$

As in standard field theory

$$\epsilon(x) = \theta(x) - \theta(-x)$$

$$\Delta_1(x) = i[\Delta_+(x) + \Delta_+(-x)], \quad \Delta(x) = \Delta_+(x) - \Delta_+(-x). \tag{6.39}$$

The field $\phi^j(x)$ satisfies the Klein-Gordon equation

$$(\Box + m^2)\phi^j_\sigma(x) = 0. \tag{6.40}$$

In momentum space, the propagator (6.37) takes the form

$$D^j(q) = \int d^4x e^{iq\cdot x} D^j(x) = -i(m)^{-2j}\Pi^j(q)/(-q^2 + m^2 - i\epsilon), \tag{6.41}$$

where the scalar function $\Pi^j(q)$ has the form

$$\Pi^j(q) = (q^2)^j \exp[-2\theta(q)\widehat{q}\cdot\mathbf{J}^{(j)}]. \tag{6.42}$$

The functions $\Pi^j(q)$ and $\bar{\Pi}^j(q)$ for the representations $(j,0)$ and $(0,j)$, respectively, are connected by $\Pi(q)\bar{\Pi}(q) = (q^2)^{2j}$. For integer j and arbitrary q, we have

$$\Pi^j(q) = (q^2)^j + [(q^2)^{(j-1)}/2!](2\mathbf{q}\cdot\mathbf{J}^{(j)})(2\mathbf{q}\cdot\mathbf{J}^{(j)} - 2q^0) + \ldots\ . \tag{6.43}$$

The series terminates after $j+1$ terms. For half-integer j and arbitrary q:

$$\Pi^j(q) = (q^2)^{(j-1/2)}[q^0 - 2\mathbf{q}\cdot\mathbf{J}^{(j)}] + \ldots\ . \tag{6.44}$$

In the case of massless particles, the momentum-space propagators are

$$D^j(q) = \int d^4x e^{iq\cdot x} D^j(x) = -i\Pi^j(q)/(-q^2 - i\epsilon). \tag{6.45}$$

We shall also introduce the fields $\psi^j(x)$ given by

$$\psi^j(x) = (2\pi)^{-3/2} \int d^3\mathbf{p}\left(\frac{m}{E(\mathbf{p})}\right)^{1/2} \sum_{\sigma'} [\mathcal{D}^{(j)}_{\sigma\sigma'} b(\mathbf{p},\sigma') w(\mathbf{p}) e^{-ip\cdot x}$$

$$+\{\mathcal{D}^{(j)}C^{-1}\}_{\sigma\sigma'} d^*(\mathbf{p},\sigma') v(\mathbf{p}) e^{ip\cdot x}], \tag{6.46}$$

where w and v are Dirac spinors. The field $\psi^j(x)$ satisfies the Dirac equation

$$(-i\gamma^\mu \partial_\mu + m)\psi^j(x) = 0 \tag{6.47}$$

and the b's and d's satisfy the commutation and anticommutation rules

$$[b(\mathbf{p},\sigma), b^\dagger(\mathbf{p}',\sigma')]_\pm = \delta(\mathbf{p}-\mathbf{p}')\delta_{\sigma\sigma'},$$

$$[d(\mathbf{p},\sigma), d^\dagger(\mathbf{p}',\sigma')]_\pm = \delta(\mathbf{p}-\mathbf{p}')\delta_{\sigma\sigma'}. \tag{6.48}$$

A calculation of the covariant propagator gives

$$S^j_{\sigma\sigma'}(x-y) = < T\{\psi^j_\sigma(x)\psi^{j\,\dagger}_{\sigma'}(y)\} >_0$$

$$= (2\pi)^{-3}(-im)^{-2j}(i\gamma^\mu\partial_\mu + m)t^{\mu_1\mu_2\ldots\mu_{2j}}_{\sigma\sigma'}\,\partial_{\mu_1}\partial_{\mu_2}\ldots\partial_{\mu_{2j}}\Delta^C(x-y). \tag{6.49}$$

In momentum space, the propagator becomes

$$S^j(q) = \int d^4x e^{iq\cdot x} S^j(x) = -i(m)^{-2j}\Pi^j(q)/(m - q\cdot\gamma - i\epsilon). \tag{6.50}$$

Weinberg has also constructed a $2(2j+1)$-component field $\Psi^j(x)$ corresponding to the $(j,0)$ field $\phi_\sigma(x)$ and the $(0,j)$ field $\chi_\sigma(x)$ that satisfies a "Dirac-type" equation of motion

$$(\gamma^{\mu_1\mu_2\ldots\mu_{2j}}\partial_{\mu_1}\partial_{\mu_2}\ldots\partial_{\mu_{2j}} + m^{2j})\Psi^j(x) = 0, \tag{6.51}$$

where the generalized Dirac matrices $\gamma^{\mu_1\mu_2\ldots}$ are defined by the $2(2j+1)$ -dimensional matrices

$$\gamma^{\mu_1\mu_2\ldots\mu_{2j}} = -i^{2j}\begin{pmatrix} 0 & t^{\mu_1\mu_2\ldots\mu_{2j}} \\ \bar{t}^{\mu_1\mu_2\ldots\mu_{2j}} & 0 \end{pmatrix} \tag{6.52}$$

and

$$\gamma_5 = \begin{pmatrix} 1 & 0 \\ 0 & -1 \end{pmatrix}, \tag{6.53}$$

$$\beta = \begin{pmatrix} 0 & 1 \\ 1 & 0 \end{pmatrix}. \tag{6.54}$$

The $\Psi^j(x)$ field transforms according to the $(j,0) \oplus (0,j)$ representation. The causal propagator for the $\Psi^j(x)$ field in momentum space is given by

$$\mathcal{S}^j(q) = -i^{-2j}[\wp(q) + m^{2j}]/(-q^2 + m^2 - i\epsilon), \tag{6.55}$$

where

$$\wp^j(q) = -i^{2j}\gamma^{\mu_1\mu_2\ldots\mu_{2j}}\, q_{\mu_1} q_{\mu_2} \ldots q_{\mu_{2j}}\,. \tag{6.56}$$

An evaluation of the functions $\wp(q)$ by Weinberg[64] for integer j gives

$$\wp^j(q) = (q^2)^j \beta + \sum_{n=0}^{j-1} \frac{(q^2)^{j-1-n}}{(2n+2)!}[(2\mathbf{q}\cdot\mathcal{J})^2 - (2\mathbf{q}^2)][(2\mathbf{q}\cdot\mathcal{J})^2 - (4\mathbf{q})^2]\ldots$$

$$[(2\mathbf{q}\cdot\mathcal{J})^2 - (2n\mathbf{q})^2][2\mathbf{q}\cdot\mathcal{J}\beta - (2n+2)q^0\gamma_5\beta], \tag{6.57}$$

where

$$\mathcal{J} = \begin{pmatrix} \mathbf{J}^{(j)} & 0 \\ 0 & \mathbf{J}^{(j)} \end{pmatrix}. \tag{6.58}$$

In the last paper in Weinberg's series of three papers on arbitrary spin fields[64], he constructed the most general propagator for a particle of spin j and mass m, off the mass shell, created by a field $\phi(x)$ of type (A_2, B_2) and destroyed by a field $\phi(x)$ of type (A_1, B_1). The result is

$$D_{(A_1B_1,A_2B_2)}(q;j,n) = N_{(A_1B_1,A_2B_2)}(q;j,m,n)/(-q^2 + m^2 - i\epsilon), \tag{6.59}$$

where

$$N_{(A_1B_1,A_2B_2)}(q;j,m,n)$$

$$= \sum_{\nu,\mu} W(A_1B_1, A_2B_2; j, n)\Pi^n_{\nu-\mu}(q) m^{-2n}(-)^{A_1-B_2-j+\mu} C_{A_1A_2}(n,\nu) C_{B_1B_2}(n,-\mu) \tag{6.60}$$

Here W is the standard Racah coefficient and the sums over ν and μ run over all integers or half-integers for which W and the Clebsch-Gordan coefficients C do not vanish. Π is given as before by (6.42), (6.43) and (6.44). In what follows, we shall restrict ourselves to the more limited forms of the fields and propagators given by (6.37) and (6.49).

We have now arrived at the important step of introducing the field operators

$$\phi(x) = \sum_{j=0}^{\infty} (j!)^{-1/2} \phi^j(x) \tag{6.61}$$

and

$$\psi(x) = \sum_{j=0}^{\infty} (j!)^{-1/2} \psi^j(x) \tag{6.62}$$

that describe the infinite-component fields in our scheme. Our D propagator now becomes for the sum over all j:

$$D(q) = \frac{1}{i} \sum_{j=0}^{\infty} (m)^{-2j} (j!)^{-1} \Pi^j(q)/(-q^2 + m_1^2 j + m_0^2 - i\epsilon), \tag{6.63}$$

where we have introduced an infinite mass spectrum of the form

$$\widehat{m}^2 = m_1^2 j + m_0^2 \tag{6.64}$$

similar to the mass spectrum that occurs in string theory[10]. The S propagator is

$$S(q) = \frac{1}{i} \sum_{j=0}^{\infty} (m)^{-2j} (j!)^{-1} \Pi^j(q)/(m_1 j + m_0 - q \cdot \gamma - i\epsilon). \tag{6.65}$$

By using (6.43) and (6.44), we obtain for large values of q^2 the results

$$D(q) \simeq \frac{1}{i} \sum_{j=0}^{\infty} (j!)^{-1} (q^2/m^2)^j / (-q^2 + m_1^2 j + m_0^2 - i\epsilon), \tag{6.66}$$

and

$$S(q) \simeq \frac{1}{i} \sum_{j=0}^{\infty} (j!)^{-1} (q^2/m^2)^j / (m_1 j + m_0 - q \cdot \gamma - i\epsilon). \tag{6.67}$$

VII. *Infinite-Parameter Gauge Transformations*

We must now return to the question of the consistency of our field theory for sums over all spin states and masses. Much effort has been devoted in recent years to develop a consistent field theory scheme that incorporates arbitrary spin states. This work began with the papers of Fierz and Pauli, Chang, Schwinger and Fronsdal[(65)–(70)]. More recent work has been stimulated by string theory ideas and supergravity where higher spins play an important role[(71)–(73)].

In the case of the strings, the free sector is descibed by an infinite set of local fields with increasing spin, the mass spectrum of which is that of the first quantized string. In the critical dimension $D = 26$ (bosonic string) or $D = 10$ (superstring), the theory exhibits enough local symmetry to gauge away all the ghosts. The gauge symmetry of these free-field theories as well as their associated BRST structure are by now well understood. However, it is not yet understood how these symmetries fix uniquely the interacting strings, although several proposals have been put forward to solve this problem.

The string theory program can be viewed as an attempt to solve the long-standing problem of finding consistent actions and gauge invariances for local fields of arbitrarily high spin. The classical boson string can be viewed as a particular collection of high spin local fields $\phi^j(x)$.

An important result has been obtained by Ouvry and Stern[(72)]. They represent the field by a vector $|\phi(x)>$ in the space spanned by an infinite set of covariant oscillators

$$[a_n^\mu, a_m^{\dagger\nu}] = \delta_{mn}\eta^{\mu\nu}. \quad n, m = 1, 2, 3.... \tag{7.1}$$

The local fields $\phi^{\mu_1\cdots\mu_n}$ are coefficients in the expansion of $|\phi(x)>$ given by

$$|\phi(x)>= \phi(x)|0> +\phi_n^\mu(x)a_\mu^{\dagger n}|0> +\phi_{nm}^{\mu\nu}(x)a_\mu^{\dagger n}a_\nu^{\dagger m}|0> +.... \tag{7.2}$$

They treat the whole infinite set of oscillators with tensors $\phi^{\mu_1\cdots\mu_n}(x)$ with any permutation symmetry of the indices. They demand that the theory be invariant under the infinite-parameter gauge transformation

$$\delta|\phi(x)>= \sum_{n=0}^{\infty}(a_n^\dagger\cdot\partial)|\Lambda_n(x)>, \tag{7.3}$$

where $a_n \cdot \partial = a_n^\mu \partial_\mu$ and $|\Lambda_n(x) >$ is an infinite set of gauge field parameters expanded in the manner of (7.2). They prove that *independently of the space-time dimension D, this local gauge symmetry is sufficient to gauge away all the ghosts associated with the time-like modes of (7.2).* As in the case of massive string theory, supplementary fields crop up and on shell there exists a gauge in which these supplementary fields vanish and all the physical fields $\phi^{\mu_1 \cdots \mu_n}(x)$ become transverse with

$$\partial_{\mu_1} \phi^{\mu_1 \cdots \mu_n}(x) = 0 \tag{7.4}$$

with all indices. It is also shown that the massless sector exhibits a BRST structure formally identical to the one obtained by Siegel[74] for massive string theory. In the case of the massive string, the BRST structure follows directly from the existence of a nilpotent operator Q associated with the Virasoro algebra, and due to the central charge this nilpotent operator exists only for $D = 26$. In the work of Ouvry and Stern[72], the BRST structure can be determined in exactly the same way beginning with the nilpotent operator associated with the Heisenberg algebra of the longitudinal components of the creation and annihilation operators $a_n^\dagger$ and a_n. The infinite-parameter gauge invariance is manifested by an infinite chain of local transformations of the gauge parameters $|\Lambda >$ that leaves $|\phi(x) >$ invariant.

It must be stressed that this result holds in $D = 4$ space-time and is not restricted to some higher-dimensional space-time with a critical dimension as in the case of strings. We consider this a distinct advantage, since we carry out our measurements in $D = 4$- dimensional space-time. The substantial problems associated with Kaluza-Klein or string compactification can also be avoided. We shall adopt the point of view that the infinite gauge transformations associated with gauge fields $\phi^j(x)$ of arbitrarily high spin remove all unphysical modes associated with our infinite-component fields defined by (6.61) and (6.62). In particular, this will hold also for vector fields $\phi_\mu(x)$ and symmetric and skew-symmetric tensor fields $\phi_{(\mu\nu)}$ and $\phi_{[\mu\nu]}$, so that we can develop an infinite-component field theory of quantum electrodynamics, weak interactions, Yang-Mills color gauge theory and gravity. More work is required to investigate further the properties of infinite-parameter gauge invariances of infinite-component fields to firmly establish the consistency of these theories.

Let us now consider specifically the infinite-component fields

$$A_\mu(x) = \sum_{j=0}^{\infty} (j!)^{-1/2} A^j_\mu(x), \tag{7.5}$$

$$h_{(\mu\nu)}(x) = \sum_{j=0}^{\infty} (j!)^{-1/2} h^j_{(\mu\nu)}(x), \tag{7.6}$$

$$h_{[\mu\nu]} = \sum_{j=0}^{\infty} (j!)^{-1/2} h^j_{[\mu\nu]}(x), \tag{7.7}$$

where the h's are the small quantities defined by the expansion in (2.29). We define the infinite-parameter gauge transformations

$$\delta A_\mu(x) = \partial_\mu \Lambda(x), \tag{7.8}$$

$$\delta h_{(\mu\nu)} = \partial_\mu \xi_\nu(x) + \partial_\nu \xi_\mu(x), \tag{7.9}$$

and

$$\delta h_{[\mu\nu]} = \partial_\nu a_\mu(x) - \partial_\mu a_\nu(x). \tag{7.10}$$

The gauge parameters $\Lambda(x), \xi_\mu(x)$ and $a_\mu(x)$ are expanded as

$$\Lambda(x) = \sum_{n=0}^{\infty} c_n \Lambda^n(x), \tag{7.11}$$

$$\xi_\mu(x) = \sum_{n=0}^{\infty} c_n \xi^n_\mu(x), \tag{7.12}$$

$$a_\mu(x) = \sum_{n=0}^{\infty} c_n a^n_\mu(x). \tag{7.13}$$

The $\psi(x)$ field in (6.62) satisfies the infinite-parameter gauge transformation

$$\psi \to \psi e^{ie\Lambda(x)}, \quad \bar{\psi} e^{-ie\Lambda(x)}, \tag{7.15}$$

where $\Lambda(x)$ is defined by (7.11) and e is a coupling constant.

In this way, we can extend the infinite-component field theory formalism developed here to all the know physical fields including the gravitational field, and obtain a consistent scheme free of ghosts.

VIII. *Ultraviolet Finite Quantum Field Theory*

Since all conventional attempts to produce a consistent ultraviolet finite quantum gravity theory in four dimensions based on point particles have failed, and because theories based on strings have potential problems with compactification, finding perturbative solutions to the string equations, and deriving a consistent two-dimensional field theory of strings, it appears that a radical modification of quantum field theory may be required to solve these problems. Attempts to formulate non-local quantum field theories based on extended objects have met with difficulties, although the attempt made by Efimov(75)(76) does not suffer from the usual problems of microscopic causality or violation of unitarity. However, the introduction by him of a non-local description of the photon propagator suffers from a degree of arbitrariness. In the following, we shall describe a theory formulated in terms of infinite-component fields that leads in a natural way to the local ineractions of non-local, extended fundamental objects, which possess ultraviolet finite Feynman diagrams and S-matrix elements. These methods can then be used to give a finite quantum gravity theory based on Einstein's GR or NGT.

Let us consider Quantum Electrodynamics (QE). In the past, form factors have been introduced into Feynman diagram propagators to measure the effects of non-locality. One method has been to write the photon propagator as

$$D_{\mu\nu}(q^2) = \delta_{\mu\nu}\left[\frac{1}{-q^2 - i\epsilon} - \frac{1}{-q^2 + M^2 - i\epsilon}\right], \tag{8.1}$$

where M is a large mass associated with an elementary length l as $l = 1/M$. These Pauli-Villars type of theories produce additional unwanted singularities in the amplitudes, giving rise to violations of causality and unitarity and difficulties with the gauge invariances of the theory. To save this situation, authors have introduced indefinite metrics but these methods apply only to lowest order perturbation theory. The Pauli-Villars regularization techniques based on (8.1) use analytic mereomorphic functions that lead to difficulties with quantum field theory. On the other hand, analytic *entire* functions do not have singularities in any finite region of the complex plane, and can be used to describe the non-local fundamental objects with a radius l that interact locally with one another and produce a finite consistent quantum theory.

We begin with a theory that contains a vector field $A_\mu(x)$, represented as an infinite-component field (7.5) and a spinor field given by (6.62). The free Lagrangian takes the form

$$L_0 = -\frac{1}{2}\partial_\mu A^\nu(x)\partial^\mu A_\nu(x) - \bar{\psi}(x)\Big(-i\gamma^\mu\partial_\mu + m\Big)\psi(x), \tag{8.2}$$

while the interaction Lagrangian is

$$L_I = eA^\mu(x)\bar{\psi}(x)\gamma_\mu\psi(x). \tag{8.3}$$

The S-matrix can be written in the form of a T-product

$$S = T\text{exp}\int d^4x[A^\mu\bar{\psi}(x)\gamma_\mu\psi(x)]. \tag{8.4}$$

L is invariant under the infinite-parameter gauge transformations (7.8) and (7.15). The invariance of the S-matrix is only guaranteed, as in standard field theory, by using a suitable regularization procedure. We expand the S-matrix into an infinite series with respect to the coupling constant e and use the N product of the field operators ψ and A_μ according to the Wick theorem. We define the T-symbol in terms of a regularization procedure which permits us to construct S-matrix elements in any order of perturbation theory.

The massless particle propagator has the form for large q^2:

$$D(q) \simeq \frac{1}{i}\sum_{j=0}^{\infty}(j!)^{-1}\frac{(q^2/\mu^2)^j}{-q^2+\mu^2 j - i\epsilon}, \tag{8.5}$$

where we have included a small mass μ^2 that vanishes at the end of the calculation. From (8.5) it follows that

$$D(q) = \sum_{j=0}^{\infty}\frac{1}{\mu^2}\frac{(q^2/\mu^2)^j}{j!}\int_0^1 dx x^{-q^2/\mu^2+j-1}$$

$$= \frac{1}{\mu^2}\int_0^1 dx \exp(\frac{q^2}{\mu^2}x)x^{-q^2/\mu^2-1}. \tag{8.6}$$

If we go to the Euclidean region $k^2 = -q^2$ and let $k^2 \to +\infty$, then a calculation gives[77]

$$D(k^2)_{k^2\to+\infty} = \frac{1}{i}(\frac{\pi\mu^2}{k^2})^{1/2}\exp(-k^2/\mu^2). \tag{8.7}$$

A similar behaviour results for the propagator S associated with the $\psi(x)$ field.

All the Feynman diagrams obtained from our S-matrix expansion are ultraviolet finite. The theory yields a causal and unitary S-matrix. The imaginary part of the propagator (8.7) is positive and is the same as the imaginary part of the "point-like" propagator given by

$$D'(q) = \sum_{j=0}^{\infty}\frac{1}{j!}\frac{j^j}{(-q^2+\mu^2 j)}. \tag{8.8}$$

The two propagators D and D' are the same up to an entire function $F(q^2)$:

$$D(q^2) = D'(q^2) + F(q^2). \tag{8.9}$$

The proof of causality and unitarity can be performed using the regularization techniques of Efimov[76], which was done for a nonlocal field theory model.

The regularization of the propagator is carried out in the following way. We define

$$\text{Reg}D_{\mu\nu}(x) = \frac{\delta_{\mu\nu}}{(2\pi)^4 i}\sum_{j=0}^{\infty}(m)^{-2j}(j!)^{-1}\int\frac{d^4q\Pi^j(q)R^\delta(q^2)}{-q^2+m_1^2 j+m_0^2-i\epsilon}e^{-iq\cdot x}, \tag{8.10}$$

where

$$R^\delta(z) = \exp[-\delta(z+iM^2)^{1/2+\nu}e^{-i\pi\sigma}]. \tag{8.11}$$

Here $0<\nu<\sigma<1/2$ and M^2 is a positive parameter. We have

$$|R^\delta(z)| \simeq \begin{cases}\exp[-\delta|z|^{1/2+\nu}] & -\pi a_2 < \arg z < \pi(1+a_1)\\ \exp[+\delta|z|^{1/2+\nu}]; & \pi(1+a_1) < \arg z < 2\pi(1-a_2/2),\end{cases} \tag{8.12}$$

where for $|z| \to \infty$:

$$a_1 = \frac{2(\sigma-\nu)}{1+2\nu}, \quad a_2 = \frac{1-2\sigma}{1+2\nu}. \tag{8.13}$$

The function $R^\delta(z)$ is analytic and decreases like an exponential function of the order $\rho_1 = 1/2 + \nu < 1$ for z in the upper half-plane. The integral in $\mathrm{Reg}\, D(x)$, in (8.10), is convergent for $\delta > 0$ and determines $\mathrm{Reg}\, D(x)$ for an entire function involving $\Pi(q)$ when we choose a ν and a σ such that $\rho < \rho_1 < 1$. We rotate the contours of integration over the time components q_0 by an angle $\pi/2$ and then take the limit $\delta \to 0$. After rotating the variable $q_0 \to ik_4$, the integral is taken over the Euclidean four-dimensional momenta and we can go to the limit $\delta = 0$, because as shown in (8.7) the integral vanishes exponentially fast as $k^2 \to +\infty$. Thus, the Efimov regularization technique guarantees the possibility of performing the transition to the Euclidean metric. We can apply the same procedure to the propagator $S(x)$ obtained from (6.65).

The propagators $D(q)$ and $S(q)$ are *nonlocal* having been "smeared-out" by the sum over the infinite spins of the mass spectrum. The potentials obtained from these propagators will be finite at short distances. A particle has a finite dimension determined by the fundamental length $l = 1/m$.

If we perform a renormalization of the coupling constant and the mass i.e., introduce the physical coupling constant e_r and the physical mass m_r, then the renormalization will be finite. The terms in the perturbation theory that depend logarithmically on the elementary length l i.e., $\ln(1/l^2m^2)$ will disappear from the renormalized perturbation theory for the S-matrix. It means that any correction descibed by using our infinite-component fields will be of order $O(p^2l^2)$. Provided the fundamental length is very small $l << 10^{-17}$ cm, then the extended nature of the fundamental "particle" will be well hidden from current experimental physics. Infinite-component field theory allows an interpretation of field theory without introducing infinite counter terms and yields a finite connection between the initial and physical parameters of the theory, i.e., $Z_1 < \infty, Z_2 < \infty, Z_3 < \infty$.

A study of field theory anomalies using infinite-component field theory could shed some new light on the physical nature of anomalies. Since the Feynman graphs are finite, we expect that the anomaly graphs will be finite and this facilitates a comparison with the usual field theory results for these graphs and their physical interpretation.

These methods can be extended to Einstein's theory of gravity and NGT by using an expansion of the field equations in flat space and writing the gravitational Lagrangians in terms of the infinite-component fields $h_{(\mu\nu)}$ and $h_{[\mu\nu]}$, given by (7.6) and (7.7) with their infinite-parameter gauge transformations (7.9) and (7.10)[78]. An analysis of the Feynman graphs shows that they are ultraviolet finite and that the S-matrix is unitary and causal to all orders of perturbation theory when an appropriate regularization technique is employed. However, it must be kept in mind that if we adopt Einstein's GR instead of NGT, then we are faced with the problem of degenerate masses in the representations of the Lorentz group $SO(3,1)$ unless we break Lorentz symmetry. Another question to investigate is the nature of the finite renormalization of gravitation within the scheme and the predictive power of the finite theory.

IX. *Discussion*

Motivated by the circumstance that the local gauge group $GL(4,R)$, of NGT, only has infinite-component spinor fields that can describe matter, we have formulated an infinite-component field theory in which fields are defined as infinite sums over the spins and masses of the particle spectrum. The particles form irreducible representations that are infinite sums of finite, non-unitary representations of the Lorentz group. This avoids the problems associated with causality , statistics and space-like solutions of the infinite-component wave equations.

A quantum field theory of these infinite-component fields is based on the S-matrix and its expansion to all orders in a coupling constant. Infinite-parameter gauge transformations are introduced to remove any problems associated with ghost poles of the fields. The nonlocal propagators are associated with extended objects that are defined in terms of a fundamental length $l = 1/m$. The infinite sums over the spins of the particle spectrum "smear-out" the point-like structure of the particles. This property of the theory must have a deep connection with Regge-poles in the complex angular momentum plane. The Feynman graphs are all ultraviolet finite and the S-matrix is gauge invariant and unitary to all orders of perturbation theory, provided a suitable regularization procedure is employed. The formalism can be applied to Einstein's gravity theory or NGT by expanding the gravitational fields about flat-space and defining the fields $h_{(\mu\nu)}$ and $h_{[\mu\nu]}$ in terms of infinite sums over the spins of massless particles. The S-matrix will yield Feynman graphs that are ultraviolet finite and the theory will be causal and unitary.

We have not specified in our field theory the nature of the fundamental extended object that describes the particles in the theory. It could be a string or a membrane or some quite complicated topological object. The formalism and its solution in terms of perturbation theory does not depend on knowing the exact nature of the extended object. This is clearly an advantage when coupled with the fact that we can work in four-dimensional space-time. The field theory and its vacuum states are all defined in four-dimensional space-time. This is the situation that one would expect to have in a phenomenologically sensible world.

It would be interesting to investigate the fundamental problem of explaining the smallness of the cosmological constant in gravity theory; the resolution of this problem may be found in a theory in which particles are described as nonlocal objects that interact locally with one another. Moreover, a reformulation of the standard Yang-Mills model based on $SU(3) \times SU(2) \times U(1)$, using the infinite-component field theory formalism, could possibly lead to new fundamental insights.

Another interesting feature of infinite-component field theory is the statistical nature of the particle spectrum at high energies. The statistics and thermodynamic properties of an infinite particle spectrum have been worked out by Hagedorn[(79)-(82)], and this subject has received some attention in the literature.

Acknowledgements

This work was supported in part by the Natural Sciences and Engineering Research Council of Canada, and in part by Contract NAS 8-36125. I thank the organizers of the Summer Institute, Professor F. C. Khanna, Professor G. Kunstatter, Professor H. C. Lee and Professor H. Umezawa, for their kind hospitality during my stay at the CAP-NSERC Summer Institute, University of Alberta. I also thank Professor C. W. F. Everitt and Professor W. M. Fairbank for their kind hospitality during my stay at Stanford University.

References

1. A. Einstein, *The Meaning of Relativity* (Princeton University Press, U.S.A., 1956, sixth edition, Appendix II).
2. J. W. Moffat, *Phys. Rev.* D**19**, 3554 (1979).
3. J. W. Moffat, *J. Math. Phys.* **21**, 1798 (1980).
4. J. W. Moffat, *Found. of Phys.* **14**, 1217 (1984).
5. J. W. Moffat, *Phys. Rev.* D**35**, 3733 (1987).
6. J. W. Moffat, Stanford University Preprint, September, 1987.
7. J. W. Moffat and E. Woolgar, University of Toronto Preprints, August and September, 1987.
8. C. Brans and R. H. Dicke, *Phys. Rev.* **124**, 925 (1961); S. Weinberg, *Gravitation and Cosmology* (John Wiley & Sons, New York, 1972.)
9. P. van Nieuwenhuizen, *Phys. Rep.* **68**, 191 (1981).
10. M. B. Green, J. H. Schwarz, and E. Witten, *Superstring Theory* (Cambridge University Press, Vols. I & II, 1987.)
11. J. W. Moffat, Can. J. Phys. **64**, 561 (1986).
12. K. S. Narain, Phys. Letts. **169B**, 41 (1986).
13. W. Lerche, D. Lüst, and A. N. Schellekens, **B287**, 477 (1987); S. J. Gates, Jr. Proc. of the CAP-NSERC Summer Institute on Two-Dimensional Field Theories, University of Alberta, July 1987, World Scientific.
14. J. G. Taylor, Proc. of the CAP-NSERC Summer Institute on Two -Dimensional Field Theories, University of Alberta, July 1987, World Scientific.
15. N. Seiberg, Weismann Institute Preprint, WIS-87/42/ June PH. 1987.
16. R. D. Reasenberg *et al.*, Astrophys. J. **234**, L219 (1979).
17. H. Hill and R. D. Rosenwald, Proc. NATO Advanced Workshop: "Mathematical Aspects of Gravity and Supergravity", Utah State University (1986).
18. C. M. Will, *Theory and Experiment in Gravitational Physics* (Cambridge University Press, 1981).
19. G. Kunstatter, J. W. Moffat, and J. Malzan, *J. Math. Phys.* **24**, 886 (1983).
20. Y. Ne'eman, *Ann. Inst. Henri Poincaré* A **28**, 369 (1978).
21. E. Cartan, *Ann. Ec. Norm. Sup.* **40**, 325 (1923).
22. D. Vincent, *Class. Quantum Grav.* **2**, 409 (1985).
23. M. Kalb and P. Ramond, *Phys. Rev.* D**9**, 2273 (1974).
24. E. Cremmer and J. Scherk, *Nucl. Phys.* B**72**, 117 (1974).

25. J. W. Moffat, *Phys. Rev.* D**23**, 2870 (1981).
26. R. B. Mann and J. W. Moffat, *J. Phys.* A**14**, 2367 (1981); **15**, 1055 (E) (1982).
27. R. B. Mann, J. W. Moffat, and J. G. Taylor, *Phys. Letts.* **97B**, 73 (1980).
28. R. B. Mann and J. W. Moffat, *Phys. Rev.* D**26**, 1858 (1982).
29. R. B. Mann and J. W. Moffat, *Phys. Rev.* D**31**, 2488 (1985).
30. P. Kelly and R. B. Mann, Class. Quantum Grav. **3**, 705 (1986); University of Toronto Preprints, 1987.
31. E. F. Guinan and F. P. Maloney, *Astron. J.* **90**, 1519 (1985).
32. J. W. Moffat, *Astrophs. J. Letts.* **287**, L77 (1984).
33. J. W. Moffat, *J. Math. Phys.* **25**, 347 (1984).
34. J. W. Moffat, *J. Math. Phys.* **26**, 528 (1985).
35. R. B. Mann, *Class. Quantum Grav.* **1**, 561 (1984).
36. N. Salingaros, *J. Math. Phys.* **22**, 2096 (1981).
37. Zai-Zhe Zhong, *J. Math. Phys.* **26**, 404 (1985).
38. W. K. Clifford, *Proc. London Math. Soc.* **4**, 381 (1873).
39. W. K. Clifford, *Amer. J. Math.* **1**, 350 (1878).
40. F. G. Frobenius, *J. Reine Angew. Math.* **84**, 59 (1877).
41. Y. Ne'eman and Dj. Šijački, *Ann. Phys.* (NY) **120**, 292 (1979); *Proc. Nat. Acad. Sci.* (USA) **76**, 561 (1979).
42. Y. Ne'eman, *Found. of Phys.* **13**, 467 (1983).
43. J. W. Moffat, "Spinor Fields in the Nonsymmetric Gravitation Theory", *Proc. of the 2nd Canadian Conf. on General Relativity and Relativistic Astrophysics* (to be published by World Scientific, 1987).
44. J. W. Moffat, University of Toronto Preprint, April 1987.
45. E. Majorana, *Nuovo Cimento* **9**, 335 (1932).
46. D. T. Stoyanov and I. T. Todorov, *J. Math. Phys.* **9**, 2146 (1968).
47. A. O. Barut, *Springer Tracts in Modern Physics, Ergebnisse der Exakten Naturewiss* (Springer, Berlin, Vol. 50, pp. 1-28, 1969.)
48. C. Fronsdal, *Phys. Rev.* **156**, 1653 (1967); **156**, 1665 (1967).
49. Y. Nambu, *Phys. Rev.* **160**, 1171 (1967).
50. B. G. Wybourne, *Classical Groups for Physicists* (Wiley-Interscience Publ. John Wiley & Sons, New York, 1974.)
51. G. Feldman and P. T. Matthews, *Ann. Phys.* (N.Y.) **40**, 19 (1966); *Phys. Rev.* **151**, 1176 (1966); **154**, 1241 (1967).
52. E. Abers, L. T. Grodsky, and R. E. Norton, *Phys. Rev.* **159**, 1222 (1967).
53. H. D. J. Abarbanel and Y. Frishman, *Phys. Rev.* **171**, 1442 (1968).

54. I. T. Grodsky and R. F. Streater, *Phys. Rev. Letts.* **20**, 695 (1968).
55. A. I. Oksak and I. T. Todorov, *Phys. Rev.* D**1**, 3511 (1970).
56. A. Chodos, *Phys. Rev.* D**1**, 2937 (1970).
57. Dj. Šijački and Y. Ne'eman, *J. Math. Phys.* **26**, 2457 (1985).
58. A. Kihlberg, *Ark. Fys.* **32**, 241 (1966).
59. B. Speh, *Mat. Ann.* **258**, 113 (1981).
60. J. L. Friedman and R. D. Sorkin, *J. Math. Phys.* **21**, 1269 (1980).
61. V. K. Dobrev and O. Ts. Stoytchev, *J. Math. Phys.* **27**, 883 (1986).
62. Y. Ne'eman and Dj. Šijački, *Phys. Letts.* **157B**, 275 (1985).
63. Y. Ne'eman and Dj. Šijački, *Phys. Letts.* **157B**, 267 (1985).
64. S. Weinberg, *Phys. Rev.* **133**, B1318 (1964); **134**, B882 (1964); **181**, 1893 (1969).
65. M. Fierz and W. Pauli, *Proc. Roy. Soc.* A**173**, 211 (1939).
66. J. Schwinger, *Particles, Sources and Fields* (Addison-Wesley, Reading, MA, 1970).
67. S. J. Chang, *Phys. Rev.* **161**, 1306 (1967).
68. L. P. S. Singh and C. R. Hagen, *Phys. Rev.* D**9**, 898 (1974).
69. C. Fronsdal, *Phys. Rev.* D**18**, 3624 (1978).
70. J. Fang and C. Fronsdal, *Phys. Rev.* D**18**, 3630 (1978).
71. T. Curtwright, *Phys. Letts.* **B85**, 219 (1979); *Phys. Letts.* **B165**, 304 (1985).
72. S. Ouvry and J. Stern, *Phys. Letts.* **B177**, 335 (1986).
73. N. A. Doughty and R. A. Arnold, University of Canterbury, New Zealand, Preprints, September (1987).
74. W. Siegel, *Phys. Letts.* **B151**, 391 (1985); **B151**, 396 (1985).
75. G. V. Efimov, *Commun. Math. Phys.* **5**, 42 (1967); **7**, 138 (1968).
76. G. V. Efimov, *Ann. of Phys.* **71**, 466 (1972).
77. N. V. Krasnikov, *Phys. Letts.* **195**, 377 (1987).
78. J. W. Moffat, in preparation.
79. R. Hagedorn, *Nuovo Cimento Suppl.* **3**, 147 (1965); **6**, 311 (1968).
80. R. Hagedorn, *Nuovo Cimento* **52A**, 1336 (1967); **56A**, 1027 (1968).
81. R. Hagedorn and J. Ranft, *Nuovo Cimento Suppl.* **6**, 169 (1968).
82. K. Huang and S. Weinberg, *Phys. Rev. Letts.* **25**, 895 (1970).

Quantum Gravity via Non-Linear Sigma Models

J. D. Gegenberg

Department of Mathematics and Statistics
University of New Brunswick
P. O. Box 4400
Fredericton, N.B., E3B 5A3, Canada

ABSTRACT

The non-linear sigma model form of the vacuum Einstein equations with one killing vector field is recalled. The resulting model, when quantized is a field theoretic version of minisuperspace quantization. The techniques involved in computing the one-loop effective action for a non-linear sigma model on a curved space-time are reviewed, and the result is applied to the aforementioned version of minisuperspace.

Quantum field theories are usually constructed by applying some quantization *ansatz*, e.g. canonical quantization or the path integral formalism, to an appropriate classical field theory. In some cases, for example QED, QCD, Weinberg-Salam theory, etc., the resulting quantum field theory is well-behaved in that physical predictions can be obtained from perturbation theory via some renormalization procedure. Quantum gravity is an unfortunate exception to this state of affairs. Quantum theories constructed from General Relativity theory [1] or related theories such as Kaluza-Klein theory [2] or Supergravity [3] are non-renormalizable and hence are not amenable to the usual perturbation theory techniques. String theories may yield a well-behaved quantum gravity theory [4], but we are probably a long way from being able to compute quantum gravity effects directly from string theory.

In the meantime, we may want, for pedagogical reasons, or in order to better understand such problems as gravitational collapse or early universe cosmology, to quantize some classical gravity theory and examine lowest order effects or qualitative features which we suspect will survive a more complete treatment. Minisuperspace quantization is just such a program [5]. The idea is to freeze out some of the dynamical degrees of freedom in the space-time metric. For example, if one fixes the sign of the curvature of space, then the only dynamical degree of freedom in an FRW metric is the time-dependent scale-factor. The resulting quantum theory is actually quantum *mechanical* rather than a true quantum field theory, in this simple case. The analysis is relatively simple and yields interesting qualitative information about quantum gravity in general [5], and perhaps physical results in the arena of quantum cosmology [6].

What I propose to discuss here is a version of minisuperspace quantization in which some of the dynamical degrees of freedom are frozen out, but those remaining are non-trivial functions of at least one spatial coordinate (as well as of the time-like coordinate), so that a true quantum field theory results. In order to freeze out some of the degrees of freedom, it will be assumed that the four-dimensional space-time metric admits at least one smooth space-like killing vector field $\vec{\xi}$ in some region R. (If $\vec{\xi}$ were time-like i.e. if space-time were stationary, then the dynamics would be trivial.) We choose coordinates in R such that $\vec{\xi} = \partial/\partial x^3$ in which case the line-element in R has the form [7]:

$$ds^2 = e^{-\omega} g_{\mu\nu} dx^\mu dx^\nu + e^\omega (A_\mu dx^\mu + dx^3)^2 \quad , \tag{1}$$

where $\mu, \nu, \ldots \in \{0, 1, 2\}$ and the quantities ω, $g_{\mu\nu}$ and A_μ are functions of x^μ only. The metric $g_{\mu\nu}(g^{\mu\nu})$ has Lorentzian signature $(-++)$ and is used to lower (raise) all greek indices. The four-dimensional region R can be thought of as a fiber bundle over a three-dimensional space-time M_3 with metric $g_{\mu\nu}$. The fiber bundle has a connection given locally by A_μ. Hence, the metric (1) can be considered as the starting point of 3 + 1 - dimensional Kaluza-Klein theory [2]. It should be mentioned here that (1) is invariant under both the group of diffeomorphisms on M_3 and under the Abelian gauge group given by $x^3 \mapsto x^3 + f(x)$, $A_\mu(x) \mapsto A_\mu(x) - f,_\mu(x)$. Henceforth, (x) denotes functional dependence on x^μ and $f,_\mu \equiv \partial f/\partial x^\mu$.

If the vacuum Einstein equations are imposed, i.e. if the metric (1) is required to be Ricci-flat, than the tensors (on M_3) ω, A_μ and $g_{\mu\nu}$ must satisfy the following field equations [8]:

$$R_{\mu\nu} + \frac{1}{2}e^{-2\omega}\,[\Omega,_\mu \Omega,_\nu - (e^\omega),_\mu (e^\omega),_\nu] = 0\,, \tag{2}$$

$$\underline{\nabla}^2(e^\omega) - e^{-\omega}(|\underline{\nabla}\Omega|^2 + |\underline{\nabla}e^\omega|^2) = 0\,, \tag{3}$$

$$\underline{\nabla}^2\Omega - 2e^{-\omega}\,\underline{\nabla}\Omega \cdot \underline{\nabla}e^\omega = 0\,, \tag{4}$$

where $R_{\mu\nu}$ is the Ricci-tensor on M_3 with respect to $g_{\mu\nu}$ and the operators $\underline{\nabla}^2$, $\cdot$ and $|\ |^2$ are defined, respectively, by
$\underline{\nabla}^2\Omega \equiv \underline{\nabla} \cdot \underline{\nabla}\Omega = g^{\mu\nu}\,\Omega_{|\mu\nu}$, $\underline{A} \cdot \underline{B} \equiv g^{\mu\nu} A_\mu B_\nu$, and $|\underline{A}|^2 \equiv \underline{A} \cdot \underline{A}$.
A vertical bar denotes covariant differentiation on M_3 with respect to $g_{\mu\nu}$. The scalar field Ω, the "twist" or "nut" potential [8], depends on A_μ via

$$\Omega,_\mu = \frac{1}{2}\,e^{2\omega}\,\eta_{\mu\nu\sigma}\,f^{\nu\sigma}\,, \tag{5}$$

where $\eta_{\mu\nu\sigma} \equiv \sqrt{-g}\,\varepsilon_{\mu\nu\sigma}$ is the completely anti-symmetric tensor on M_3 and $f_{\mu\nu} \equiv A_{\nu,\mu} - A_{\mu,\nu}$.

It is possible to derive the field equations (2) - (4) from the following action principle on M_3:

$$\begin{aligned} S_{\text{Total}}[g,\omega,\Omega] &= \int_{M_3} d^3x\,\sqrt{-g}\,\left[R - \frac{1}{2}(|\underline{\nabla}\omega|^2 - e^{-2\omega}|\nabla\Omega|^2)\right] \\ &\equiv S_{\text{grav}}[g] + S[\omega,\Omega] \end{aligned} \tag{6}$$

where R is the scalar curvature on M_3 and $S_{\text{grav}}[g]$ is the Einstein-Hilbert action on M_3. The remaining term, $S[\omega,\Omega]$, is the action for a non-linear sigma model over M_3[9]. To see this, define the two-component field $\phi^i(x)$ over M_3 by $\phi^1(x) \equiv \omega(x)$, $\phi^2(x) = \Omega(x)$, with $i, j, \ldots \in \{1, 2\}$. Furthermore define a 2×2 symmetric matrix $G_{ij}(\phi)$

by:

$$[G_{ij}(\phi)] = \begin{bmatrix} 1 & 0 \\ 0 & -e^{-2\phi^1} \end{bmatrix} \tag{7}$$

Then

$$S[\phi] = S[\omega, \Omega] = -\frac{1}{2}\int_{M_3} d^3x \sqrt{-g}\, G_{ij}\,(\phi(x))\phi^i{}_{,\mu}(x)\phi^j{}_{,\nu}(x)\, g^{\mu\nu}(x) \ . \tag{8}$$

The target space for the sigma model is a two-dimensional manifold, denoted N_2 . Corresponding to each point in M_3 with coordinates x^μ , there is a copy of N_2 , whose points are coordinatized by the $\phi^i(x)$, which satisfy the field equations (2) - (4) at x^μ . In addition, N_2 is provided with a fixed metric, namely G_{ij} , given by (7), which turns out to have signature (- +) and constant curvature.

The version of minisuperspace is obtained by treating the fields $\phi^i(x)$, whose classical dynamics is given by (8), as quantum fields over a fixed curved three-dimensional space-time M_3 with metric $g_{\mu\nu}$.

The remainder of this talk consists of, first, a review of the procedure for calculating the effective action (to one-loop order) of a general sigma model over a fixed curved space-time. This will then be applied to the particular model under consideration here.

Since we will employ dimensional regularization and since it comes at virtually no cost, the model discussed above is generalized to arbitrary dimensions for both space-time and the target space. Hence the classical action (8) is generalized to:

$$S[\phi] = -\frac{1}{2}\int_{M_m} d^mx \sqrt{-g}\, G_{ij}(\phi(x))\, \phi^i{}_{,\mu}(x)\, \phi^j{}_{,\nu}(x)\, g^{\mu\nu}(x) \ . \tag{9}$$

Now the indices $i, j, \ldots \in\{1, 2, \ldots, n\}$ and $\mu, \nu, \ldots \in\{0, 1, \ldots, m - 1\}$. The metric $g_{\mu\nu}$ over m-dimensional space-time M_m is a fixed Lorentzian-signature metric, while G_{ij} is a fixed n-dimensional metric (with arbitrary signature) over the target space N_n. The ϕ^i's are the local expressions of a map $\phi: M_m \to N_n$.

The computation of the effective action begins with the generating functional Z given by

$$Z = \int [d\phi]\, e^{iS[\phi]} \quad , \tag{10}$$

where $S[\phi]$ is given by (9) and $[d\phi]$ is the appropriate measure for the path integral [10]. As usual Z is computed perturbatively. To effect this, one begins by splitting the field $\phi^i(x)$ into a "classical part" $\phi_0^i(x)$ and quantum fluctuations about the $\phi_0^i(x)$ [11]. The classical part $\phi_0^i(x)$ obeys the classical equations of motion, i.e. equations (3) and (4), or, in sigma model notation:

$$\nabla^2 \phi_0^i + \Gamma^i_{jk}(\phi)\, \phi_0{}^j{}_{,\mu}\, \phi_0{}^k{}_{,\nu}\, g^{\mu\nu} = 0 \quad , \tag{11}$$

where ∇^2 is the usual Laplacian in $(M_m, g_{\mu\nu})$ (see the discussion following equation (4)), while $\Gamma^i_{jk}(\phi)$ is the Christoffel symbol in N_n with respect to G_{ij}:

$$\Gamma^i{}_{jk} \equiv \frac{1}{2} G^{i\ell} (G_{j\ell,k} + G_{k\ell,j} - G_{jk,\ell}) \; . \tag{12}$$

In general, one cannot define the quantum fluctuations as $\phi^i(x) - \phi_0{}^i(x)$ since adding coordinates at different points of N_n is not possible except in the special case where G_{ij} is flat [11]. Instead one resorts to differential geometry and defines the quantum fluctuations $\sigma_0{}^j(x)$ as the components of the tangent vector to the unique geodesic $\lambda^j(x, \tau)$ connecting $\phi_0{}^i(x)$ to $\phi^i(x)$. See Figure 1.

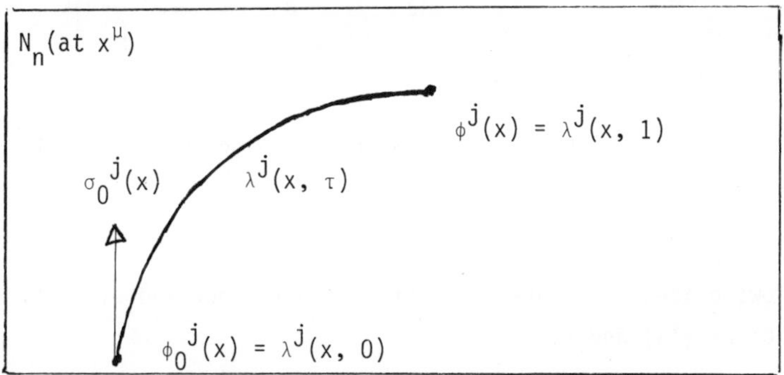

Figure 1. Covariant representation of quantum fluctuations

The geodesic $\lambda^j(x, \tau)$ is unique if $\phi^j(x)$ is sufficiently close to $\phi_0{}^j(x)$. Put another way, the $\sigma_0{}^j(x)$ are the normal coordinates of $\phi^j(x)$ with the origin at $\phi_0{}^j(x)$[12]. The $\sigma_0{}^j(x)$ will be the integration variables when the path integration in (10) is performed.

Consider the one-parameter family of actions $S(\tau)$ obtained from (9) via the replacement $\phi^i(x) \mapsto \lambda^i(x, \tau)$. Expand $S(\tau)$ in a Taylor series in τ about $\tau = 0$ and evaluate at $\tau = 1$ to obtain [11], [13], [14]:

$$S[\phi] = S(1) = S_0[\phi_0] + \frac{d}{d\tau} S(\tau)\Big|_{\tau=0} + \frac{1}{2}\frac{d^2}{d\tau^2} S(\tau)\Big|_{\tau=0} + \dots \quad . \tag{13}$$

The first term $S_0[\phi_0] \equiv S(0)$ is the classical action. The second term vanishes via the classical equations of motion (11). The third term is the one-loop correction to the classical action. It turns out that:

$$S_2[\phi_0, \sigma_0] \equiv \frac{1}{2}\frac{d^2}{d\tau^2} S(\tau)\Big|_{\tau=0} = \int_{M_m} d^m x \; L_2 \quad , \tag{14}$$

where

$$L_2 = -\frac{1}{2}\sqrt{-g(x)}\; \sigma_0{}^j(x)[-G_{jk}(\phi_0(x))D^2(x) + U_{jk}(x)]\sigma_0{}^k(x). \tag{15}$$

The operator $D^2(x) \equiv D^\mu(x)\, D_\mu(x)$, with

$$D_\mu(x)\; Y^j(\phi(x)) \equiv Y^j{}_{|k}(\phi(x))\; \phi^k{}_{,\mu}(x) \; , \tag{16}$$

for an arbitrary vector field Y^j on N_n, where $Y^j{}_{|k}$ denotes the covariant derivative on N_n with respect to G_{ij}. The N_n tensor field U_{jk} is given by:

$$U_{jk} \equiv -R_{jrks}\; \phi_0{}^r{}_{,\mu}\; \phi_0{}^s{}_{,\nu}\; g^{\mu\nu} \quad . \tag{17}$$

It is worth noting at this point that the classical equations of motion (11) can be written in terms of D_μ as:

$$D_\mu\, (\phi^i{}_{0,\nu}\; g^{\mu\nu}) = 0 \quad . \tag{18}$$

The following identities are straightforward to show and simplify the derivation of (14) and (15):

$$D_\mu\, \sigma_0{}^i = \phi_0{}^i{}_{,\mu|j}\; \sigma_0{}^j \; , \tag{19}$$

$$D_\mu G_{ij} = 0 \quad , \tag{20}$$

$$[D_\mu , D_\nu] Y^j(\phi) = i F_{\mu\nu}{}^j{}_k Y^k(\phi) \quad , \tag{21}$$

where

$$F_{\mu\nu}{}^j{}_k \equiv -i R^j{}_{krs} \phi^r{}_{0,\mu} \phi^s{}_{0,\nu} \quad . \tag{22}$$

The details of the calculations above can be found in Boulware and Brown [13], Howe *et. al.*[14], or Mukhi [15] .

To one-loop order, the path integral (10) is Gaussian, and one obtains

$$\ln Z = i S_o[\phi_0] + i W^{(1)} [\phi_0] + \ldots \quad , \tag{23}$$

where $W^{(1)}[\phi_0]$ is the effective action, and it is formally given in terms of the functional determinant of the operator H ,

$$H^i_j \equiv -\delta^i_j D^2 + U^i{}_j \quad . \tag{24}$$

In particular,

$$i W^{(1)} [\phi_0] \equiv \ln \mathrm{Det}^{-1/2} H \equiv -\frac{1}{2} \mathrm{Tr}(\ln H) . \tag{25}$$

The quantity $W^{(1)}$ is generally infinite and hence must be regulated [11], [13]. To accomplish this, the proper-time representation is employed [16], [11]. One proceeds as follows: Formally, one may represent the inverse H^{-1} of H by the integral

$$H^{-1} = i \int_0^\infty ds\, e^{-isH} \quad . \tag{26}$$

If $W^{(1)} [\phi_0]$, given by (25) is varied with respect to $g_{\mu\nu}$, one can show that identically,

$$\delta W^{(1)} = \delta\left[-\frac{i}{2}\int_0^\infty i \frac{ds}{is} \mathrm{Tr}\, e^{-isH}\right] \quad , \tag{27}$$

which can be solved to yield (up to an additive constant) [11] :

$$W^{(1)} = \frac{-i}{2}\int_0^\infty \frac{i ds}{is} \mathrm{Tr}\, e^{-isH} \quad . \tag{28}$$

A matrix representation $\langle x, k | e^{-isH} | x', k' \rangle$ for e^{-isH} can be defined such that

$$\mathrm{Tr}\, e^{-isH} = \int d^m x \langle x, k | e^{-isH} | x, k \rangle \ . \tag{29}$$

Alternatively, the matrix $\langle x, k | e^{-isH} | x', k' \rangle$ satisfies the proper time Schrödinger equation

$$-\frac{\partial}{i\partial s} \langle \cdots \rangle = H \langle \ldots \rangle \ , \tag{30}$$

together with the initial condition

$$\langle x, k | x', k' \rangle = \delta^{(m)} (x - x') \, \delta^k{}_{k'} \ . \tag{31}$$

The indices k and k' appearing, respectively, in the bra and ket are contra- and co- variant.

The dominant singular behaviour of $\mathrm{Tr}\, e^{-isH}$ resides in the "free-field" part of the solution of (30). Hence, one may write [11], [13]:

$$\langle x, k | e^{-isH} | x', k' \rangle = \left[\frac{i}{(4\pi i s)^{m/2}} \tilde{\Delta}^{1/2}(x,x') e^{i\Omega(x,x')/2s} \right] \Psi^k{}_{k'}(x, x'; is) \ . \tag{32}$$

The factor in brackets is the solution of (30) with $D_\mu = \nabla_\mu$ (i.e. the ordinary covariant derivative on M_m) and $U^j{}_k = 0$. The remaining factor $\Psi^k{}_{k'}$ (x, x'; is) is regular for all s and approaches zero exponentially as $s \to \infty$ [13]. The two-point functions $\Omega(x, x')$ and $\tilde{\Delta}(x, x')$ are defined, respectively, by

$$\frac{\partial}{\partial x^\mu} \Omega(x, x') \frac{\partial}{\partial x^\nu} \Omega(x, x') \, g^{\mu\nu}(x) = 2\, \Omega(x, x') \ , \tag{33}$$

$$m\, \tilde{\Delta}^{1/2}(x, x') = \Big[\tilde{\Delta}^{1/2}(x, x') \nabla_\mu(x) \nabla_\nu(x) \Omega(x, x') + 2 \nabla_\mu(x) \tilde{\Delta}^{1/2}(x,x') \nabla_\nu(x) \Omega(x, x') \Big] g^{\mu\nu}(x) \ . \tag{34}$$

The two-point function Ω is Synge's world-function [17], while $\tilde{\Delta}(x, x')$ is the curved space-time generalization [11] of the Van Vleck determinant [18] $\Delta(x, x')$, i.e.

$$\tilde{\Delta}(x, x') \equiv \sqrt{-g(x)}\, \Delta(x, x') \sqrt{-g(x')} \ . \tag{35}$$

Useful summaries of the properties of Ω and Δ are given by De Witt [11] and by Brown [19]. The reason for the appearance of these quantities here is, first, that $\Omega(x, x')$ is half the square of the geodesic distance from x^{μ} to $x^{\mu'}$, and hence generalizes $(x^{\mu} - x^{\mu'})(x^{\nu} - x^{\nu'})\eta_{\mu\nu}$, which appears in Schwinger [16], while $\tilde{\Delta}(x, x')$ is essentially the Jacobian of the transformation from the variables x^{μ}, $x^{\mu'}$ to the new variables x^{μ}, $\Omega(x, x')$. Before proceeding, it should be noted that in the limit as $x^{\mu'} \to x^{\mu}$, $\Omega(x, x') \to 0$, $\tilde{\Delta}^{1/2}(x, x') \to 1$.

Using (32) in (28), one obtains for the effective action [13]:

$$W^{(1)}[\phi_0] = \int d^m x \quad \mathcal{L}^{(1)}_{eff}(x; m) \quad , \tag{36}$$

where $\mathcal{L}^{(1)}_{eff}(x; m)$, the effective Lagrangian density in m space-time dimensions, is given by

$$\mathcal{L}^{(1)}_{eff}(x; m) = \frac{1}{2}(4\pi)^{-m/2} \int_0^{\infty} \frac{i\, ds}{(is)^{1+m/2}} \Psi(x; is) \quad , \tag{37}$$

with $\Psi(x, is) \equiv \sum_{k=1}^{n} \Psi^k{}_k(x, x; is)$.

The ultraviolet infinities come from those terms in (37) which are singular in the $s \to 0$ limit [11], [13], [19]. (Singularities at the upper limit, $s \to \infty$, are infrared divergences.) To isolate and regulate the ultraviolet infinities, one continues to the physical dimension (in this case, m = 3) of space-time from a region of the complex m-plane in which the potentially divergent terms at s = 0 are finite [13]. To isolate these terms (for the case m = 3), one integrates (37) by parts twice. Because m is odd, no potentially divergent terms occur, and one obtains a manifestly finite expression for $\mathcal{L}^{(1)}_{eff}(x; 3)$:

$$\mathcal{L}^{(1)}_{eff}(x; 3) = \frac{1}{12\pi^{3/2}} \int_0^{\infty} i\, ds (is)^{-1/2} \left(\frac{\partial}{\partial is}\right)^2 \Psi(x; is) \quad . \tag{38}$$

This agrees formally with the results obtained by De Witt [11], using zeta-function regularization.

An exact result for the effective Lagragian density could now be obtained from (38) if (30) and (31) could be solved for the $\Psi^k{}_{k'}$. This is a formidable task, and one usually resorts to an approximation. In particular, for very small values of the parameter s, one can approximate $\Psi^k{}_{k'}$ by a power series in (is) [11], [13], [19], [20]:

$$\Psi^k{}_{k'}(x, x'; is) = \sum_{\alpha=0}^{\infty} \psi_\alpha{}^k{}_{k'}(x, x')\,(is)^\alpha \quad , \tag{39}$$

where the coefficients satisfy the recursion relations:

$$-\alpha\psi_\alpha{}^k{}_{k'} = \Omega_{|\mu} D^\mu \psi_\alpha{}^k{}_{k'} + \tilde{\Delta}^{-1/2} H(\tilde{\Delta}^{1/2}\psi_{\alpha-1}{}^k{}_{k'}) \quad , \tag{40}$$

for $\alpha \geq 1$ and

$$\psi_0{}^k{}_{k'}(x, x') = \delta^{(3)}(x - x')\,\delta^k{}_{k'} \quad , \tag{41}$$

as follows from substitution of (39) into (30) and (31). In the limit as $x' \to x$, writing $\psi_\alpha(x) = \lim_{x'\to x} \sum_k \psi_\alpha{}^k{}_k(x, x')$, it turns out that [20], [21]

$$\psi_0(x) = n \quad , \tag{42}$$

$$\psi_1(x) = \frac{n}{6} R(x) + R_{ij}(\phi_0)\,\phi^i_{0,\mu}\,\phi^j_{0,\nu}\, g^{\mu\nu} \quad , \tag{43}$$

$$\psi_2(x) = \frac{1}{360}(12n\,\nabla^2 R + 5n R^2 - 2R_{\mu\nu}R^{\mu\nu} + 2R_{\mu\nu\pi\sigma}R^{\mu\nu\pi\sigma} + 60\,RU$$

$$+ 180\,U^2 - 60\,R\,\nabla^2 U - 30\,F^2) \quad , \tag{44}$$

and so on for $\psi_3(x)$, etc. In (44), the notation $U \equiv \sum_k U^k{}_k$ etc. and $F^2 \equiv \sum_k F_{\mu\nu}{}^k{}_\ell\, F^{\mu\nu\ell}{}_k$ has been used. The expansion (39) is called the asymptotic expansion for $\Psi^k{}_{k'}$.

If the asymptotic expansion is used for $\Psi^k{}_{k'}$ in (38), then as expected the integrals over each term are finite in the limit $s \to 0$, but blow up at the upper limit $s \to \infty$. This is an infrared divergence, and one avoids it here by simply putting in a cut-off. In particular, replace $\Psi^k{}_{k'}$ in (38) by $e^{-iM^2 s}\Psi^k{}_{k'}$ for arbitrary real M. The result is that the coefficients of $\psi_0(x)$ and $\psi_1(x)$ vanish as $M \to 0$, while the coefficients of $\psi_2(x)$, etc., tend to infinity.

Hence, it is concluded that the lowest order (in powers of the curvature on M_3 and powers of gradients of $\phi_0{}^i$) corrections to the classical Lagrangian are:

$$\mathcal{L}^{(1)}_{eff}(x) = \frac{\sqrt{-g}}{12\pi^{3/2}} C(M^2)\,\psi_2(x) + \ldots \quad . \tag{45}$$

The constant $C(M^2)$ is cut-off dependent.

Under current investigation is the application of the above to a simple model of spherically symmetric gravitational collapse. Quantum gravity effects become relevant inside the event horizon near the singularity. The time-like killing vector field becomes space-like inside the horizon and it turns out that, using the latter for $\partial/\partial x^3$, the metric $g_{\mu\nu}$ of three-space is a three dimensional FRW model. The result of these investigations will be reported elsewhere.

REFERENCES

1. For a review, see Isham, C. J., in Quantum Gravity 2, ed. by Isham, C. J., et al. (Oxford, Clarendon Press, 1981).
2. An Introduction To Kaluza-Klein Theory, ed. Lee, H. C. (Singapore, World Scientific, 1984).
3. van Nieuwenhuizen, P., Phys. Reps. 68, 189 (1981).
4. Green, M. B., Schwarz, J. H., and Witten, E., Superstring Theory, V. 1 (Cambridge, Cambridge Univ. Press, 1987).
5. De Witt, B. S., Phys. Rev. 186, 1113 (1967).
6. Hawking, S. W., Nuc. Phys. B239, 257 (1984).
7. For a review, see Kramer, D., et al., Exact Solutions of Einsteins Field Equations (Cambridge, Cambridge Univ. Press, 1980).
8. Israel, W. and Wilson, G. A., J. Math Phys. 13, 865 (1972); Gibbons, G. W. and Hawking, S. W., Comm. Math. Phys. 66, 291 (1979).

9. Sanchez, N.,"Harmonic Maps in General Relativity and Quantum Field Theory", CERN preprint 1986.

10. Ramond, P., Field Theory: A Modern Primer (Reading, MS, Benjamin/Cummings, 1981).

11. De Witt, B. S., Dynamical Theory of Groups and Fields (New York, Gordon and Breach, 1965); in General Relativity: An Einstein Centenary Volume ed. by Hawking, S. W. and Israel, W., (Cambridge, Cambridge Univ. Press, 1979).

12. Misner, C. W., Thorne, K. S. and Wheeler, J. A., Gravitation (San Francisco, Freeman & Co., 1973).

13. Boulware D. G. and Brown, L. S., Ann. of Phys. 138, 392 (1982).

14. Howe, P. S., Papadopoulos, G. and Stelle, K. S., "The Background Field Method and the Non-Linear σ-Model", preprint, 1986.

15. Mukhi, S.,"Non-Linear σ-Models, Scale Invariance and String Theories: A Pedagogical Review", preprint 1986.

16. Schwinger, J., Phys. Rev. 82, 664 (1951).

17. Synge, J. L., Relativity: The General Theory (Amsterdam, North Holland Press, 1960).

18. Van Vleck, J.H., Proc. Nat. Acad. Sci (USA) 14, 178 (1928).

19. Brown, L. S., Phys. Rev. D15, 1469 (1977).

20. Gilkey, P., J. Diff. Geom. 10, 601 (1975).

21. Coquereaux, R., "Comments About the Geometry of Non-Linear Sigma Models", CERN preprint 1986.

ENLARGING THE SU(n+1) SYMMETRY GROUP OF THE EUCLIDEAN CP^n MODELS TO SL(n+1, C)†

G. Arsenault
CRM and Département de physique

M. Jacques
CRM and Laboratoire de physique nucléaire

Y. Saint-Aubin
CRM and Département de mathématiques et de statistique
Université de Montréal, C.P. 6128, succ. A, Montréal, QUÉ, Canada H3C 3J7

Relying on our previous study of infinite dimensional Lie algebras acting on the solution space of various sigma models, we consider the case of the finite action solutions of the Euclidean CP^n *models. Here the symmetry algebra turns out to be finite dimensional and we give a closed form for the corresponding group action. The symmetry group is identified to be SL(n+1,C), thus enlarging the well-known SU(n+1) symmetry. Some remarks concerning the orbit structure of the solution space under the SL(n+1,C) action are also made.*

1. INTRODUCTION

During the last few years, a great deal of work [1,2] has been devoted to the study of infinite dimensional Lie algebras acting on the solution space of miscellaneous non-linear field theoretic models. The reasons of this extensive study lie in the hope that such an infinite symmetry might help to better understand the dynamics of these complicated models. An extremely useful example in that respect is obtained when we consider the sigma models defined on two-dimensional space-time. The general case where the field takes its values in an arbitrary Riemannian symmetric space has been completely solved [3].

On the other hand, much more is known in the particular case of the Euclidean **CP**^{n} models. Indeed, due to the work of Din and Zakrzewski [4], it is possible to characterize the structure of the space of finite action solutions

† Supported in part by the Natural Sciences and Engineering Research Council of Canada and the "Fonds FCAR", Québec.

of these models. All these solutions may be obtained from the instantons (i.e. the solutions of the self-duality condition) by applying some differential operator a finite number of times. Hence, the instantons of a given degree k and the solutions gotten from them by the above scheme are described by a finite number of parameters. We thus expect the infinite dimensional symmetry algebra to collapse into a finite dimensional one if the symmetry generators map k-instantons into k-instantons.

This problem has been addressed elsewhere [5]. The result is that the symmetry algebra indeed collapses into a finite dimensional Lie algebra with the structure of sl(n+1,C). It has also been noticed [5] that it is possible to exponentiate some particular generators of sl(n+1,C). Here we want to present the general case, namely an explicit closed form for the action of the group SL(n+1,C) on the space of finite action solutions of the $\mathbf{CP}^n$ models. This in turn allows us to draw some conclusions on the SL(n+1,C) orbits in the solution space.

The content of the paper is the following. We begin by recalling some facts about $\mathbf{CP}^n$ models and organizing them in a way suitable for our later purposes. In particular, we describe the field of the model alternatively in terms of a unit vector in $\mathbf{C}^{n+1}$, the associated rank one projector and a collection of rank p projectors ($0 \leq p \leq n+1$) satisfying additional constraints. We also briefly review the construction of Din and Zakrzewski [4]. The next section is the central part of the paper. Starting with the action of the Lie algebra sl(n+1,C), we show how to exponentiate it for an arbitrary element in the group SL(n+1,C). We then prove that the result indeed provides a symmetry of the model, i.e. that it preserves all the required properties of the objects upon which it acts. The closing section contains further comments on how the group action splits the solution space into strata and orbits.

2. COLLECTING MATERIAL ABOUT THE $\mathbf{CP}^n$ MODELS

We consider models defined on Euclidean two-dimensional space with real coordinates (x_1, x_2) or , alternatively, complex coordinates (x_+, x_-) where $x_\pm \equiv x_1 \pm i x_2$. The field of the $\mathbf{CP}^n$ model is a vector $z \in \mathbf{C}^{n+1}$ of unit length: $|z|=1$. We fix the dynamics by introducing the following Lagrangian density:

$$\mathcal{L} \equiv 2\,[\,|D_+ z|^2 + |D_- z|^2\,] \qquad (1)$$

where $D_\pm$ are the covariant derivatives associated with the connection $(z^\dagger \partial_\pm z)$:

$$D_\pm \equiv \partial_\pm - (z^\dagger\, \partial_\pm z) \qquad (2)$$

The field equation obtained from (1) reads :

$$D_- D_+ z + (z^\dagger D_- D_+ z)\, z = 0 \qquad (3)$$

The (anti-) self-duality condition takes the simple form :

$$D_{\underset{(+)}{-}} z = 0 \qquad (4)$$

and one sees immediately that any solution of (4) is a solution of (3). In the sequel, we will consider the solutions of (3) that have finite action. The finite action solutions of (4) are called (anti-) instantons.

The remarkable fact about the **CP**^{n} model is that it is possible to give an algorithm to compute explicitly all its finite action solutions. This construction is due to Din and Zakrzewski [4] and here is a brief account of it. First, define the following differential operators :

$$P_\pm f \equiv \partial_\pm f - \left(\frac{f^\dagger\, \partial_\pm f}{|f|^2} \right) f \qquad \text{for any } f(x_+, x_-) \in \mathbf{C}^{n+1} \qquad (5)$$

Let now z be any finite action solution of the **CP**^{n} model. Then [4],

- there exist i and j such that : $P_-^{i+1} z = P_+^{j+1} z = 0$ with $i+j \leq n$
- every element of the set :

$$\left\{ \frac{P_-^i z}{|P_-^i z|}, \frac{P_-^{i-1} z}{|P_-^{i-1} z|}, \ldots, \frac{P_- z}{|P_- z|}, z, \frac{P_+ z}{|P_+ z|}, \ldots, \frac{P_+^{j-1} z}{|P_+^{j-1} z|}, \frac{P_+^j z}{|P_+^j z|} \right\} \qquad (6)$$

is a finite action solution of the CP^n model

- the elements of the set (6) are mutually orthogonal.

By definition of i and j, it is obvious that $P_-^i z$ ($P_+^j z$) is an (anti-) instanton. The situation is thus rather simple: any finite action solution belongs necessarily to a set like (6) and is thus of the form $z_l = P_+^l z / |P_+^l z|$ where $z \equiv z_0$ is an instanton. Of course, an equivalent statement holds for the anti-instanton.

Moreover, all the instantons of the CP^n model are known. They are given by a set of (n+1) polynomials $p_i(x_+)$ of x_+ only ($0 \le i \le n$), with no common root ($z_{0,i}$ denotes the i-th component of the vector z_0):

$$z_{0,i} = \frac{p_i(x_+)}{\sqrt{\sum_{j=0}^{n} |p_j(x_+)|^2}} \tag{7}$$

and z_0 is called a k-instanton if k is the maximum degree among the degrees of the p_i's. Putting together (6) and (7) gives a complete description of the space of finite action solutions of the CP^n model.

We call the set (6) the *family* of the solution z and denote it by {z}. With every family {z} we associate an integer m(z) which is the number of elements in {z} minus one; thus, $m(z) \equiv i+j \le n$. If $m(z) = n$, we call the family {z} *generic* and, if $m(z) < n$, we call the family {z} *degenerate.*

We now turn to the projector description of the solution space, because the action of the symmetry group (to be introduced in the next section) is easier to obtain when expressed in these terms. With every solution z_l of the family {z} we associate the rank one hermitian projector P_l which projects onto the complex line generated by z_l:

$$P_l \equiv z_l z_l^\dagger \quad , \; P_l^2 = P_l^\dagger = P_l \qquad (0 \le l \le m(z)) \tag{8}$$

$$P_l P_{l'} = \delta_{l,l'} P_l \qquad (0 \le l, l' \le m(z)) \tag{9}$$

For later convenience, we still need another equivalent formulation. Define a set of (m(z)+2) matrices Σ_k $(0 \leq k \leq m(z)+1)$ as:

$$\Sigma_0 \equiv 0 \qquad \text{and} \qquad \Sigma_k \equiv \sum_{l=0}^{k-1} P_l \qquad (1 \leq k \leq m(z)+1) \tag{10}$$

It can be shown [4,5] that asking the vectors z_l of unit length defined by:

$$P_l z_l = z_l \tag{11}$$

with

$$P_l = \Sigma_{l+1} - \Sigma_l \tag{12}$$

to be solutions of (3) is fully equivalent to the following conditions on the set $\{\Sigma_k, 0 \leq k \leq m(z)+1\}$:

- the Σ_k's are rank k hermitian projectors:

$$\Sigma_k^2 = \Sigma_k^\dagger = \Sigma_k \qquad \text{with rank } \Sigma_k = k \tag{13}$$

- the Σ_k's satisfy a condition which is recognized as the self-duality condition for the Grassmannian sigma models:

$$\Sigma_k \, \partial_+ \Sigma_k = 0 \tag{14}$$

- the range of Σ_i is a subspace of the range of Σ_j if $i \leq j$:

$$\Sigma_i \, \Sigma_j = \Sigma_j \, \Sigma_i = \Sigma_i \qquad \text{if } i \leq j \tag{15}$$

- the Σ_k's satisfy the additional differential constraints:

$$\Sigma_{k+1} \, \partial_+ \Sigma_k = \partial_+ \Sigma_k \qquad (0 \leq k \leq m(z)) \tag{16}$$

Before closing this section, we note that the last projector in the family

$\Sigma_{m(z)+1}$ is a constant projector:

$$\partial_{\pm} \Sigma_{m(z)+1} = 0 \tag{17}$$

(In the generic case, this is obvious since m(z) = n implies $\Sigma_{m(z)+1} = \Sigma_{n+1} = 1$. That the result remains true in the degenerate case requires however a detailed proof [6].)

From now on, we consider finite action solutions of the $\mathbf{CP}^n$ model through their equivalent formulation in terms of a family of projectors $\{\Sigma_k\}$ satisfying the conditions (13-16).

3. FROM SU(n+1) TO SL(n+1,C)

All sigma models with values in a Riemannian symmetric space enjoy an infinite set of symmetry transformations acting on their solution space [3]. In a previous publication [5], we showed that, in the case of the finite action solutions of the Euclidean $\mathbf{CP}^n$ model, this infinite dimensional symmetry algebra collapses into a finite dimensional one and the algebraic structure was identified to be that of sl(n+1,C). Explicitly, we got the following transformation laws:

$$\Sigma_k \longrightarrow \Sigma_k + \Delta^T \Sigma_k \qquad \text{for } 0 \leq k \leq m(z)+1 \text{ , } T \in sl(n+1,C) \tag{18a}$$

where

$$\Delta^T \Sigma_k = [\, T, \Sigma_k \,] \qquad \text{if } T^\dagger = -T \tag{18b}$$

$$\Delta^T \Sigma_k = \{\Sigma_k \, , \, T\} - 2\, \Sigma_k \, T \, \Sigma_k \qquad \text{if } T^\dagger = T \tag{18c}$$

Take now an arbitrary generator $T \in sl(n+1,C)$ and decompose it into its hermitian and anti-hermitian parts:

$$H = \frac{1}{2}(T + T^\dagger) \qquad\qquad H = H^\dagger$$

$$A = \frac{1}{2}(T - T^\dagger) \qquad\qquad A = -A^\dagger$$

Thus, the general infinitesimal transformation law is, by linearity:

$$\begin{aligned}\Delta^T \Sigma_k &= [A, \Sigma_k] + \{\Sigma_k, H\} - 2\Sigma_k H \Sigma_k \\ &= T\Sigma_k + \Sigma_k T^\dagger - \Sigma_k (T+T^\dagger)\Sigma_k \end{aligned} \tag{19}$$

As already noticed [5] for special cases, it is possible to exponentiate the law (19). Indeed, (19) is the lowest order of an infinitesimal linear fractional transformation:

$$\Sigma_k \longrightarrow [(1+T)\Sigma_k]\,[1 - T^\dagger + (T+T^\dagger)\Sigma_k]^{-1} \tag{20a}$$

which can be expressed in matrix form as:

$$\begin{pmatrix} 1 \\ \Sigma_k \end{pmatrix} \longrightarrow \begin{pmatrix} (1-T^\dagger) & (T+T^\dagger) \\ 0 & (1+T) \end{pmatrix} \begin{pmatrix} 1 \\ \Sigma_k \end{pmatrix} \tag{20b}$$

by imposing on $C^{2(n+1)\times(n+1)}$ the equivalence relation:

$$\begin{pmatrix} 1 \\ \Sigma_k \end{pmatrix} \approx \begin{pmatrix} A \\ \Sigma_k A \end{pmatrix} \qquad \text{for all invertible } A \tag{20c}$$

Hence, the strategy is to exponentiate the $2(n+1) \times 2(n+1)$ matrix in (20.b) and use (20.c) afterwards. Let now $g \equiv \exp T \in SL(n+1,C)$. One easily checks that:

$$\exp\begin{pmatrix} -T^\dagger & T+T^\dagger \\ 0 & T \end{pmatrix} = \begin{pmatrix} g^{\dagger-1} & g - g^{\dagger-1} \\ 0 & g \end{pmatrix} \tag{21}$$

and the $SL(n+1,C)$ action on the solution space finally reads:

$$\Sigma_k' = g\,\Sigma_k\,[\,g^{\dagger -1}\,(\mathbb{1}-\Sigma_k) + g\,\Sigma_k]^{-1} = g\,\Sigma_k\,[(\,g^\dagger g)^{-1}\,(\mathbb{1}-\Sigma_k) + \Sigma_k]^{-1}\,g^{-1} \qquad (22)$$

$$\text{for } 0 \leq k \leq m(z)+1,\ g \in SL(n+1,C)$$

Up to now, we do not know still if (22) provides a symmetry transformation at all. We thus have to prove that the relation (22) is well-defined (i.e. that the inverse makes sense) and that the properties (13-16) which define solutions of the $\mathbf{CP}^n$ model are satisfied by the new Σ_k' if they were by the old Σ_k. For some of these proofs, it is useful to write down matrices in the basis where Σ_k is diagonal. Note that, by (15), all the Σ_k commute among themselves and can thus be diagonalized by the same matrix Λ:

$$\Sigma_k = \Lambda^{-1} \begin{pmatrix} \mathbb{1}_k & 0 \\ 0 & 0 \end{pmatrix} \Lambda \qquad (23)$$

Using an obvious block decomposition, we also introduce the matrix $(g^\dagger g)$ and its inverse in the basis where Σ_k is diagonal as:

$$(g^\dagger g)^{-1} \equiv \Lambda^{-1} \begin{pmatrix} \tau_{k1} & \tau_{k2} \\ \tau_{k3} & \tau_{k4} \end{pmatrix} \Lambda \qquad g^\dagger g \equiv \Lambda^{-1} \begin{pmatrix} \tilde{\tau}_{k1} & \tilde{\tau}_{k2} \\ \tilde{\tau}_{k3} & \tilde{\tau}_{k4} \end{pmatrix} \Lambda \qquad (24)$$

We now turn to the list of these properties, together with their proofs.

(P1) $[(g^\dagger g)^{-1}\,(\mathbb{1}-\Sigma_k) + \Sigma_k\,]$ is always invertible.

Indeed,

$$(g^\dagger g)^{-1}\,(\,\mathbb{1}-\Sigma_k\,) + \Sigma_k = \Lambda^{-1} \begin{pmatrix} 1 & \tau_{k2} \\ 0 & \tau_{k4} \end{pmatrix} \Lambda$$

and is thus invertible as soon as τ_4 is. But, $(g^\dagger g)^{-1}$ is a positive definite matrix and τ_4 , being one of its submatrix, is also positive definite and, hence, invertible.

(P2) $\Sigma_k'^{\dagger} = \Sigma_k'$

$$\Sigma_k'^{\dagger} = g^{-1\dagger}\,[(g^{\dagger}g)^{-1}\,(1-\Sigma_k) + \Sigma_k]^{-1\dagger}\,\Sigma_k\,g^{\dagger}$$

$$= g^{-1\dagger}\,[(1-\Sigma_k)(g^{\dagger}g)^{-1} + \Sigma_k]^{-1}\,\Sigma_k\,[(1-\Sigma_k)+(g^{\dagger}g)\Sigma_k]\,[(1-\Sigma_k)+(g^{\dagger}g)\Sigma_k]^{-1}\,g^{\dagger}$$

$$= g[(1-\Sigma_k)+\Sigma_k(g^{\dagger}g)]^{-1}\Sigma_k[(1-\Sigma_k)+(g^{\dagger}g)\Sigma_k][g^{\dagger-1}(1-\Sigma_k)+g\Sigma_k]^{-1}$$

$$= g\Lambda^{-1}\begin{pmatrix} 1_k & 0 \\ 0 & 0 \end{pmatrix}\begin{pmatrix} 1_k & -\tau_{k2}\tau_{k4}^{-1} \\ 0 & \tau_{k4}^{-1} \end{pmatrix}\begin{pmatrix} 1_k & 0 \\ 0 & 0 \end{pmatrix}\Lambda[(g^{\dagger}g)^{-1}(1-\Sigma_k)+\Sigma_k]^{-1}g^{-1}$$

$$= g\Sigma_k[(g^{\dagger}g)^{-1}(1-\Sigma_k)+\Sigma_k]^{-1}g^{-1} = \Sigma'_k$$

(P3) $\Sigma_k'^2 = \Sigma_k'$

$$\Sigma_k'^2 = g\Sigma_k[(g^{\dagger}g)^{-1}(1-\Sigma_k)+\Sigma_k]^{-1}\Sigma_k[(g^{\dagger}g)^{-1}(1-\Sigma_k)+\Sigma_k]^{-1}g^{-1}$$

$$= g\Lambda^{-1}\begin{pmatrix} 1_k & 0 \\ 0 & 0 \end{pmatrix}\begin{pmatrix} 1_k & -\tau_{k2}\tau_{k4}^{-1} \\ 0 & \tau_{k4}^{-1} \end{pmatrix}\begin{pmatrix} 1_k & 0 \\ 0 & 0 \end{pmatrix}\Lambda[(g^{\dagger}g)^{-1}(1-\Sigma_k)+\Sigma_k]^{-1}g^{-1}$$

$$= g\Sigma_k[(g^{\dagger}g)^{-1}(1-\Sigma_k)+\Sigma_k]^{-1}g^{-1} = \Sigma'_k$$

(P4) rank $\Sigma_k' = k$

$$\text{rank}\ \Sigma_k' = \text{tr}\ \Sigma_k' = \text{tr}\ \Sigma_k[(g^{\dagger}g)^{-1}(1-\Sigma_k)+\Sigma_k]^{-1} = \text{tr}\begin{pmatrix} 1_k & -\tau_{k2}\tau_{k4}^{-1} \\ 0 & 0 \end{pmatrix} = k$$

(P5) Let v_k be an eigenvector of Σ_k with eigenvalue 1, then gv_k is an eigenvector of Σ_k' with eigenvalue 1.

$\Sigma_k v_k = v_k$ means $\Lambda v_k = [v'_{k1}, \dots, v'_{kk}, 0, \dots, 0]^T$

Then,

$$\Sigma_k'(gv_k) = g\Sigma_k[(g^\dagger g)^{-1}(1-\Sigma_k)+\Sigma_k]^{-1}v_k = g\Lambda^{-1}\begin{pmatrix} 1_k & -\tau_{k2}\tau_{k4}^{-1} \\ 0 & 0 \end{pmatrix}\Lambda v_k$$

$$= g\Lambda^{-1}\Lambda v_k = gv_k$$

(P6) $\Sigma_k' \partial_+ \Sigma_k' = 0$

Observe first that:

$$\partial_+\Sigma_k' = g\partial_+\Sigma_k[(g^\dagger g)^{-1}(1-\Sigma_k)+\Sigma_k]^{-1}g^{-1}$$

$$- \Sigma_k' g(1-(g^\dagger g)^{-1})\partial_+\Sigma_k[(g^\dagger g)^{-1}(1-\Sigma_k)+\Sigma_k]^{-1}g^{-1}$$

and thus,

$$\Sigma_k'\partial_+\Sigma_k' = \Sigma_k' g^{\dagger-1}\partial_+\Sigma_k[(g^\dagger g)^{-1}(1-\Sigma_k)+\Sigma_k]^{-1}g^{-1}$$

where we used (P3). But,

$$\Sigma_k' g^{\dagger-1} = g\Sigma_k[(g^\dagger g)^{-1}(1-\Sigma_k)+\Sigma_k]^{-1}(g^\dagger g)^{-1} = g\Sigma_k[(1-\Sigma_k)+(g^\dagger g)\Sigma_k]^{-1}$$

$$= g\Lambda^{-1}\begin{pmatrix} 1_k & 0 \\ 0 & 0 \end{pmatrix}\begin{pmatrix} \tilde{\tau}_{k1} & 0 \\ \tilde{\tau}_{k3} & 1 \end{pmatrix}^{-1}\Lambda = g\Lambda^{-1}\begin{pmatrix} \tilde{\tau}_{k1}^{-1} & 0 \\ 0 & 0 \end{pmatrix}\Lambda = g\Lambda^{-1}\begin{pmatrix} \tilde{\tau}_{k1}^{-1} & 0 \\ 0 & 0 \end{pmatrix}\begin{pmatrix} 1_k & 0 \\ 0 & 0 \end{pmatrix}\Lambda$$

$$= g\Lambda^{-1}\begin{pmatrix} \tilde{\tau}_{k1}^{-1} & 0 \\ 0 & 0 \end{pmatrix}\Lambda\Sigma_k$$

Hence,

$$\Sigma_k' \, \partial_+\Sigma_k' = g\Lambda^{-1}\begin{pmatrix} \tilde{\tau}_{k1}^{-1} & 0 \\ 0 & 0 \end{pmatrix}\Lambda\Sigma_k\partial_+\Sigma_k[(g^\dagger g)^{-1}(\mathbb{1}-\Sigma_k)+\Sigma_k]^{-1}g^{-1} = 0 \qquad \text{by (14)}$$

(P7) $\Sigma_i'\Sigma_j' = \Sigma_j'\Sigma_i' = \Sigma_j' \qquad \text{if } j \le i$

$$\Sigma_i'\Sigma_j' = g\Sigma_i[(g^\dagger g)^{-1}(\mathbb{1}-\Sigma_i)+\Sigma_i]^{-1}\Sigma_j[(g^\dagger g)^{-1}(\mathbb{1}-\Sigma_j)+\Sigma_j]^{-1}g^{-1}$$

But,

$$\Sigma_i[(g^\dagger g)^{-1}(\mathbb{1}-\Sigma_i)+\Sigma_i]^{-1}\Sigma_j = \Lambda^{-1}\begin{pmatrix} \mathbb{1}_i & 0 \\ 0 & 0 \end{pmatrix}\begin{pmatrix} \mathbb{1}_i & -\tau_{i2}\tau_{i4}^{-1} \\ 0 & \tau_{i4}^{-1} \end{pmatrix}\begin{pmatrix} \mathbb{1}_j & 0 \\ 0 & 0 \end{pmatrix}\Lambda$$

$$= \Lambda^{-1}\begin{pmatrix} \mathbb{1}_j & 0 \\ 0 & 0 \end{pmatrix}\Lambda = \Sigma_j$$

because $j \le i$. Thus,

$$\Sigma_i'\Sigma_j' = \Sigma_j' \qquad \text{for } j \le i$$

The other equality follows by taking the hermitian conjugate and using (P2).

(P8) $\Sigma_{k+1}' \, \partial_+\Sigma_k' = \partial_+\Sigma_k'$

$$\partial_+\Sigma_k' = g\partial_+\Sigma_k[(g^\dagger g)^{-1}(\mathbb{1}-\Sigma_k)+\Sigma_k]^{-1}g^{-1}$$

$$- \Sigma_k' g(\mathbb{1}-(g^\dagger g)^{-1})\partial_+\Sigma_k[(g^\dagger g)^{-1}(\mathbb{1}-\Sigma_k)+\Sigma_k]^{-1}g^{-1}$$

and, by (P7),

$$\Sigma'_{k+1}\,\partial_+\Sigma'_k = \Sigma'_{k+1} g\partial_+\Sigma_k[(g^\dagger g)^{-1}(\mathbb{1}-\Sigma_k)+\Sigma_k]^{-1}g^{-1}$$

$$- \Sigma'_k g(\mathbb{1}-(g^\dagger g)^{-1})\,\partial_+\Sigma_k[(g^\dagger g)^{-1}(\mathbb{1}-\Sigma_k)+\Sigma_k]^{-1}g^{-1}$$

But,

$$\Sigma'_{k+1} g\partial_+\Sigma_k = g\Sigma_{k+1}[(g^\dagger g)^{-1}(\mathbb{1}-\Sigma_{k+1})+\Sigma_{k+1}]^{-1}\Sigma_{k+1}\partial_+\Sigma_k$$

$$= g\Lambda^{-1}\begin{pmatrix} \mathbb{1}_{k+1} & 0 \\ 0 & 0 \end{pmatrix}\begin{pmatrix} \mathbb{1}_{k+1} & \tau_{k+1,2} \\ 0 & \tau_{k+1,4} \end{pmatrix}^{-1}\begin{pmatrix} \mathbb{1}_{k+1} & 0 \\ 0 & 0 \end{pmatrix}\Lambda\partial_+\Sigma_k$$

$$= g\Lambda^{-1}\begin{pmatrix} \mathbb{1}_{k+1} & 0 \\ 0 & 0 \end{pmatrix}\Lambda\partial_+\Sigma_k = g\Sigma_{k+1}\partial_+\Sigma_k = g\partial_+\Sigma_k$$

where, in the first and last equalities, we used (16). Finally,

$$\Sigma'_{k+1}\partial_+\Sigma'_k = \partial_+\Sigma'_k$$

All the required properties are thus satisfied and (22) indeed gives an SL(n+1,C) action on the space of finite action solutions of the $\mathbf{CP}^n$ model. Before sketching the consequences of this new symmetry, note that the transformation law (22) *cannot* be written down easily for the vectors z_l of the family {z}. Indeed, one should be careful with the content of property (P5) : it does *not* mean that the vectors z_l are transformed according to $z_l \to gz_l$. This is true only in the case of the rank one projector $\Sigma_1 = P_0$, i.e. the instanton. So the best way to give the transformation law for the vectors z_l is through the action on the instanton z_0 of the family to which they belong:

$$z'_l = \frac{P^l_+(gz_0)}{|P^l_+(gz_0)|} \qquad g\in SL(n+1,C)\,,\ 0\le l\le m(z_0) \tag{25}$$

In general: $P_+^1(gz_0) \neq g\,P_+^1(z_0)$. Of course, if g belongs to the unitary subgroup of SL(n+1,C), everything boils down to the usual symmetry of the model:

$$z_l' = gz_l \qquad g \in SU(n+1)\,,\, 0 \leq l \leq m(z_0) \qquad (26)$$

4. DISCUSSION

Let us now briefly review some consequences of the transformation law (22) [6]. First of all, we would like to know if the symmetry transformations can change a degenerate family into a generic one. The answer is no and the proof [6] is essentially contained in property (P4) and the fact that the law is invertible. In fact, the number of elements in a family is an invariant of the law (22): $m(gz_0) = m(z_0)$ for all $g \in SL(n+1,C)$.

One can deduce even more. Take for instance an instanton in $\mathbf{CP}^n$ such that the vector representing it has vanishing (n-k) last components. It is obvious that this instanton z_0 belongs to a degenerate family and $m(z_0) \leq k$. This form with vanishing (n-k) last components is typical of a degenerate family with $m(z) < n$ [6]. Indeed, because of (17), the projector $\Sigma_{m(z)+1}$ can be diagonalized by a constant unitary matrix and it can be shown that, in the basis where $\Sigma_{m(z)+1}$ is diagonal, all the vectors of the family {z} have vanishing (n-m(z)) last components. A degenerate family in $\mathbf{CP}^n$ is thus always on the SU(n+1) orbit of a generic family in $\mathbf{CP}^m$ trivially imbedded in $\mathbf{CP}^n$.

So what is the full SL(n+1,C) symmetry good for? The full group SL(n+1,C) can be used to put the instantons into a canonical form [6]. By this, we mean that a k-instanton is represented by polynomials (see (7)) such that $k = \deg p_0 > \deg p_1 > \ldots > \deg p_n$. An immediate consequence of the existence of this canonical form is that no k-instanton in $\mathbf{CP}^n$ with $k < n$ can be generic. Another consequence is that the isotropy subgroup of a particular solution under the action (22) (i.e. the set of elements in SL(n+1, C) leaving it unchanged) is easily readable. It turns out [6] that the isotropy subgroup of a family characterized by an integer m is $SL(n-m,C) \times \{C \backslash \{0\}\}$ if $m<n$ (if $m=n$, the isotropy subgroup is trivial). Each orbit of the action (22) has thus the structure of $SL(n+1,C)/(SL(n-m,C) \times \{C\backslash\{0\}\})$ and the strata of the action (22) (i.e. the collections of orbits with equivalent isotropy subgroup) are hence

labeled by this integer m.

REFERENCES

1] Dolan, L., Phys. Rev. Lett. **47**, 1371 (1981)
2] Wu, Y.-S., Nucl. Phys. **B211**, 160 (1983)
3] Jacques, M. , Saint-Aubin, Y. ,J. Math. Phys. (1987) in press
4] Din, A. , Zakrzewski, W. , Nucl. Phys. **B174**, 397 (1980)
5] Arsenault, G. , Jacques, M. , Saint-Aubin, Y. , *Collapse and exponentiation of infinite symmetry algebras of Euclidean projective and Grassmannian σ models* , Preprint CRM-1458 (Montréal, 1987)
6] Arsenault, G., Jacques, M., Saint-Aubin, Y., *SL(n+1)-strata and orbits in the solution space of Euclidean CP^n models,* Preprint CRM-1504 (Montréal, 1987)

GROUP THEORY AND THE CANONICAL QUANTIZATION OF STRINGS ON TORI[†]

C.J.ISHAM

The Blackett Laboratory
Imperial College
South Kensington
London SW7 2BZ

ABSTRACT

The quantization of particles and strings moving on a torus is discussed within the framework of a group theoretic scheme which respects the global topological structure of the system. It is shown how the representation of the disconnected part of the string's canonical group leads naturally to the extra canonical variables that have previously been introduced *ad hoc*. This extra circular degree of freedom is combined group theoretically with the constant string modes to give a well-defined left-right split.

1. INTRODUCTION

This talk is based on joint work performed in collaboration with N. Linden [1]. The topic is certain topological problems that arise in the quantization of a bosonic string propagating in a torodially compactified spacetime of the form $\mathbb{R}^4 \times T^n$ where $\mathbb{R}^4$ and T^n are Minkowski space and the n-torus respectively. A canonical, light-cone gauge, analysis of the classical theory yields a system whose configuration space is $C(S^1, \mathbb{R}^2 \times T^n) \approx C(S^1, \mathbb{R}^2) \times C(S^1, T^n)$, where $C(X,Y)$ denotes the space of smooth maps from the manifold X into the manifold Y. The quantization of loops in $\mathbb{R}^2$ (the transverse part of Minkowski space) involves standard quantum field theory and produces no surprises. However, the configuration space $C(S^1,T^n)$ is (i) disconnected, and (ii) not a vector space, and this might be expected to cause problems in any careful consideration of the quantum programme. Such difficulties do indeed arise but we shall show that they can be handled successfully using a global, group-theoretic approach to quantization. These techniques can be generalized to handle the more complicated general non-linear σ-model with configuration space $C(\Sigma, G/H)$ where Σ and G/H are the domain space and homogeneous target space respectively.

To simplify the discussion (and with no great loss of generality) we will restrict our attention to the case $n=1$. The usual starting point for the quantization of this system is a mode expansion of $X \epsilon C(S^1, T^1)$

$$X(\sigma) = q + \omega\sigma + \sum_{n\neq 0} (a_n \cos n\sigma + b_n \sin n\sigma) \tag{1}$$

in which q is the constant mode and ω is the winding number (the number of times the loop wraps itself around the target T^1). Note that, since X takes its values in the circle T^1, it is necessary to identify values of $X(\sigma)$ that differ by $2\pi n$ where n is any integer (ie. we identify T^1 with the quotient space $\mathbb{R}/2\pi\mathbb{Z}$).

It is then assumed that there is a canonical momentum $P(\sigma)$ conjugate to $X(\sigma)$ with the usual commutation relations:

$$[\hat{X}(\sigma), \hat{P}(\sigma')] = i\hbar\delta(\sigma, \sigma') \tag{2}$$

and with $[\hat{X}(\sigma), \hat{X}(\sigma')] = [\hat{P}(\sigma), \hat{P}(\sigma')] = 0$. The light-cone gauge equations of motions are linear and the Heisenberg operator equations generated with the aid of (2) can be solved in the familiar form

$$\hat{X}(\sigma,\tau) = \hat{q} + \hat{J}\tau + \hat{\omega}\sigma + \sum_{n\neq 0} ({}_L\hat{\alpha}_n e^{-in(\sigma+\tau)} + {}_R\hat{\alpha}_n e^{-in(\tau-\sigma)}) \tag{3}$$

in which $\hat{J}$, ${}_L\hat{\alpha}_n$ and ${}_R\hat{\alpha}_n$ are constant operators.

In the analysis of the heterotic string by Gross et al [2], an important role is played by a decomposition of the string modes into left- and right-movers. The part of (3) that lies within the sum splits naturally in this way with respect to the coefficients ${}_L\hat{\alpha}_n$ and ${}_R\hat{\alpha}_n$. However, it is by no means clear how one should proceed with the "$\hat{q} + \hat{J}\tau + \hat{\omega}\sigma$" term. The authors of [2] simply assumed that there were operators $\hat{q}_L$, $\hat{q}_R$, $\hat{J}_L$ and $\hat{J}_R$ with

$$\hat{q} + \hat{J}\tau + \hat{\omega}\sigma = (\tfrac{1}{2}\hat{q}_L + \hat{J}_L(\tau+\sigma)) + (\tfrac{1}{2}\hat{q}_R + \hat{J}_R(\tau-\sigma)) \tag{4}$$

and which satisfied the commutation relations

$$[\hat{q}_L, \hat{J}_L] = i\hbar = [\hat{q}_R, \hat{J}_R] \; ; \qquad [\hat{q}_L, \hat{J}_R] = 0 = [\hat{q}_R, \hat{J}_L] \tag{5}$$

Substituting the expansion (3) into the basic commutation relations (2) gives the equation

$$[\hat{q}, \hat{J}] = i\hbar \tag{6}$$

for the constant mode operators $(\hat{q},\hat{J})$ and it is easy to check that this is reproduced by (4) and (5). The operators $\hat{q}_L$, $\hat{q}_R$, $\hat{J}_L$ and $\hat{J}_R$ can be constructed if it is assumed [2,3] that there is an operator $\hat{V}$ which is conjugate to the winding number

$$[\hat{V}, \hat{\omega}] = i\hbar \tag{7}$$

and which commutes with the remaining string modes; one simply defines

$$\begin{aligned} \hat{q}_L &:= \hat{q} + \hat{V}\,; \qquad & \hat{J}_L &:= \tfrac{1}{2}(\hat{J} + \hat{\omega}) \\ \hat{q}_R &:= \hat{q} - \hat{V}\,; \qquad & \hat{J}_R &:= \tfrac{1}{2}(\hat{J} - \hat{\omega}) \end{aligned} \tag{8}$$

This is the received wisdom on the implementation of the left-right split for the constant- and winding-number modes. However, a number of objections can be raised against the procedure that has been followed. Specifically:

(a) Is it really true that we have commutation relations of the form (6) for the constant mode q ? The problem is that q is essentially an *angular* variable (ie. its values are constrained to lie between 0 and 2π) whereas it is well known that the spectrum of any (continuously exponentiable) self-adjoint operator satisfying (6) is necessarily unbounded. Put another way, if (6) is true we have the uncertainty relations

$$\Delta q \, \Delta J \geq \hbar \tag{9}$$

and hence the dispersion Δq can be made arbitrarily large by choosing a state for which ΔJ is sufficiently small. Once again this contradicts the bounded nature of the angular variable q.

Another problem is that (6) yields a continuous spectrum for the operator $\hat{J}$. But $\hat{J}$ is an angular momentum (it is conjugate to an angular variable) and therefore its spectrum should be discrete. It is clear from these remarks that the quantization of the constant string mode should be discussed within the framework of a proper quantum theory for a system whose configuration space is a circle, not the more familiar real line to which (6) is applicable.

(b) Can there exist any operator $\hat{V}$ that is conjugate to $\hat{\omega}$ in the sense of (7)? One obvious objection is that $\hat{\omega}$ is the winding number and so its spectrum should presumably be discrete whereas, as remarked above in the context of (6), commutation relations of the Heisenberg form (7) imply a continuous spectrum.

Another serious difficulty arises in trying to find an explicit expression for $\widehat{V}$ in terms of the operators $\widehat{X}(\sigma)$ and $\widehat{P}(\sigma)$. Since $\widehat{\omega}$ commutes with $\widehat{X}(\sigma)$, the momentum $\widehat{P}(\sigma)$ must be used in some essential way; for example via the formal relation

$$e^{i\widehat{P}(f)}\,\widehat{\omega}\,e^{-i\widehat{P}(f)} = \widehat{\omega} + \frac{\hbar}{2\pi}\int_0^{2\pi}\frac{df(\sigma)}{d\sigma}\,d\sigma \tag{10}$$

where $\widehat{P}(f) := \frac{1}{2\pi}\int_0^{2\pi} f(\sigma)\widehat{J}(\sigma)\,d\sigma$ is the operator $\widehat{P}(\sigma)$ smeared with the test function f. Since any test-function should be differentiable, the integral on the right hand side of (10) is in fact zero, and we see that $\widehat{\omega}$ commutes with $\widehat{P}(f)$ as well as with $\widehat{X}(\sigma)$. However, it was proposed in [3] to construct the desired operator $\widehat{V}$ by employing (10) with the test function $f(\sigma) = -\sigma$:

$$\widehat{V} := -\frac{1}{2\pi}\int_0^{2\pi}\sigma\widehat{P}(\sigma)\,d\sigma \tag{11}$$

Formally this does indeed satisfy (7). However, the angular coordinate σ on the domain circle is *not* a continuous function and (11) is exceptionally ill-defined.

(c) There is another serious objection to granting operator status to angular variables. Suppose $\widehat{q}_1, \widehat{q}_2, \ldots \widehat{q}_n$ is a set of such operators for a n-torus T^n, with conjugate variables $\widehat{p}_1, \widehat{p}_2, \ldots \widehat{p}_n$. Then the operators $\widehat{J}_{ij} := \widehat{q}_i\,\widehat{p}_j - \widehat{q}_j\,\widehat{p}_i$ apparently generate continuous rotations from one angular direction on T^n to another. But this is incompatible with the global topological structure of the torus (although of course it *is* allowed on the $\mathbb{R}^n$ tangent space). Serious mistakes may be made if it is assumed that operators of the type $\widehat{J}_{ij}$ exist. For example, it can lead to an overestimation of the symmetries possessed by the theory.

2. QUANTIZATION OF NON-LINEAR SYSTEMS

2.1 The General Problem

We will tackle the problems discussed above within the framework of the rather general question of how one should approach the quantization of an arbitrary classical system whose state space is a non-linear symplectic manifold $\mathfrak{S}$. The observables of such a system belong to the function space $C(\mathfrak{S}, \mathbb{R})$ and the simplest approach to canonical quantization is to seek a linear map $f \longrightarrow \widehat{f}$ which associates to each $f \in C(\mathfrak{S}, \mathbb{R})$ a self-adjoint operator $\widehat{f}$ on some Hilbert space $\mathfrak{H}$ such that, for all $f, g \in C(\mathfrak{S}, \mathbb{R})$,

$$\{f, g\} = \frac{1}{i}[\hat{f}, \hat{g}] \tag{12}$$

The classical observables form a complete set and this is reflected in the quantum theory by requiring the operator representation (12) of the classical Poisson bracket algebra to be irreducible. This guarantees that any quantum observable can be written as a function of the operators $\hat{f}$ representing the classical observables $f \in C(\mathfrak{S}, \mathbb{R})$. If the representation is not irreducible, there will exist quantum operators that cannot be written in this form and which describe the decomposition of the Hilbert space $\mathfrak{H}$ into irreducible subspaces. This might correspond to the existence of superselection sectors (or "internal" quantum numbers) and, as such, could be quite acceptable. However, it cannot be regarded as simply being a quantization of the given classical system alone, and that is what concerns us at the moment.

From this perspective, the non-dynamical aspects of quantization are reflected in the problem of finding unitary, irreducible, representations of the symplectic group of canonical transformations of $\mathfrak{S}$. This is an elegant way of formulating the quantization programme and therefore it is rather unfortunate that it cannot be implemented. More precisely, there is an old theorem of Groenwold and Van Hove [4,5] showing that, in the example $\mathfrak{S} = \mathbb{R}^{2n}$, there are *no* irreducible representations of (12) which reproduce the usual wavefunction representation of the Heisenberg relations

$$[\hat{x}^i, \hat{p}_j] = i\hbar\delta^i_j \qquad i,j = 1....n \tag{13}$$

for the subset of observables $\mathfrak{U} = (x^1, x^2, ... x^n; p_1, p_2, ... p_n)$. Since, modulo pathologies with unbounded operators, the Stone-Von Neumann theorem shows that every irreducible representation of (13) is equivalent to this particular one, a better statement of the Groenwold-Van Hove theorem is that there is no irreducible unitary representation of the full symplectic group of $\mathfrak{S} = \mathbb{R}^{2n}$ which remains irreducible when restriced to the Heisenberg-Weyl subgroup generated by $\mathfrak{U}$.

Thus the classical Poisson bracket algebra can be preserved at the quantum level only for some subset of observables and, for any particular symplectic space $\mathfrak{S}$, the central question is how this preferred subset is to be selected. In the familiar example $\mathfrak{S} = \mathbb{R}^{2n}$, the usual choice is simply the subset of globally defined coordinate functions $\mathfrak{U}$ satisfying the commutation relations (13). In an irreducible representation of this algebra, every operator can be written in terms of this basic set $(\hat{x}^1, \hat{x}^2, ... \hat{x}^n ; \hat{p}_1, \hat{p}_2, ... \hat{p}_n)$ and this reflects the obvious fact that every classical observable $f \in C(\mathfrak{S}, \mathbb{R})$ can be written as a function of the basic set of globally-defined coordinate functions $(x^1, x^2, \ldots x^n; p_1, p_2, \ldots p_n)$. For a general symplectic manifold $\mathcal{S}$, the analogue would be to find a subset $\mathcal{D} \subset C(\mathcal{S}, \mathbb{R})$ of classical observables such that

(i) $\mathcal{D}$ is a Lie subalgebra of $C(\mathcal{S}, \mathbb{R})$

(ii) $\mathcal{D}$ is big enough to generate $C(\mathcal{S}, \mathbb{R})$ as a ring, ie. every smooth function on $\mathcal{S}$ can be written as (a suitable limit of) sums and products of elements of $\mathcal{D}$.

(iii) There is no subalgebra $\mathcal{D}'$ of $\mathcal{D}$ which is also large enough to generate $C(\mathcal{S}, \mathbb{R})$ but which is such that there exists an irreducible self-adjoint representation of $\mathcal{D}$ which becomes reducible when restricted to $\mathcal{D}'$.

'Quantization' can then be defined to be the construction of irreducible, self-adjoint representations of this special subalgebra $\mathcal{D}$ of classical observables. Condition (ii) justifies choosing representations that are *irreducible*, while (iii) is a statement of the absence of a Groenwold-Van Hove phenomenon. It is not a trivial matter to show that an operator on an infinite-dimensional Hilbert space is genuinely self-adjoint and, in the context of a Lie algebra, the most efficient way is to construct a unitary representation of an associated Lie group. Thus quantization means finding unitary irreducible representations of a Lie group $\mathcal{G}$ whose Lie algebra is isomorphic to $\mathcal{D}$. In the case $\mathcal{G} = \mathbb{R}^{2n}$, the Stone-Von Neumann theorem guarantees that there is only one such representation. However, there is no particular reason why this should be so for a general $\mathcal{S}$ and different representations will correspond to genuinely inequivalent quantizations of the classical system. This phenomenon is familiar enough in conventional quantum field theory (where $\mathcal{S}$ is an infinite-dimensional vector space) but it is important to appreciate that, even for a finite-dimensional system, it might also arise if the topological structure of $\mathcal{S}$ is not that of a vector space $\mathbb{R}^{2n}$. There may be more than one possible choice for the preferred subset $\mathcal{S}$ and, again, this corresponds to the existence of inequivalent ways of quantizing the system.

2.2 Construction of $\mathcal{D}$

For any given classical state space $\mathcal{S}$, the critical question now is whether or not it is possible to find a subset $\mathcal{D} \subset C(\mathcal{S}, \mathbb{R})$ which satisfies the three conditions above. For any differentiable manifold $\mathfrak{M}$, the natural source of Lie algebras is the set of vector fields on $\mathfrak{M}$ which form an infinite-dimensional Lie algebra under the vector field commutator operation (modulo technical niceties, this is the Lie algebra of the diffeomorphism group of $\mathfrak{M}$). In particular, for a symplectic space $\mathfrak{S}$ there is a well-known map $f \rightarrow \xi(f)$ which associates to each $f \in C(\mathfrak{S}, \mathbb{R})$ a vector field on $\mathfrak{S}$ defined by

$$\xi(f) := \sum_{i=1}^{n} \left(\frac{\partial f}{\partial q^i} \frac{\partial}{\partial p_i} - \frac{\partial f}{\partial p_i} \frac{\partial}{\partial q^i} \right) \tag{14}$$

which satisfies

$$\xi(\{f,g\}) = [\,\xi(f), \xi(g)\,] \qquad \text{for all } f, g \in C(\mathfrak{S}, \mathbb{R}) \tag{15}$$

where the right hand side denotes the commutator of the vector fields $\xi(f)$ and $\xi(g)$. Thus the map $f \rightarrow \xi(f)$ is a homomorphism from the Poisson bracket algebra of $C(\mathfrak{S}, \mathbb{R})$ into the Lie algebra of vector fields on $\mathfrak{S}$. It is clear from (14) that the kernel of this map is simply the constant functions on $\mathfrak{S}$ and is hence isomorphic to $\mathbb{R}$.

Subalgebras of $C(\mathfrak{S}, \mathbb{R})$ can then be studied via the following steps [6]:

(i) Find sets of vector fields $\{\gamma^A \mid A \in \mathfrak{L}\}$ whose commutators close on some Lie algebra $\mathfrak{L}$, ie.

$$[\,\gamma^A, \gamma^B] = \gamma^{[AB]} \qquad \text{for all } A, B \in \mathfrak{L} \tag{16}$$

where $[AB]$ denotes the Lie bracket of A and B.

(ii) For each vector field γ^A, $A \in \mathfrak{L}$, find a function $f^A \in C(\mathfrak{S}, \mathbb{R})$ such that $\gamma^A = \xi(f^A)$. Then (16) implies that $[\xi(f^A), \xi(f^B)] = \xi(f^{[AB]})$ and hence, by (15),

$$\xi(f^{[AB]}) = \xi(\{f^A, f^B\}) \tag{17}$$

Since $f \rightarrow \xi(f)$ has kernel $\mathbb{R}$, this implies that

$$f^{[AB]} = \{f^A, f^B\} + \mathfrak{z}(A,B) \tag{18}$$

where the "two-cocycle" $\mathfrak{z}(A,B) \in \mathbb{R}$ may be changed to some extent by adjusting the arbitrary real numbers associated with the choices of f^A.

Thus implementing these two steps gives a subalgebra of $C(\mathfrak{S}, \mathbb{R})$ that is isomorphic to $\mathfrak{L}$ or, if the two-cocycle $\mathfrak{z}$ cannot be removed, to a central extension of $\mathfrak{L}$. By these means the problem has been recast into one of finding subalgebras of vector fields on $\mathcal{S}$ which lie in the image of the map $f \rightarrow \xi(f)$. If $\mathcal{S}$ is simply connected (we shall consider below an example where $\pi_1(\mathcal{S}) \neq 0$), such vector fields are best found by looking for a group $\mathcal{G}$ that acts on $\mathcal{S}$ as a group of canonical transformations (ie. it preserves the symplectic two-form). Each element A in the Lie algebra $\mathfrak{L}(\mathcal{G})$ of such a group generates a vector field γ^A on $\mathcal{S}$ which lies in the image of the map $f \rightarrow \xi(f)$ and which satisfies $[\gamma^A, \gamma^B] = \gamma^{[AB]}$ for all $A, B \in \mathfrak{L}(\mathcal{G})$. Our ultimate intention is that $\mathcal{G}$ shall play the same role for $\mathcal{S}$ as does the canonical Heisenberg-Weyl group for the special case $\mathcal{S} = \mathbb{R}^{2n}$.

(iii) There are some rather uninteresting ways of following the procedure outlined so far. For example, any complete vector field on $\mathcal{S}$ corresponds to the action of a one-parameter abelian group $\mathcal{G}$. Such trivial examples are excluded by imposing the requirement that the set of functions $\{f^A \mid A \in \mathfrak{L}(\mathcal{G})\}$ should be large enough to generate all elements in the classical ring $C(\mathcal{S}, \mathbb{R})$. It is relatively straightforward to show [6] that this is essentially equivalent to requiring that the action of $\mathcal{G}$ on $\mathcal{S}$ should be *transitive*.

(iv) The final restriction on $\mathcal{G}$ is that it should not lead to a Groenwold-Van Hove phenomenon in which there exists some proper subalgebra $\mathcal{D}'$ of $\{f^A \mid A \in \mathfrak{L}(\mathcal{G})\}$ which also generates the ring $C(\mathcal{S}, \mathbb{R})$ and which is represented reducibly in some irreducible representation of $\mathfrak{L}(\mathcal{G})$. The simplest way of satisfying this requirement is to make sure that $\mathcal{G}$ has no proper subgroups whose action on $\mathcal{S}$ is also transitive. Thus, given any transitively acting group $\mathcal{G}$, one should select the smallest subgroups whose action is still transitive. Each such group is a possible choice as the analogue of the canonical Heisenberg-Weyl group of $\mathcal{S} = \mathbb{R}^{2n}$.

2.3 Quantum Theory on S^1

A useful example for illustrating these ideas is the quantization of the system whose classical configuration space $\mathcal{Q}$ is the circle S^1 [1,6]. The state space $\mathcal{S}$ is the cotangent bundle $T^*S^1 \approx S^1 \times \mathbb{R}$ and the obvious choice for the canonical group $\mathcal{G}$ is $\mathbb{R} \times SO2$ which acts transitively and symplectically on T^*S^1 by

$$\tau_{(r,\varphi')}(\varphi, p) := ((\varphi + \varphi') \bmod 2\pi,\ p - r) \tag{19}$$

where $(r, \varphi) \in \mathbb{R} \times SO2$ and $(\varphi, p) \in S^1 \times \mathbb{R} \approx T^*S^1$. The vector fields corresponding to the two generators T, J are $\gamma^T = \partial/\partial p$ and $\gamma^J = -\partial/\partial\varphi$ respectively and it easy to see that $\gamma^J = \xi(f^J)$ where $f^J(\varphi, p) := p$. One is tempted to write $\gamma^T = \xi(f^T)$ where $f^T(\varphi, p) := \varphi$, but this is not allowed since f^T is not a *continuous* function on S^1.

This means that the vector field γ^T is not in the image of the map $f \to \xi(f)$ and has arisen because $\pi_1(S^1)$ is non-trivial. Basically, this is why it is not possible to use the conventional canonical commutation relations (6) for quantum theory on a circle.

The heart of the problem is that the angle φ is not a globally-defined smooth function on S^1 ; in fact, the smallest number of smooth functions needed to specify the position of a point on a circle is *two*, with the natural choice being the pair $(\cos\varphi, \sin\varphi)$. This suggests that perhaps it is these functions, rather than φ itself, that should be used to construct a group action. Specifically, we define (cf. (19))

$$\tau_{(r_1, r_2, \varphi')}(\varphi, p) := ((\varphi + \varphi') \bmod 2\pi,\ p + r_1 \sin(\varphi + \varphi') - r_2 \cos(\varphi + \varphi')) \quad (20)$$

It is straightforward to show that this does indeed define a transitive and symplectic action of a group $\mathcal{G}$ on $\mathcal{S} = T^*S^1$ and that $\mathcal{G}$ is isomorphic to the Euclidean group $E_2 \approx \mathbb{R}^2 \circledS SO2$. The vector fields corresponding to the three generators $(u,v;J)$ of E_2 are

$$\gamma^u = -\sin\varphi \frac{\partial}{\partial p}, \qquad \gamma^v = \cos\varphi \frac{\partial}{\partial p}, \qquad \gamma^J = -\frac{\partial}{\partial \varphi} \quad (21)$$

and these satisfy the commutation relations

$$[\gamma^J, \gamma^u] = \gamma^v \qquad [\gamma^J, \gamma^v] = -\gamma^u, \qquad [\gamma^u, \gamma^v] = 0\,. \quad (22)$$

Unlike γ^T above, both γ^u and γ^v lie in the image of the map $f \to \xi(f)$ with $\gamma^u = \xi(f^u)$, $f^u(\varphi,p) := \cos\varphi$, and $\gamma^v = \xi(f^v)$, $f^v(\varphi,p) := \sin\varphi$. Explicit calculation gives the basic Poisson brackets

$$\{f^J, f^u\} = f^v, \qquad \{f^J, f^v\} = -f^u, \qquad \{f^u, f^v\} = 0 \quad (23)$$

which show in particular that there is no two-cocycle. Thus it is possible to choose the group E_2 itself as the canonical group for quantum theory on a circle and the operator equivalent of the classical equations (23) is

$$[\hat{J}, \hat{u}] = i\hat{v}, \qquad [\hat{J}, \hat{v}] = -i\hat{u}, \qquad [\hat{u}, \hat{v}] = 0 \quad (24)$$

The group E_2 is a regular semi-direct product and hence all of its irreducible representations can be obtained from induced representation theory. This shows that all such representations are realised on the Hilbert space $L^2(S^1, d\varphi)$ and are of the form

$$(\hat{J}\psi)(\varphi) = -i\frac{d\psi}{d\varphi}(\varphi), \quad (\hat{u}\psi)(\varphi) = R\cos\varphi\, \psi(\varphi), \quad (\hat{v}\psi)(\varphi) = R\sin\varphi\, \psi(\varphi) \quad (25)$$

where R is some positive real number. Note that:

(i) The operator $\hat{u}^2 + \hat{v}^2 = R^2\hat{\mathbf{1}}$ is the Casimir operator for the group E_2 and representations with different values of R are unitarily inequivalent. The statement that this is a multiple of the unit operator $\hat{\mathbf{1}}$ can be regarded as the quantum equivalent of the classical identity on trigonometrical functions, $(\cos\varphi)^2 + (\sin\varphi)^2 = 1$. A more detailed analysis [6] shows that the parameter R is related to Planck's constant.

(ii) The spectrum of $\hat{J}$ in (25) is always the integers, irrespective of the value of R. Thus the incorrect spectral property associated with the Heisenberg commutation relations (6) has been remedied.

(iii) The use of E_2 also solves the problem associated with the uncertainty relation and the bounded nature of the angular variable. For example, if $\Delta_\psi J$ and $\Delta_\psi u$ are the dispersions for the observables J and u in the quantum state ψ, then the generalized uncertainty relation (applicable to an arbitary pair of non-commuting operators) gives

$$\Delta_\psi J\, \Delta_\psi u \geq \tfrac{1}{2}|\langle\psi, [\hat{J},\hat{u}]\psi\rangle| = \tfrac{1}{2}|\langle\psi, \hat{v}\psi\rangle| \quad (26)$$

There is no longer any contradiction with the bounded nature of u since a state that makes $\Delta_\psi J$ small will do the same for the expectation value of v and (26) will still be satisfied.

Another interesting feature of this group-theoretic approach to quantum theory on a circle is the natural way in which "ϑ-structure" enters. The key observation is that if a Lie group $\mathcal{G}$ acts transitively and symplectically on a classical state space $\mathcal{S}$, then so does its universal covering group $\mathcal{G}$ (simply let the kernel of the homomorphism $\tilde{\mathcal{G}} \to \mathcal{G}$ act trivially) and it induces the same Lie algebra as $\mathcal{G}$. Hence it is in order to extend the concept of quantization to include the irreducible representations of this covering group. In the case where $\mathcal{S} = T^*S^1$ we have $\pi_1(E_2) \approx \mathbf{Z}$ and the representations of the group $\tilde{E}_2 \approx \mathbb{R}^2 \circledS \mathbb{R}$ can again be studied with induced representation theory. This shows that the most general representation involves cross-sections of a complex line bundle over S^1 associated with the principal $\mathbf{Z}$-bundle $\mathbf{Z} \to \mathbb{R} \to S^1$ via a character $\chi_\vartheta(n) := e^{in\vartheta}$ of the group $\mathbf{Z}$ where $0 \leq \vartheta \leq 2\pi$.

An equivalent way of writing such a cross-section is as a function defined on $\mathbb{R}$ satisfying $\psi(x+2\pi) = e^{i\vartheta}\psi(x)$. The operator $\hat{J}$ becomes

$$(\hat{J}\psi)(x) = \left(-i\frac{d}{dx} + \vartheta\right)\psi(x) \tag{27}$$

and its spectrum is the integers shifted by the parameter ϑ.

3. QUANTUM THEORY ON LT[1]

3.1 The Canonical Group

We can now use this group-theoretic quantization scheme to tackle the bosonic string problems discussed in the Introduction. We are restricting our attention to the case where the target space is the one-torus T^1, and so the configuration space of the system is $\mathcal{Q} = C(S^1, T^1)$. This is a special case of the general non-linear σ-model $\mathcal{Q} = C(\Sigma, Y)$ and the first task is consider what is meant by a 'tangent' or cotangent' vector on an infinite-dimensional space of this type. Of the several possibilities, the simplest for our purposes is the "L^2-dual" in which the tangent and cotangent spaces to a map $X \in C(\Sigma, Y)$ are defined as [6,7]

$$T_X C(\Sigma, Y) := \{\, \tau : \Sigma \rightarrow TY \mid \tau(\sigma) \in T_{X(\sigma)}Y \,\} \tag{28}$$

$$T_X^* C(\Sigma, Y) := \{\, \lambda : \Sigma \rightarrow T^*Y \mid \lambda(\sigma) \in T^*_{X(\sigma)}Y \,\} \tag{29}$$

and the inner product between $\tau \in T_X C(\Sigma, Y)$ and $\lambda \in T_X^* C(\Sigma, Y)$ is

$$\langle \lambda, \tau \rangle := \int_\Sigma \langle \lambda(\sigma), \tau(\sigma) \rangle_{X(\sigma)}\, d\vartheta_\sigma \tag{30}$$

where $\langle\,,\rangle_y$ denotes the usual pairing between elements of T_y^*Y and T_yY at the point $y \in Y$. The integral in (30) is performed with respect to some background d-form on Σ (where $d = Dim\,\Sigma$); alternatively, the maps in $T_X^*C(\Sigma, Y)$ may be defined to lie in the cotangent vector *densities* on Y.

Several points should be kept in mind when using infinite-dimensional manifolds of this type. In particular:

(i) If T^*Y is a non-trivial bundle then, in general, $T^*C(\Sigma, Y)$ will not be isomorphic to $C(\Sigma, T^*Y)$.

(ii) If Y is a homogenous space G/H, there is a natural action of $C(\Sigma, G)$ on $C(\Sigma, G/H)$ but (unlike the action of G on G/H) this is *not* necessarily transitive. For example, the orbit passing though a point X in the connected component of $C(\Sigma, G/H)$ consists of those smooth functions from Σ to G/H that can be lifted in the fibre bundle $H \rightarrow G \rightarrow G/H$ to give a smooth function from Σ to G. [8] Thus there may be topological obstructions to writing the configuration space $\mathcal{Q} = C(\Sigma, G/H)$ of the G/H-valued non-linear σ-model as $C(\Sigma, G)/C(\Sigma, H)$.

Fortunately, neither of these potential difficulties arises in the present case of interest where $\mathcal{Q} = LT^1 = C(S^1, T^1)$. It is easy to show that $T^*T^1 \approx T^1 \times \mathbb{R}$ and hence that $T^*(LT^1) \approx L(T^*T^1) \approx LT^1 \times L\mathbb{R}$. In particular, this implies that $T^*(LT^1)$ has a countable number of components: $\pi_0(T^*(LT^1)) \approx \pi_0(LT^1) \times \pi_0(L\mathbb{R}) \approx \pi_0(LT^1) \approx \mathbb{Z}$. Problem (ii) is also absent since it is obvious that the action of the group $LSO2$ on LT^1 by pointwise multiplication is transitive. This suggests that a possible canonical group $\mathcal{G}$ for the system with the classical state space $T^*(LT^1)$ is the group of loops taking their values in the canonical group for the system with state space T^*T^1. We discussed the latter in Section 2.3 and showed that an appropriate choice was the Euclidean group $E_2 \approx \mathbb{R}^2 \circledS SO2$. Thus the indicated canonical group for the bosonic string propogating on the one-torus T^1 is the loop group LE_2.

It is straightforward to check that this group does indeed act transitively and symplectically on $T^*(LT^1)$ and it is therefore a *bona fide* possible choice for the canonical quantization of this particular system. The important task is to see whether it can be used to solve the problems discussed in the Introduction that arise in a more naive approach to the quantum theory that neglects the global structure of the classical configuration space.

The first point to note is that the semi-direct product structure $\mathbb{R}^2 \circledS SO2$ of E_2 carries across to LE_2 which is isomorphic to $L\mathbb{R}^2 \circledS LSO2$. The loop group $LSO2$ has a countable number of components and hence $LE_2 \approx (L\mathbb{R}^2 \circledS (LSO2)_0) \circledS \mathbb{Z}$ where $(LSO2)_0$ is the connected component of $LSO2$. Thus a unitary representation of the canonical group $\mathcal{G} \approx LE_2$ will yield (i) a self-adjoint representation of the generators of $L\mathbb{R}^2 \circledS (LSO2)_0$, and (ii) a unitary representation of the integers $\mathbb{Z}$ suitably intertwined with the rest of the group. The former is a "current-algebra" version of the commutation relations (24) of E_2 :

$$[\hat{J}(h), \hat{u}(\sigma)] = i\hat{v}(\sigma), \quad [\hat{J}(h), \hat{v}(\sigma)] = -i\hat{u}(\sigma), \quad [\hat{u}(\sigma), \hat{v}(\sigma)] = 0 \qquad (31)$$

where $\hat{u}(\sigma)$ and $\hat{v}(\sigma)$ are the generators of $L\mathbb{R}^2$ and $\hat{J}(h) = \frac{1}{2\pi}\int_0^{2\pi} h(\sigma)\,\hat{J}(\sigma)\,d\sigma$ is the (smeared) generator of $(LSO2)_0$. (The generators $\hat{u}(\sigma)$ and $\hat{v}(\sigma)$ can also be smeared if desired).

The unitary representation operators $\hat{U}_n$, $n \in \boldsymbol{Z}$, of the integers will obey

$$\hat{U}_n \hat{U}_m = \hat{U}_{n+m} \tag{32}$$

$$\hat{U}_n (\hat{u}(\sigma) - i\hat{v}(\sigma))\hat{U}_n^{-1} = e^{-in\sigma}(\hat{u}(\sigma) - i\hat{v}(\sigma)) \tag{33}$$

$$\hat{U}_n \hat{J}(h) \hat{U}_n^{-1} = \hat{J}(h) \tag{34}$$

where the phase factor $e^{-in\sigma}$ on the right hand side of (33) appears because the group $\boldsymbol{Z}$ is combined with the connected component $(LE_2)_0 \approx L\mathbb{R}^2 \circledS (LSO2)_0$ in the form of a semi-direct product.

Note that (31) and (33) imply that, at least formally, $\hat{u}^2(\sigma) + \hat{v}^2(\sigma)$ is a Casimir operator for the group LE_2. Hence, in any irreducible representation of LE_2 in which this operator is well-defined, we must have

$$\hat{u}^2(\sigma) + \hat{v}^2(\sigma) = \rho(\sigma)\hat{\mathbb{1}} \tag{35}$$

where $\hat{\mathbb{1}}$ is the unit operator and ρ is some positive and, apparantly, arbitrary function. However, we recall that in the light-cone gauge, the original invariance of the bosonic string theory under the diffeomorphism group of the domain circle S^1 is lost with the exception of the $SO2$ subgroup of rigid rotations. An elegant way of incorporating this residual gauge invariance into the quantum scheme would be to take the classical configuration space $(LT^1)/SO2$ and find a new canonical group that acts directly on the cotangent bundle $T^*((LT^1)/SO2)$. Unfortunately, I do not know whether this can be done and so we resort to the well-tried Dirac procedure in which the physical states are defined as vectors $|phys\rangle$ in the Hilbert space of the LE_2 representation which satisfy the additional property

$$\hat{D}_\sigma \,|\,phys\rangle = |\,phys\rangle \tag{36}$$

where the unitary operators $\hat{D}_\sigma$, $0 \leq \sigma \leq 2\pi$, carry the representation of $SO2 \subset Diff\, S^1$. The $SO2$ action on S^1 implies that

$$\hat{D}_\sigma \hat{u}(\sigma')\hat{D}_\sigma^{-1} = \hat{u}(\sigma' - \sigma), \quad \hat{D}_\sigma \hat{v}(\sigma')\hat{D}_\sigma^{-1} = \hat{v}(\sigma' - \sigma) \tag{37}$$

which is clearly incompatible with (35) unless ρ is a constant. The implication is that an irreducible representation of the canonical group $\mathfrak{G} = LE_2$ can be extended to a representation of the semidirect product $\mathfrak{G} \circledS SO2$ (with $SO2 \subset \mathrm{Diff}\, S^1$) only if the Casimir operator in (35) is equal to R^2 where R is some constant number.

3.2 The Left-Right Split

In an irreducible representation of LE_2 in which $\hat{u}^2(\sigma) + \hat{v}^2(\sigma) = R^2\hat{\mathbf{1}}$, the winding number operator can be defined as

$$\hat{\omega} := \frac{1}{2\pi R^2}\int_0^{2\pi} d\sigma\left(\hat{u}\frac{d\hat{v}}{d\sigma} - \hat{v}\frac{d\hat{u}}{d\sigma}\right) \tag{38}$$

and has the following commutation relations with the operators of the LE_2 representation :

$$[\hat{\omega}, \hat{u}(\sigma)] = [\hat{\omega}, \hat{v}(\sigma)] = [\hat{\omega}, \hat{J}(h)] = 0 \tag{39}$$

$$\hat{U}_n\hat{\omega}\hat{U}_n^{-1} = \hat{\omega} + n\hat{\mathbf{1}}. \tag{40}$$

If the spectrum of $\hat{\omega}$ is the integers (as one might anticipate) then the Hilbert space $\mathfrak{H}$ carrying the irreducible representation of the canonical group LE_2 will decompose into a direct sum

$$\mathfrak{H} \approx \bigoplus_{n\in\mathbf{Z}}\mathfrak{H}_n \qquad \text{where } \hat{\omega}\psi = n\psi \text{ for all } \psi\in\mathfrak{H}_n. \tag{41}$$

It follows from (39) that the connected component $(LE_2)_0$ maps each eigenspace $\mathfrak{H}_n$ of $\hat{\omega}$ into itself and hence the representation of $(LE_2)_0$ on $\mathfrak{H}$ is reducible. Irreducibility for the full group LE_2 is restored by (40) which shows that the set of operators $\hat{U}_n$, $n\in\mathbf{Z}$, act as intertwining operators with $\hat{U}_n$ mapping $\mathfrak{H}_m$ isomorphically onto $\mathfrak{H}_{m-n}$.

We can now confront the problems discussed in the Introduction. These arose from the bounded nature of the constant string mode q and from the attempt to regards $\hat{\omega}$ as part of a canonical pair of operators $(\hat{V}, \hat{\omega})$ satisfying the Heisenberg commutation relations (7). We recall that the difficulty with the latter lay in the discrete nature of the spectrum of $\hat{\omega}$ and the ill-defined nature of the expression (11) purporting to define $\hat{V}$ as a function of the canonical string variables. In terms of the discussion above, it is clear how we should proceed:

(i) The canonical group for the string theory should be the loop group LE_2 rather than the conventional Heisenberg-Weyl group of (2). Consequently, the constant mode will now correspond to an E_2 subgroup of LE_2 and, as discussed in Section 2.3 in the context of quantum theory on S^1, the uncertainty relations are now of the form (26) and are perfectly compatible with the boundedness of the classical observables.

(ii) The winding number operator $\hat{\omega}$ should be made into the $SO2$ generator of an E_2 canonical group, rather than the Heisenberg-Weyl group of (7). This will guarantee that its spectrum is discrete. It should also be possible to perform a proper left-right split by combining this new "topological" E_2 group with the E_2 group of the constant string modes.

Clearly, the critical step is to construct a new "topological" E_2 group containing $\hat{\omega}$ as the $SO2$ generator. The new translation generators $\hat{\boldsymbol{u}},\hat{\boldsymbol{v}}$ must be constructed from the existing variables in the theory (since the representation of the canonical group LE_2 is irreducible) and (39) shows that they cannot be built from the generators of the connected part $(LE_2)_0$. Thus the only possibility is to employ the unitary operators $\{\hat{U}_n \mid n\in\boldsymbol{Z}\}$ which represent the disconnected part of the canonical group. The crucial observation is that the unitarity relation $\hat{U}_n\hat{U}_n^\dagger=\hat{\mathbf{1}}$, coupled with the group law (32), implies that $\hat{U}_n^\dagger = \hat{U}_{-n}$, and hence the two operators $\hat{\boldsymbol{u}},\hat{\boldsymbol{v}}$ defined by

$$\hat{\boldsymbol{u}} := (\hat{U}_{-1}+\hat{U}_1)/2, \qquad \hat{\boldsymbol{v}} := (\hat{U}_{-1}-\hat{U}_1)/2i \tag{42}$$

are *hermitian*. Explicit calculation then gives

$$[\hat{\boldsymbol{u}},\hat{\boldsymbol{v}}] = 0 \tag{43}$$

as desired. Furthermore, the intertwining relations (40) show that $[\hat{\omega},\hat{U}_n\ ] = -n\hat{U}_n$, and hence

$$[\hat{\omega},\hat{\boldsymbol{u}}] = i\hat{\boldsymbol{v}}, \qquad [\hat{\omega},\hat{\boldsymbol{v}}] = -i\hat{\boldsymbol{u}}\ . \tag{44}$$

Equations (43) and (44) show that we have succeeded in embedding the winding number operator as part of a new E_2 group whose generators are formed from the elements of the canonical group LE_2. Note that the Casimir operator for this new E_2 takes on the special value

$$\hat{\boldsymbol{u}}^2+\hat{\boldsymbol{v}}^2 = (\hat{U}_{-2}+\hat{U}_2+2)/4 - (\hat{U}_{-2}+\hat{U}_2-2)/4 = \hat{\mathbf{1}} \tag{45}$$

The final step is to perform the "left-right" split by combining together the "topological" E_2 group and the E_2 associated with the constant string mode. The combination of the $SO2$ generators is assumed to be same as that proposed by Gross et al [2]:

$$\hat{J}_L := \tfrac{1}{2}(\hat{J}+\hat{\omega}), \qquad \hat{J}_R := \tfrac{1}{2}(\hat{J}-\hat{\omega}) \tag{46}$$

where $\hat{J}$ is the $SO2$ generator of the constant mode E_2. The critical step is to construct the left- and right-versions of the translation generators of the respective E_2 groups. We define

$$\hat{u}_L - i\hat{v}_L := (\hat{u} - i\hat{v})(\hat{\boldsymbol{u}} - i\hat{\boldsymbol{v}}), \quad \hat{u}_R - i\hat{v}_R := (\hat{u} + i\hat{v})(\hat{\boldsymbol{u}} - i\hat{\boldsymbol{v}}) \tag{47}$$

where $\hat{u}, \hat{v}$ are the generators of the translation ($\mathbb{R}^2$) subgroup of the constant mode E_2 group. With some calculation, it can be shown that the sets of operators $(\hat{u}_L, \hat{v}_L, \hat{J}_L)$ and $(\hat{u}_R, \hat{v}_R, \hat{J}_R)$ generate two, independent, E_2 Lie algebras. The non-constant modes can be decomposed more or less as in (3), and thus we have succeeded in giving a rigourous and well-defined left-right split for the T^1-valued bosonic string.

4. CONCLUSIONS

We have seen how topological problems arising in the quantization of the toroidally compactified string can be surmounted with the aid of a global, group-theoretical, approach to quantization. The discussion can be extended readily to include the case where the target space is a general torus T^n ; the details are given in [1]. Note that the discussion in Section 2.3 of quantum theory on a circle showed that the covering group $\tilde{E}_2$ of E_2 was also a candidate for the canonical group for this system and that the origin of the ϑ-angle could be understood in this way. This applies in particular to the constant mode of the string and, in turn, suggests that there may also be a ϑ-angle associated with the E_2 canonical group for the winding number. This gives the rather peculiar result that the spectrum of this operator may not be the integers, but rather the integers shifted by a ϑ-parameter. *A priori*, there is no reason why the ϑ-angles for the two groups should be related and (46) shows that there will be an asymmetry in the spectra of the left- and right-operators $\hat{J}_L$ and $\hat{J}_R$ respectively.

Of the various ways in which the work above might be extended, one of the most interesting is to the case where the target space is something other than a torus. String compactifications on orbifolds might well be treatable with these methods but the case where the target manifold is a Lie group G is more problematical. The quantization of a system whose configuration space is G can certainly be handled using the group theoretic scheme. However, a difficulty arises in implementing the left-right split since, in general, there is no obvious way in which the "winding number" in $\pi_0(LG) \approx \pi_1(G)$ can be associated with a generator of a second copy of the canonical group for the system with $\mathcal{Q} = G$. It appears that the case $G = T^n$ is rather special in this respect.

We have concentrated on the constant- and winding-number modes of the string and have handled the topological problems using the appropriate E_2 canonical groups. However, the theory also suggests that the correct canonical group for the full string theory is the loop group LE_2 and this has implications in its own right. In particular, it is not the case that every representation of this group can be related to a representation of the Weyl-Heisenberg group normally used to describe the quantization of the infinite set of modes that remain after removing the constant- and winding-number degrees of freedom. This suggests that there may be some intriguing and unexpected effects in the non-perturbative quantization of the string theory. This, and the zero-mode results, could have interesting implications for some of the recent attempts to construct string field theories for toroidally compactified strings [9].

References

1. Isham, C.J. and Linden, N. To appear *Class. Quan. Grav.*

2. Gross, D.J., Harvey, J.A., Martinec, E. and Rohm, M. *Phys. Rev. Letts.* **54**, 502 (1985); *Nucl. Phys.* **B256**, 253 (1985).

3. Ezawa, Z.F., Nakamura, S. and Tezuka, A. Toyko University preprint TU/86/298.

4. Groenwold, H.J., *Physica* **12**, 405 (1946).

5. Van Hove, L.V., *Acad. Roy. Belg. Bull. CL. Sci.*, **37**, 610 (1951).

6. Isham, C.J., in *Relativity, Groups and Topology II. Proceedings, 1983 Les Houches Summer School*, eds. B.S. DeWitt and R. Stora. North-Holland, Amsterdam (1984).

7. Fischer, A.E. and Marsden, J.E. in *General Relativity - An Einstein Centenary Survey*, eds. S.W. Hawking and W. Israel. Cambridge University Press, Cambridge (1979).

8. Percacci, R. *Geometry of Nonlinear Field Theories*, World Scientific Press, Singapore (1986).

9. Bresslov, P., Restuccia, A. and Taylor, J.G., to appear in *Journal of Modern Physics A.*

(Oriented) Smooth Strings on Genus-One Surface

K.S. VISWANATHAN and ZHOU XIAOAN

Department of Physics
Simon Fraser University
Burnaby, B.C., CANADA V5A 1S6

ABSTRACT

The methods of Polchinski, and Burgess and Morris for evaluating bosonic string path integral are used and extended to smooth strings on genus-one surfaces. This extension is nontrivial since the smooth string action contains higher derivative terms (the extrinsic curvature term). An additional (on-shell) invariance, the H-invariance, is shown to be possessed by the smooth string action, which removes the longitudinal modes of the strings. The results are exciting in that the smooth string theory can be made tachyon free. It is also graviton free and the static potential for open smooth strings in dimension four has the preferred coefficient for the 1/R term for large R.

1. INTRODUCTION

Polyakov[1] has suggested that for a realistic string model of hadrons one must add to the Nambu-Goto action a term that depends on the extrinsic curvature of the world sheet. Extrinsic curvature term has been known in the investigations on membranes in fluids by

Helfrich[2] and others.[3] Such a term suppresses those string configurations which are sharply kinked, favoring those that are smooth. This theory has become a focus of interest since it shares with QCD not only the property of asymptotic freedom[1],[2],[3] (for the extrinsic coupling α_0), but also the property of infrared confinement.[4]

Since Polyakov's work, there have been two main directions in understanding the physics of smooth strings. These are: (a) to calculate the effect of the curvature term on the static quark-antiquark potential.[4] This involves studying open strings with quark and antiquark at end points. These calculations extend the results obtained for the Nambu-Goto action[5]; and (b) to study the renormalization group behavior of the extrinsic coupling α_0.[6] In addition to these studies, a number of authors[7] have studied classical solutions to this action. In studies quoted above involving loop corrections, their authors have considered only flat two-dimensional topology. In this article we undertake the study of the partition function for smooth strings on genus-one surfaces. Our calculations follow that of Polchinski[8] for closed smooth strings and that of Burgess and Morris[9] for open smooth strings with some important modifications which will be specified in the sequel. Particularly, as pointed out by Kleinert in Refs. (2) and (3), the theory of strings with extrinsic curvature may contain additional negative norm states due to the higher derivative terms in the action. If this happens then it will be difficult to interpret the quantum theory of smooth strings. This argument is based on our experience with higher derivative gravity theory.[10] To solve this problem, a key observation made here is the invariance of the extrinsic curvature term in the action (in the second-order form) under normal variations of the world sheet $X_\mu(\sigma)$ in the direction of the mean curvature vector provided that the Riemann two-surfaces of genus-p of the world sheet are H-stationary.[11] H-invariance of the action introduces an additional Faddeev-Popov (F-P) determinant in the path integral which removes the additional longitudinal modes of smooth strings.

Following Polyakov,[1] the first-order formalism is used in the paper. The Euclidean path integral over string world sheet is expressed as an integral over Teichmüller parameter τ. From the path integral we calculate the free energies of a gas of closed and open smooth strings. This is compared with the free energy of a collection of non-interacting particles. We find that there exists a critical coupling $\alpha_0 T L_1{}^2$ for closed smooth strings or open ones, where L_1 is the length of the strings, above which the mass spectrum does not contain a tachyon. The fact that smooth strings may be tachyon free has been conjectured by Braaten, Pisarski, and Tze[4] (see also, Pisarski[12]). Another interesting property that we obtain is that there are no zero mass excitations of spin-2. This agrees with the fact that the smooth string theory does not have modular invariance.[13] Curtright, Ghandour, Thorn and Zachos[7] have previously conjectured on the basis of the shape of the classical Regge trajectory that smooth strings may have no massless excitations.

The path integral for smooth strings is not modular invariant. This implies that the region of integration over the Teichmüller parameter τ is not restricted to the fundamental domain[14] $-\frac{1}{2} < \tau_1 < \frac{1}{2}$, $\mathrm{Im}\ \tau_2 > 0$; $|\tau| > 1$. Instead the region of integration is $\mathrm{Im}\ \tau_2 > 0$, $-\frac{1}{2} < \tau_1 < \frac{1}{2}$. The restriction $-\frac{1}{2} < \tau_1 < \frac{1}{2}$ arises from the fact that the path integral is still invariant under translations $\tau \to \tau + 1$.

For open smooth strings, the static quark-antiquark potential is also extracted from the path integral. The surprising result is that the coefficient of the Lüscher term (1/R term) in large R expansion of the potential is twice as large as the former results[4],[5] calculated on a flat topology.

The present article is organized as follows. Section II shows that the smooth string action has an additional (on-shell) invariance, the H-invariance. Section III evaluates the path integrals, the free energy, and the mass spectrum for both closed and open smooth strings. In Section IV the static quark-antiquark potential is extracted from the path integral of open smooth strings.

II. H-INVARIANCE

The action for Polyakov's model of smooth strings may be written in the following form[1] (in the first-order formalism)

$$S = \mu \int \sqrt{g}\, d^2\sigma + \frac{1}{2\alpha_0} \int d^2\sigma \left[\sqrt{g}\, (\Delta X^\mu)^2 + \lambda^{ab}(\partial_a X^\mu \partial_b X_\mu - \rho \hat{g}_{ab}(\tau))\right] \quad (2\text{-}1)$$

where the conformal gauge $g_{ab} = \rho\hat{g}_{ab}(\tau)$ has been used. $\hat{g}_{ab}(\tau)$ depends on the Teichmüller parameter τ. The string coordinate $X^\mu(\sigma)$ spans a d-dimensional space-time. ΔX^μ is given by

$$\Delta X^\mu = \frac{1}{\sqrt{g}}\, \partial_a(\sqrt{g}\, g^{ab}\, \partial_b X^\mu) \quad . \quad (2\text{-}2)$$

The first term in (2-1) is the usual Nambu-Goto action and can be also considered as a cosmological term in the first-order formalism.[8] The second term in (2-1) is the extrinsic curvature term. In the third term λ^{ab} is a Lagrange multiplier field which fixes the metric to be the induced metric, namely $g_{ab} = \partial_a X^\mu \partial_b X_\mu$ on the surfaces. The extrinsic coupling α_0 is dimensionless and is asymptotically free.[1],[2],[3] The action (2-1) has reparametrization invariance. Both the cosmological and the extrinsic curvature terms are not invariant under conformal transformations

$$g_{ab} \to \rho(\sigma) g_{ab} \quad , \quad X^\mu \to X^\mu \quad (2\text{-}3)$$

where $\rho(\sigma)$ is an arbitrary function of σ. Varying the action (2-1) with respect to λ^{ab}, we find $g_{ab} = \partial_a X \partial_b X$. If we further set $\mu = 0$ at the classical level as in Ref. (8), the action (2-1) is on-shell equivalent to the extrinsic curvature term in the second-order form. It has been shown[1],[15] that Polyakov's action (the extrinsic curvature term) classically correspond to a particular topological sector of the $G_{2,n}$ σ-model.

We now show that the extrinsic curvature term possess an additional invariance, called H-invariance, on stationary world sheet surfaces. Because of the length limitation of the article, we can only give a brief discussion here and refer the reader to Refs. (11) and (16) for details.

An equivalent form for the extrinsic curvature term is given by (up to a total derivative)

$$\frac{1}{2\alpha_0}\int_M d^2\sigma\,\sqrt{g}\,(\Delta X)^2 = \frac{1}{2\alpha_0}\int_M d^2\sigma\,\sqrt{g}\,h^{ra}_{a}\,h^{rb}_{b} \qquad (2\text{-}4)$$

$$(a,b = 1,2;\ r = 3,\ldots,d)$$

where $h_{ab}^{\ r}$'s are components of the second fundamental form defined by

$$\partial_a\partial_b X = \Gamma_{ab}^{\ c}\,\partial_c X + h_{ab}^{\ r}\,e_r \qquad (2\text{-}5a)$$

$$e_r\cdot e_s = \delta_{rs} \qquad (2\text{-}5b)$$

$$e_r\cdot\partial_a X = 0 \qquad (2\text{-}5c)$$

e_r $(r = 3,\ldots,d)$ are d-2 orthonormal normal vectors to the world sheet $X(\sigma)$ and $\Gamma_{ab}^{\ c}$ are the connection coefficients determined by the induced metric $g_{ab} = \partial_a X \partial_b X$. The Weingarten formula takes the form

$$\partial_a e_r = -h^{rb}_{a}\,\partial_b X + \ell^{s}_{ra}\,e_s\ . \qquad (2\text{-}6)$$

Let M be compact n-dimensional submanifold (with or without boundary) of a Euclidean R^d (n=2 for the string theory). Let X denote the position vector of M in R^d. Then $X = X(\sigma_1,\ldots,\sigma_n)$, where $\sigma_1,\ldots,\sigma_n$ are local coordinates of M. Consider the following variation of X:

$$X'(\sigma_1,\ldots,\sigma_n,t) = X(\sigma_1,\ldots,\sigma_n) + t\Phi(\sigma_1,\ldots,\sigma_n)\xi(\sigma_1,\ldots,\sigma_n), \qquad (2\text{-}7)$$

where $\Phi(\sigma_1,\ldots,\sigma_n)$ is a differentiable function and t an infinitesmal parameter. $\xi(\sigma_1,\ldots,\sigma_n)$ is a unit normal vector field of M in R^d. We require that both Φ and $\partial_a\phi$ vanish identically on the boundary ∂M of M. For a submanifold without boundary, there are no requirements on Φ. The variation in (2-7) is called a normal variation and some of its consequences are discussed by Kleinert.[3)]

If ξ is a unit normal vector field which is in the direction of the mean curvature vector H, then the variation (2-7) is called an H-variation of M in R^d. The mean curvature vector is defined by

$$H = h^r_{ab}\, g^{ab} e_r \,, \tag{2-8}$$

where e_r's are defined in equations (2-5a-c). The submanifold M is called stationary if

$$\delta \int_M \alpha^n dv = 0 \tag{2-9}$$

for all normal variations of M. In (2-9)

$$\delta = \left.\frac{\partial}{\partial t}\right|_{t=0} , \tag{2-10}$$

α is the mean curvature

$$\alpha = \frac{1}{n} (H\cdot H)^{1/2} = \frac{1}{n}\left((h^{ra}_{a})^2\right)^{1/2} \tag{2-11}$$

and

$$dv = \sqrt{g}\, d\sigma_1 \wedge d\sigma_2 \wedge \ldots \wedge d\sigma_n \,. \tag{2-12}$$

M is called H-stationary if (2-9) is true for all H-variations of M. We sketch here the proof but leave out the details (see Ref. 16). Now $\partial_1 X, \ldots, \partial_n X, e_{n+1}, \ldots, e_d$ define the natural orientation of R^d. Let us choose $e_{n+1} = \xi$. From (2-7) and (2-10) it follows that

$$\delta X = \Phi e_{n+1} \,. \tag{2-13}$$

The change in the induced metric is

$$\delta g_{ab} = -2\Phi h^{n+1} \tag{2-14a}$$

$$\delta g^{ab} = 2\Phi h^{n+1\, ab} \tag{2-14b}$$

$$\delta\sqrt{g} = -\Phi(\mathrm{tr}\, h^{n+1})\sqrt{g} \tag{2-14c}$$

Using (2-13), (2-14) and the formulas of Gauss (Eq. (2-5a) and Weingarten (Eq. (6)), a long calculation yields

$$\delta \int_M \alpha^n dv = \int_M \Phi[\Delta\alpha^{n-1} - \alpha^{n-1}\ell^2 - n\alpha^{n+1} + \|A_{n+1}\|^2]dv \qquad (2\text{-}15)$$

where

$$\|A_{n+1}\|^2 = (h^{n+1}_a{}^b)(h^{n+1}_b{}^a) \qquad (2\text{-}16)$$

$$\ell^2 = \ell^r_{n+1,t}\, \ell^{rt}_{n+1} , \qquad (2\text{-}17)$$

and

$$\ell^r_{n+1,t} = e_r \partial_t e_{n+1} = -\ell^{n+1}_{r,t} . \qquad (2\text{-}18)$$

Thus the action for extrinsic curvature is H-invariant if and only if the mean curvature α satisfies (for $n = 2$).

$$\Delta\alpha - \alpha(\ell^2 + 2\alpha^2 - \|A_3\|^2) = 0 , \qquad (2\text{-}19)$$

where A_3 denotes the Weingarten map with respect to the unit vector in the direction of H. It is shown in Ref. (17) that there exist H-invariant immersions of M of arbitrary genus-p in R^d. Chen[16] shows, for example, that a 2-torus in R^3 is H-stationary if and only if $a = \sqrt{2}\, b$, where a and b are radii of the circles generating the torus.

Since the action (2-1) is on-shell H-invariant (we have set $\mu = 0$ at the classical level), in evaluating the string path integral, it is necessary to use the F-P procedure to determine the Haar measure with respect to H-variations in addition to the usual F-P determinants associated with the reparametrization invariance. The "H-gauge" fixing condition is taken in the form

$$\mathrm{tr}\, h^{(3)} \equiv h^{(3)}_{ab}\, g^{ab} = 0 . \qquad (2\text{-}20)$$

With this "gauge fixing" choice, the F-P determinant is readily seen to be

$$\Delta_{F-P} = \det(-\Delta - \|A_3\|^2 + \ell^2) \tag{2-21}$$

where $\|A_3\|^2$ depends on X through the fourth order derivative (2-21), while ℓ^2 (see Eq. (2-17)) only depends on the local coordinates σ. In the next section, we shall show that the F-P determinant in Eq. (2-21) arising from H-invariance plays a crucial role in cancelling the longitudinal modes in the mass spectrum of smooth strings if we replace ℓ^2 by a suitable constant and ignore the high derivative term $\|A_3\|^2$. In Ref. (10), we showed that the above "H-gauge" fixing procedure introduces a pair of ghost and antighost into the theory which are scalar Grassmannian fields and contribute to the conformal anomaly.

III. PATH INTEGRAL ON GENUS-ONE SURFACES[18),19)]

Following Polyakov,[1] we split all fields into slow and fast parts and integrate out the fast components

$$\begin{aligned} x &= x_0 + x_1 \\ \rho &= \rho_0 + \rho_1 \\ \lambda^{ab} &= \lambda_0^{ab} + \lambda_1^{ab} \end{aligned} \tag{3-1}$$

where the wave vectors of fast quantities lie between $\tilde{\Lambda}$ and Λ. In the one-loop approximation, one obtains the renormalized action in the form (for details, see Ref. 1)

$$S = \mu^2 \int d^2\sigma \sqrt{g_0} + \frac{1}{2\alpha_0} \int d^2\sigma \left\{ (1 - \frac{\alpha_0}{2\pi} \ln \frac{\Lambda}{\tilde{\Lambda}}) \sqrt{g_0}\, (\Delta x_0)^2 + \lambda_0^{ab} \left[\partial_a x_0 \partial_b x_0 - \rho_0 \hat{g}_{ab}(\tau)(1 - \frac{D-2}{4\pi} \alpha_0 \ln \frac{\Lambda}{\tilde{\Lambda}})\right] \right\} \tag{3-2}$$

After renormalization of the x field

$$x_0 \to z^{1/2} x_0 \qquad \lambda_0^{ab} \to z^{-1} \lambda_0^{ab}$$
$$z = 1 - \frac{D-2}{4\pi} \alpha_0 \ln \frac{\Lambda}{\tilde{\Lambda}} \tag{3-3}$$

one obtains

$$S = \mu \int d^2\sigma \sqrt{g_0} + \frac{1}{2\tilde{\alpha}_0} \int d^2\sigma \left[\sqrt{g_0}\, (\Delta x_0)^2 + \lambda_0^{ab} (\partial_a x_0 \partial_b x_0 - \rho_0 \hat{g}_{ab}(\tau))\right] \tag{3-4}$$

where

$$\frac{1}{\tilde{\alpha}_0} = \frac{1}{\alpha_0} - \frac{D}{2} \frac{1}{2\pi} \ln \frac{\Lambda}{\tilde{\Lambda}} , \quad g_0^{ab} = \rho_0 \hat{g}^{ab}(\tau) . \tag{3-5}$$

The quantum theory of smooth strings involves evaluating the path integral over the space of all metrics of a given topology and over all configurations x_0^μ. In our evaluation of the path integral we will ignore the effect of conformal fields that may fluctuate on the world sheet. We shall choose the space of metrics on a torus to be conformal to the metric with $\rho_0 = 1$. The metric can then be put in the form[8),9)]

$$\hat{g}_{ab} = \begin{bmatrix} 1 & \tau_1 \\ \tau_1 & \tau_1^2 + \tau_2^2 \end{bmatrix} \quad \text{for a torus} \tag{3-6a}$$

and

$$\hat{g}_{ab} = \begin{bmatrix} 1 & 0 \\ 0 & \tau_2^2 \end{bmatrix} \quad \text{for a cylinder} \tag{3-6b}$$

where $\tau \equiv (\tau_1, \tau_2)$ is a complex Teichmüller parameter that characterizes the torus. Since the smooth string theory is not conformally invariant we introduce length scales (L_1, L_2) explicitly. We proceed to treat λ_0^{ab} as a background field[6] in the form

$$\lambda_0^{ab} = T\tilde{\alpha}_0 \sqrt{g_0}\, \hat{g}_{ab}(\tau) \tag{3-7}$$

where T is the bare string tension, and choose $\mu = T$ at the quantum level. The path integral over torus can be written as

$$W_{torus} = \int \frac{[dg_{ab}][dx^{\mu}]}{V_{G.C.}} \exp(-S_{S.S.}) \, . \tag{3-8}$$

We work in Euclidean space. $V_{G.C.}$ is the volume of the general coordinate group. Then a straightforward calculation yields the following[8)]

$$W_M = T^{d/2}\left(\prod_{\mu=1}^{d} L^{\mu}\right) \int d^2\tau/2 \; J(\tau)\left\{(2\pi L_1 L_2/\int\sqrt{g}d^2\sigma)\det{}'(-\Delta)\det{}'(-\frac{\Delta}{\tilde{\alpha}_0 T}+1)\right\}^{-d/2} \tag{3-9}$$

where L^{μ} ($\mu = 1,2,\ldots d$) denote the lengths of the box in d-dimension and $J(\tau)$ is the Jacobian defined by

$$[dg] = (d\xi)' \, d^2\tau \; J(\tau) \, . \tag{3-10}$$

The Jacobian can be written as

$$J(\tau) = (\det f_{ij})^{1/2}\left(\frac{\det Q_{AB}}{V_T{}^c}\right)^{-1/2} (\det{}' P_1 P_1)^{1/2} \tag{3-11}$$

where

$$V_M = \frac{1}{L_1 L_2} \int d^2\sigma \sqrt{g}$$

$$c = 2 \text{ for a torus}$$

$$= 1 \text{ for a cylinder}$$

$$(\det{}' P_1 P_1)^{1/2} = [\det{}'(-2\delta_c{}^d \, \Delta)]^{1/2}$$

$$= \frac{1}{2}\det{}' (-\Delta) \, . \tag{3-12}$$

The primes over determinants in (3-9) and (3-12) indicate that the zero modes are to be excluded. Furthermore,

$$f_{ij} = g^{ac} g^{bd} f_{ab,i} f_{cd,j}$$

$$f_{ab,i} = \frac{\partial g_{ab}}{\partial \tau_i} - \frac{1}{2} g_{ab} g_{cd} \frac{\partial g_{cd}}{\partial \tau_i} \tag{3-13}$$

and

$$Q_{AB} = \frac{1}{L_1 L_2} \int d^2\sigma \sqrt{g}\, g_{ab}\, \xi^a_A\, \xi^b_B \tag{3-14}$$

where ξ^a_A (A = 1,2,...c) represent c independent conformal killing vectors on the manifold M.

Now we take the F-P determinant (2-23) into account, ignore $\|A_3\|^2$ term and choose $\ell^2 = T\tilde{\alpha}_0$. The complete Jacobian takes the form

$$\begin{aligned} \tilde{J}(\tau) &= J(\tau)\cdot\det{}'(-\Delta + T\tilde{\alpha}_0) \\ &= J(\tau)\cdot\det{}'(-\frac{1}{T\tilde{\alpha}_0}\Delta + 1) \end{aligned} \tag{3-15}$$

where a constant factor $1/T\tilde{\alpha}_0$ has been dropped and $J(\tau)$ is determined by (3-11)-(3-14).

For a torus, we have

$$\begin{aligned} \sqrt{g} &= \tau_2 \\ V_T &= \frac{1}{L_1 L_2}\int d^2\sigma \sqrt{g} = \tau_2 \\ \det(f_{ij}) &= 4/\tau_2{}^4 \\ \det Q_{AB} &= V_T{}^c \det g_{ab} = \tau_2{}^4 . \end{aligned} \tag{3-16}$$

Therefore

$$\tilde{J}_T(\tau) = \frac{1}{\tau_2{}^3} \det{}'(-\Delta)\det{}'(-\frac{1}{\tilde{\alpha}_0 T}\Delta + 1) \quad . \tag{3-17}$$

The path integral then takes the form

$$W_{torus} = T^{d/2} \left(\prod_\mu L^\mu\right) \frac{d^2\tau}{4\pi\tau_2{}^2} (2\pi\tau_2)^{-(d/2 - 1)} \left\{ \det{}'(-\Delta) \times \det{}'\left(-\frac{1}{\alpha_0 T}\Delta + 1\right) \right\}^{1-d/2} \tag{3-18}$$

Det'(-Δ) has been calculated in Ref. 8.

$$\det{}'(-\Delta) = \tau_2{}^2 e^{-\pi\tau_2/3} \left| f(e^{2\pi i\tau}) \right|^4 \tag{3-19a}$$

where

$$f(e^{2\pi i\tau}) = \prod_{n=1}^{\infty} (1 - e^{2\pi i n\tau}) . \tag{3-19b}$$

The determinant of $(-\frac{1}{\tilde{\alpha}_0 T}\Delta + 1)$ can be evaluated in a similar way[18]

$$\det{}'\left(-\frac{1}{\tilde{\alpha}_0 T}\Delta + 1\right) = \exp\left(\frac{\tau_2 a^2}{4\pi} + 2\pi\tau_2 I(a)\right) \left| f(e^{2\pi i W_+}) \right|^4 \times (1 - e^{-\tau_2 a})^2 \tag{3-20}$$

where

$$I(a) = 4 \int_0^\infty \frac{dy}{1 - e^{2\pi(y + a/2\pi)}} y^{1/2} (y + \frac{a}{\pi})^{1/2} \tag{3-21}$$

$$W_+ = n\tau_1 + i\tau_2\sqrt{n^2 + a^2/4\pi^2} \tag{3-22a}$$

and

$$a^2 = \tilde{\alpha}_0 T L_1{}^2 . \tag{3-22b}$$

I(a) has the following limiting values

$$I(a) = -\frac{1}{6} \quad \text{as} \quad a \to 0 \qquad\qquad = 0 \quad \text{as} \quad a \to \infty \tag{3-23}$$

Substituting (3-18) and (3-20) into (3-17), we find

$$W_{torus} = T^{d/2} (\prod_{\mu} L^{\mu}) \int \frac{d^2\tau}{4\pi\tau_2{}^2} (2\pi\tau_2)^{1-d/2} \left|f(e^{2\pi i\tau})\right|^{2(2-d)}$$

$$\times \left|f(e^{2\pi i W_+})\right|^{2(2-d)} (1 - e^{-\tau_2 a})^{2-d}$$

$$\times \exp\left\{(\frac{d}{2} - 1) \frac{\pi\tau_2}{3} \left(1 - \frac{3a^2}{4\pi^2} - 6I(a)\right)\right\} . \quad (3\text{-}24)$$

Equation (3-24) is invariant under the transformation $\tau \to \tau + 1$. This invariance requires that τ_1 is restricted to the region chosen to be $-\frac{1}{2} < \tau_1 < \frac{1}{2}$ and $\tau_2 > 0$. In the usual string theory the path integral is also invariant under $\tau \to -1/\tau$. This makes the path integral modular invariant where modular transformations are of the form

$$\tau \to \frac{\alpha\tau + \beta}{\gamma\tau + \delta}$$

α, β, γ, δ are integers and $\alpha\delta - \beta\gamma = 1$. This requires that τ be chosen to be in the fundamental domain defined by $-\frac{1}{2} < \tau_1 < \frac{1}{2}$, $\tau_2 > 0$ and $|\tau| > 1$. Extrinsic curvature term breaks modular invariance of the path integral. Note that (3-24) is not invariant under $\tau \to -1/\tau$. The free energy $F(\beta)$ for a gas of strings is given by

$$F(\beta) = -(\prod_{\mu} L^{\mu})^{-1} W_{connected} \cdot \quad (3\text{-}25)$$

Following Polchinski's procedure[8] we find that

$$F(\beta) = -T^{d/2} \int_0^{\infty} \frac{d\tau_2}{2\pi\tau_2{}^2} \int_{-1/2}^{1/2} d\tau_1 (2\pi\tau_2)^{1-d/2}$$

$$\times \left|f(e^{2\pi i\tau})\right|^{2(2-d)} \left|f(e^{2\pi i W_+})\right|^{2(2-d)}$$

$$\times (1 - e^{-\tau_2 a})^{2-d} \exp\left\{4\pi\tau_2 \left(1 - \frac{3a^2}{4\pi^2} - 6I(a)\right) \frac{d-2}{24}\right\}$$

$$\times \sum_{\gamma=1}^{\infty} e^{-\gamma^2\beta^2 T/2\tau_2} . \quad (3\text{-}26)$$

In order to understand the content of (3-26) we compare it with the free energy for a collection of free particles whose energy spectrum is $\omega_k = \sqrt{k^2 + m^2}$.

$$F(\beta,m^2) = \frac{1}{\beta}\int \frac{d^{d-1}k}{(2\pi)^{d-1}} \ln\left(1 - e^{-\beta\omega_k}\right)$$

$$= \int_0^\infty \frac{ds}{s} (2\pi s)^{-d/2} \sum_{\gamma=1}^{\infty} e^{-m^2 s/2 - \gamma^2\beta^2/2s} . \qquad (3\text{-}27)$$

In terms of occupation numbers of transverse oscillators $N^{(1)}_{n_i}$, $\tilde{N}^{(1)}_{n_i}$, $N^{(2)}_{\tilde{n}_i}$, $\tilde{N}^{(2)}_{\tilde{n}_i}$, and $N^{(3)}_i$, the spectrum is given by

$$m^2(a) = 4\pi T \left\{ \left(-2 + \frac{3a^2}{2\pi^2} + 12I(a)\right)\frac{d-2}{24} + \sum_{i=1}^{d-2}\sum_{n=1}^{\infty} n\left(N^{(1)}_{n_i} + \tilde{N}^{(1)}_{n_i}\right)\right.$$

$$\left. + \sum_{i=1}^{d-2}\sum_{n=1}^{\infty} \tilde{n}\left(N^{(2)}_{\tilde{n}_i} + N^{(2)}_{\tilde{n}_i}\right) + \frac{a}{2\pi}\sum_{i=1}^{d-2} N^{(3)}_i \right\} \qquad (3\text{-}28)$$

subject to the closed string constraints

$$\sum_{i=1}^{d-2}\sum_{n=1}^{\infty} n\left(N^{(1)}_{n_i} - \tilde{N}^{(1)}_{n_i}\right) = 0$$

$$\sum_{i=1}^{d-2}\sum_{n=1}^{\infty} n\left(N^{(2)}_{\tilde{n}_i} - \tilde{N}^{(2)}_{\tilde{n}_i}\right) = 0 \qquad (3\text{-}29)$$

where

$$\tilde{n} = \sqrt{n^2 + \frac{a^2}{4\pi^2}} . \qquad (3\text{-}30)$$

Summing (3-27) over oscillator spectrum (3-28) subject to constraints (3-29) yields exactly the free energy given in (3-26). The τ_1 integral enforces (3-29) while $\tau_2 = sT$. Thus (3-28) and (3-29) replace the usual spectrum for the Nambu-Goto theory. There are several interesting features to take note of.

From (3-28), we see that there are only transverse modes in the mass spectrum. The longitudinal modes of $N^{(1)}_{n_i}$ and $\tilde{N}^{(1)}_{n_i}$ have been cancelled by the factor det'$(-\Delta)$ in the Jacobian (19) while those of $N^{(2)}_{\tilde{n}_i}$, $N^{(2)}_{\tilde{n}_i}$ and $N^{(3)}_i$ by the factor det'$(-1/\tilde{\alpha}_0 T\ \Delta + 1)$ in (3-15) which comes from "H-gauge" fixing procedure.

In Eq. (3-28) the number operators $N^{(1)}_{n_i}$, $N^{(2)}_{\tilde{n}_i}$ represent independent right movers while $N^{(1)}_{\tilde{n}_i}$ and $\tilde{N}^{(2)}_{\tilde{n}_i}$ are the corresponding left movers. These are commuting operators. $N^{(3)}_i$ arises due to zero point fluctuations of the smooth string. The constraints (3-29) are the usual ones; namely, the number of left moving degrees of freedom coincide with those of the right moving degrees of freedom. Consider first the Tachyonic state. The $(\text{mass})^2$ for this state in the presence of extrinsic curvature is given by

$$m^2(a) = \left(\frac{d-2}{24}\right) 4\pi T \left(-2 + \frac{3a^2}{2\pi^2} + 12I(a)\right) . \qquad (3\text{-}31)$$

As $a \to 0$ and $I(a) \to -1/6$ and $m^2(a) = (d - 2/24)\ (-16\ \pi T)$ is just twice the mass square of the Tachyon of the Nambu-Goto theory. As $a \to 0$, the extrinsic curvature term dominates and in this case the theory has a tachyon. For non-vanishing a, clearly, there exists a critical value of a for which $m^2(a_c) = 0$. This implies that the smooth strings can be free of tachyons for $a \geq a_c$. It should be remarked here that our expression for the path integral and the free energy are not valid for $a^2 \to \infty$. The reason for this lies in our regularization scheme which is valid as long as a is finite.[19]

Turning now to the excited states, we note that there are two sets of oscillator states described by $N^{(1)}, \tilde{N}^{(1)}$ and $N^{(2)}$ and $\tilde{N}^{(2)}$. This is reminiscent of the situation in classical non-relativistic stiff strings.[20] In the non-relativistic case the equations of motion are fourth order in space and each normal mode, for a clamped string, is two-fold degenerate. According to (3-28) we see that for the quantum relativistic stiff string, this degeneracy is lifted as n and $\tilde{n}$ have different values. It can be seen from (3-28) that there are no zero mass spin 2 states in the excitation spectrum of the smooth strings. This agrees with the fact that smooth string theory does not have modular invariance.[13] Finally, we note that the operator $N_i^{(3)}$ denotes zero point excitation of smooth strings.

For a cylinder, we have[9]

$$\begin{aligned} \sqrt{g} &= \tau_2 \\ V_c &= \frac{1}{L_1 L_2} \int d^2\sigma\sqrt{g} = \tau_2/2 \\ \det f_{22} &= 2/\tau_2{}^2 \end{aligned} \tag{3-32}$$

and

$$\det Q_{AB} = \det Q_{22} = V_C^c \det g_{ab} = \tau_2^3/2 . \tag{3-33}$$

Combining all these pieces of information we can now write the path integral on the cylinder as

$$W_c = (T)^{d/2} (\Pi\, L^\mu) \int_0^\infty \frac{d\tau_2}{\tau_2^2} \sqrt{2} \left(\frac{\tau_2}{4\pi}\right)^{d/2} (\det{}'\ P_1^+ P_1)_C^{1/2}$$

$$\times \left[\det{}'\ (-\Delta) \det{}'\ (-\frac{1}{\tilde{\alpha}_0 T}\ \Delta + 1)\right]_C^{-\,d/2} \tag{3-34}$$

In (3-34) subscripts C's denote that these determinants are to be evaluated over spectrum of these operators defined on the cylinder. Further use of the results in Ref. 9 yields

$$(\det{}'\ -\Delta)_C = \tau_2\ (Z_T^{++}(\tau))^{1/2} \Big|_{\tau=i\tau_2} \tag{3-35}$$

where

$$Z_T^{++}(\tau) = \tau_2^2 \exp\left(-\frac{L_2\pi\tau_2}{3L_1}\right) \left|\left|\prod_{n=1}^{\infty}\right| \left[1 - \exp(2\pi i n\tau L_2/L_1)\right]\right|^4 \tag{3-36}$$

$\mathrm{Det}'(-\frac{1}{\tilde{\alpha}_0 T}\Delta + 1)_C$ is given by[19]

$$\left[\det{}'(-\frac{1}{\tilde{\alpha}_0 T}\ \Delta + 1)\right]_C$$

$$= (1 - e^{-\sqrt{\tilde{\alpha}_0 T}\ \tau_2 L_2}) \exp(\sqrt{\tilde{\alpha}_0 T}\ \tau_2 L_2) \left[Z_T^{++}(\tau, \tilde{\alpha}_0 T)\right]\Big|_{\tau = i\tau_2}^{1/2} \tag{3-37}$$

In (3-34) and (3-35) (++) denotes the periodic boundary conditions for $X^\mu(\sigma)$ in the two directions (σ^1, σ^2). $Z_T^{++}(\tau, \tilde{\alpha}_0 T)$ was evaluated in Ref. 19 (see (3-20)-(3-22)):

$$Z_T^{++}(\tau,\tilde{\alpha}_0 T) = \left| \prod_{n=1}^{\infty} (1 - e^{2\pi i \frac{L_2}{L_1} W_+}) \right|^4 \times (1 - e^{-\tau_2 \sqrt{\tilde{\alpha}_0 T} L_2})^2$$

$$\times \exp \{\frac{L_1 L_2 \tau_2 \tilde{\alpha}_0 T}{4\pi} + 2\pi \frac{L_2}{L_1} \tau_2 I(\tilde{\alpha}_0 T L_1^2)\} \tag{3-38}$$

where $I(\tilde{\alpha}_0 T L_1{}^2)$ and W_+ are given by (3-21) and (3-22) respectively

$(\mathrm{Det}'\ P_1^\dagger P_1)_C$ is given by[19]

$$(\det{}' P_1^\dagger P_1)_C = \frac{1}{2} \{Z_T^{++}(\tau)\ Z_T^{++}(\tau, \tilde{\alpha}_0 T)\}\Big|_{\tau = i\tau_2} \tag{3-39}$$

In Eqs. (3-35), (3-37) and (3-39), we note that the determinants of the Laplacian operators that occur in the theory for the cylinders are related to those for the torus not linearly but by the exponent of half. Our final expression for the path integral for open smooth strings in one loop is given by

$$W_C = T^{d/2}\Omega \int_0^\infty \frac{d\tau_2}{2(4\pi\tau_2^2)} \left[16\pi^2\ Z_T^{++}(\tau)\ Z_T^{++}(\tau,\tilde{\alpha}_0 T)\right]^{(2-d)/4}_{\tau = i\tau_2}$$

$$\times \{[1 - \exp(-\sqrt{\tilde{\alpha}_0 T}\ \tau_2 L_2)]\ \exp(\sqrt{\tilde{\alpha}_0 T}\ \tau_2 L_2/2\pi)\}^{-d/2+1}$$

$$= T^{d/2}\Omega \int_0^\infty \frac{d\tau_2}{2(4\pi\tau_2)^2} (4\pi\tau_2)^{(2-d)/2} \left| \prod_{n=1}^{\infty} (1 - \exp - 2\pi n\ \tau_2 L_2/L_1) \right|^{2-d}$$

$$\times \left| \prod_{n=1}^{\infty} (1 - \exp - 2\pi\tau_2 L_2 (n^2 + L_1^2/4\pi^2)^{1/2}/L_1) \right|^{2-d} (1 - \exp - \sqrt{\tilde{\alpha}_0 T}\ \tau_2 L_2)^{2-d}$$

$$\times \exp \frac{\pi\tau_2 L_2}{L_1} \left[\left(1 - \frac{3\tilde{\alpha}_0 T L_1^2}{4\pi^2} - 6\ I\ (\tilde{\alpha}_0 T L_1^2)\right) \frac{d-2}{12} - \frac{\sqrt{\tilde{\alpha}_0 T}\ L_1}{4\pi^2}\ d\right]$$

$$= T^{d/2}\Omega' \int_0^\infty \frac{d\tilde{\tau}_2}{2\tilde{\tau}_2} (2\pi\tilde{\tau}_2)^{-d/2} \left| \prod_{n=1}^\infty (1 - \exp - 2\pi n\tilde{\tau}_2)\right|^{2-d}$$

$$\times \left| \prod_{n=1}^\infty \left(1 - \exp - 2\pi\tilde{\tau}_2(n^2 + a^2/4\pi^2)^{1/2}\right)\right|^{2-d} (1 - \exp - a\tilde{\tau}_2)^{2-d} \times$$

$$\times \exp \pi\tilde{\tau}_2 \left[\left(1 - \frac{3a^2}{4\pi^2} - 6\, I(a^2)\right)\frac{d-2}{12} - \frac{a}{4\pi^2} d\right] . \qquad (3\text{-}40)$$

In the last step, we have set

$$\prod_{\mu=1}^{d} L^\mu = \Omega$$

$$a^2 = \tilde{\alpha}_0 T L_1^2$$

$$\tilde{\tau} = \frac{L_2 \tau}{L_1}$$

and

$$\Omega' = \left(\frac{L_2}{2L_1}\right)^{d/2} \Omega . \qquad (3\text{-}41)$$

From the path integral (3-40), it is a straightforward exercise to evaluate the free energy of a gas of open smooth strings. The procedure is the same as that used for closed smooth strings and is explained in Polchinski's paper.[8] The result is

$$F(\beta) = - (T)^{d/2} \int_0^\infty \frac{d\tilde{\tau}_2}{4\pi\tilde{\tau}_2^2} (2\pi\tilde{\tau}_2)^{\frac{2-d}{2}} \left| \prod_{n=1}^\infty (1 - \exp^{-2\pi n\tilde{\tau}_2})\right|^{2-d}$$

$$\times \left| \prod_{n=1}^\infty (1 - \exp - 2\pi\tau_2(n^2 + a^2/4\pi^2)^{1/2})\right|^{2-d} \times (1 - \exp^{-\tilde{\tau}_2 a})^{2-d}$$

$$\times \exp \left\{\pi\tilde{\tau}_2\left[\left(1 - \frac{3a^2}{4\pi^2} - 6I(a^2)\right) \frac{d-2}{12} - \frac{a}{4\pi^2} d\right]\right\} \sum_{r=1}^\infty e^{-r^2\beta^2 T^2/2\tilde{\tau}_2} \qquad (3\text{-}42)$$

This expression can be compared with the expression (3-27) for a collection of free particles. Let the mass spectrum in (3-27) be given by

$$m^2(\alpha_0 T) = 2\pi T \, [(-2 + \frac{3a^2}{2\pi^2} + 12\, I\,(a^2))\, \frac{d-2}{24} + \frac{a}{4\pi^2}\, d$$

$$+ \sum_{i=1}^{d-2} \sum_{n=1}^{\infty} (nN_{n_i}^{(1)} + \tilde{n}\, N_{n_i}^{(2)}) + \frac{a}{2\pi} \sum_{i=1}^{d-2} N_i^{(3)} \,] \tag{3-43}$$

with

$$\tilde{n} = \left(n^2 + \frac{a^2}{4\pi^2}\right)^{1/2} \tag{3-44}$$

$N_{n_i}^{(1)}$, $N_{n_i}^{(2)}$, and $N_i^{(3)}$ are number operators and commute with each other.

Summing (3-27) over the spectrum (3-43) reproduces (3-42) with the identification $ST = \tilde{\tau}_2$.

The ground state is one in which the number operators have zero eigenvalues. We find that

$$m_0^2\,(a^2) = 2\pi T \, [(-2 + \frac{3a^2}{2\pi^2} + 12\, I\,(a^2))\, \frac{d-2}{24} + \frac{a}{4\pi^2}\, d] \tag{3-45}$$

From the known behavior of $I(a^2)$ given by (3-23), it is seen that there exists a critical coupling a_c above which the tachyon disappears. The critical coupling constant as determined from (3-45) for open smooth strings is different from the one for closed strings given by Eq. (3-31) and has a lower value.

IV. THE STATIC POTENTIAL

To extract the static potential from the path-integral of the open smooth strings, we set $\mu = -T$ in (3-2) at the quantum level or equivalently add a factor $\exp(TL_1L_2\tau_2)$ to the path integral (3-34).

In order to compare our result with those in Refs. (4) and (5) we introduce a mass scale M^2 in such a way that

$$[\det'(-\frac{1}{\tilde{\alpha}_0 T}\,\Delta + 1)]_C = \frac{\tilde{\alpha}_0 T}{M^2}\,[\det'\,\frac{1}{M^2}\,(-\Delta + \tilde{\alpha}_0 T)]_C \tag{4-1}$$

$[\mathrm{Det}'\,\frac{1}{M^2}\,(-\Delta + \tilde{\alpha}_0 T)]_C$ can be evaluated in the same way as in evaluating $[\det'(-\frac{1}{\tilde{\alpha}_0 T}\,\Delta + 1)]_C$ given by (3-37). We find

$$\left[\det{}' \frac{1}{M^2} (- \Delta + \widetilde{\alpha}_0 T)\right]_C = (1 - e^{-\tau_2\sqrt{\widetilde{\alpha}_0 T}\, L_2}) \exp(\tau_2\sqrt{\widetilde{\alpha}_0 T}\, L_2/2\pi)$$

$$\times \left|(Z_T^{++}(\tau, M^2, \widetilde{\alpha}_0 T)^{1/2}\right|_{\tau = i\tau_2} \qquad (4\text{-}2)$$

with

$$Z_T^{++}(\tau, M^2, \widetilde{\alpha}_0 T) = \left|\prod_{n=1}^{\infty} (1 - e^{2\pi i W_+ \frac{L_2}{L_1}})\right|^4 (1 - e^{-\tau_2\sqrt{\widetilde{\alpha}_0 T}\, L_2})^{1/2}$$

$$\times \exp\left\{\frac{L_1 L_2 \tau_2 \widetilde{\alpha}_0 T}{4\pi} (1 - \ln \frac{\widetilde{\alpha}_0 T}{M^2}) + 2\pi\tau_2 I(\widetilde{\alpha}_0 T L_1^2) \frac{L_2}{L_1}\right\} \qquad (4\text{-}3)$$

where $I(\widetilde{\alpha}_0 T L_1{}^2)$ and W_+ are given by (3-21) and (3-22) respectively. $(\det{}' P_1{}^+ P_1)_C$ is given by

$$(\det{}' P_1 P_1)_C = \frac{1}{2} \left[\det{}'(-\Delta) \det{}' \frac{1}{M^2} (-\Delta + \widetilde{\alpha}_0 T)\right]_T . \qquad (4\text{-}4)$$

The remaining procedure is the same as that leading to (3-40) and we find

$$W_C' = (T)^{d/2} \left(\prod_{\mu=1}^{d} L''_\mu\right). \int_0^\infty \frac{d\tau_2}{8\pi\tau_2^2} (4\pi\tau_2)^{\frac{2-d}{2}} \exp\Big\{- L_2\Big[(T_r(M) L_1$$

$$- \frac{1}{L_1} \frac{(d-2)\pi}{12} (1 - 6 I(\widetilde{\alpha}_0 T L_1^2)) - \frac{\sqrt{\widetilde{\alpha}_0 T}}{4\pi} d)\tau_2$$

$$- \frac{d-2}{L_2} \sum_{n=1}^{\infty} \ln(1 - \exp - \frac{2\pi\tau_2}{L_1} L_2 (n^2 + \widetilde{\alpha}_0 T L^2/4\pi^2)^{1/2})$$

$$\times (1 - \exp - 2\pi\tau_2 L_2 n/L_1) - \frac{d-2}{L_2} \ln (1 - \exp^{-\sqrt{\widetilde{\alpha}_0 T}\, \tau_2 L_2})\Big]\Big\}$$

$$= (T)^{d/2} \left(\prod_{\mu=1}^{d} L''^\mu\right) \int_0^\infty \frac{d\tau_2}{8\pi\tau_2^2} (4\pi\tau_2)^{2-d/2} \exp[-L_2\tau_2 V(\tau_2, L_1, L_2)]$$

$$(4\text{-}5a)$$

$$T_r(M) = T\left[1 - \frac{d-2}{16\pi^2} \widetilde{\alpha}_0 (\ln \frac{\widetilde{\alpha}_0 T}{M^2} - 1)\right] \qquad (4\text{-}5b)$$

is the renormalized string tension,

$$L''^{\mu} = L^{\mu}(\frac{\sqrt{\tilde{\alpha}_0 T}}{M}) \tag{4-5c}$$

and

$$V(\tau_2.L_1,L_2) = T_r(M)L_1 - \frac{d}{4\pi}\sqrt{\tilde{\alpha}_0 T} - \frac{1}{L_1}\ \frac{(d-2)\pi}{12}(1 - 6I(\tilde{\alpha}_0 TL_1^2))$$

$$- \frac{d-2}{L_2\tau_2}\sum_{n=1}^{\infty} \ln(1 - \exp - \frac{2\pi\tau_2}{L_1} L_2\ (n^2 + \tilde{\alpha}_0 TL^2/4\pi^2)^{1/2})$$

$$\times\ (1 - \exp - 2\pi\tau_2 T_2 L_2/L_1) - \frac{d-2}{L_2\tau_2}\ln\ (1 - \exp^{-\sqrt{\tilde{\alpha}_0 T}\ \tau_2 L_2}) \tag{4-6}$$

The static potential may be identified as the following quantity[4)]

$$V(L_1) = \lim_{L_2 \gg L_1} V(\tau_2,L_1 L_2)$$

$$\approx T_r(M)L_1 - \frac{d-2}{4\pi}\sqrt{\tilde{\alpha}_0 T} - \frac{1}{L_1}\ \frac{(d-2)\pi}{12}(1 - 6I(\tilde{\alpha}_0 TL_1^2)) \tag{4-7}$$

The static potential has been evaluated by a number of authors.[4), 5)] These authors evaluated the determinants of the relevant operators on scalar functions which vanish on the boundary of a flat rectangular 2-dimensional surface. This requires that the determinants det'(- Δ) and det' $(- \frac{1}{\tilde{\alpha}_0 T}\Delta + 1)$ contain sum only over eigenvalues $m,n > 0$. As we have seen in Refs. 18 and 19 by considering the topology of genus one surfaces properly we find that in the case of closed string the sum is over eigenvalues $m,n\ \varepsilon Z$. In other words, the degeneracy of each eigenvalue is 4 for closed strings. For open strings on a cylinder, only (σ^1, o) is identified (σ_1, L_2). In this case n runs over positive integers while m is unrestricted. Therefore the degeneracy of each eigenvalue is 2 for open strings. This is the topological source of the crucial factor 2 for Lucher term in the static potential. Thus the coefficient of the 1/R term in the static potential for the Nambu-Goto model is in fact $-\frac{\pi}{6}$ (d = 4) instead of $-\frac{\pi}{12}$. The effect of the extrinsic curvature term to the 1/R term is give by $I(\tilde{\alpha}_0 TL_1{}^2)$ and for large L_1

$$I(\tilde{\alpha}_0 T L_1^2) \sim e^{-\sqrt{\tilde{\alpha}_0 T} L_1} . \tag{4-8}$$

We conclude that the coefficient of the "Luscher term" is universal for large L_1. It should be pointed out that even though the coefficient of the Luscher term is $-\pi/6$ even in the Nambu model on a cylinder, the absence of tachyons depends on the extrinsic curvature.

REFERENCES

1. Polyakov, A.M., Nucl. Phys. B268, 406 (1986).
2. Helfrich, W., Z. Naturforsch. C, 28, 693 (1973); Helfrich, W., J. de Phys. 46, 1263 (1985).
3. Peliti, L. and Leibler, S., Phys. Rev. Lett. 54, 1690 (1985); Förster, D., Phys. Lett. 114A, 115 (1986); Kleinert, H., Phys. Lett. B 174, 335 (1986).
4. Braaten, E., Pisarski, R.D. and Tze, Sze-man, Phys. Rev. Lett. 58 93 (1987); Kleinert, H., Phys. Rev. Lett. 58, 1915 (1987); Oleson, P. and Yang, Sung-Kil, Nucl. Phys. B283, 73 (1987); Bagan, E., Phys. Lett. 192B, 420 (1987).
5. Lüscher, M., Symanzik, K. and Weisz, P., Nucl. Phys. B173, 365 (1980); Alvarez, O., Phys. Rev. D24 (1981) 440; Arvis, J.F., Phys. Lett. 27B, 106 (1983); Olesen, P., Phys. Letts. B160, 144 (1985).
6. Alonso, F. and Espriu, D., Nucl. Phys. B283, 393 (1987); David, F., Europhys. Lett. 2, 577 (1986).
7. Curtright, T.L., Ghandour, G.I., Thron, C.S. and Zachos, C.K., Phys. Rev. Lett. 57, 799 (1986); Phys. Rev. D34, 3811 (1986).
8. Polchinski, J., Comm. Math. Phys. 104, 37 (1986).
9. Burgess, C.P. and Morris, T.R., Princeton Preprint (1986); Nucl. Phys. B29, 256 (1987).
10. Julve, J. and Torin, M., Nuovo Cim. 46B, 137 (1981); Fradkin, E.S. and Tseytlin, A.A., Phys. Lett. B 104, 377 (1981).
11. Viswanathan, K.S. and Xiaoan, Zhou, SFU Preprint.
12. Pisarski, R., Fermi Lab. Preprint, Fermi Lab. Conf. 86/171-T, Phys. Rev. Lett. 58, 1300 (1987).
13. Antoniadis, I., Bachas, C.P. and Kounnas, C., Nucl. Phys. B289, 8 (1987).
14. Nelson, Philip, HUTP 86/A047. Lectures on Strings and Moduli Space.
15. Carfora, M., Martellini, M., Marzuoli, A. and Silvotti, R., Preprint.
16. Chen, Bang-Yan, Total Mean Curvature and Submanifolds of Finite Type, World Scientific, Singapore (1984).
17. Wilmore, T.J., Total Curvature in Riemannian Geometry, Ellis Horword Ltd., England (1982).
18. Viswanathan, K.S. and Xiaoan, Zhou, SFU Preprint.
19. Viswanathan, K.S. and Xiaoan, Zhou, SFU Preprint.
20. Morse, P.M., Vibration and Sound, McGraw-Hill (1948).

SCALAR FIELD AND THE POLYAKOV STRING

Z.Y. Zhu[1,2] and H.C. Lee[1]

[1]Theoretical Physics Branch
Chalk River Nuclear Laboratories
Atomic Energy of Canada Limited
Chalk River, Ontario, Canada K0J 1J0

and

[2]Institute of Theoretical Physics
Academia Sinica, Beijing, China

Some time ago Polyakov [1] suggested that in string theories the world-sheet metric tensor components should be considered as independent fields. In this program the partition function includes the integration over all distinct world-sheet surfaces as well as the functional integration over the string coordinates. Polyakov showed that for surfaces without handles, these integrals can be reduced to the functional integral of a single scalar field which has the lagrangian of the quantum Liouville theory, and that for string theories in less than critical space-time dimensions ($d_{critical} = 26$ for bosonic strings) the dynamics of the scalar field must be taken into account if conformal invariance is to be maintained. Since the scalar field is proportional to the logarithm of the Weyl factor of the world-sheet metric in the conformal gauge, we shall call it the Weyl field.

Polyakov's idea has spawned several research directions including intensive studies of properties of Riemann surfaces and their relation to string theories [2], and studies of the quantum Liouville theory [3]. In yet another direction, Marnelius [4] showed that Polyakov's result (for world-sheets without handles) can be alternatively derived from an action including a world-sheet and space-time scalar field $\phi(\xi)$ in addition to the string $X_\mu(\xi)$ and the two-dimensional metric $g_{ab}(\xi)$

$$S = \int d^2\xi \sqrt{-g} \left[\frac{1}{2} g^{ab}\partial_a X^\mu \partial_b X_\mu + N\left(-\frac{1}{2} g^{ab}\partial_a\phi\partial_b\phi + R\phi + \mu^2\right)\right]$$

$$\equiv \int d^2\xi \left[\mathcal{L}_0(X) - N\mathcal{L}_0(\phi) + N\sqrt{-g}\,(R\phi + \mu^2)\right] \tag{1}$$

where ϕ is not free but is constrained by its field equation. We use the Minkowski signatures $\eta_{ab} = (+,-)$ for the world-sheet and $\eta_{\mu\nu} = (-,+,+,\cdots)$ for space-time. This is analogous to the reduction of the string lagrangian of Brink, di Vecchia and Howe [5] to the Nambu-Goto lagrangian when the world-sheet metric is constrained by its field equations. In the case of (1), it was shown that the dynamic Liouville mode is constrained to be a linear combination of the Weyl field and ϕ. The action (1) has an interesting interpretation in the quantum field theory context. The new terms proportional to N can be viewed as counterterms, the value of the coefficient N being determined to exactly cancel the conformal anomaly of the string lagrangian $\mathcal{L}_0(X)$.

Marnelius' action is a very special case of a general constrained system involving the fields X^μ, g_{ab} and ϕ. Here we report the investigation of the general case, where the scalar field is allowed to enter into the theory nonlinearly, subject only to the requirements that the theory be (i) world-sheet reparametrization invariant; (ii) space-time Poincaré invariant; (iii) renormalizable. Briefly, our result is as follows. If the scalar field is not coupled to the world-sheet curvature R, then the theory reduces to either the d-dimensional string theory, or to a special case of nonlinear sigma model on a (d+1)-dimensional target space, with a metric that is flat in the $d+1^{th}$(i.e., the ϕ) dimension but has a ϕ-dependent Weyl factor in the other d dimensions. If the scalar is coupled to the world-sheet curvature, then the theory in general has ghosts, except for special cases, when the ghosts decouple. The special cases exclude direct coupling between ϕ and the string, but include nonlinear coupling between ϕ and g_{ab}. Remarkably, all the ghost-free cases (with ϕ-curvature coupling) can be reduced by field redefinition to the Liouville theory. Boundary terms needed for covariant quantization are also given.

We wish to remark that the scalar field $\phi(\xi)$ is not the same as the scalar field $\Phi(X)$ commonly known as the dilaton in nonlinear sigma-models [6]. In the standard approach to such models space-time has the critical number of dimensions and the dynamics of the Weyl field associated with the world-sheet is ignored. Nevertheless, it will be evident that the two scalar fields $\phi(\xi)$ and $\Phi(X)$ play remarkably similar roles in conformal anomaly cancellation, at least at the lowest order of perturbation theory. Since all Φ's are functions of ξ but not all ϕ's are functions of X, it may be that ϕ is a necessary generalization of Φ for non-critical dimensions. In this paper we do not pursue this line of thought any further and will forget the dilaton from now on.

The most general lagrangian with ϕ coupled nonlinearly to X^μ and g_{ab} while satisfying the conditions (i) to (iii) above has the form

$$S = \int d^2\xi \left[C(\phi)\mathcal{L}_o(X^\mu) + D(\phi)\mathcal{L}_o(\phi) + \sqrt{-g}(RB(\phi) + A(\phi)) \right] \tag{2}$$

where A, B, C and D are arbitrary scalar functions of ϕ, R is the scalar curvature which, in terms of the world sheet variables $\xi^0 = \tau$ and $\xi^1 = \sigma$ and the metric with $\alpha = g_{00}$, $\beta = g_{11}$ and $\gamma = g_{01}$, is explicitly

$$eR = \partial_\tau\left(\frac{1}{e} R_o\right) + \partial_\sigma\left(\frac{1}{e} R_1\right) \qquad (e \equiv \sqrt{-g}) \tag{3}$$

where

$$R_o \equiv -\dot{\beta} + \gamma' + \frac{\gamma}{2}\ell', \qquad R_1 \equiv -\alpha' + \dot{\gamma} - \frac{\gamma}{2}\dot{\ell},$$

$$\ell \equiv \ln(\alpha/\beta), \qquad \dot{\beta} = \partial_\tau \beta, \qquad \gamma' = \partial_\sigma\gamma, \qquad \text{etc.}$$

In the following, anticipating the conformal gauge, we shall use instead of α and β the linear combinations

$$\rho \equiv \rho_- = \frac{1}{2}(\beta-\alpha), \quad \Delta \equiv \rho_+ = \frac{1}{2}(\alpha+\beta) \tag{4}$$

The independent fields are now X^μ, ϕ, ρ, Δ and γ, with the corresponding canonical momenta

$$P_\mu = \frac{-C}{e}\left[(\Delta+\rho)\dot{X}_\mu - \gamma X'_\mu\right] \tag{5a}$$

$$P_\phi = \frac{-D}{e}\left[(\Delta+\rho)\dot{\phi} - \gamma\phi'\right] - \frac{1}{e} R_0\, \partial_\phi B \tag{5b}$$

$$P_\rho = \frac{1}{e}\left[\dot{B} + \gamma\Delta B'/(\rho^2-\Delta^2)\right] \tag{5c}$$

$$P_\Delta = +\frac{1}{e}\left[\dot{B} - \gamma\rho B'/(\rho^2-\Delta^2)\right] \tag{5d}$$

$$P_\gamma = \frac{-1}{e} B' \tag{5e}$$

The variations of the Lagrangian with respect to the metric are given by ($\rho_\gamma \equiv \gamma$)

$$\delta_g \mathcal{L}_o(\phi) \equiv \frac{1}{2eg} \sum_{j=+,-,\gamma} \psi_j^\phi \, \delta\rho_j \tag{6a}$$

$$\psi_\mp^\phi = (\mp\rho\Delta \mp \rho_\pm^2+\gamma^2)\,\dot{\phi}^2+(-\rho\Delta + \rho_\pm^2 \mp \gamma^2)\phi'^2 \pm 2\rho_\mp\gamma\dot{\phi}\phi' \tag{6b}$$

$$\psi_\gamma^\phi = -\gamma(\Delta+\rho)\,\dot{\phi}^2 - \gamma(\Delta-\rho)\phi'^2 - 2(\rho^2-\Delta^2)\dot{\phi}\phi' \tag{6c}$$

$$\delta_g(eRB) \equiv \frac{1}{2eg} \sum_{j=+,-,\gamma} \psi_j^R \, \delta\rho_j \tag{7a}$$

$$\psi_\mp^R = 2g(-\ddot{B} \pm B'') \mp \left[(2\rho_\mp R_o \mp \dot{g}) + \frac{\rho_\pm}{\rho^2-\Delta^2}(\gamma g' - 2g\gamma')\right]\dot{B}$$

$$\mp \left[(2\rho_\mp R_1 + g') - \frac{\rho_\pm}{\rho^2-\Delta^2}(\gamma\dot{g} - 2g\dot{\gamma})\right]B' \tag{7b}$$

$$\psi_\gamma^R = -(2\gamma R_0 + g' + g\ell')\dot{B} - (2\gamma R_1 + \dot{g} - g\dot{\ell})B' + 4g\dot{B}' \tag{7c}$$

Note that the trace of ψ^ϕ vanishes, while that of ψ^R is proportional to the Laplacian Δ_g of B:

$$\mathrm{Tr}\psi^R = \rho_+\psi_+^R + \rho_-\psi_-^R + 2\gamma\,\psi_\gamma^R \sim \Delta_g B \equiv \frac{1}{e}\,\partial_a(eg^{ab}\partial_b B)$$

The field equations are therefore

$$\Delta_g(X^\mu C) = 0 \tag{8a}$$

$$\mathcal{L}_o(X)\partial_\phi C - \partial_a(eDg^{ab}\partial_b\phi) + eR\partial_\phi B + e\partial_\phi A = 0 \tag{8b}$$

$$\Delta_g B - A = 0 \tag{8c}$$

$$\phi_\Delta \equiv C\phi_+^x + D\phi_+^\phi + \phi_+^R - 2g\Delta A = 0 \tag{8d}$$

$$\phi_\gamma \equiv C\phi_\gamma^x + D\phi_\gamma^\phi + \phi_\gamma^R + 2g\gamma A = 0 \tag{8e}$$

The absence of $\dot{\gamma}$ in (5) suggests then γ is not a dynamical variable, and can therefore be constrained to have any value. We choose $\gamma=0$. In this case $P_\rho = +P_\Delta$, so that either ρ or Δ is an independent dynamical variable, but not both. Having already chosen the Minkowski signature, we now choose $\chi \equiv \ell n|\rho|$ to be the dynamical variable, and set $\Delta = 0$. As is well known, the fixed values $\gamma = \Delta = 0$ precisely define the conformal gauge. (Had we chosen the Euclidean signature, the roles of Δ and ρ would inter-change, and conformal gauge would mean $\gamma = \rho = 0$.) In this gauge $g_{ab} \to e^\chi \eta_{ab}$, $\sqrt{-g} \to e^\chi$, $R \to e^{-\chi} \Box\chi$, $\Delta_g \to e^{-\chi} \Box$, and the canonical momenta are

$$P_\mu = C\dot{X}_\mu \tag{9a}$$

$$P_\phi = D\dot{\phi} - \partial_\phi B\dot{\chi} \tag{9b}$$

$$P_\chi = -\dot{B} = -\partial_\phi B\dot{\phi} \tag{9c}$$

The gauge fixing also elevates the two equations (8e,f) to the status of constraints that must be satisfied at all times. With (7), they become (to within inessential factors)

$$\phi_\Delta = C(\dot{X}^2+X'^2) + D(\dot{\phi}^2+\phi'^2) + 4B'' - 2e^\chi A - 2\dot{B}\dot{\chi} - 2B'\chi' = 0 \tag{10a}$$

$$\phi_\gamma = C\dot{X}X' + D\dot{\phi}\phi' - \dot{B}\chi' - B'\dot{\chi} + 2\dot{B}' = 0. \tag{10b}$$

In the following we examine special cases.

<u>B is a constant</u>. Then, since R is a total derivative, the B term in (2) is reduced to a nondynamical surface term. From (8c) it follows that $A(\phi)=0$, and the resulting theory depends on whether or not A is a function of ϕ. If not, then A must be a vanishing constant. (In other words, it is not consistent to have a nonvanishing cosmological term without B being a function of ϕ.) In this case D can be scaled away by

a redefinition of ϕ: $\phi \to \phi D^{-1/2}$, and the Lagrangian reduces to that of a nonlinear sigma model with a (d+1)-dimensional target space, on which the metric is flat in the ϕ-direction, but has a Weyl factor equal to (the rescaled) $C(\phi)$ in the remaining d dimensions. On the other hand, if A is a function of ϕ, then $A(\phi)=0$ becomes a constraint on ϕ, so that the $D(\phi)\mathcal{L}_0(\phi)$ term becomes nondynamical and $C(\phi)$ can be reduced to a constant factor, and the theory reduces to that of a d-dimensional string. In short, the case B = constant is not very interesting.

B is a function of ϕ. In this case the Lagrangian can be simplified by redefining B as a normalization constant N times a new scalar field ϕ, and understanding A, C and D to be functions of (the new) ϕ. The Lagrangian can be further simplified by the field redefinitions

$$\chi = \psi + F(\phi), \quad \partial_\phi F = \frac{1}{2N} D(\phi) \tag{11}$$

so that in terms of the new fields

$$\mathcal{L} = \frac{1}{2} C(\phi)(\partial X^\mu)^2 - N\partial\phi\partial\psi + e^\psi \tilde{A}(\phi) + B_1 \tag{12}$$

where $\tilde{A}(\phi) \equiv e^{F(\phi)} A(\phi)$ and $B_1 = N\partial_a[\phi(\partial_a\psi + \frac{1}{2N} D\partial_a\phi)]$ is a boundary term. It is easy to see that, by a linear transformation, the second term in (12) can be separated into the decoupled kinetic terms of two scalars, one of which will have negative norm and is therefore a ghost. The ghost will be coupled to the string unless C is a constant. We will therefore consider only the case when C is a constant. In this case, the ghost will still in general be coupled to the other scalar through the A term. However, when A has the form

$$\tilde{A}(\phi) = c\, e^{-s\phi}, \qquad s > 0 \tag{13}$$

the ghost is decoupled. This is revealed by the linear transformation

$$\Psi_\pm = \psi \mp s\phi \tag{14}$$

under which the lagrangian becomes, to within boundary terms to be discussed later,

$$\mathcal{L} = \frac{1}{2}(\partial X^\mu)^2 + \eta[\frac{1}{2}(\partial\Psi_+)^2 - \mu^2 e^{\Psi_+}] - \frac{\eta}{2}(\partial\Psi_-)^2, \quad \eta \equiv \frac{N}{2s}, \quad \mu^2 \equiv -c/\eta \tag{15}$$

with field equations and momenta

$$\Box x^{\mu} = 0 \ , \quad \Box \Psi_{-} = 0 \tag{16}$$

$$\Box \Psi_{+} + \mu^{2} e^{\Psi_{+}} = 0 \tag{17}$$

$$P_{\mu} = \dot{x}_{\mu} \tag{18}$$

$$P_{\Psi_{\pm}} = \pm \eta \dot{\Psi}_{\pm} \tag{19}$$

and the constraints (10) can be more conveniently re-expressed as

$$\Phi_{\pm} = \frac{1}{4}(P \pm X')^{2} + \frac{1}{4}\sum_{j=+,-}\left[\frac{j}{\eta}(P_{\Psi_{j}} \pm j\eta\Psi_{j}')^{2} \mp 4(P_{\Psi_{j}} \pm j\eta\Psi_{j}')^{-}\right] - \frac{1}{2} Nce^{\Psi_{+}} = 0 \tag{20}$$

If $\phi = X_{\mu}$, Ψ_{+}, Ψ_{-} and their corresponding momentum P_{ϕ} have canonical Poisson brackets, then the algebra of $\Phi_{\pm}$ is known. Define

$$F_{\phi} \equiv P_{\phi} + u\, \partial_{\sigma}\phi \tag{21}$$

$$G_{\phi} \equiv F_{\phi}^{2} - 4u\, \partial_{\sigma}F_{\phi} + v\, e^{\phi} \ , \qquad u,v \text{ constants.} \tag{22}$$

Then

$$\{F_{\phi}^{2}(\sigma),\ F_{\phi'}^{2}(\sigma')\} = 4u\, \delta_{\phi\phi'}\big(F_{\phi}^{2}(\sigma) + F_{\phi}^{2}(\sigma')\big)\partial_{\sigma}\delta(\sigma-\sigma') \tag{23}$$

$$\{\partial_{\sigma}F_{\phi}(\sigma),\ \partial_{\sigma'}F_{\phi'}(\sigma')\} = -2u\, \delta_{\phi\phi'}\partial^{3}\delta(\sigma-\sigma') \tag{24}$$

$$\{G_{\phi}(\sigma), G_{\phi'}(\sigma')\} = 4u\, \delta_{\phi\phi'}\big[\big(G_{\phi}(\sigma)+G_{\phi}(\sigma')\big)\partial_{\sigma} - 8u^{2}\partial_{\sigma}^{3}\big]\delta(\sigma-\sigma') \tag{25}$$

One recognizes the last term in (25), originating from the commutator (24), as a Schwinger term. In the Poisson brackets for $\Phi_{\pm}$, the Schwinger terms arising respectively from the two modes Ψ_{+} and Ψ_{-} cancel, so that

$$\{\Phi_{i}(\sigma), \Phi_{j}(\sigma')\} = j\delta_{ij}\big(\Phi_{j}(\sigma) + \Phi_{j}(\sigma')\big)\partial_{\sigma}\delta(\sigma-\sigma'), \qquad i,j = +,- \tag{26}$$

Thus, reparametrization invariance is preserved as long as both $\Psi_{\pm}$ are kept.

Of course a physical theory must not admit ghosts. Since the ghost in (15) is decoupled, one may excise it from the theory. This can be achieved by demanding Ψ_- operated on any physical state vanishes

$$\Psi_- |\text{phys.}\rangle = 0 \tag{27}$$

The price one pays for this is the breaking of reparametrization invariance. Let $\Phi_\pm$ be the constraints (20) with the contributions from Ψ_- discarded, then their algebra

$$\{\tilde{\Phi}_i(\sigma),\ \tilde{\Phi}_j(\sigma')\} = j\delta_{ij}[(\tilde{\Phi}_j(\sigma) + \tilde{\Phi}_j(\sigma')e)_\sigma - 2\eta\partial_\sigma^3]\ \delta(\sigma-\sigma') \tag{28}$$

exhibits the familiar conformal anomaly in the last term.

By a suitable choice of η, this anomaly can be used to cancel the string anomaly. This is easily shown with the GGRT method [7] of examining Lorentz invariance in the light-cone gauge. In this gauge, P^+ is a constant, $X^+ \propto P^+\tau$, and X^- and P^- are given by the two constraints in which only the Liouville mode $\Psi \equiv \Psi_+$ is kept,

$$\begin{aligned} P^- &= \frac{1}{2P^+}(P^iP^i + X^{i\prime}X^{i\prime} + \frac{1}{\eta}P_\Psi^2 + \eta\Psi'^2 - 4\eta\Psi'' - 2ce^\Psi) \\ X^{-\prime} &= \frac{1}{P^+}(P^iX^{i\prime} + P_\Psi\Psi' - 2P_\Psi') \end{aligned} \tag{29}$$

The dynamical transverse modes X^i can then be quantized in the usual way by expansion in terms of oscillators a_n^i satisfying commutation relations

$$[a_n^i,\ a_m^{j\dagger}] = \delta^{ij}\delta_{mn}$$

If one assumes the commutation relation of the quantized Ψ and P_Ψ is unchanged from the canonical Poisson bracket

$$[P_\Psi(\tau,\sigma),\ \Psi(\tau,\sigma')] = i\delta(\sigma-\sigma') \tag{30}$$

then the Lorentz noninvariance is given by

$$[M^{i-},M^{j-}] = \frac{2\pi}{(p^+)^2}\sum_{h=1}^{\infty}[\alpha_0 - \frac{d-2}{24} + n^2(2\eta\pi - \frac{d-2}{24} - 1)]\ (a_n^{i\dagger}a_n^j - a_n^ja_n^{i\dagger}) \tag{31}$$

where

$$M^{i-} = \int_0^\pi d\sigma(X^i P^- - X^- P^i) \tag{32}$$

is the angular momentum operator of the string. The parameter α_0 arises from uncertainty in the ordering of operators. It is seen that Lorentz invariance is restored by setting

$$\begin{aligned} \eta &= \frac{26-d}{48\pi} \\ \alpha_0 &= \frac{d-2}{24} \end{aligned} \tag{33}$$

The first condition is precisely the value needed for the Polyakov string.

So far the role played by the string tension α' has been ignored. It is revealed by the transformation

$$\begin{aligned} X^\mu &\to X^\mu/\sqrt{\alpha'} \\ P^\mu &\to \sqrt{\alpha'}\, P^\mu \\ \eta &\to \alpha'\eta = \frac{26-d}{48\pi}\,\alpha' \end{aligned} \tag{34}$$

after which the Lagrangian (without the ghost) is

$$\mathcal{L} = \frac{1}{2\alpha'}(\partial X^\mu)^2 + \alpha'\left(\frac{26-d}{48\pi}\right)\left[\frac{1}{2}(\partial\Psi)^2 - \mu^2 e^\Psi\right] \tag{35}$$

since α' is the coupling constant for loop expansion on the world-sheet, this expression emphasizes the fact that the conformally noninvariant Liouville action cancels at the tree level the one-loop conformal anomaly of the string, when $d \neq 26$.

We now discuss boundary conditions for the open string. Other than the term B_1 in (12) arising from partial integration of $\mathcal{L}$, boundary terms involving the scalar field satisfying the conditions of renormalizability and reparametrization invariance have the general form

$$\int_M d^2\xi \sqrt{-g}\, D_a(E(\phi)D^a\phi) + 2\int_{\partial M} ds\, \kappa_g G(\phi) + \int_{\partial M} ds\, H(\phi) \tag{36}$$

where D_a is the covariant derivative on the world sheet M with boundary ∂M, ds is the line element and κ_g is the geodesic curvature of the boundary. In the conformal gauge,

$$\begin{aligned} ds &= e^{\chi/2} d\ell \\ \kappa_g &= e^{-\chi/2}[-\sigma'(t)\tau''(t) + \tau'(t)\sigma''(t)] + \partial_n e^{-\chi/2} \end{aligned} \tag{37}$$

where $d\ell$ is the line element on the flat σ-τ space, t is parametric variable defining the boundary and ∂_n is the inward normal derivative on the tangent plane of the world-sheet. The quantity in the square bracket can be interpreted as the curvature of the boundary on the flat σ-τ parameter space, which vanishes whenever at least one of the parameters is a consant on the boundary. This latter condition is always satisfied for open strings, which on the flat parameter space is just the infinite strip M defined by

$$M: \quad \sigma = 0 \text{ and } \pi, \qquad \tau = \pm\infty . \tag{38}$$

Therefore, for the strip, and with the assumption that contributions on the $\tau = \pm\infty$ boundaries vanish, the second term in (36) reduces in the conformal gauge to

$$-\int_{\partial M} d\ell\, G\partial_n\chi = \int_{-\infty}^{\infty} d\tau\, G\chi' \Big|_{\sigma=0}^{\sigma=\pi} \tag{39}$$

The other terms are similarly simplified. The sum of the boundary terms, including a contribution from the B_1 term in (12), is

$$\begin{aligned} B.T. = &\int_{-\infty}^{\infty} d\tau[E\phi' + (G+N\phi)\chi' + e^{\chi/2}H]_{\sigma=\pi} \\ &-\int_{-\infty}^{\infty} d\tau[E\phi' + (G+N\phi)\chi' - e^{\chi/2}H]_{\sigma=0} \end{aligned} \tag{40}$$

Upon transformation to Ψ_+ and Ψ_-, it is seen that the ghost can be decoupled only if

$$G(\phi) = m_2 - N\phi$$

$$E(\phi) = m_1 - \frac{1}{2N} m_2\, D(\phi) \qquad (41)$$

$$H(\phi) = m_3 e^{-(F(\phi)+s\phi)/2}, \qquad \partial_\phi F = \frac{1}{2N} D(\phi)$$

where m_i are arbitrary constants. The condition (27) then leads to ($\Psi \equiv \Psi_+$)

$$B.T. = -\frac{1}{2}\left(\frac{m_1}{s} - m_2\right) \int d\tau\left[(\Psi' - me^{\Psi/2})_{\sigma=\pi} - (\Psi' + me^{\Psi/2})_{\sigma=0}\right] \qquad (42)$$

where

$$m = 2m_3/(m_1/s - m_2) \qquad (43)$$

Thus the variation of the action on the boundary vanishes provided the Liouville mode satisfies the boundary conditions

$$\partial_\sigma \Psi = m' e^{\Psi/2}, \qquad \sigma = \pi,$$

$$= -m' e^{\Psi/2}, \qquad \sigma = 0$$

$$m' = 2m_3/\left[\frac{2-m_1}{s} + m_2\right] \qquad (44)$$

which are precisely the condition given by Gervais and Neveu [8].

We now give some examples. First we note that N and s always appear in the constraints as $\eta = N/2s$, so without loss of generality, we can set $s = 1/2$. Then, for anomaly cancellation, $N = (26-d)/48\pi$.

(i) $D(\phi) = -N$. A solution for F is $F(\phi) = -\phi/2$. This is the case considered by Marnelius. The generalized Polyakov action including boundary terms is

$$S = \int_M d^2\xi\left[\mathcal{L}_0(x) - N\mathcal{L}_0(\phi) + \sqrt{-g}\left(NR\phi - N\mu^2 + D_a\left[(m_1 + \frac{m_2}{2})D^a\phi\right]\right)\right]$$

$$+ \int_{\partial M} ds\left[\kappa_g(m_2 - N\phi) + m_3\right] \qquad (45)$$

Note that in this case the kinetic term for the scalar field already appears with a "wrong" sign, suggesting the existence of a ghost in the system.

(ii) $D(\phi) = 0$. A solution for F is $F(\phi) = 0$. Then

$$S = \int_M d^2\xi[\mathcal{L}_o(X) + \sqrt{-g}(NR\phi - N\mu^2 e^{-\phi/2} + D_a(m_1 D^a\phi))]$$

$$+ \int_{\partial M} ds[\kappa_g(m_2 - N\phi) + m_3 e^{-\phi/4}] \tag{46}$$

(iii) $D(\phi) = N$. A solution for F is $F(\phi) = \phi/2$. Then

$$S = \int_M d^2\xi[\mathcal{L}_o(X) + N\mathcal{L}_o(\phi) + \sqrt{-g}(NR\phi - N\mu^2 e^{-\phi} + D_a[(m_1 - \frac{m_2}{2})D^a\phi]))]$$

$$+ \int_{\partial M} ds[\kappa_g(m_2 - N\phi) + m_3 e^{-\phi/2}] \tag{47}$$

(iv) $D(\phi) = N\phi^r$. A solution for F is $F(\phi) = a\phi^{(r+1)/2(r+1)}$, and

$$S = \int_M d^2\xi[\mathcal{L}_o(x) + N\phi^r\mathcal{L}_o(\phi) + \sqrt{-g}\,(NR\phi + N\mu^2 e^{-(\frac{\phi^r}{n+1} + 1)\phi/2}$$

$$+ D_a((m_1 - \frac{m_2}{2}\phi^r)D^a\phi)] + \int_{\partial M} ds[\kappa_g(m_2 - N\phi) + m_3 e^{-(\frac{\phi^r}{r+1} + 1)\phi/4}] \tag{48}$$

In conclusion, we have seen that a general class of two-dimensional theories with a constrained scalar field reduces to the Polyakov string at the semiclassical level, in which the string variables obey properly quantized commutation relations but commutation relations for the Liouville mode are taken to have the same form as the corresponding classical Poisson brackets. An interesting observation is that the insistence of the decoupling of the ghost and the existence of ghost-free physical states guarantees the reduction to the Liouville model. It was shown the semiclassical theory corresponds to a field theory where the string variables are taken to the one-loop order but the Liouville mode remains at the tree order. An obvious question is whether the semiclassically equivalent theories remain equivalent at higher loop orders. As yet we have not discovered any symmetry that will guarantee the relations (11) and (41) needed for the preservation of this equivalence.

References

1. A.M. Polyakov, Phys. Lett. 103B(1981)207.
2. O. Alvarez, Nucl. Phys. B216(1983)125; for a review, see P. Nelson, Phys. Rep. 149(1987)337.
3. See A. Bilal and J.-L. Gervais, Nucl. Phys. B293(1987)1, and references therein.
4. R. Marnelius, Nucl. Phys. B211(1983)14.
5. L. Brink, P. di Vecchia and P. Howe, Nucl. Phys. B118(1977)76.
6. E.S. Fradkin and A.A. Tseytlin, Phys. Lett. 158B(1985)316;
   C.G. Callan, D. Friedan, E. Martinec and M.J. Perry, Nucl. Phys. B262(1985)593;
   C. Hull, "Lectures in Sigma-Models", in Super Field Theories, Eds. H.C. Lee, et al. Plenum, New York) 1987;
   R.R. Metsaev and A.A. Tseytlin, Nucl. Phys. B293(1987)385.
7. P. Goddard, J. Goldstone, C. Rebbi and C.B. Thorn, Nucl. Phys. B56(1973)109.
8. J.-L. Gervais and A. Neveu, Nucl. Phys. B194()1982)59.

MODULAR INVARIANT PARTITION FUNCTIONS FROM CONFORMAL SUBALGEBRAS*

Mihaela Niculescu Sanielevici

Center for Theoretical Physics
Laboratory for Nuclear Science
and Department of Physics
Massachusetts Institute of Technology
Cambridge, MA 02139 U.S.A.

ABSTRACT

The modular invariance and conformal invariance constraints, as well as the asymptotic behaviour of characters are used to decompose explicitely highest weight representations of Kač-Moody algebras with respect to their conformal subalgebras. The explicit decompositions are used to construct modular invariant combinations of characters. Thus a whole class of modular invariant partition functions is obtained. The SU(2) partition functions classified be Cappelli-Itzykson-Zuber[1] are also obtained as a particular case. The new partition functions describe WZW string theories with arbitrary symmetry group.

Conformal and modular invariance play a crucial role in string theory as well as in conformal field theory in two dimensions. There is justified hope that by imposing modular invariance constraints for a mathematically consistent compactification one can select from the too numerous low energy string theories. In the context of conformal field theory for statistical physics, Cardy[2] was the first one to study how the modular invariance requirements restrict the field content of a conformal field theory. Comparing the transformation properties of the characters under the modular group, Gepner[3] realized that the Virasoro characters can be obtained from a pair of $A_1^{(1)}$ Kač–Moody characters. Thus the minimal conformal models are equivalent to the product of two $SU(2)$ WZW

* This work is supported in part by funds provided by the U. S. Department of Energy (D.O.E.) under contract #DE-AC02-76ER03069 and by the Natural Sciences and Engineering Research Council of Canada.

theories and all such theories are classified by the modular invariant combinations of $SU(2)_L \times SU(2)_R$ characters which can be written at a given level k. The complete list of the $A_1^{(1)}$ modular invariants of level $k+2 \leq 100$ is summarized below:[1]

(1a) $k \geq 1, \qquad Z = \sum_{j=1}^{k+1} |\chi_j|^2$

(1b) $k = 4p, \qquad Z = \sum_{j\,\mathrm{odd}=1}^{2p-1} |\chi_j + \chi_{4p+2-j}|^2 + 2\,|\chi_{2p+1}|^2$

(1c) $k = 4p-2, \; Z = \sum_{j\,\mathrm{odd}=1}^{4p-1} |\chi_j|^2 + |\chi_{2p}|^2 + \sum_{j\,\mathrm{even}=2}^{2p-2} \left(\chi_j \chi_{4p-j}^* + \chi_j^* \chi_{4p-j}\right)$

(1d) $k = 10, \qquad Z = |\chi_1 + \chi_7|^2 + |\chi_4 + \chi_8|^2 + |\chi_5 + \chi_{11}|^2$

(1e) $k = 16, \qquad Z = |\chi_1 + \chi_{17}|^2 + |\chi_5 + \chi_{13}|^2 + |\chi_7 + \chi_{11}|^2 + |\chi_9|^2$
$\qquad\qquad + (\chi_3 + \chi_{15})\,\chi_9^* + \chi_9\,(\chi_3^* + \chi_{15}^*)$

(1f) $k = 28, \qquad Z = |\chi_1 + \chi_{11} + \chi_{19} + \chi_{29}|^2 + |\chi_7 + \chi_{13} + \chi_{17} + \chi_{23}|^2$

In the above, the characters χ_j are labeled by the dimension j of the $A_1^{(1)}$ representation at level k:

$$j = \dim L\left((k - j + 1)\,\Lambda_0 + (j-1)\Lambda_1\right)$$

The dimensions of the left-right symmetric representations in the $A_1^{(1)}$ partition functions listed above are the exponents of the simply-laced simple Lie algebras as follows:

$k \geq 1$, the algebra A_{k+1} whose dual Coxeter number is $g = k+2$.

$k = 4p$, the algebra D_{2p+2} with $g = 4p+2 = k+2$.

$k = 4p-2$, the algebra D_{2p+1} with $g = 4p = k+2$

$k = 10,\ 16,\ 28$ the algebras E_6, E_7, E_8 with $g = k+2 = 12,\ 18,\ 30$.

The $A_1^{(1)}$ modular invariant partition functions listed above describe also the propagation of a string on a group manifold[4] and the infinite series A_{k+1}, D_{2p+2} and D_{2p+1} were first found by Gepner and Witten as partition functions for the $SU(2)$ and $SO(3)$ WZW models, respectively. The partition functions for $A_1^{(1)}$ given in (1a) – (1f) were classified by Cappelli–Itzykson–Zuber and are referred to as the CIZ classification. Cappelli[5] extended the classification to the superconformal case. Besides the $A_1^{(1)}$ modular invariant partition functions of the CIZ series no other non-trivial partition functions were known.

Here we use the decomposition of a highest weight representation $L(\Lambda)$ of an affine algebra $\mathbf{g}$ into representations of its conformal subalgebra $\mathbf{p} \subset \mathbf{g}$ in order to generate *new* modular invariant partition functions. In this way we obtain partition functions for strings on the group manifold associated with the subalgebra $\mathbf{p}$ and at the level $k = I$, the Dynkin index of the embedding. This corresponds to a much larger class of theories than the $SU(2)$ and $SO(3)$ WZW models considered by Gepner and Witten. In the cases when the conformal subalgebra is $A_1^{(1)}$ or is semisimple and contains $A_1^{(1)}$ as a factor, our partition functions reproduce the CIZ ones (or linear combination of them) when the extra symmetries are gauged away. In Ref[6] we list the decompositions of the exceptional affine algebras with respect to their conformal subalgebras. Now we summarize the important features of the conformal subalgebras and explain the algorithm used for decompositions. For more details on the decomposition algorithm or on modular properties of the representations of Kač-Moody algebras we send the reader to Ref.[6,13,12].

We are interested in decompositions of a highest weight representation of an affine algebra $\mathbf{g}$ into representations of its affine conformal subalgebra $\mathbf{p}$. By conformal subalgebras we mean that the central charges of the Sugawara construction on $\mathbf{g}$ and $\mathbf{p}$ are equal, $c(\mathbf{g}) = c(\mathbf{p})$. This insures preservation of conformal invariance in string compactification from $\mathbf{g}$ to $\mathbf{p}$. Because of their relevance to string compactification, several authors compiled complete lists of conformal pairs $(\mathbf{g}, \mathbf{p})$.[7]

To a finite-dimensional simple Lie algebra $\bar{\mathbf{g}}$, $[T^a, T^b] = i f^{abc} T^c$, $a, b, c = 1, 2, \cdots, \dim \bar{\mathbf{g}}$, one associates the affine Kač–Moody algebra $\mathbf{g}$ via the loop algebra of $\bar{\mathbf{g}}$, $\mathcal{L}(\bar{\mathbf{g}}) = \mathbf{C}[t, t^{-1}] \otimes \bar{\mathbf{g}}$. Here, $\mathbf{C}[t, t^{-1}]$ is the ring of Laurent polynomials in t and an element of the loop algebra is $T_n^a = t^n T^a$, $n \in \mathbf{Z}$, $a = 1, 2, \cdots, \dim \bar{\mathbf{g}}$. The affine algebra $\mathbf{g}$ is defined as

$$\mathbf{g} = \mathcal{L}(\bar{g}) \oplus \mathbf{C}k \oplus \mathbf{C}d$$

with $\mathbf{C}k$ the one-dimensional central extension (called also the affine charge or the Schwinger term in the current algebra) determined by the two-cocycle on $\mathcal{L}(\bar{\mathbf{g}})$, and $\mathbf{C}d$ is the derivation acting on $\mathcal{L}(\bar{\mathbf{g}})$ as $t\frac{d}{dt}$, *i.e.*, it counts the degree in t of the generators T_n^a.

The commutation relations of the Kač–Moody algebra are

$$[T^a_m, T^b_n] = if^{abc}T^c_{m+n} + km\delta^{ab}\delta_{m,-n}$$

$$[d, T^a_n] = nT^a_n \ ; \qquad [d, k] = 0 \ .$$

Similar to the decomposition of $\bar{g}$ into Cartan subalgebras $\bar{h}$ and step operators $\bar{n}_\pm$, $\bar{g} = \bar{n}_+ + \bar{h} + \bar{n}_-$, the affine algebra g has the triangular decomposition

$$g = n_+ + h + n_-$$

$$n_\pm = \bar{n}_\pm + \sum_{k>0} t^{\pm k} \otimes \bar{g}$$

$$h = \bar{h} + Ck + Cd \ .$$

Based on this decomposition, one defines for a dominant weight Λ of g an irreducible, highest weight representation as the integrable highest weight g-module $L(\Lambda)$ with highest weight Λ as follows: $L(\Lambda)$ contains a highest weight vector $|v\rangle$, the "vacuum", which is killed by n_+ part of g, $n_+|v\rangle = 0$ and $|v\rangle$ is an eigenvector for the Cartan subalgebra h of g: $H|v\rangle = \Lambda(H)|v\rangle$ for all $H \in h$. The step generators in n_- acting on the vacuum produce all the states of the representation. The eigenvalue of the affine charge is called the level of the representation $m = \Lambda(k)$ and the dominant weight Λ is said to be of level m. The fundamental weights of the affine algebra are $\Lambda_0, \Lambda_1, \cdots \Lambda_\ell$: $\Lambda_i = \bar{\Lambda}_i + a_i^V \Lambda_0$ where $\bar{\Lambda}_i$ are the fundamental weights of $\bar{g}$.

The highest root is a linear combination of simple roots $\theta = \sum_{i=1}^{\ell} a_i \alpha_i$, where α_i are the simple roots. For algebras of type $A - D - E$ (simply laced), $a_i^V = a_i$. For the non-simply lased case a_i^V are the components of the highest coroot $\theta^V = \sum a_i^V \alpha_i^V$ in terms of the simple coroots α_i^V. Put $\bar{\rho} = \sum_{i=1}^{\ell} \bar{\Lambda}_i$ the sum of fundamental weights of $\bar{g}$, then $\rho = \bar{\rho} + g\Lambda_0$, g is the dual Coxeter number of $\bar{g}$, $g = 1 + \sum_{i=1}^{\ell} a_i^V$. The affine charge acts on all vectors of $L(\Lambda)$ as multiplication by m.

We are interested in conformal invariance and study the affine algebras together with their Virasoro algebra partner given by the Sugawara construction. The generators of the Virasoro algebra are

$$L^g_n = \frac{1}{2(m+g)} \sum_{k \in \mathbf{Z}} : \sum_{a=1}^{\dim \bar{g}} T^a_{k+n} T^a_{-k} : \ .$$

Here g is the dual Coxeter number for $\bar{\mathbf{g}}$ and m is the level of the representation $L(\Lambda)$. The dots denote normal ordering and thus the T_n^a's of positive n are moved to the right. Since $n_+|v\rangle = 0$ this ensures finiteness when L_n^g act on $|v\rangle$ as an infinite sum over $k \in \mathbf{Z}$. The L_n^g's satisfy the Virasoro algebra:

$$[L_m, L_n] = (m-n)L_{m+n} + \frac{1}{12}c(m^3 - m)\delta_{m,-n}$$

$$[T_n^a, L_m] = nT_{n+m}^a \quad ; \qquad [k, L_m] = 0$$

with the central term (conformal charge or conformal anomaly) given by

$$c = \frac{k \dim \bar{\mathbf{g}}}{k+g} \tag{2}$$

where k is the affine charge (if viewed as an operator, the level m is its eigenvalue) and g is the dual Coxeter number of $\bar{\mathbf{g}}$.

For physical applications we require that the energy (or mass) spectrum to be bounded below (or to be non-negative). In quantum field theory in two dimensions, one has two commuting Virasoro algebras L_n and $\bar{L}_n$ (components of θ_{++} and θ_{--}) and the total energy is $H \simeq L_o + \bar{L}_0$. In string theory L_0 gives the mass spectrum. It is understandable then to require the spectrum of L_0 to be bounded below. A representation for which L_0 has a lower bound is called a positive energy representation and the highest weight g-module $L(\Lambda)$ defined above is such a highest weight representation. The representations of the Virasoro algebra L_n^g are labeled by the conformal charge c and by h, the minimal eigenvalue of L_0; Vir$\,(c,h)$. Unitarity requires $T_n^{a+} = T_{-n}^a$ and $L_n^+ = L_{-n}$ and the states obtained from $|v\rangle$ by applying the step operators L_{-n} and T_{-n}^a must have non-negative norm. Using the Kač formula for determinants of matrices of inner products of vectors of a given L_0-eigenvalue, Friedan, Qiu and Shenker[8] have shown that $c \geq 0$, $h \geq 0$ is a necessary condition for unitarity of a highest weight representation Vir$\,(c,h)$. No further restrictions are needed for $c \geq 1$, $h \geq 0$ and in this case there exists a continuum of unitary representations, for each value of c there is an infinity of values of h which are allowed. For $0 \leq c \leq 1$ the unitary representations form a discrete sequence with the permitted values given by

$$c = 1 - \frac{6}{(m+2)(m+3)} \,, \qquad m = 0,\, 1,\, \cdots$$
$$h_{p,q} = \frac{[(m+3)p - (m+2)q]^2 - 1}{4(m+2)(m+3)} \,, \qquad \begin{array}{l} p = 1,\, 2,\, \cdots,\, m+1 \\ q = 1,\, 2,\, \cdots,\, p \end{array} \tag{3}$$

Consider the affine embedding $\mathbf{p} \subset \mathbf{g}$. The unitary irreducible highest weight (HW) representation $L(\Lambda)$ of $\mathbf{g}$ with HW Λ at level m is reducible into representations $L(\dot{\Lambda})$ of $\mathbf{p}$ of HW $\dot{\Lambda}$ at level $\dot{m} = mI$. In general, the decomposition $\mathbf{g} = \mathbf{p} \oplus \mathbf{v}$ has an infinity of terms:

$$L(\Lambda) = \bigoplus_{j=1}^{\infty} L(\dot{\Lambda}_j) \times \text{Vir}\,(c(\mathbf{g}) - c(\mathbf{p}), h_j) \quad . \tag{4}$$

Here, $\text{Vir}(c,h)$ is the Virasoro algebra corresponding to the coset $\mathbf{g}/\mathbf{p}$ in the GKO[9] construction. Its generators K_n and central charge are :

$$K_n = L_n^q - L_n^p \,;\; c \equiv c(\mathbf{g}) - c(\mathbf{p}) = \frac{m \dim \bar{\mathbf{g}}}{m+g} - \frac{\dot{m} \dim \bar{\mathbf{p}}}{\dot{m} + \dot{g}} \tag{5}$$

(here and throughout the paper the dot reffers to the subalgebra and the bar means restriction to the finite dimensional algebra). When $0 \le c \le 1$, the unitarity restricts the representations to the "discrete series" [8] and the decomposition (4) is finitely reducible,*i.e.* the sum on the right hand side has a finite number of terms (the ones allowed by unitarity). Conversely,[2] finite reducibility implies $c \le 1$. For $c = 0$, $\text{Vir}(c, h_j)$ is trivial, all generators vanish identically $K_n \equiv 0$ and $L(\Lambda)$ is finitely reducible with respect to $\mathbf{p}$ alone

$$L(\Lambda) = \bigoplus_{j}^{N} L(\dot{\Lambda}_j) \,. \tag{6}$$

It follows that an explicit decomposition as (6) can be written down only for $c_k = 0$, $\text{Vir}(c_k, h) \equiv 0$. In this case $c(g) = c(p)$ and we then say that the conformal factor $f = \frac{c(p)}{c(g)}$ equals one.

$$f = \frac{c(p)}{c(g)} = \frac{\dim \bar{\mathbf{p}}}{\dim \bar{\mathbf{g}}} \, \frac{m+g}{\dot{m}+\dot{g}} \, \frac{\dot{m}}{m}$$

The conformal factor is ≤ 1 and we are interested in its maximal value. Let's see the dependence of f on the level m

$$f = \frac{\dim \bar{\mathbf{p}}}{\dim \bar{\mathbf{g}}} \cdot \frac{\dot{m}}{m} \cdot \frac{m+g}{\dot{m}+\dot{g}} = \text{constant} \cdot \frac{m+g}{m+\dot{g}/I}$$

$$\frac{\partial f}{\partial m} \sim \left(\frac{\dot{g}}{I} - g\right)$$

Here we used $\dot{m} = mI$, I is the embedding index $I(g/p) = I_2^p(R)/I_2^g(R)$ where I_2^p, I_2^g are the second index of $\mathbf{p}$ and $\mathbf{g}$ in the representation R (see Ref. [14]). By the second index sum rule $Ig > \dot{g}$ and f decreases with the level m of $\mathbf{g}$. Therefore, Kač–Moody algebra with $m > 1$ *do not* have conformal subalgebras.[9]

Obviously the decomposition (6) can be used to write nontrivial modular invariant partition functions for the subalgebra. The condition $c(\mathbf{g}) = c(\mathbf{p})$ occurs only for levels $m = 1$ (see the second of Ref.[9]) and $\dot{m} = I$, therefore one starts with the level one trivial partition function of $\mathbf{g}$ and substitutes the $\mathbf{p}$-content of each representation. Nahm and Bouwknegt[10] used the fact that the Virasoro generators obtained from $\mathbf{g}$ and $\mathbf{p}$ via Sugawara construction are identified (the "quantum equivalence theorem"[11]) and, in particular, $L_0^{\mathbf{g}} = L_0^{\mathbf{p}}$. On a HW vector $|\Lambda\rangle$ this reads:

$$\frac{\Omega_\Lambda}{2(m+g)} + n = \frac{\Omega_\lambda}{2(\dot{m}+\dot{g})} \quad , \qquad n \in \mathbf{Z}_+ \tag{7}$$

where Ω_Λ and Ω_λ are the eigenvalues of the second Casimir on $L(\Lambda)$ and $L(\lambda)$, respectively. They compute the Ω_Λ and Ω_λ for all the representations of level $m = 1$ and $\dot{m} = I$, respectively, and find the content of $L(\Lambda)$ by matching (7). This matching is equivalent to the "trace condition" we impose: the trace anomaly $\dot{h}_\lambda$ equals the trace anomaly h_Λ modulo an integer[13]:

$$h_\Lambda - \dot{h}_\lambda \in \mathbf{Z}. \tag{8}$$

In very simple cases the trace condition alone determines the decomposition. In our examples we use first the trace condition to cut down the number of eligible representations and then determine the multiplicities by matching the asymptotics ($\tau \to 0$) of the Kač characters. In some cases even this matching does not solve the decomposition and one finds the multiplicities by solving some matrix equation.[6] To find the decompositions we use the modular properties of the characters.

The character of the g-module $L(\Lambda)$ is defined:

$$ch_\Lambda(\tau,z) = q^{-c/24}\,\mathrm{tr}_{\,L(\Lambda)}q^{L_0}e^{2\pi i z} \tag{9}$$

$q = e^{2\pi i\tau}$ and we denote an element in the affine CSA by $(z,t,\tau) = 2\pi i(z + tk - \tau d), \tau, t \in \mathrm{C}$, $z \in \overline{CSA}$, the Cartan subalgebra of the finite dimensional Lie algebra, k is the affine charge and $d = -L_0$. The spectrum of L_0 is bounded below, therefore $Im\tau > 0$. Under the action of the modular group $SL_2(\mathbf{Z})$, the complex linear span of characters for level m representations $\Lambda \in P^{(m)}$, form an invariant subspace.[12]

$$ch_\Lambda\left(-\frac{1}{\tau},z\right) = \sum_{M\in P^{(m)}} a(\Lambda,M)ch_M(\tau,z) \tag{10a}$$

$$ch_\Lambda(\tau+1,z) = e^{2\pi i S_\Lambda}ch_\Lambda(\tau,z). \tag{10b}$$

$S_\Lambda = h_\Lambda - \frac{1}{24}c$ is the modular anomaly and h_Λ the trace anomaly. Since $S_\Lambda - \dot{S}_\lambda = (h_\Lambda - \dot{h}_\lambda) - \frac{1}{24}(c_\mathrm{g} - c_\mathrm{p})$, for conformal pairs $c_\mathrm{g} = c_\mathrm{p}$ and the cancellation of the modular anomaly requires the cancellation of the trace anomaly, Eq.(7) or (8).

The complex matrix $a(\Lambda,M)$ is symmetric, unitary and invariant under automorphisms of the Dynkin diagram.[6,12,13] From (10a) follows the asymptotic behavior[12,13]:

$$\text{as } \tau \to 0 \ , \quad ch_\Lambda(\tau,0) \sim a(\Lambda)\cdot e^{2\pi i c/24\tau} \tag{11}$$

where $a(\Lambda) \equiv a(\Lambda, m\Lambda_0)$ are positive real numbers which play the role of dimensions of $L(\Lambda)$ and we call them "asymptotic dimensions". Unitarity of $a(\Lambda,M)$ implies $\sum_\Lambda a(\Lambda)^2 = 1$, when the sum is taken over all representations at a given level.

The branching functions b_λ^Λ are defined:

$$b_\lambda^\Lambda(\tau) = q^{-c/24}\ \mathrm{tr}_{\mathrm{Vir}\,(c,h)}q^{(L_0^\mathrm{g} - L_0^\mathrm{p})} \tag{12}$$

For $c = 0$, $L_0^\mathrm{g} - L_0^\mathrm{p} = 0$ they are constants and equal the multiplicities with which $L(\lambda)$ appear in the decomposition of $L(\Lambda)$:

$$ch_\Lambda(\tau,z) = \sum_{\lambda\in P^{(\dot{m})}} b_\lambda^\Lambda ch_\lambda(\tau,\dot{z}). \tag{13}$$

The branching functions b_λ^Λ have similar transformation properties under the modular group as the characters. The asymptotic behavior of the branching functions is obtained from (11) and (13):

$$\sum_\lambda \lim_{\tau\to 0} b_\lambda^\Lambda(\tau)\cdot \dot{a}(\lambda) = a(\Lambda)\, e^{\pi i(c_{\mathrm{g}}-c_{\mathrm{p}})/12\tau} \tag{14}$$

For $c_{\mathrm{p}} = c_{\mathrm{g}}$, the branching functions are constants anyways and they equal their asymptotic limit, $b_\lambda^\Lambda = mult_\Lambda(\lambda)$. Then (14) becomes:

$$a(\Lambda) = \sum_{\lambda\in P^{(I)}} mult_\Lambda(\lambda)\dot{a}(\lambda) \tag{15}$$

which is the matching of the asymptotic dimensions which we use to solve for the multiplicities. We calculate the real numbers $a(\Lambda)$ and $\dot{a}(\lambda)$ with the formula:

$$a(\Lambda) = \frac{1}{\sqrt{\det A_{\mathrm{g}}}}\frac{1}{(m+g)^{\ell/2}}\prod_{\alpha\in\bar{\Delta}_+} 2\sin\frac{\pi(\bar{\Lambda}+\bar{\rho}|\alpha)}{(m+g)} \quad . \tag{16}$$

Here $\det A_{\mathrm{g}}$ is the determinant of the Cartan matrix for the algebra $\bar{\mathrm{g}}$, ℓ the rank of $\bar{\mathrm{g}}$ and the product is over all positive roots. If $\Lambda = (k_0, k_1, \ldots, k_l)$ and s_i are the components of the positive roots $\alpha = \sum_{i=1}^\ell s_i\alpha_i$, $(\bar{\Lambda}+\bar{\rho}|\alpha) = \sum_{i=1}^\ell (k_i+1)s_i$. In solving (15) we restrict the search to the representations $L(\lambda)$ which obey the "trace condition" of Eq.(8). For $\Lambda = k_0\Lambda_0 + \sum_{i=1}^\ell k_i\bar{\Lambda}_i$ the trace anomaly is given by:

$$h_\Lambda = \frac{(\Lambda+2\rho|\Lambda)}{2(m+g)} = \sum_{i=1}^\ell\sum_{j=1}^\ell (k_i+2)k_j\cdot(\bar{\Lambda}_i|\bar{\Lambda}_j)\Big/ 2(m+g) \tag{17}$$

The matrix $(\bar{\Lambda}_i|\bar{\Lambda}_j)$ is listed in Ref. [14] as the quadratic form matrix.

As examples of decomposition, we use the conformal pairs (g, p) with g the affine algebras associated with the exceptional Lie algebras. The subalgebras p are simple or semi-simple and there are examples of both regular (same rank and index of embedding $= 1$) and non-regular (lower rank and index ≥ 1). We list them all in Table 1 and the explicit decompositions are listed in Ref.[6].

g	p	Index	g	p	Index
		(a)			
E_6	$A_5 \oplus A_1$	(1,1)	E_8	$E_7 \oplus A_1$	(1,1)
E_6	$A_2 \oplus A_2 \oplus A_2$	(1,1,1)	E_8	$E_6 \oplus A_2$	(1,1)
E_7	$D_6 \oplus A_1$	(1,1)	E_8	A_8	1
E_7	$A_5 \oplus A_2$	(1,1)	E_8	$A_4 \oplus A_4$	(1,1)
E_7	A_7	1	F_4	$C_3 \oplus A_1$	(1,1)
		(b)			
E_6	C_4	1	E_7	A_2	21
E_6	A_2	9	E_8	C_2	12
E_6	G_2	3	E_8	$G_2 \oplus F_4$	(1; 1)
E_6	$A_2 \oplus G_2$	(2; 1)	E_8	$A_1 \oplus A_2$	(16; 6)
E_7	$A_1 \oplus F_4$	(3; 1)	F_4	$A_3 \oplus A_2$	(2; 1)
E_7	$G_2 \oplus C_3$	(1; 1)	F_4	$G_2 \oplus A_1$	(1; 8)
E_7	$A_1 \oplus G_2$	(7; 2)	G_2	$A_1 \oplus A_1$	(3; 1)

Table 1: Maximal conformal subalgebras (a) regular, (b) non-regular of affine exceptional algebras.

In the cases when the subalgebra **p** is semi-simple and has, for example, two pieces $\mathbf{p} = \mathbf{p}_1 + \mathbf{p}_2$ with indexes (I_1, I_2), we consider the representations $\dot{L}(\lambda_1)$ of $\mathbf{p}_1$ of level $\dot{m}_1 = I_1$ and the representations $\ddot{L}(\lambda_2)$ of $\mathbf{p}_2$ of level $\ddot{m}_2 = I_2$. The trace condition becomes

$$h_\Lambda = \dot{h}_{\lambda_1} + \ddot{h}_{\lambda_2} \qquad \text{modulo } \mathbf{Z} \ .$$

The numbers $\dot{a}(\lambda)$ and $\dot{a}(\lambda,\mu)$ for $\mathbf{p} = \mathbf{p}_1 + \mathbf{p}_2$ are replaced by the products $\dot{a}(\lambda_1) \cdot \ddot{a}(\lambda_2)$ and $\dot{a}(\lambda_1,\mu_1) \cdot \ddot{a}(\lambda_2,\mu_2)$, respectively.

For strings propagating on a group manifold G the one-loop amplitude is the partition function of the theory. The modular invariance of the partition function is nothing else but the requirement that the amplitude is invariant under the global reparametrizations of the torus. The general expression for such a partition function is

$$Z(\tau) = \sum_{R,R'} M_{R,R'} ch_R ch^*_{R'} \tag{18}$$

where R, R' are representations of G for the left and right movers and the coefficients $M_{R,R'}$ (the "mass-matrix") specify the field content of the theory. Since the characters at a given level form a modular invariant subspace, an obvious modular invariant partition function is the trivial one, with $M_{R,R'} = \delta_{R,R'}$. Starting with the level 1 trivial partition functions we substitute the subalgebra content. Since the decompositions preserve the conformal and modular anomalies, the resulting partition functions are modular invariant. Unfortunately we are restricted to the levels $k = I$ (the level of embedding of the conformal subalgebra) and to the algebras which appear in the list of conformal subalgebras.[9] Still, this is a much larger class of modular invariants than the $SU(2)$ ones classified by Cappelli-Itzykson-Zuber.

From the embeddings $SU(N) \supset SO(N)$, of index $I = 2$ given in Ref.[13] we obtain the partition functions $Z^{k=2}_{SO(N)}$.

a) $N = 2\ell+1$, $\Lambda_0 = (2\dot{\Lambda}_0) + (2\dot{\Lambda}_1)$, $\Lambda_1 = (\dot{\Lambda}_0 + \dot{\Lambda}_1)$, $\Lambda_j = \dot{\Lambda}_j$, $2 \leq j \leq \ell - 1$, $\Lambda_\ell = (2\dot{\Lambda}_\ell)$ and given the $\mathbf{Z_2}$ symmetry of $A_{2\ell}$, $\Lambda_{\ell+1} = \Lambda_\ell, \ldots, \Lambda_{2\ell} = \Lambda_1$ such that

$Z^{k=1}_{SU(2\ell+1)} = \sum_{i=0}^{2\ell} |\chi_{\Lambda_i}|^2 = |\chi_{\Lambda_0}|^2 + 2\sum_{i=1}^{\ell} |\chi_{\Lambda_i}|^2$ becomes

$Z^{k=2}_{SO(2\ell+1)} = |\chi_{2\dot{\Lambda}_0} + \chi_{2\dot{\Lambda}_1}|^2 + 2\left(\chi^2_{\dot{\Lambda}_0+\dot{\Lambda}_1} + \chi^2_{2\dot{\Lambda}_\ell} + \sum_{j=2}^{\ell-1} \chi^2_{\dot{\Lambda}_j}\right)$. Example:

$Z^{(2)}_{SO(7)} = |\chi_{(000)} + \chi_{(200)}|^2 + 2\chi^2_{(100)} + 2\chi^2_{(010)} + 2\chi^2_{(002)}$.

b) $N = 2\ell$, $\Lambda_0 = (2\dot{\Lambda}_0) + (2\dot{\Lambda}_1)$, $\Lambda_1 = (\dot{\Lambda}_0 + \dot{\Lambda}_1)$, $\Lambda_j = \dot{\Lambda}_j$, $2 \leq j \leq \ell - 2$, $\Lambda_{\ell-1} = (\dot{\Lambda}_{\ell-1} + \dot{\Lambda}_\ell)$, $\Lambda_\ell = (2\dot{\Lambda}_{\ell-1}) + (2\dot{\Lambda}_\ell)$ and, by symmetry, $\Lambda_{\ell+1} = \Lambda_{\ell-1}, \ldots, \Lambda_{2\ell-1} = \Lambda_1$. Then,

$Z^{k=1}_{SU(2\ell)} = \sum_{i=0}^{2\ell-1} |\chi_{\Lambda_i}|^2 = |\chi_{\Lambda_0}|^2 + |\chi_{\Lambda_\ell}|^2 + 2\sum_{i=1}^{\ell-1} |\chi_{\Lambda_i}|^2$ becomes

$$Z^{k=2}_{SO(2\ell)} = |\chi_{2\dot{\Lambda}_0} + \chi_{2\dot{\Lambda}_1}|^2 + |\chi_{2\dot{\Lambda}_{\ell-1}} + \chi_{2\dot{\Lambda}_\ell}|^2 + 2\left(\chi^2_{\dot{\Lambda}_0+\dot{\Lambda}_1} + \chi^2_{\dot{\Lambda}_{\ell-1}+\dot{\Lambda}_\ell} + \sum_{j=2}^{\ell-2} \chi^2_{\dot{\Lambda}_j}\right).$$ Example:

$$Z^{k=2}_{SO(8)} = |\chi_{(0000)} + \chi_{(2000)}|^2 + |\chi_{(0020)} + \chi_{(0002)}|^2 + 2\left(\chi^2_{(1000)} + \chi^2_{(0011)} + \chi^2_{(0100)}\right).$$

Besides these infinite series we can start with the level 1 exceptional partition functions $Z_{E_8} = |\chi_{\Lambda_0}|^2$, $Z_{E_7} = |\chi_{\Lambda_0}|^2 + |\chi_{\Lambda_6}|^2$, $Z_{E_6} = |\chi_{\Lambda_0}|^2 + |\chi_{\Lambda_1}|^2 + |\chi_{\Lambda_5}|^2$ and use the explicit decompositions given in Ref.[6]. When the index of embedding is 1, the decompositions do not seem very useful: the new partition function contains all level one characters and equals the trivial partition function. This happens for:

$E_8 \supset E_7 + A_1;\ E_6 + A_2;\ A_4 + A_4;\ G_2 + F_4$,

$E_7 \supset D_6 + A_1;\ A_5 + A_2;\ G_2 + C_3;$

$E_6 \supset A_5 + A_1;\ A_2 + A_2 + A_2$ and $F_4 \supset C_3 + A_1$.

In the above the subalgebra has two components. If we gauge away one of the symmetries (disregard it), we obtain the trivial partition function for the remaining algebra. We choose the example of $E_8 \supset E_6 + A_2 : \Lambda_0 = (\dot{\Lambda}_0, \ddot{\Lambda}_0) + (\dot{\Lambda}_1, \ddot{\Lambda}_1) + (\dot{\Lambda}_5, \ddot{\Lambda}_2)$, or, $\Lambda_0 = (\dot{\Lambda}_0, \ddot{\Lambda}_0) + (\dot{\Lambda}_1, \ddot{\Lambda}_2) + (\dot{\Lambda}_5, \ddot{\Lambda}_1)$. Gauging away E_6 (A_2) we obtain from $Z_{E8}^{k=1} = |\chi_{\Lambda_0}|^2$ the trivial partition functions for A_2 (E_6). The above decomposition acts as a change of basis in the character space. We also wish to remark here that whenever the decompositions admit a double solution, both solutions produce the same partition function.

Rather surprising are the partition functions from:

$E_8 \supset A_8$, $\Lambda_0 = (\dot{\Lambda}_0) + (\dot{\Lambda}_3) + (\dot{\Lambda}_6)$, $Z_{A_8} = |\chi_{\dot{\Lambda}_0} + \chi_{\dot{\Lambda}_3} + \chi_{\dot{\Lambda}_6}|^2$

$E_8 \supset D_8$, $\Lambda_0 = (\dot{\Lambda}_0) + (\dot{\Lambda}_7)$ or $(\dot{\Lambda}_0) + (\dot{\Lambda}_8)$, $Z_{D_8} = |\chi_{\dot{\Lambda}_0} + \chi_{\dot{\Lambda}_{7(8)}}|^2$

$E_7 \supset A_7$, $\Lambda_0 = (\dot{\Lambda}_0) + (\dot{\Lambda}_4); \Lambda_6 = (\dot{\Lambda}_2) + (\dot{\Lambda}_6)$, $Z_{A_7} = |\chi_{\dot{\Lambda}_0} + \chi_{\dot{\Lambda}_4}|^2 + |\chi_{\dot{\Lambda}_2} + \chi_{\dot{\Lambda}_6}|^2$ which contain only one-third or a half of level one representations which we expect by triviality. This reflects the Z_3 and Z_2 symmetries of $A_8^{(1)}$ and $D_8^{(1)}$, $A_7^{(1)}$. Since we are still at level 1 we can further substitute for the conformal subalgebras $A_8 \supset SO(9)$ and $A_7 \supset D_4$ to obtain more nontrivial partition functions.

From decompositions of Ref.[6] with higher index of embedding we write down higher level partition functions.

Using $E_8 \supset A_1 + A_2$ we can write $Z_{A_1+A_2}^{(16,6)}$. Disregarding the A_2, one obtains a level $k = 16$ $A_1^{(1)}$ partition function already in the CIZ series. Indeed, one gets in this way $Z_{A_1}^{k=16} = Z(D_{10}) + 2Z(E_7)$ where $Z(D_{10})$ is the one in (1b) with $p = 4$ and $Z(E_7)$ is the one in (1e). From the same decomposition we can write the $SU(3)$ level 6 partition function:

$$Z_{A_2}^{k=6} = |\chi_{(00)} + \chi_{(06)} + \chi_{(60)}|^2 + |\chi_{(03)} + \chi_{(30)} + \chi_{(33)}|^2$$
$$+|\chi_{(41)} + \chi_{(14)} + \chi_{(11)}|^2 + 3|\chi_{(22)}|^2$$

From the conformal subalgebra $E_8 \supset C_2$ follow the partition function:

$$Z_{C_2}^{k=12} = |\chi_{(0,0)} + \chi_{(0,12)} + \chi_{(2,3)} + \chi_{(2,7)} + \chi_{(8,1)} + \chi_{(8,3)} + 2\chi_{(4,4)} + \chi_{(6,0)} + \chi_{(6,6)}|^2$$

From the conformal subalgebras of E_7 we obtain the following higher level MIPF's

$$Z_{A_1+G_2}^{(7,2)} = |\chi_1\chi_{(00)} + \chi_7\chi_{(10)} + \chi_3\chi_{(02)} + \chi_5\chi_{(01)}|^2 +$$

$$+|\chi_8\chi_{(00)} + \chi_2\chi_{(10)} + \chi_6\chi_{(02)} + \chi_4\chi_{(01)}|^2$$

$Z^{(3,1)}_{A_1+F_4} = |\chi_1\chi_{\ddot{\Lambda}_0} + \chi_3\chi_{\ddot{\Lambda}_4}|^2 + |\chi_4\chi_{\ddot{\Lambda}_0} + \chi_2\chi_{\ddot{\Lambda}_4}|^2$ from which, gauging away F_4, we obtain the $k = 3$ CIZ partition function of (1a).

Another example of conformal subalgebra of E_7 is:

$$Z^{k=21}_{A_2} = |\chi_{(0,0)} + \chi_{(21,0)} + \chi_{(0,21)} + \chi_{(10,10)} + \chi_{(10,1)} + \chi_{(1,10)} + \chi_{(4,4)} + \chi_{(13,4)} + \chi_{(4,13)} + \chi_{(6,6)} + \chi_{(9,6)} + \chi_{(6,9)}|^2 + |\chi_{(6,0)} + \chi_{(0,6)} + \chi_{(15,0)} + \chi_{(0,15)} + \chi_{(15,6)} + \chi_{(6,15)} + \chi_{(7,4)} + \chi_{(4,7)} + \chi_{(10,4)} + \chi_{(4,10)} + \chi_{(10,7)} + \chi_{(7,10)}|^2$$

From decompositions of E_6 with respect to its conformal subalgebras we obtain the following MIPF's:

$$Z^{k=9}_{A_2} = |\chi_{(0,0)} + \chi_{(9,0)} + \chi_{(0,9)} + \chi_{(4,4)} + \chi_{(4,1)} + \chi_{(1,4)}|^2 + \\ +2|\chi_{(2,2)} + \chi_{(5,2)} + \chi_{(2,5)}|^2.$$

$$Z^{k=3}_{G_2} = |\chi_{(00)}\chi_{(11)}|^2 + 2|\chi_{(02)}|^2$$

These are just a few examples. More modular invariant partition functions can be calculated using the lists of conformal subalgebras of any Kač-Moody algebra and the algorithm described in Ref.[6] to perform the explicit decomposition. When looking for nontrivial modular invariant partition function, one actually looks for an invariant subspace in the space of characters at a certain level. As already shown on a few examples, the method recovers the CIZ classification when the group is $SU(2)$. A further example is the decomposition $F_4 \supset G_2 + A_1$, $I = (1,8)$.

$$\Lambda_0 = (\dot{\Lambda}_0, 8\ddot{\Lambda}_0) + (\dot{\Lambda}_0, 8\ddot{\Lambda}_1) + (\dot{\Lambda}_2, 4\ddot{\Lambda}_0 + 4\ddot{\Lambda}_1),$$
$$\Lambda_4 = (\dot{\Lambda}_0, 4\ddot{\Lambda}_0 + 4\ddot{\Lambda}_1) + (\dot{\Lambda}_2, 6\ddot{\Lambda}_0 + 2\ddot{\Lambda}_1) + (\dot{\Lambda}_2, 2\ddot{\Lambda}_0 + 6\ddot{\Lambda}_1)$$

yields the partition function

$$Z^{k=1}_{F_4} = |\chi_{\Lambda_0}|^2 + |\chi_{\Lambda_4}|^2 =$$
$$Z^{(1,8)}_{G_2+A_1} = |\chi_{\dot{\Lambda}_0}(\chi_1 + \chi_9) + \chi_{\dot{\Lambda}_2}\chi_5|^2 + |\chi_{\dot{\Lambda}_0}\chi_5 + \chi_{\dot{\Lambda}_2}(\chi_3 + \chi_7)|^2.$$

If we disregard the G_2 part we obtain

$Z^{k=8}_{SU(2)} = |\chi_1 + \chi_9|^2 + |\chi_3 + \chi_7|^2 + 2|\chi_5|^2$, the D_6-CIZ of Eq.(1b) with $p = 2$.

It seems that all $SU(2)$ or Virasoro modular invariants in the list of Cappelli-Itzykson-Zuber can be obtained from decompositions with respect to conformal subalgebras.

If all modular invariants correspond to decompositions with respect to a conformal subalgebra, then one can use the already known MIPF's (for example those in the CIZ classification) to deduce conformal subalgebras decompositions. For instance, in the list of conformal subalgebras appears $G_2 \supset A_1$ index 28. From Eq.(1f) we know Z_{A_1} at $k = 28$ and this is also the level one trivial partition function of G_2, $Z_{G_2}^{k=1} = |\chi_{\Lambda_0}|^2 + |\chi_{\Lambda_2}|^2$, therefore we read the content of Λ_0 and Λ_2 from (1f):

$$\Lambda_0 = (28\dot{\Lambda}_0) + (28\dot{\Lambda}_1) + (18\dot{\Lambda}_0 + 10\dot{\Lambda}_1) + (10\dot{\Lambda}_0 + 18\dot{\Lambda}_1) \text{ and}$$

$$\Lambda_2 = (22\dot{\Lambda}_0 + 6\dot{\Lambda}_1) + (6\dot{\Lambda}_0 + 22\dot{\Lambda}_1) + (16\dot{\Lambda}_0 + 12\dot{\Lambda}_1) + (12\dot{\Lambda}_0 + 16\dot{\Lambda}_1).$$

All the conformal subalgebras decompositions yield modular invariant partition functions bilinear in the Kač-Moody characters. Note that all coefficients are non-negative in the partition functions. Gepner[3], searching for combinations (14) which are modular invariant found the E_6 and E_7-CIZ solutions but also some modular invariant combinations with a few negative coefficients, which he interpreted as "fermionic" terms. These partition functions failed to satisfy the consistency of the operator algebra and they were dropped. By our method there is no need for consistency checks, all the solutions are valid. It remains now to use these modular invariants to classify string models algebraically and further to consider the supersymmetric extension of this work, *i.e.*, the modular invariant partition functions from decompositions of super-conformal subalgebras.

REFERENCES

1. A. Cappelli, C. Itzykson and B. Zuber, *Nucl. Phys.* **B280** [FS18], 445 (1987).

2. J. L. Cardy, *Nucl. Phys.* **B270** [FS16] (1986) 186.

3. D. Gepner and E. Witten, *Nucl. Phys.* **B278**, 493 (1986); D. Gepner, "On the Spectrum of 2-D Conformal Field Theory", preprint(1986)

4. E. Witten, *Commun. Math. Phys.* **92**, 455 (1984) .

5. A. Cappelli, *Phys. Lett.* **185B**, 82 (1987) .

6. M. Niculescu Sanielevici, "String Functions from Conformal Subalgebras", MIT-CTP# 1470, (1987); V. G. Kač and M. N. Sanielevici, in preparation.

7. F. A. Bais and P. G. Bouwknegt, *Nucl. Phys.* **B279** , 561 (1987) ; A. N. Schellekens and N. P. Warner, *Phys. Rev.* **D34**, 3092 (1986) ; R. C. Arcuri, J. F. Gomez and D. Olive, "Conformal Subalgebras and Symmetric Spaces," preprint 1987.

8. D. Friedan, Z. Qiu and S. Shenker,*Phys. Rev. Lett.* **52** , 1575 (1984) .

9. P. Goddard and D. Olive, *Nucl. Phys.* **B257** [FS14], 226 (1985) ; P. Goddard, A. Kent and D. Olive, *Phys. Lett.* **152B** , 88 (1985) .

10. P. Bouwknegt and W. Nahm, *Phys. Lett.* **184B** , 359 (1987).

11. J. F. Gomes, *Phys. Lett.* **171B**, 75 (1986).

12. V. G. Kač and D. H. Peterson, *Adv. Math* **53**, 125 (1984); *Bull. Amer. Math. Soc.* **3**, 1057 (1980).

13. V. G. Kač and M. Wakimoto, "Modular and Conformal Invariance Constraints in Representation Theory of Affine Algebras." MIT preprint, April 1987 ; V. G. Kač and M. Wakimoto, "Unitarizable Highest Weight Representations of the Virasoro Neveu-Schwarz and Ramond Algebras," preprint 1986.

14. W. McKay and J. Patera, *Tables of Dimensions, Indices and Branching Rules* (Decker, New York, 1981).

MORE FOUR DIMENSIONAL (SUPER) STRING THEORIES?

S. James Gates, Jr. *

Department of Physics and Astronomy
University of Maryland, College Park, MD 20742

Abstract

We discuss the superfield lagrangian formulation for the NSR representation of four dimensional string theories. After a review of previous results, we study the possible existence of a class of anomaly-free conformal theories with manifest (1,0) and (1,1) supersymmetry and world-sheet gauge vectors. This may lead to many new string and superstring theories.

1. INTRODUCTION

This discussion might be subtitled, "There may be more four dimensional (super) string theories than Yau ever imagined." Soon after the construction of the heterotic string [1], it was suggested [2] that compactification of the $E_8 \times E_8$ version on a "Calabi-Yau" manifold might lead to an acceptable low energy theory. Of course by the whimsical subtitle, we really mean to raise an important question about four dimensional string theories. Over the last few years, there has been a considerable effort made to use string theories to go beyond the "Kaluza-Klein-Supergravity-GUT" scenario and to provide modifications of the standard model. Although some models have come tantalizingly close, to my knowledge no **compelling** three-generation model has emerged. So perhaps at this point it is not unreasonable to ask, "Are there more four dimensional superstring theories?" I won't provide a complete answer to this question by the end of my discussion. However, I will provide some very suggestive evidence that the class of four dimensional string and superstring theories is much larger than previously thought. Our work suggests that there may be an infinite number of such theories!

---

*Work supported by National Science Foundation Grant PHY 86-19077.

Most of the progress in understanding superstrings has occurred with the use of the NSR representation. Since this representation makes use of a supersymmetry (either (1,0) or (1,1) in the cases of interest) on the world-sheet, the use of superspace techniques manifest this supersymmetry at each stage. These have been known for some time [3] for the spinning strings and have more recently [4] been developed for heterotic strings. A second important requirement to impose on two dimensional field theories, which also describe strings, is the absence of ABJ-type [5] anomalies in the two dimensional Weyl and Lorentz symmetries. Using these two requirements, we will study the structure and build consistent field theories. Our evidence is only suggestive because the final requirement of a string is the absence of global anomalies (modular invariance). We are presently studying this question.

2. UNIDEXTEROUS SUPERSPACE SUPERGRAVITY

To banish the tachyon from string theories, the introduction of supersymmetry seems essential. In the NSR representation this supersymmetry is first seen on the world sheet of the string as a two dimensional supersymmetry. We may assemble the coordinates (τ, σ) of the string world sheet into light-cone coordinates $\sigma^{++} \equiv \sigma + \tau$, $\sigma^{--} \equiv \sigma - \tau$. Along with a single real Grassmann number, ζ^{+}, these make up the "supercoordinate" $z^{M} \equiv (\zeta^{+}, \sigma^{++}, \sigma^{--})$. A superfield is just a function of z^{M}. Assuming analyticity in the Grassmann direction, we have functions such as $X(z)$ given by

$$X(z) = X(\sigma^{++}, \sigma^{--}) + \zeta^{+}\psi_{+}(\sigma^{++}, \sigma^{--}) \ . \tag{2.1}$$

As further definitions we take

$$\partial_{++} \equiv \frac{\partial}{\partial\sigma} + \frac{\partial}{\partial\tau} \ , \quad \partial_{--} \equiv \frac{\partial}{\partial\sigma} - \frac{\partial}{\partial\tau} \ . \tag{2.2}$$

so that $\partial_{++}X$ and $\partial_{--}X$ are also superfunctions. Furthermore, as is customary in supersymmetric theories, we require a spinorial covariant derivative $D_{+} \equiv \frac{\partial}{\partial\zeta^{+}} + i\zeta^{+}\partial_{++}$ which implies that $D_{+}X$ is a superfunction.

In string theories, the reparametrization group of the world-sheet coordinates plays an important role. In a field theory, this implies that we must consider a supergravity theory to admit arbitrary reparametrization $\left(z^{M} \rightarrow f^{M}(z)\right)$ invariance of the theory. Following familiar procedures, we define a supergravity covariant derivative ∇_{A} [4] by

$$\nabla_{A} \ = \ E_{A}{}^{M} D_{M} + \omega_{A}\mathcal{M} \ \ , \ \ D_{M} \equiv (D_{+}, \partial_{++}, \partial_{--}) \ , \tag{2.3}$$

where $E_A{}^M(z) \equiv$ super vielbein and $\omega_A(z) \equiv$ superspin connection. The quantity M is the Lie algebra generator of $SO(1,1)$. Given $SO(1,1)$ tensors of various diferent types $(t_+, t_-, t^+, t^-, t_{++}, t_{--}, t^{++}, t^{--})$, the action of $\mathcal{M}$ is defined by

$$[\mathcal{M}, t_+] = \tfrac{1}{2}t_+ \ , \quad [\mathcal{M}, t_-] = -\tfrac{1}{2}t_- \ , \quad [\mathcal{M}, t_{++}] = t_{++} \ , \quad [\mathcal{M}, t_{--}] = -t_{--} \ ,$$

$$[\mathcal{M}, t^+] = -\tfrac{1}{2}t^+ \ , \quad [\mathcal{M}, t^-] = \tfrac{1}{2}t^- \ , \quad [\mathcal{M}, t^{++}] = -t^{++} \ , \quad [\mathcal{M}, t^{--}] = t^{--} \ . \tag{2.4}$$

In other words, $\mathcal{M}$ just assigns a weight of plus one-half for each lower plus index and minus one-half for each lower minus index. An upper plus (minus) index is equivalent to lower minus (plus) index. Our notation is such that a single plus or minus index is a spinorial index. Usually any quantity with an even (odd) number of such indices is bosonic (fermionic). Torsion and curvature super tensors are defined by graded commutation $[\nabla_A, \nabla_B\} \equiv T_{AB}{}^C \nabla_C + R_{AB}\mathcal{M}$. These are subject to constraints [4] which imply,

$$[\nabla_+, \nabla_+\} = i2\nabla_{++} \quad , \quad [\nabla_+, \nabla_{++}\} = 0 \ ,$$

$$[\nabla_+, \nabla_{--}\} = -i2\Sigma^+\mathcal{M} \quad , \quad [\nabla_{++}, \nabla_{--}\} = -(\ \Sigma^+\nabla_+ + \mathcal{R}\mathcal{M}) \quad . \tag{2.5}$$

These constraints have been solved [4,6] in terms of five independent "prepotential" superfields: $H_+{}^{--}$, $H_{--}{}^{++}$, H^+, L, and Ψ. The first two of these $(H_+{}^{--}, H_{--}{}^{++})$ represent the superconformal multiplet, the next one (H^+) is the local supersymmetry compensator, the following one (L) is the local Lorentz compensator, and the final (Ψ) is the local Weyl compensator. Classically, all of the compensators can be "gauged" to zero by local supersymmetry, local Lorentz, and local scale transformations, respectively. Alternatively, if this statement breaks down in a quantum theory, it is a sign of an anomaly in the classical symmetries. The linearized transformation laws of the superconformal multiplet are $\delta H_+{}^{--} \simeq D_+K^{--}$, $\delta H_{--}{}^{++} \simeq \partial_{--}K^{++}$. Locally these can be used to set the conformal multiplet to zero, but globally there may be obstructions. It was first noted in ref. [7] that $D = 2$ superconformal multiplets should **not** be "gauged away" owing to the inherent non-locality of this process. These obstructions lead to the notion of supermoduli [8]. We may characterize these by a set of Teichmuller parameters $\tilde{u}$ [8]. Since we have the complete solution [6], it is easy to deduce the dependence of the (1,0) superdeterminant on the Teichmuller parameters through the superconformal multiplet $H_+{}^{--}(\tilde{u})$ and $H_{--}{}^{++}(\tilde{u})$. We note (defining $E \equiv sdet\left(E_A{}^M\right))$ that

$$E^{-1} = \Psi^{\frac{3}{2}}(1 \cdot e^{\frac{1}{2}H^+\ \overleftarrow{D}_+})e^{\frac{1}{2}L}\hat{E}^{-1}(\tilde{u}) \ ,$$

$$\hat{E}(\tilde{u}) = [1 + iH_{--}{}^{++}(\tilde{u})(D_+H_+{}^{--}(\tilde{u}) + H_+{}^{--}(\tilde{u})\partial_{--}H_+{}^{--}(\tilde{u}))\ ] \ . \tag{2.6}$$

The path integral over $E_A{}^M$ is defined in terms of its independent prepotentials.

$$\int [DE_A{}^M] \equiv \sum_{(\tilde{u})} \int [DH_+{}^{--}][DH_{--}{}^{++}][DH^+][DL][D\Psi] \quad , \qquad (2.7)$$

where the sum of $\tilde{u}$ covers all the different topologies.

The result in (2.6) may be compared to the bosonic case. In the bosonic theory we may define "conformal" zwiebeins $\hat{e}_{++} \equiv \partial_{++} + h_{++}{}^{--}\,\partial_{--}$ and $\hat{e}_{--} \equiv \partial_{--} + h_{--}{}^{++}\,\partial_{++}$. The usual "Poincaré" zwiebeins are defined by $e_{++} = \psi e^{\ell}\,\hat{e}_{++}$ and $e_{--} = \psi\, e^{-\ell}\,\hat{e}_{--}$. The fields ψ and ℓ have the interpretation of scale and Lorentz compensators, respectively. A short calculation of $det\,(e_a{}^m)$ yields

$$e^{-1} = \psi^2 \hat{e}^{-1}(\tilde{u}) \quad , \quad \hat{e}(\tilde{u}) = 1 - h_{--}{}^{++}(\tilde{u}) h_{++}{}^{--}(\tilde{u}) \quad . \qquad (2.8)$$

The similarities are now obvious. The dependence on the Teichmuller parameter is restricted to the determinant or superdeterminant of the conformal $(\hat{e})$ or superconformal $(\hat{E})$ zwiebeins. Following earlier discussions [6], the reduction to components can be found. A key ingredient in this is the identity

$$\int d\zeta^- E^{-1}\mathcal{L}_- = [det(e_a{}^m)]^{-1}((\nabla_+ - i\psi_{++}{}^+)\mathcal{L}_-)| \equiv e^{-1}((\nabla_+ - i\psi_{++}{}^+)\mathcal{L}_-)| \quad (2.9)$$

where | indicates the limit as ζ^+ approaches zero. The actual component fields of (1,0) supergravity are the vielbein $e_a{}^m\,(\sigma^{++},\sigma^{--})$ and the gravitino $\psi_a{}^+(\sigma^{++},\sigma^{--})$. Starting from superspace [6], we have derived the component "tensor calculus" for these theories. The reader is referred to our earlier works for details.

3. MATTER MULTIPLETS

As in all theories of gravitational interactions, there are arbitrary numbers of matter field multiplets. These have been enumerated previously, up to and including matter gravitino multiplets [4,6]. In order of increasing spin these are: (a) the scalar multiplet, (b) the minus spinor multiplet, (c) the righton multiplet, (d) the lefton multiplet, (e) the vector multiplet, (f) the plus matter gravitino multiplet, and (g) two minus matter gravitino multiplets. Each multiplet may be described in terms of superfields or components.

The first two (1,0) matter multiplets and rigid (1,0) superspace were described by Sakamoto [9]. These are the scalar multiplet (SM) and minus spinor multiplet (MSM). The scalar multiplet $X(z)$ has component fields defined by $X\,(\sigma^{++},\sigma^{--}) \equiv$

$X(z) \mid$ and $\psi_+(\sigma^{++},\sigma^{--}) \equiv (\nabla_+ X(z)) \mid$. The action for d free scalar multiplets coupled to supergravity is

$$\begin{aligned} S_{SM} &= \int d^2\sigma d\zeta^- E^{-1}\eta_{\underline{a}\underline{b}}[\, i\tfrac{1}{2}(\nabla_+ X^{\underline{a}})(\nabla_{--}X^{\underline{b}})\,] \;, \\ &= -\tfrac{1}{2}\int d^2\sigma e^{-1}\eta_{\underline{a}\underline{b}}[\,(e_{++}X^{\underline{a}})(e_{--}X^{\underline{b}}) + i\psi_+{}^{\underline{a}}\mathcal{D}_{--}\psi_+{}^{\underline{b}} + 2\psi_{--}{}^{+}\psi_+{}^{\underline{a}}(e_{++}X^{\underline{b}})\,] \;, \end{aligned} \tag{3.1}$$

where the indices $\underline{m},\underline{n}$ take on values $0,\ldots,d-1$ and $\eta_{\underline{m}\underline{n}}$ is the spatial Minkowski metric. A number, N_F, of free minus spinor multiplets $\eta_-^{\hat{I}}$ $\left(\hat{I}=1,\ldots,N_F\right)$ may be coupled to (1,0) supergravity via the action

$$\begin{aligned} S_{MSM} &= -\tfrac{1}{2}\int d^2\sigma d\zeta^- E^{-1}\delta_{\hat{I}\hat{J}}[\,\eta_-^{\hat{I}}\nabla_+\eta_-^{\hat{J}}\,] \;, \\ &= \int d^2\sigma e^{-1}\tfrac{1}{2}[\, i\lambda_-^{\hat{I}}\mathcal{D}_{++}\lambda_-^{\hat{I}} + F^{\hat{I}}F^{\hat{I}}\,] \;. \end{aligned} \tag{3.2}$$

where the component fields are defined by $\lambda_-^{\hat{I}} \equiv \eta_-^{\hat{I}}(z) \mid$ are $F^{\hat{I}} \equiv -i\left(\nabla_+\eta_-^{\hat{I}}(z)\right) \mid$. As can be seen from the action, the component fields, $F^{\hat{I}}$, are auxiliary fields with algebraic equations of motion that set them to zero. It is one of the unusual features of (1,0) supersymmetry that off-shell there are equal numbers of bosonic and fermionic degrees of freedom, but on-shell this equality is absent.

The next multiplet is the right-moving chiral boson (or "righton") multiplet. The superfield formulation [10] of this multiplet provides the off-shell version of the "bosonic formulation" [1] of the heterotic string. In fact, the righton multiplet represents the superspace bosonization of two minus spinor multiplets. This bosonization is not surprising since the right-handed sectors of (1,0) theories are not **on-shell** supersymmetric. The linear action for N_R such multiplets coupled to supergravity takes the form

$$\begin{aligned} S_R &= i\tfrac{1}{2}\int d^2\sigma d\zeta^- E^{-1}[\nabla_+\Phi_R{}^{\hat{a}}\nabla_{--}\Phi_R{}^{\hat{a}} \;+\; \Lambda_{--}{}^{++}{}_{\hat{a}\hat{b}}\nabla_+\Phi_R{}^{\hat{a}}\nabla_{++}\Phi_R{}^{\hat{b}}] \;, \\ &= -\int d^2\sigma e^{-1}\tfrac{1}{2}\{\, e_{++}\phi_R^{\hat{a}}e_{--}\phi_R^{\hat{a}} \;+\; i\chi_+^{\hat{a}}\mathcal{D}_{--}\chi_+^{\hat{a}} \;+\; 2\psi_{--}{}^{+}\chi_+^{\hat{a}}e_{++}\phi_R^{\hat{a}} \\ &\qquad -\lambda_{--}{}^{+}{}_{\hat{a}\hat{b}}\chi_+^{\hat{a}}e_{++}\phi_R^{\hat{b}} \;+\; \lambda_{--}{}^{++}{}_{\hat{a}\hat{b}}[e_{++}\phi_R^{\hat{a}}e_{++}\phi_R^{\hat{b}} \\ &\qquad + i\chi_+^{\hat{a}}\mathcal{D}_{++}\chi_+^{\hat{b}} \;+\; 2\psi_{++}{}^{+}\chi_+^{\hat{a}}e_{++}\phi_R^{\hat{b}}]\,\} \;, \end{aligned} \tag{3.3}$$

where $\hat{a}=1,\ldots,N_R$, $\phi_R^{\hat{a}} \equiv \Phi_R^{\hat{a}} \mid$, $\chi_+^{\hat{a}} \equiv \left(\nabla_+\Phi_R^{\hat{a}}\right) \mid$, $\lambda_{--}{}^{++}{}_{\hat{a}\hat{b}} \equiv \Lambda_{--}{}^{++}{}_{\hat{a}\hat{b}} \mid$, and $\lambda_{--}{}^{+}{}_{\hat{a}\hat{b}} \equiv -i(\nabla_+\Lambda_{--}{}^{++}{}_{\hat{a}\hat{b}}) \mid$. The component action contains the usual righton action for $\phi_R^{\hat{a}}$. Remarkably, the lagrange multiplier terms also imply that $\mathcal{D}_{++}\chi_+^{\hat{a}} = 0$. This together with χ's equation of motion means that $\chi_+^{\hat{a}}$ vanishes on-shell! It is an auxiliary field like $F^{\hat{I}}$. In a sense $\chi_+^{\hat{a}}$ is the fermionization of $F^{\hat{I}}$ since neither propagates any physical degrees of freedom.

Even though (1,0) superspace has a preferred handedness, it also admits a supersymmetric generalization of a left-moving chiral boson ("lefton"). It is isomorphic to the (1,1) lefton [11]. But this is as expected, since on-shell the left-handed sectors of (1,0) and (1,1) superspace are isomorphic. The linear action for N_L such multiplets coupled to supergravity is [12] (with $\hat{\alpha} = 1, \ldots, N_L$)

$$\begin{aligned} S_L &= i\tfrac{1}{2}\int d^2\sigma d\zeta^- E^{-1}[\nabla_+\Phi_L^{\hat{\alpha}}\nabla_{--}\Phi_L^{\hat{\alpha}} \;+\; \Lambda_+{}^{--}{}_{\hat{\alpha}\hat{\beta}}\nabla_{--}\Phi_L^{\hat{\alpha}}\nabla_{--}\Phi_L^{\hat{\beta}}] \;, \\ &= -\int d^2\sigma e^{-1}\tfrac{1}{2}\{\, e_{++}\phi_L^{\hat{\alpha}}e_{--}\phi_L^{\hat{\alpha}} \;+\; i\beta_+^{\hat{\alpha}}\mathcal{D}_{--}\beta_+^{\hat{\alpha}} \;+\; 2\psi_{--}{}^+\beta_+^{\hat{\alpha}}e_{++}\phi_L^{\hat{\alpha}} \\ &\qquad - \lambda_{++}{}^{--}{}_{\hat{\alpha}\hat{\beta}}[e_{--}\phi_L^{\hat{\alpha}}e_{--}\phi_L^{\hat{\beta}} \;+\; 2\psi_{--}{}^+\beta_+^{\hat{\alpha}}e_{--}\phi_L^{\hat{\beta}}] \\ &\qquad + i\lambda_+{}^{--}{}_{\hat{\alpha}\hat{\beta}}[2\,\mathcal{D}_{--}\beta_+^{\hat{\alpha}}e_{--}\phi_L^{\hat{\beta}} \;+\; \psi_{--}{}^+\beta_+^{\hat{\alpha}}\mathcal{D}_{--}\beta_+^{\hat{\beta}} \\ &\qquad + i\psi_{--}{}^+e_{++}\phi_L^{\hat{\alpha}}e_{--}\phi_L^{\hat{\beta}} \;-\; i\psi_{++}{}^+e_{--}\phi_L^{\hat{\alpha}}e_{--}\phi_L^{\hat{\beta}} \\ &\qquad + i\psi_{--}{}^+\psi_{++}{}^+\beta_+^{\hat{\alpha}}e_{--}\phi_L^{\hat{\beta}}]\,\}\;. \end{aligned} \tag{3.4}$$

The component fields here are defined in a manner analogous to those in (3.3).

Unidexterous superfields are like all other superfields. There are additional multiplets that represent higher spins. However, in two dimensions, all fields of spin greater than one-half do not propagate physical degrees of freedom. The (1,0) superspace generalization of these fields has been discussed elsewhere [6].

4. WHY USE RIGHTON AND LEFTON MULTIPLETS?

In a large number of the discussions [13] of four dimensional strings, the notion of a purely fermionic representation of $D = 2$ supersymmetry has been used. This concept can be traced to a work by Di Vecchia, Kniznik, Petersen, and Rossi who first suggested it. (It was "rediscovered" later by Antoniadis, Bachas, Kounnas, and Windey.) It has been stated that these free fermions provide an example of a nonlinear representation of supersymmetry. We will now argue that what was really found [14] should more properly be described as an "anomalous" representation of supersymmetry. By this we mean that the free fermion "representation" is **not** supersymmetric as a classical theory. (This fact was explicitly stated previously [14] in the first work on free fermion representations.) However, due to the presence of a quantum mechanical anomaly, it appears to provide a consistent free-field representation of supersymmetry as a quantum theory at least on a trivial genus surface.

To support our contention, we first review the $D = 2$ Volkov-Akulov model [15,16]. This model actually does provide a nonlinear realization of supersymmetry and will be useful for comparison with the anomalous representation [14]. The $D = 2$ Volkov-Akulov model is defined by

$$S_{V-A} = a^{-2} \int d^2\sigma det(f_a{}^b) \ ,$$

$$f_a{}^b \equiv \begin{pmatrix} 1 - ia^2\chi_+\partial_{--}\chi_+ & ia^2\chi_+\partial_{++}\chi_+ \\ -ia^2\chi_-\partial_{--}\chi_- & 1 + ia^2\chi_-\partial_{++}\chi_- \end{pmatrix} \ , \tag{4.1}$$

where $\chi_+(\tau,\sigma)$ and $\chi_-(\tau,\sigma)$ are Majorana-Weyl, one component spinor fields and a is a dimensionful constant with the units of length. This action changes by a total derivative under the variations

$$\delta_Q(\epsilon)\chi_+ = ia^{-1}\epsilon^- - a(\epsilon^+\chi_-\partial_{++} + \epsilon^-\chi_+\partial_{--}\ )\chi_+ \ ,$$

$$\delta_Q(\epsilon)\chi_- = ia^{-1}\epsilon^+ - a(\epsilon^+\chi_-\partial_{++} + \epsilon^-\chi_+\partial_{--}\ )\chi_- \ . \tag{4.2}$$

This provides a genuine nonlinear representation of supersymmetry since a direct calculation reveals

$$[\ \delta_Q(\epsilon_1)\ ,\ \delta_Q(\epsilon_2)\ ]\chi_+ = i2(\epsilon_1^+\epsilon_2^+\partial_{++} + \epsilon_1^-\epsilon_2^-\partial_{--})\chi_+ \ ,$$

$$[\ \delta_Q(\epsilon_1)\ ,\ \delta_Q(\epsilon_2)\ ]\chi_- = i2(\epsilon_1^+\epsilon_2^+\partial_{++} + \epsilon_1^-\epsilon_2^-\partial_{--})\chi_- \ . \tag{4.3}$$

All of the above discussion is applicable to classical considerations where the spinor fields are anticommuting quantities.

Now we may ask whether the free fermions [14] are also a nonlinear representation of supersymmetry. Anomaly-freedom requires a triplet of Majorana-Weyl spinors $\psi_+^{\hat{I}}(\tau,\sigma)$, $\chi_+^{\hat{I}}(\tau,\sigma)$, and $\lambda_+^{\hat{I}}(\tau,\sigma)$ $\left(\hat{I} = 1,\ldots,N_G\right)$ each of which (on-shell) satisfies a Dirac equation. The supersymmetry variation of each spinor takes the form (c_o, constant)

$$\delta_Q'\chi_+^{\hat{I}} = c_o f^{\hat{I}}{}_{\hat{J}\hat{K}}\epsilon^+\chi_+^{\hat{J}}\chi_+^{\hat{K}} \ . \tag{4.4}$$

The commutator of two such transformations yields

$$\begin{aligned} \left[\delta_Q'(\epsilon_1)\,,\,\delta_Q'(\epsilon_2)\right]\chi_+^{\hat{I}} &= -4c_o(\epsilon_1^+\epsilon_2^+)f^{\hat{I}}{}_{\hat{J}\hat{K}}f^{\hat{J}}{}_{\hat{L}\hat{M}}\chi_+^{\hat{K}}\chi_+^{\hat{L}}\chi_+^{\hat{M}} \\ &= -\tfrac{4}{3}(\epsilon_1^+\epsilon_2^+)[\ f^{\hat{I}}{}_{\hat{J}\hat{K}}f^{\hat{J}}{}_{\hat{L}\hat{M}} + f^{\hat{I}}{}_{\hat{J}\hat{L}}f^{\hat{J}}{}_{\hat{M}\hat{K}} \\ &\qquad + f^{\hat{I}}{}_{\hat{J}\hat{M}}f^{\hat{J}}{}_{\hat{K}\hat{L}}\ ]\chi_+^{\hat{K}}\chi_+^{\hat{L}}\chi_+^{\hat{M}} \ , \end{aligned} \tag{4.5}$$

where in obtaining the last line we have used the anti-commutativity of $\chi_+^{\hat{I}}$, etc. If $f_{\hat{J}\hat{K}}{}^{\hat{I}}$ are the structure constants for a Lie algebra then the right hand side of (4.5) vanishes since it is the Jacobi identity for those constants. Therefore, we conclude that **classically** the transformation in (4.4) is more similar to a BRST-like transformation $\left[\delta'_{Q_1}, \delta'_{Q_2}\right] = 0$, than to a supersymmetry transformation. Quantum mechanically, it is well-known that the current algebra with generators $J_{++}{}^{\hat{I}} \equiv i f_{\hat{J}\hat{K}}{}^{\hat{I}} \psi_+^{\hat{J}} \psi_+^{\hat{K}}$ is anomalous. It has been argued [14] that the presence of this anomaly will modify (4.5) to produce the correct form as in (4.3). Thus, we see that the anomaly is responsible for modifying the algebra of a symmetry, at the quantum level, that was already present as a BRST-like symmetry at the classical level!

This is certainly not the usual situation with anomalies. Usually anomalies break classical symmetries. The fermionic representation [14] may work well at the level of current algebra, but we have not been able to find a **local supercovariant two dimensional field theory action** which leads to the appropriate currents.

Using the Noether procedure, it is relatively simple to see the problem. The rigid action for $\chi_+^{\hat{I}}$ is simply the Dirac action: $S = -i\frac{1}{2}\int d^2\sigma \chi_+^{\hat{I}} \partial_{--} \chi_+^{\hat{I}}$. Next we make a "supersymmetry" variation (4.4) to derive the "supercurrent" $S_{++}{}^{-} = \frac{2}{3} c_o f_{\hat{I}\hat{J}\hat{K}} \chi_+^{\hat{I}} \chi_+^{\hat{J}} \chi_+^{\hat{K}}$. Following the usual Noether rules, we couple this to the gravitino $\psi_{--}{}^{+}$ and minimally covariantize with respect to gravity. In this way we reach the intermediate action

$$S = \int d^2\sigma e^{-1} [\, -i\tfrac{1}{2}(\chi_-^{\hat{I}} \mathcal{D}_{--} \chi_-^{\hat{I}} + \psi_{--}{}^{+} S_{++}{}^{-}) + \dots\,] \ . \tag{4.6}$$

On taking the "supersymmetry" variation of this, the $(\mathcal{D}_{--}\epsilon^+)$-term cancels (as it should) but there remains a term linear in $\psi_{--}{}^{+}$. We have been unable to find any terms, which are represented by the ellipses, that can cancel this term. At this stage, we have conjectured that no such terms exist.

There are several known four dimensional precedents for this situation. Some time ago, the "double antisymmetric tensor model" was suggested [17] as a representation of four dimensional $N = 1$ supersymmetry. It was later shown that this model could **not** be consistently coupled to supergravity. We believe the free fermion model falls into the same category. On the other hand, the Volkov-Akulov model, which does provide an example of a nonlinear representation of supersymmetry, can be consistently coupled to supergravity [16] in two dimensions. (This is also the case for higher dimensions.) There is also the example of a single scalar field together with a Majorana or Weyl spinor field [18]. It is possible to define a "phony" supersymmetry invariance of their combined actions.

This brings us to the use of righton and lefton supermultiplets. These manifestly provide representations of supersymmetry since they are expressible in terms of superfields. In the (1,0) case, the component lefton multiplet (on-shell) corresponds to a left-moving chiral boson $\phi_L^{\hat{\alpha}}(\tau,\sigma)$ and a left-moving spinor $\beta_+^{\hat{\alpha}}(\tau,\sigma)$. These provide a simple linear representation of supersymmetry and their coupling to supergravity is straightforwardly given by equation (3.4). In the (1,1) case, lefton and righton multiplets are isomorphic to the (1,0) lefton multiplet. So the answer to the question raised at the beginning of this section is that righton and lefton multiplets realize supersymmetry both quantum mechanically and classically in two dimensional field theories. Additionally, these linear realizations allow for both Polyakov-type string formulations and "β-function" type studies of four dimensional string theories. The fact that the free fermion model cannot be coupled classically to supergravity suggests that it has serious difficulties on topologically nontrivial surfaces. In comparison to ordinary bosonization, super manifold bosonization may be more restricted or even forbidden.

5. CLASSICAL WEYL INVARIANCE

Up to this point, we have discussed the formulation of any (1,0) supersymmetric field theory. In searching for string candidates, we will now impose the restriction of local Weyl or scale invariance. This requirement along with local supersymmetry implies full local superconformal invariance.

As in all supergravity theories, a local scale transformation is generated by a redefinition of the scale compensator Ψ in the supergravity covariant derivative ∇_A. An infinitesimal change of $\delta\Psi$ given by $\delta\Psi = S\Psi$, via the solution to constraints [6], implies the following changes in ∇_A [19]:

$$\begin{aligned} \delta_S \nabla_+ &= \tfrac{1}{2} S \nabla_+ + (\nabla_+ S)\mathcal{M} , \\ \delta_S \nabla_{++} &= S\nabla_{++} - i(\nabla_+ S)\nabla_+ + (\nabla_{++} S)\mathcal{M} , \\ \delta_S \nabla_{--} &= S\nabla_{--} - (\nabla_{--} S)\mathcal{M} , \end{aligned} \tag{5.1}$$

where S is the parameter of superscale transformations. From (5.1) we may deduce that E^{-1} transforms as $\delta_S E^{-1} = -\frac{3}{2} S\, E^{-1}$. Note that although $E_A{}^M$ is not scale covariant, E^{-1} transforms covariantly with the law $\delta_S E^{-1} = w\, S\, E^{-1}\left(w = -\frac{3}{2}\right)$. Any superfield F that transforms like E^{-1} but with arbitrary weight w is scale covariant. Below we list some of the scale weights for multiplets discussed in section three.

Superfield(s)	w
$X^{\underline{a}}$	0
$\eta_-^{\hat{I}}$	$\frac{1}{2}$
$\Phi_R^{\hat{a}}$	0
$\Lambda_{--}{}^{++}$	0
$\Phi_L^{\hat{\alpha}}$	0
Λ_+^{--}	$-\frac{1}{2}$

Table I

Any scale covariant lagrangian $\mathcal{L}_-$ constructed from (1,0) superfields in such a way that its scale weight $w = \frac{3}{2}$ is classically superscale invariant. This is seen by the following consideration

$$\begin{aligned}\delta_S S_{\mathcal{L}} &= \int d^2\sigma d\zeta^- \delta_S(E^{-1}\mathcal{L}_-) \\ &= \int d^2\sigma d\zeta^-(-\tfrac{3}{2}S\, E^{-1}\mathcal{L}_- + E^{-1}\delta_S\mathcal{L}_-) \\ &= \int d^2\sigma d\zeta^- E^{-1}(\ w\ - \tfrac{3}{2})S\mathcal{L}_-\ \ .\end{aligned} \tag{5.2}$$

For $w = \frac{3}{2}$, $\delta_S\, S_{\mathcal{L}}$ vanishes and is therefore superscale invariant at the classical level.

6. BUILDING ANOMALY-FREE MODELS

By starting from the two-dimensional field theory viewpoint there emerges a simple well-defined way [10,20] to construct candidate superstring theories. The steps are:

(i.) Consider any classically scale invariant (1,0) superfield action.

(ii.) Quantize the action via the path integral and for convenience use the background field method. After appropriate gauge fixing, derive propagators and three point vertices.

(iii.) Calculate the renormalized background effective action by integrating over quantum fields and ghosts. Since anomalies in two dimensions are found from two point functions, we need only consider one-loop diagrams.

(iv.) Check to see if the effective action depends on background gauge fields in a non-gauge invariant manner. If the non-gauge invariant dependence cannot be removed by the introduction of local counter terms (Adler-Bardeen method) then we have found an anomaly.

In particular, to isolate the Weyl and Lorentz anomalies it suffices to calculate the contribution to the effective action of the form of $\Gamma_{Anom.}$ where

$$\Gamma_{Anom.} = \frac{1}{96\pi}\int d^2\sigma d\zeta^- \{ \nu_R H_+{}^{--}[\frac{(\partial_{--})^4}{\Box} D_+ H_+{}^{--}] + i\nu_L H_{--}{}^{++}[\frac{(\partial_{++})^3}{\Box} D_+ H_{--}{}^{++}] \} \ , \tag{6.1}$$

ν_R and ν_L are some constants and $H_+{}^{--}$ and $H_{--}{}^{++}$ are background supergravity prepotentials. We note that the purely bosonic part of (6.1) reduces to

$$\Gamma_{Anom.} = -\frac{1}{96\pi}\int d^2\sigma \{ \nu_R h_{++}{}^{--}[\frac{(\partial_{--})^4}{\Box} h_{++}{}^{--}] + \nu_L h_{--}{}^{++}[\frac{(\partial_{++})^4}{\Box} h_{--}{}^{++}] \} \ . \tag{6.2}$$

In this expression $h_{++}{}^{--}(\sigma^{++}, \sigma^{--})$ and $h_{--}{}^{++}(\sigma^{++}, \sigma^{--})$ are the linearizations of the background zwiebein $e_a{}^m$ (i.e. $e_a{}^m = \delta_a^m + h_a{}^m$). Equation (6.2) can also be derived in a purely bosonic theory and thus define ν_R and ν_L in such a theory.

The actual values of ν_R and ν_L are found from one loop supergraph calculations. Typically the Feynman supergraphs required for their determination have the structures seen below

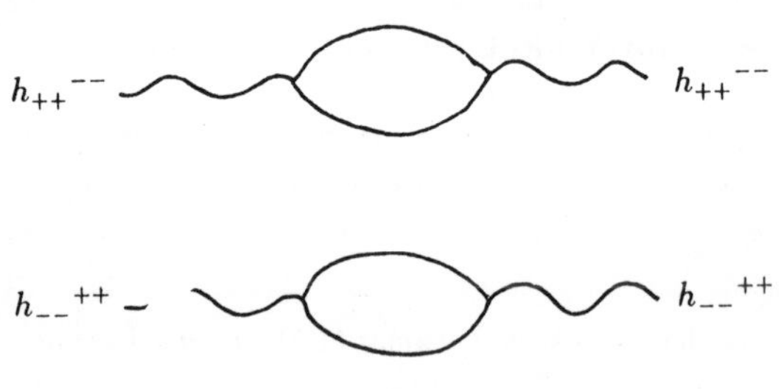

figure I

With our normalizations, ν_R and ν_L are always either integer or half odd integer valued. In fact, ν_R and ν_L are related to the Weyl anomaly (w) and the Lorentz anomaly (ℓ) as $\nu_R = \frac{1}{2}(w + \ell)$ and $\nu_L = \frac{1}{2}(w - \ell)$. In the loops above, only quantum fields or ghosts make contributions. For two dimensional gauge fields it is only the F-P ghosts which make contributions. (This statement is only true if it is classically possible to set the gauge field identically to zero modulo topologically nontrivial gauge orbit pieces.)

Before we give the values of ν_R and ν_L for the various multiplets, it is useful to know these numbers for arbitrary two dimensional $SO(1,1)$ representations. We have previously deduced these [20] and will restate our results here. In (1,0) superspace there are two basic types of gauge superfields. Denoting these by $\hat{\Gamma}_+^{(q)}$ and $\hat{\Gamma}_{--}^{(q)}$, their linearized gauge transformations take the forms

$$\delta_G \hat{\Gamma}_+^{(q)} = D_+ \lambda^{(q)} \quad , \quad \delta_G \hat{\Gamma}_{--}^{(q)} = \partial_{--} \lambda'^{(q)} \quad , \tag{6.3}$$

$\lambda^{(q)}$ and $\lambda'^{(q)}$ are the parameters of the gauge transformations. Some examples of these are (1,0) Yang-Mills theories where $\hat{\Gamma}_+^{(q)} = \Gamma_+$ and $\hat{\Gamma}_{--}^{(q)} = \Gamma_{--}$ or (1,0) supergravity where $\hat{\Gamma}_+^{(q)} = H_+{}^{--}$ and $\hat{\Gamma}_{--}^{(q)} = H_{--}{}^{++}$. The (q)-index is the "excess spin" of the gauge connection above that required to match the index of the derivative. In particular this implies

$$\begin{aligned} [\, \mathcal{M} \, , \, \hat{\Gamma}_+^{(q)} \,] &= -\tfrac{1}{2}(q-1)\hat{\Gamma}_+^{(q)} \quad , \\ [\, \mathcal{M} \, , \, \hat{\Gamma}_{--}^{(q)} \,] &= -\tfrac{1}{2}(q+2)\hat{\Gamma}_{--}^{(q)} \quad . \end{aligned} \tag{6.4}$$

Therefore q is equal to plus one for each plus index and minus one for each minus index on the generalized connections $\hat{\Gamma}_+^{(q)}$ and $\hat{\Gamma}_{--}^{(q)}$.

On choosing a gauge wherein these connections vanish, we are led to consider two classes of ghost actions

$$S_{Ghost}^{(1)\ (q)} = \int d^2\sigma d\zeta^- E^{-1}[B_{--}{}^{(-q)} \nabla_+ C^{(q)}] \quad , \tag{6.5}$$

$$S_{Ghost}^{(2)\ (q)} = \int d^2\sigma d\zeta^- E^{-1}[-i(B_+{}^{(-q)} \nabla_{--} C'^{(q)})\,] \quad . \tag{6.6}$$

In these expressions the B's are FP anti-ghosts and the C's are FP ghosts arising by the usual replacement of the gauge parameters. For (6.5) we have

$$\nu_L = 0 \quad , \quad \nu_R = -(-)^q [\, 3(q-1)^2 - 1] \quad , \tag{6.7}$$

while for (6.6) we obtain

$$\nu_L = -(-)^q 3(2q+1) \quad , \quad \nu_R = 0 \quad . \tag{6.8}$$

The asymmetry of these values is due to the fact that the right-handed sector of (1,0) theories is not supersymmetric but the left-handed sector is supersymmetric.

We can now efficiently find the values of ν_R and ν_L for all (1,0) supermultiplets from the scalar multiplet up to and including the (1,0) supergravity multiplet. These are summarized below.

Anomalies and (1,0) Superfield $SO(1,1)$ Reps

$(1,0)$ S Field	Comp. Fields	ν_L	ν_R
X	(X, ψ_+)	$\frac{3}{2}$	1
η_-	(λ_-, F)	0	$\frac{1}{2}$
Φ_R	(ϕ_R, χ_+)	0	1
Φ_L	(ϕ_L, β_-)	$\frac{3}{2}$	0
Λ_+	$(v_{++}, 0)$	0	-2
Λ_{--}	(v_{--}, λ_-)	-3	0
$\Psi_+{}^-$	$(\psi_{++}{}^-, 0)$	0	11
$\Psi_{--}{}^-$	$(\psi_{--}{}^-, S)$	-3	0
$\Psi_+{}'^-$	$\left(\psi'_{++}{}^-, v_{++}\right)$	0	9
$\Psi_+{}^+$	$(\psi_{++}{}^+, 0)$	0	-1
$\Psi_{--}{}^+$	$(\psi_{--}{}^+, A_{--})$	9	0
$E_A{}^M$	$(e_a{}^m, \psi_a{}^+)$	-15	-26

Table II

Since the right hand column is nonsupersymmetric, we can use it to deduce the analogous results for component fields alone. As a check on this reasoning we can use the fact that the number defined by the Feynman graphs obey a linearity rule: the values of ν_R and ν_L for a multiplet is the sum of the values for the components within the multiplet. This implies that the nonsupersymmetric values of are obtained simply by interchanging plus sign with minus signs and right for left. We must also take care to properly account for fields which only move in one direction. Thus for component fields we may obtain the following table by a variety of methods.

Anomalies and Component Field SO(1,1) Reps

Comp field	ν_R	ν_L
X	1	1
λ_-	0	$\frac{1}{2}$
β_+	$\frac{1}{2}$	0
ϕ_R	0	1
ϕ_L	1	0
v_{++}	0	-2
v_{--}	-2	0
$e_a{}^m$	-26	-26

Table III

These two tables provide the secrets for looking for superstring and string candidates. The point is that any two dimensional action which describes a string must be free of world-sheet Weyl and Lorentz anomalies. This can be accomplished by using an anomaly-free set. By this we mean a set of fields such that the conditions $\sum \nu_R = \sum \nu_L = 0$ are satisfied when the sums are taken over all of the fields in the set. For bosonic string candidates table III provides the basic building blocks for constructing such sets. For all superstrings candidates, both spinning and heterotic, table II is our starting point.

7. BUILDING UP TO SUPERSTRINGS

Using the information discussed in the previous section we can now easily understand the structure of the known string theories. We believe that these arguments provide the simplest explanation concerning the structure of string theories.

We begin by considering the bosonic string. One question often asked by "non-stringy" physicists is, "Why does the bosonic string exist only in 26 dimensions?" The action is well known to be

$$S_{Bosonic} = -\tfrac{1}{2}\int d^2\sigma e^{-1}\eta_{\underline{ab}}\,(e_{++}X^{\underline{a}})(e_{--}X^{\underline{b}}) \quad . \tag{7.1}$$

On the basis of table III we see that the set $(e_a{}^m, X^{\underline{a}})$ is an anomaly-free set if $\underline{m} = 0,\ldots,25$. On the other hand, this is clearly not the **only** way to obtain an anomaly-free set. The set $\left(e_a{}^m, X^{\underline{a}}, \lambda_-^{\hat{I}}, \beta_+^{\hat{I}}\right)$ with $\underline{a} = 0, \ldots, d-1$, $\hat{I} = 1, \ldots, N_F$ is anomaly-free if $d + \frac{1}{2}N_F - 26 = 0$. An action for these fields is easy to write

$$
\begin{aligned}
S'_{Bosonic} = -\tfrac{1}{2}\int d^2\sigma e^{-1}[\ \eta_{\underline{ab}}\,(e_{++}X^{\underline{a}})(e_{--}X^{\underline{b}}) \\
+\ i\beta_+^{\hat{I}}(\mathcal{D}_{--}\beta_+^{\hat{I}}) -\ i\lambda_-^{\hat{I}}(\mathcal{D}_{++}\lambda_-^{\hat{I}})\ ]\ ,
\end{aligned}
\tag{7.2}
$$

and in doing so we have forced a toroidal compactification of the bosonic string. The question of the dimension of the spacetime in which the string is embedded is really a question about the zero modes of the two dimensional fields in the string action. If these zero modes range continuously over the real numbers, then we interpret them as the coordinates of a point in spacetime. If these modes are restricted to lie on some type of lattice or group manifold, then we cannot make such an identification. Our experience with fields that move in one direction generally indicates that the zero modes of such fields satisfy this condition.

To understand all superstrings table II may be taken as the starting point. We may add the actions in section three freely since classical superconformal symmetry is maintained. The question then becomes one of finding anomaly-free sets. Some of these are summarized below.

Set	Anomaly-Freedom Cond.	Sol'n
$\left(E_A{}^M, X^{\underline{m}}, \eta_-^{\hat{I}}\right)$	$-26 + d + \frac{1}{2}N_F = 0$	$d = 10$
	$-15 + \frac{3}{2}d = 0$	$N_F = 32$
$\left(E_A{}^M, X^{\underline{m}}, \Phi_R^{\hat{a}}\right)$	$-26 + d + N_R = 0$	$d = 10$
	$-15 + \frac{3}{2}d = 0$	$N_R = 16$
$\left(E_A{}^M, X^{\underline{m}}, \eta_-^{\hat{I}}, \Phi_L^{\hat{a}}\right)$	$-26 + d + \frac{1}{2}N_F = 0$	Not unique,
		$N_L = 6\,,\ d = 4$
	$-15 + \frac{3}{2}d + \frac{3}{2}N_L = 0$	$N_F = 44$ possible

These anomaly-free sets correspond respectively to the fermionic formulation of the heterotic string, the bosonic formulation, and compactified four dimensional heterotic strings. In the last case we have used the freedom to pick one parameter arbitrarily in order to describe four dimensional heterotic strings.

However, the utility of table II does not end with heterotic strings. **Any** string theory that possesses arbitrary type (p, q) supersymmetry $(p > 0\,,\ q > 0)$ on the world sheet may be analyzed. For example, the NSR spinning string may be regarded as a (1,0) supergravity theory coupled to (1,0) matter multiplets including a matter gravitino multiplet. Explicitly we find the NSR action takes the form,

$$\begin{aligned} S_{NSR} = S_{SM} - \tfrac{1}{2}\int d^2\sigma d\zeta^- E^{-1}\eta_{\underline{mn}}[\ \eta_-^{\underline{m}}\nabla_+\eta_-^{\underline{n}}\ ] \\ + i\int d^2\sigma d\zeta^- E^{-1}[\ \eta_{\underline{mn}}\Psi_+{}^-\eta_-^{\underline{m}}(\nabla_{--}X^{\underline{n}})\ ]\ . \end{aligned} \tag{7.3}$$

Although written in terms of $(1,0)$ superfields, this model is actually (1,1) supersymmetric. The second term is the usual minus spinor multiplet action with the replacements $\eta_-^{\hat{I}} \to \eta_-^{\underline{m}}$, $\delta_{\hat{I}\hat{J}} \to \eta_{\underline{mn}}$. The third term, however, includes a (1,0) matter gravitino $\Psi_+{}^-$! Considering anomaly-freedom of the set we have

Set	A-F Cond.	Sol'n
$\left(E_A{}^M, \Psi_+{}^-, X^{\underline{m}}, \eta_-^{\underline{m}}\right)$	$-26 + 11 + d + \frac{1}{2}d = 0$	$d = 10$
	$-15 + \frac{3}{2}d = 0$	

In a similar fashion, the compactified version of the NSR type II theory (which has also been called the type III theory) may be analyzed. The set of superfields is given by $\left(E_A{}^M, \Psi_+{}^-\right) \oplus (X^{\underline{m}}, \eta_-^{\underline{m}}) \oplus \left(\Phi_R^{\hat{a}}, \eta_-^{\hat{a}}\right) \oplus \Phi_L^{\hat{\alpha}}$ where $\underline{m} = 0, \ldots, d-1$, $\hat{a} = 1, \ldots, N_R$, $\hat{\alpha} = 1, \ldots, N_L$. The anomaly-freedom of the set implies $N_R = N_L$, $d = 10 - N_L$. This possesses a solution $N_L = 6$, $d = 4$.

Likewise, for the $N = 2$ spinning string we can note the complete superfield set must be $(E_A{}^M\ ,\ 2\Psi_+{}^-\ ,\ \Psi_{--}{}^+\ ,\ \Gamma_+) \oplus (2X^{\underline{m}}, 2\eta_-^{\underline{m}})$. The first subset is actually the $N = 2$ superconformal multiplet. The second subset is an $N = 2$ scalar multiplet. This is why the multiplicity factors of two appear in front of $X^{\underline{m}}$ and $\eta_-^{\underline{m}}$. The anomaly-freedom condition is easily found to be $d = 2$.

Quite literally, the search for all anomaly-free sets reduces to a "numbers game" in which two Diaphantine equations $\Sigma\nu_R = \Sigma\nu_L = 0$ must be solved in complete generality. However, when using extra matter gravitino multiplets there is the added complication that the extra non-manifest supersymmetries must be properly represented on the other matter superfields.

8. WORLD-SHEET SPIN-ONE FIELDS

As we saw in the last section, the addition of matter gravitino multiplets very rapidly causes the critical dimension of the string to decrease. But there is something interesting to note in the $N = 2$ spinning string that distinguishes it from the other cases. Unlike all the other models, the $N = 2$ spinning string consists of three **distinct** anomaly-free sets! It is most clearly seen if we work in terms of component fields. We may group the component fields into three sets $(e_a{}^m, \ldots) \oplus (A_{++}, \ldots) \oplus (A_{--}, \ldots)$ where each subset is itself anomaly-free. This same phenomenon would occur in the $N = 4$ spinning string which possesses seven distinct anomaly-free sets. (For the $N = 4$ theory each spin-one field A_a is replaced by a SU(2) triplet.) In the $N = 4$ case we have the $N = 4$ superconformal multiplet

$$
\begin{aligned}
(1,0)SF \qquad & (E_A{}^M, \Psi_+{}^-, 3\Psi'_+{}^-, 3\Psi_{--}{}^+) \ , \\
\Sigma\nu_R = & \ -26 + 11 + 27 + 0 \qquad = 12 \ , \\
\Sigma\nu_L = & \ -15 + 0 + 0 + 27 \qquad = 12 \ ,
\end{aligned}
$$

together with the $N = 4$ scalar multiplet

$$
\begin{aligned}
(1,0)SF \qquad & (4X^m, 4\eta_-^m) \ , \\
\Sigma\nu_R = & \ 4 + 2 \qquad = 6 \ , \\
\Sigma\nu_L = & \ 6 + 0 \qquad = 6 \ .
\end{aligned}
$$

and it is apparent that the anomaly-free condition is $12 + 6d = 0$ or $d = -2$. (Usually at this point it is said that because the critical dimension is negative, the $N = 4$ model is quantum mechanically inconsistent. However, the fact that the critical dimension is precisely minus two may also be interpreted as implying that the scalar system is equivalent to the FP ghost system of a $N = 4$ vector multiplet!) The extended spinning strings use world sheet vectors in their NSR formulations. We should therefore consider the possibility that world-sheet gauge vectors can appear in the more interesting cases of $N = \frac{1}{2}$, and $N = 1$ superconformal theories. To our knowledge, this question was first raised by Tomboulis [21] in the $N = 0$ case.

The distinguishing feature of world-sheet gauge fields is that their values of ν_R and ν_L are of the same sign as that for the graviton. In the following, we will argue the requirements of classical local superconformal and anomaly-freedom can simultaneously be fulfilled. There are also issues concerning spin-one anomalies and mixed anomalies which must be addressed.

We demand that any world-sheet spin-one field appears in such a way that it can be "gauged away." This is implicit in the derivation of ν_R and ν_L for such a field. This is automatically satisfied if v_{++} couples only to right moving currents and v_{--} couples only to left moving currents in an action. We also demand that v_{++} and v_{--} belong to anomaly-free sets. In this way we are first led to consider the component sets

$$\begin{gathered}(A_{++}, 4\psi_-{}') \; , \; (A_{++}, \phi_R{}', 2\psi_-{}') \; , \; (A_{++}, 2\phi_R{}') \; , \\ (A_{--}, 4\psi_+{}') \; , \; (A_{--}, \phi_L{}', 2\psi_+{}') \; , \; (A_{--}, 2\phi_L{}') \; , \\ (A_{++}, A_{--}, 2\psi_-{}', 2\psi_+{}', X) \; . \end{gathered} \tag{8.1}$$

As written the spinors above are assumed to be one component Majorana-Weyl spinors. Two Majorana-Weyl spinors $(2\psi'_-)$ are equivalent to a Dirac-Weyl spinor ρ_-. So we may replace the two Majorana-Weyl spinors by one Dirac spinor. We may take M_R copies of the set $\left(A_{++} \, , \, \lambda'_- \, , \, \lambda''_- \, , \, \rho_-\right)$ with ρ_- a complex spinor, M_L copies of $\left(A_{--} \, , \, \beta'_+ \, , \, \beta''_+ \, , \, \rho_+\right)$ with ρ_+ a complex spinor, and consider the action

$$\begin{aligned} S_{(M_R, M_L)} \;=\; -\tfrac{1}{2} \int d^2\sigma e^{-1} [& - i\bar{\rho}_-{}^{a'} \hat{\mathcal{D}}_{++} \rho_-{}^{a'} \;-\; i\lambda_-{}^{a''} \mathcal{D}_{++} \lambda_-{}^{a''} \\ & + \; i\bar{\rho}_+{}^{\alpha'} \hat{\mathcal{D}}_{--} \rho_+{}^{\alpha'} \;+\; i\beta_+{}^{\alpha''} \mathcal{D}_{--} \beta_+{}^{\alpha''}] \;, \end{aligned}$$

$$a' = 1, \ldots, M_R \;, \quad a'' = 1, \ldots, 2M_R \;, \quad \alpha' = 1, \ldots, M_L \;, \quad \alpha'' = 1, \ldots, 2M_L \;, \tag{8.2}$$

where the modified covariant derivative $\hat{\mathcal{D}}_a$ includes the $[U_R(1)]^{M_R} \times [U_L(1)]^{M_L}$ gauge fields $A^{a'}_{++}$ and $A^{\alpha'}_{--}$. The charges of $\rho^{a'}_-$ and $\rho^{\alpha'}_+$ are denoted by $q^{a'}$ and $q^{\alpha'}$. The action in (8.2) slightly generalizes that in ref. [21] by permitting heterodexterity between right-moving and left-moving degrees of freedom. The action $S_{(M_R, M_L)}$ may be added freely to either (7.1) or (7.2) without disturbing the anomaly-freedom of those actions.

There is also another way in which an anomaly-free vector set can be useful. It is simple to show that the set (X, A_a, ρ_+, ρ_-) is anomaly-free. But the vector cannot couple to X since it is real. So we may take p-copies of this multiplet and write the action

$$\begin{aligned} S_{(p)} \;=\; -\tfrac{1}{2} \int d^2\sigma e^{-1} [& (\partial_{++} X^{\underline{m}'})(\partial_{--} X_{\underline{m}'}) \\ & - \; i\rho_-{}^{a'} \hat{\mathcal{D}}_{++} \rho_-{}^{a'} \;+\; i\rho_+{}^{a'} \hat{\mathcal{D}}_{--} \rho_+{}^{a'}] \;, \end{aligned} \tag{8.3}$$

with $a' = 1, \ldots, p$ and $\underline{m}' = 1, \ldots, p$. Since this is anomaly-free, it can also be added to (7.1) or (7.2) without disturbing their anomaly-freedom. Now it is an interesting question to ask whether there exist supersymmetric extensions of these results.

In the case of (1,0) supersymmetric theories, it is immediately clear how to construct the analog of (8.2). We simply write

$$
\begin{aligned}
S_{(M_R,M_L)} &= i\tfrac{1}{2}\int d^2\sigma d\zeta^- E^{-1}[\, i\hat{\rho}_-{}^{a'}\hat{\nabla}_+\rho_-{}^{a'} + i\eta_-{}^{a''}\nabla_+\eta_-{}^{a''} \\
&\quad + (\nabla_+\tilde{\Phi}_L{}^{\alpha'})(\hat{\nabla}_{--}\tilde{\Phi}_L{}^{\alpha'}) + \tilde{\Lambda}_+{}^{--}{}_{\alpha'\beta'}(\hat{\nabla}_{--}\tilde{\Phi}_L{}^{\alpha'})(\hat{\nabla}_{--}\tilde{\Phi}_L{}^{\beta'}) \\
&\quad + (\nabla_+\Phi_L{}^{\alpha'})(\nabla_{--}\Phi_L{}^{\alpha'}) + \Lambda_+{}^{--}{}_{\alpha'\beta'}(\nabla_{--}\Phi_L{}^{\alpha'})(\nabla_{--}\Phi_L{}^{\beta'})\,]\ , \\
&a' = 1,\ldots,M_R\ , \qquad a'' = 1,\ldots,2M_R\ , \qquad \alpha' = 1,\ldots,M_L\ .
\end{aligned} \tag{8.4}
$$

Only the complex minus spinor multiplets $\rho_-^{a'}$ and the tilde lefton multiplets carry non-trivial $U_R(1)\times U_L(1)$ charges. Modified covariant derivatives, $\hat{\nabla}_A$, are gravitationally as well as $[U_R(1)]^{M_R}\times[U_L(1)]^{M_L}$-gauge covariant; $\hat{\nabla}_A = \nabla_A - i\Gamma_A\cdot t$. The Lie algebra valued connection Γ_+ contains M_R independent gauge superfields and Γ_{--} contains M_L independent gauge superfields. Now the (1,1) supersymmetric extension is also easy to deduce. We find the following action has all of the required properties

$$
\begin{aligned}
S_{(M_R,M_L)} &= -i\tfrac{1}{2}\int d^2\sigma d^2\zeta E^{-1}\mathcal{L}_{(M_R,M_L)}\ , \\
\mathcal{L}_{(M_R,M_L)} &= (\hat{\nabla}_+\tilde{\Phi}_R{}^{a'})(\nabla_-\tilde{\Phi}_R{}^{a'}) + \tilde{\Lambda}_-{}^{++}{}_{a'b'}(\hat{\nabla}_+\tilde{\Phi}_R{}^{a'})(\hat{\nabla}_{++}\tilde{\Phi}_R{}^{b'}) \\
&\quad + (\nabla_+\tilde{\Phi}_L{}^{\alpha'})(\hat{\nabla}_-\tilde{\Phi}_L{}^{\alpha'}) + \tilde{\Lambda}_+{}^{--}{}_{\alpha'\beta'}(\hat{\nabla}_-\tilde{\Phi}_L{}^{\alpha'})(\hat{\nabla}_{--}\tilde{\Phi}_L{}^{\beta'}) \\
&\quad + (\nabla_+\Phi_R{}^{a'})(\nabla_-\Phi_R{}^{a'}) + \Lambda_-{}^{++}{}_{a'b'}(\nabla_+\Phi_R{}^{a'})(\nabla_{++}\Phi_R{}^{b'}) \\
&\quad + (\nabla_+\Phi_L{}^{\alpha'})(\nabla_-\Phi_L{}^{\alpha'}) + \Lambda_+{}^{--}{}_{\alpha'\beta'}(\nabla_-\Phi_L{}^{\alpha'})(\nabla_{--}\Phi_L{}^{\beta'})\ .
\end{aligned} \tag{8.5}
$$

The tilded chiral boson multiplets $\tilde{\Phi}_R^{a'}$ and $\tilde{\Phi}_L^{\alpha'}$ carry the $[U_R(1)]^{M_R}\times[U_L(1)]^{M_L}$ charges. The actions in (8.4) and (8.5) have manifest supersymmetry along with superconformal symmetry, Weyl and Lorentz anomaly-freedom. These actions are not necessarily free of pure gauge anomalies, but we will address this shortly.

It is also possible to generalize (8.3) to the supersymmetric case. First for (1,0) supersymmetry, there exists the action

$$
\begin{aligned}
S_{(p)} = i\tfrac{1}{2}\int d^2\sigma d\zeta^-[\, &(\nabla_+X^{\underline{m}'})(\nabla_{--}X_{\underline{m}'}) + i\rho_-{}^{a'}\hat{\nabla}_+\rho_-{}^{a'} \\
&+ (\nabla_+\tilde{\Phi}_L{}^{a'})(\hat{\nabla}_{--}\tilde{\Phi}_L{}^{a'}) + \tilde{\Lambda}_+{}^{--}{}_{a'b'}(\hat{\nabla}_{--}\tilde{\Phi}_L{}^{a'})(\hat{\nabla}_{--}\tilde{\Phi}_L{}^{b'})\,]\ .
\end{aligned} \tag{8.6}
$$

and for (1,1) supersymmetry we write

$$
\begin{aligned}
S_{(p)} &= -i\tfrac{1}{2}\int d^2\sigma d^2\zeta E^-[\, (\nabla_+X^{\underline{m}'})(\nabla_{--}X_{\underline{m}'}) \\
&\quad + (\hat{\nabla}_+\tilde{\Phi}_R{}^{a'})(\nabla_-\tilde{\Phi}_R{}^{a'}) + \tilde{\Lambda}_-{}^{++}{}_{a'b'}(\hat{\nabla}_+\tilde{\Phi}_R{}^{a'})(\hat{\nabla}_{++}\tilde{\Phi}_R{}^{b'}) \\
&\quad + (\nabla_+\tilde{\Phi}_L{}^{a'})(\hat{\nabla}_-\tilde{\Phi}_L{}^{a'}) + \tilde{\Lambda}_+{}^{--}{}_{a'b'}(\hat{\nabla}_-\tilde{\Phi}_L{}^{a'})(\hat{\nabla}_{--}\tilde{\Phi}_L{}^{b'})\,]\ .
\end{aligned} \tag{8.7}
$$

In both of these actions the indices range as $\underline{m}' = 1, \ldots, p, a' = 1, \ldots, p$.

Having seen that supersymmetry can be realized in models with anomaly-free vector sets, we must now consider the possibility of new anomalies: spin-one gauge anomalies and mixed gravitational-gauge anomalies. For simplicity we only consider the bosonic case. The mixed gravitational-gauge anomaly is generated by diagrams like the figure in section six, where one external graviton field is replaced by a spin-one field. It follows that up to some normalizations, these anomalies take the forms

$$\begin{aligned} \hat{\Gamma}_{Anom.} = \int d^2\sigma\{ \quad & c_1\,(\Sigma q_{(a)})A_{++}^{(a)}[\frac{(\partial_{--})^3}{\Box}h_{++}{}^{--}] \\ & + c_2(\Sigma q_{(\alpha')})A_{--}^{(\alpha')}[\frac{(\partial_{++})^3}{\Box}h_{--}{}^{++}]\ \}\ . \end{aligned} \tag{8.8}$$

where the sums are performed over all species of charged fermions which couple to the spin-one connections. Clearly we can enforce the absence of these anomalies by demanding that these sums vanish. This is quite simple to achieve. All we need to do is introduce at least one anomaly-free vector set where all of the fermions are charged. In this way we obtain more species of charged fermions than spin-one fields and can easily pick the charges appropriately.

The pure gauge anomalies are more difficult to handle. Again up to some normalizations, these are given by

$$\begin{aligned} \tilde{\Gamma}_{Anom.} = \int d^2\sigma\{ \quad & c_3\,(\Sigma q_{(a)}q_{(b)})A_{++}^{(a)}[\frac{(\partial_{--})^2}{\Box}A_{++}^{(b)}] \\ & + c_4(\Sigma q_{(\alpha')}q_{(\beta')})A_{--}^{(\alpha')}[\frac{(\partial_{++})^2}{\Box}A_{--}^{(\beta')}]\ \}\ . \end{aligned} \tag{8.9}$$

The sums over charges can not be used to cancel this anomaly. However, there appear to be at least two solutions to this apparent anomaly. One of these suggestions is motivated by the coupling of spin-one fields to chiral bosons [22]. It has been suggested that this is possible, if the parameter of the gauge transformation is chiral. Thus, using this idea we propose that the gauge group of the charged spinors be consistent with

$$\begin{aligned} \delta A_{++} &= \partial_{++}\alpha_L \quad , \quad \partial_{--}\alpha_L = 0\ , \\ \delta A_{--} &= \partial_{--}\alpha_R \quad , \quad \partial_{++}\alpha_R = 0\ , \end{aligned} \tag{8.10}$$

where α_L and α_R are the gauge parameters. Under this subgroup, $\tilde{\Gamma}_{Anom}$ only changes by a surface term. Now it turns out that this seemingly ad hoc solution is in fact demanded in the supersymmetric case. The reason is that leftons and rightons can only interact with spin one fields if the conditions in (8.10) are satisfied! A more radical solution is to demand that the connections are chiral, satisfying $\partial_{--}A_{++} = 0$ and $\partial_{++}A_{--} = 0$. This constraint makes both $\hat{\Gamma}_{Anom}$ and $\tilde{\Gamma}_{Anom}$ surface terms! The cancellation of this type of anomaly has also been achieved in a slightly different context [23]. In fact, the well-known Green-Schwarz mechanism has been used for precisely this type of anomaly. Using a group manifold, as opposed to toroidal compactification, appears to allow us to implement this mechanism by slight modification to the latter work in ref. [23]. Thus, we appear to have mechanisms which avoid the new anomalies.

We have yet to prove that our suggestions are totally consistent but this is under further study. Also we are more closely studying the $N = 2$ spinning string to see how it avoids the new anomaly problems. Of course, the most important question is whether the sum of an anomaly-free vector set with an anomaly-free graviton set corresponds to a new string theory. This question is of great importance to string theories. If the answer is affirmative, then the actions in (8.2) and (8.3) along with their supersymmetric extensions imply that the notion of critical dimension for strings may be fundamentally modified. The addition of (7.1), (7.2), (8.2) and (8.3) would lead to bosonic strings in arbitrary dimensions with almost arbitrary rank gauge groups. In our view this would be a very satisfactory result. This would imply that most if not all ordinary point particle field theories could be "strung out," i.e. made to be the low energy limit of string field theories. In this way, the concept of the string or superstring would evolve into a principle that can be applied to model-building much in the same way as supersymmetry. There is a very important reason to hope this is the case. If there are an infinite number of consistent four-dimensional superstring theories, it is much more likely that our efforts to find a model that describes our world will be successful. Thus, our efforts might be better directed to build such a theory from the bottom-up instead of from the top-down.

Along this line we should mention that the modification (8.5) of the four dimensional spinning string would seem to be the most profitable theory to pursue. The reason for this is that the no-go theorem derived in ref. [24] would not apply. If modular invariance allows $M_R \neq 0$ and/or $M_L \neq 0$, then the spacetime gauge group may be enlarged enough to permit the appearance of a chiral GUT with quarks and leptons. (Note that if $M_R \neq M_L$, then the theory will be heterodexterous but not heterotic.) Such models may not be burdened with the typically large GUT groups as are four dimensional heterotic strings.

Another point of interest is to note that other anomaly-free superconformal sets may exist. The reason is that the values of ν_R and ν_L for component fields are given by $\nu_R = -(-)^q [3(q-1)^2 - 1]$, $\nu_L = 0$, and $\nu_R = 0$, $\nu_L = -(-)[3(q'-1)^2 - 1]$ for gauge fields $\Gamma^{(q)}_{++}$ and $\Gamma^{(q')}_{--}$ respectively. The quantities q and q' are the numbers of plus and minus indices, respectively, on the different gauge fields. So if we start with some values q_{max} and q'_{max} such that $\nu_R < 0$ and $\nu_L < 0$, then to find an anomaly-free set we must solve the equations

$$[3(q_{max} - 1)^2 - 1] = \Sigma a_q (-)^q [3(q-1)^2 - 1] \quad ,$$

$$[3({q'}_{max} - 1)^2 - 1] = \Sigma b_{q'} (-)^{q'} [3(q'-1)^2 - 1] \quad . \tag{8.13}$$

where a_q and $b_{q'}$ are integers and the sums are performed up to q_{max} and q'_{max}, respectively. If an action containing these fields with appropriate gauge invariances can be constructed, it will automatically be anomaly-free. The action in (8.2) is such a construction. The generalization, classification, and significance of these additional anomaly-free combinations are open questions.

Acknowledgment

I wish to thank my collaborators Roger Brooks, Marc Grisaru, Luca Mezincescu, Fuad Muhammad, and Paul Townsend for their cooperation at various stages in carrying out this research.

References

1. D. Gross, J. Harvey, E. Martinec, and R. Rohm, Phys. Rev. Lett. 54 (1985) 102; ibid. Nucl. Phys. B256 (1985) 253; ibid B267 (1986) 75.

2. P. Candelas, G. Horowitz, A. Strominger, and E. Witten, Nucl. Phys. B258 (1985) 46.

3. P. Howe, J. Phys. A12 (1979) 393; M. Brown and S. Gates, Jr., Ann. Phys., NY 122 (1979) 443.

4. R. Brooks, F. Muhammad, and S. Gates, Jr., Nucl. Phys. B268 (1986) 599; R. Brooks and S. Gates, Jr., Nucl. Phys. B287 (1987) 699; P. Nelson and G. Moore, Nucl. Phys. B274 (1986) 509; M. Evans and B. Ovrut, Phys. Lett. 171B (1986) 177; E. D'Hoker and D. Phong, Nucl. Phys. B269 (1986) 205.

5. S.L. Adler, Phys. Rev. 177 (1969) 2426; J.S. Bell and R. Jackiw, Nuovo Cimento 60A (1969) 47.

6. See first two works of ref. [4].

7. S. Gates, Jr. and H. Nishino, Class. Quant. Grav. 3 (1985) 391.

8. E. D'Hoker and D. Phong, Nucl. Phys. B278 (1986) 225; D. Friedan, E. Martinec, and S. Shenker, Nucl. Phys. B271 (1986) 93; G. Moore, P. Nelson and J. Polchinski, Phys. Lett. 169B (1986) 47.

9. M. Sakamoto, Phys. Lett. 151B (1985) 115.

10. M. Grisaru, L. Mezincescu and P. Townsend, Phys. Lett. 179B (1986) 247.

11. W. Siegel, Nucl. Phys. B238 (1984) 307.

12. S. Gates, Jr., R. Brooks, and F. Muhammad, Univ. of Md. Preprint UMDEPP 87-122, to appear in Phys. Lett.

13. H. Kawai, D. Lewellen, and S. Tye, Phys. Rev. Lett. 57 (1986) 1832; W. Lerche, D. Lüst, and A. Schellekens, CERN Preprint CERN-TH.4590/86; R. Bluhm, L. Dolan and P. Goddard, Rockefeller Preprint RU/B/187.

14. I. P. Di Vecchia, V. Knizhnik, J. L. Petersen, and P. Rossi, Nucl. Phys. B253 (1985) 701; P. Goddard, W. Nahm, and D. Olive, Phys. Lett. 160B (1985) 111; P. Goddard, A. Kent, and D. Olive, Comm. Math. Phys. 103 (1986) 105;

I. Antoniadis, C. Bachas, C. Kounnas, and P. Windey, Phys. Lett.171B (1986) 51.

15. D. Volkov and V. Akulov, Phys. Lett. 46B (1973) 109; T. Dereli and S. Deser, J. Phys. A10 (1973) L149.

16. M. Roček, Phys. Rev. Lett. 41 (1978) 451.

17. D. Freedman, Caltech preprint CALT-68-624 (1977).

18. P. West, *Introduction To Supersymmetry and Supergravity* World Scientific Pub. Co. (1986) Singapore, p. 23.

19. G. Moore, P. Nelson, and J. Polchinski, Phys. Lett. 169B (1986) 47.

20. S. Gates, Jr., M. Grisaru, L. Mezincesu, and P. Townsend, Nucl. Phys. B286 (1987) 1.

21. E. Tomboulis, "Gauge Fields on the World-sheet and New String Models," UCLA Preprint # UCLA/86/TEP/30, Aug. 1986.

22. M. Bernstein and J. Sonnenschein, "Quantization of Chiral Bosons," Weizmann Inst. preprint PPN WIS-86/47/Sept-PH. (Sept., 1986).

23. M. Green and J. Schwarz, Phys. Lett. 149B (1984) 117; C. Hull and E. Witten, Phys. Lett. 160B (1985) 398.

24. L. Dixon, V. Kaplunovsky, and C. Vafa, SLAC preprint SLAC-PUB-4282 (Mar., 1987).

DYNAMICAL FINITE SIZE EFFECT, INFLATIONARY COSMOLOGY AMD THERMAL PARTICLE PRODUCTION

B. L. Hu
University of Maryland
Department of Physics and Astronomy
College Park, Maryland 20742, USA*
and
The Institute for Advanced Study
Princeton, New Jersey 08540, USA

ABSTRACT

We propose a way to understand the symmetry behavior of quantum fields in inflationary cosmology and thermal particle production in curved spacetime as consequences of dynamically-induced constrained systems. The constraints are generated by exponential scaling transformations. Like event horizons, as a geometric entity of spacetime, they give rise to Hawking effect and have coarse-graining attributes. But unlike event horizons, they are kinematic quantities which depend on the energy of the system, the geometry and dynamics of spacetime and the interactions of the system with the background. Implications of these ideas on the quantum and statistical behavior of gravitational and non-gravitational systems are discussed.

*Permanent Address

In this talk I would like to discuss some recent thoughts I have on certain interconnecting aspects of quantum processes in curved spacetime including symmetry behavior in inflationary cosmology, thermal particle production and the entropy of fields and geometry.

Many people who witnessed in the 70's Bekenstein's[1] audacious proposal of black hole entropy and Hawking's[2] revolutionary prediction of quantum thermal radiance cannot help being bewildered by the many intriguing possibilities these discoveries bring forth, touching on many fundamental aspects of gravity, quantum physics and statistical thermodynamics. Equally interesting are the many different ways of understanding these results:[3] Davies and Fulling's analog of moving mirrors, Unruh's accelerating detector; Hartle and Hawking's periodicity in the propagator, Gibbon and Perry's thermal Green function, Candelas and Sciama's fluctuation-dissipation theory, to name just a few. With earlier interest on particle production in cosmological spacetimes following Parker's[4] pioneering footsteps, I was naturally curious about what makes thermal radiance so special and black hole spacetimes - or spacetimes with event horizons, as is commonly believed - so unique in these phenomena. As Hawking, Penrose, Wald and Page had argued,[5] the thermal character of radiance is interesting because these systems allow a one-way pure to mixed state transformation, a property believed by some to be special to and characteristic of quantum gravity. This also raises questions on whether Hawking effect can occur only in spacetimes with event horizons, whether it is special to curved spacetime, and whether gravity has some intrinsic feature other quantum fields do not have, like the loss of quantum coherence.[6] On the first question, moving mirrors and accelerated detectors are indeed systems which can acquire an event horizon not because of curved space. Parker and others showed that thermal particle production can also come from cosmological spacetimes which in general do not have event horizons. I was particulary attracted by Unruh's[7] and Sciama's[8] views. They point at simpler but more general causes - excitation of vacuum and statistical fluctuations by observers in some special state of motion. In a way these views are less welcoming, because they tend to strip gravity of its mystique and uniqueness.

These were questions in the back of my mind. When I was later working on finite temperature theory in curved spacetime and pondering on how entropy of quantum fields and geometry can be defined, I was partly driven by a desire to understand the Bekenstein-Hawking-Unruh effect and its implications in a wider context. It was not until the recent advent of inflationary universe[9] and its implications, and the fruition of our (Denjoe O'Connor and myself) studies on the symmetry behavior of curved spacetime that these questions are brought to the fore and certain leads begin to appear.

I want to emphasize that what I am about to discuss are not proven results, but are merely hints, indications, and "ideas for ideas" which I find useful to understand some of the interconnections between symmetry breaking, inflation, thermal particle production and entropy generation in certain classes of curved spacetime. In the spirit of a discussion workshop, and for the sake of lucidity, I hope you can excuse me for any imprecision of statements and looseness of presentation.

I. FINITE SIZE EFFECT AND INFRARED BEHAVIOR

Let me begin by relating to you some recent results Denjoe O'Connor and I obtained in studying the symmetry behavior of quantum fields in curved spacetime.[10-15] We are particulary interested in how gravitational effects related to global properties of spacetime like topology and geometry enter in the quantum field-theoretical description of systems undergoing symmetry breaking. From detailed analysis of the infrared behavior of quantum fields in a number of representative cosmological spacetimes (Einstein[11], Taub[12], de Sitter[13], mixmaster[14]), we soon come to realize that as much as topology and curvature effects are important in their own rights, the determinant physical factor governing the symmetry behavior of quantum systems in curved spacetime (at least for those with some compact or finite dimension) is the so-called <u>finite-size</u> <u>effect</u> (FSE). Finite size effect is not a curvature effect <u>per</u> <u>se</u>, because it exists for flat spaces with boundaries, even though curvature can bring forth finite size effect. Neither is it a topological effect _per se_ as manifest in multiply-connected manifolds or for twisted fields, because spaces with different topology can give similar finite size effect such as the three-sphere S^3 and three-torus T^3, even though topology can manifest finite size effect. We use the term finite size effect in a general sense, referring to the influence of constraints on the vacuum fluctuations of quantum fields due to the presence of non-trivial boundary conditions or topology in spacetime or the field configurations. One familiar example of finite size effect is the Casimir effect.[16] We know that for example the attractive force between two parallel plates or the repulsive force on a conducting sphere is caused by the difference in the vacuum energy in the space inside and outside of the conductors, which act as constraints on the vacuum fluctuation fields. One can ask the same question and perform similar calculations for curved space, e.g. the Einstein universe $R^1 \times S^3$ or Kaluza-Klein $M^4 \times S^7$ universe.[17] In these cases the vacuum energy is calculated by taking into account the zero-point flucutation of all modes. A technical problem which arises is the regularization of ultraviolet divergence from the high frequency modes. For symmetry breaking in curved space one needs to study how certain global features of spacetime and fields influence the infrared behavior. Since the critical point of a system is reached when the correlation length ξ or the inverse effective mass M_{eff} of the order parameter $\hat{\phi}$ field goes to infinity, the problem in focus would be how the constraints in geometry or fields enter in changing this behavior. Therefore, in contrast to the Casimir effect where the vacuum energy ρ_0 (equal to the value of the effective potential V_{eff} at the symmetric state $\hat{\phi} = 0$) is calculated with contributions from the full spectrum of the fluctuation field, the study of symmetry breaking concentrates on the infrared sector of the fluctuation field spectrum but requires a full picture of the effective potential which provides information on the location and energetics of the minimum energy states. From our recent studies of finite size effect in curved-space symmetry behavior we can make a number of observations:[10]

1) For spacetimes with some compact spatial dimensions or invariant operators (of the fluctuation field) with a discrete (or band) spectrum, the most important contribution to the infrared behavior comes from its zero mode (or band).

2) The decoupling of higher modes from the dynamics gives rise to dimensional reduction in the low energy limit.

3) The symmetry behavior of these constrained systems is determined by the value of a parameter η equal to the ratio of the correlation length ξ ($\xi = M^{-1}_{eff}, M^2_{eff} = \partial^2 V/\partial\hat{\phi}^2|\hat{\phi}_{minima}$) to the geometric scale L of the background spacetime. (For compact dimensions $L = 2\pi a$ where a is its curvature radius. For non compact dimensions $L = \infty$.) Dimensional reduction occurs for very large values of η.

4) In the low energy limit the system behaves effectively as in a lower dimension. The effective infrared dimension (EIRD) for product spaces $R^d \times B^b$ (with d non-compact dimensions and b compact dimensions) is, to a first approximation, equal to d.

5) Since the geometric and field parameters entering in the effective potential run with curvature and energy (according to a set of renormalization group equations), the correlation length at different minimum energy states can decrease or increase with curvature, changing η accordingly. Therefore the EIRD can be different at states of different symmetry.

The energy and curvature dependence of the correlation length of quantum systems in curved space and its related dimensional reduction behavior make the notion and effect of constraints a lot more interesting. Finite size effect, despite the plainness of its name, should not be viewed simply as the study of box-like structures in vacuum. We have seen that constraints in curved space not only carry geometric but also kinematic properties. As we shall show next, they could also be generated dynamically, yielding very interesting results.

In our previous work, we have discussed the symmetry behavior in product spaces with topology $R^d \times B^b$. This includes many physically interesting systems such as

1) S^4 the Euclideanized de Sitter universe
2) $R^1 \times S^3$ the Einstein, Taub and mixmaster universes
3) $R^2 \times S^2$ the Einstein-Rosen and axisymmetric solutions, and
4) $R^3 \times S^1$ the imaginary time finite temperature field theory or $M^4 \times S^1$ the Kaluza-Klein theory

We have also discussed from the first two categories the implications of finite size effect on phase transitions in the inflationary[13] and mixmaster[14] cosmologies. In this talk I will focus on the relation of dynamical effects with finite size effect. The problem is exemplified by viewing the de Sitter universe both in the $R^1 \times S^3$ Robertson-Walker (RW) and the S^4 Euclidean coordinatization. In the S^4 formulation it is easy to understand the many salient features of the infrared behavior of quantum fields in a de Sitter space. Since the fluctuation operator is discrete, one can easily identify the lowest mode contribution to the effective potential (e.g. via zeta-function method) and infer the infrared behavior in the way we have described. For example de Sitter spacetime should have EIRD=0. What is not obvious is how the effect of "finite size" in S^4 translate to in the $R^1 \times S^3$ (k=1 RW) or the $R^1 \times R^3$ (k=0 RW) coordinates. By contrast the latter are dynamical spacetimes. One can only see a FSE in the S^3 spatial dimensions, the time being an "open" dimension contributes an EIRD = 1. This is not the same FSE as in S^4. However, if FSE were a physical effect, it should exist independent of the coordinatization. How then does finite size effect manifest in these other formulations? The gist of the matter lies in finding the conditions upon which the time dimension effectively acquires a scale and appears closed by virtue of the finite size effect in that dimension. This is what we called dynamical finite size effect. We conjecture that a particular class of dynamical behavior where the scale factor a of a system undergoes exponential expansion $a \sim e^{Ht}$ would induce such a response. In particular this implies that the $R^1 \times R^3$ or the late time $R^1 \times S^3$ Robertson-Walker coordinatization (where H is the Hubble constant) would exhibit finite size effect equivalent to that in the S^4 formulation. In what follows I will give a few plausibility arguments without any formal proof.

II. GEOMETRIC DEFORMATION, SYMMETRY CHANGE AND DIMENSIONAL REDUCTION

For the cosmological problem at hand, our conjecture states that an exponential expansion of the universe ($R^1 \times S^3$) would generate a scale $L = H^{-1} = (\dot{a}/a)^{-1}$ in the time direction which exhibits finite size effect on a quantum system equivalent to that from a space with topology $S^1 \times S^3$, where $L \simeq a_1$, where a_1 is the radius of S^1. The generation of a scale renders the time dimension effectively finite and hence decreases the effective infrared dimension associated with time from 1 to 0. Let me develop these ideas in steps: First let us look at a closed spatial dimension and analyze the opposite effect of "opening" up $S^1 \to R^1$. This can be the result of extreme geometric deformation (or high field perturbation in an analogous quantum mechanical problem). In Ref. 10 by analyzing the lowest mode contribution to the ζ-function we show how a specetime can manifest very different infrared behavior and have different EIRD's under different modes of extreme deformations. An example is $S^2 \times S^1$ (with radii a_2 and a_1 respectively):

$$S^2 \times S^1 \to S^2 \times R^1 \text{ as } a_1 \to \infty, \text{ and}$$
$$\to R^2 \times S^1 \text{ as } a_2 \to \infty ,$$

where the EIRD becomes 1 and 2 in the respective cases. The arrow should be taken to mean that when the appropriate limits on the scale factor are taken, the spaces show equivalent infrared behavior. Note that in all these processes the topology of space remains unchanged. However the symmetry behavior changes qualitatively once the scale of the configuration under deformation becomes comparable to the correlation length. Finite size effect works in the opposite sense, i.e. in dimensional reduction in the infrared regime. In the above example, it would be $S^2 \times R^1 \to S^2 \times S^1$. Let us examine the eigenfunction and spectrum of these successively reduced systems $S^3 \to S^2 \times S^1 \to S^2 \times R^1$. The line element of a 3-sphere with radius a is

$$d\ell^2(S^3) = a^2 d\Omega_3{}^2 = a^2[d\chi^2 + \sin^2\chi(d\theta^2+\sin^2\theta d\phi^2)] \tag{1}$$

where $0 \le \chi \le \pi$, $0 \le \theta \le \pi$, $0 \le \phi \le 2\pi$. The Laplace-Beltrami operator on S^3 is $\Delta = L^2/a^2$, where L^2 is the Casimir operator on S^3. The eigenfunctions belonging to the eigenvalues $\kappa_N = n(n+2)/a^2$, $N = (n, \ell, m)$ with $n = 0, 1, 2, \ldots$ $\ell = 0, 1, \ldots n$ and $m = -\ell, -\ell + 1, \ldots \ell$ are the hyperspherical harmonics $Y_{\ell m}(\chi, \theta, \phi)$. It can be expressed as a product of the Gegenbauer polynomial $G_{n\ell}(\chi)$ and the spherical harmonics $Y_{\ell m}(\theta, \phi)$, i.e.

$$Y_{n\ell m}(\chi, \theta, \phi) = G_{n\ell}(\chi) Y_{\ell m}(\theta, \phi) \tag{2}$$

$G_{n\ell}$ obeys a second-order differential equation

$$\frac{d^2G}{d\chi^2} + 2 \cot\chi \frac{dG}{d\chi} - \frac{\ell(\ell+1)}{\sin\chi} G = - n(n+2)G \tag{3}$$

For $S^2 \times S^1$, with radii a_2 and a_1 respectively,

$$\Delta = \frac{L_2^2}{a_2^2} + \frac{L_z^2}{a_1^2} \quad , \tag{4}$$

where L_2^2 is the total angular momentum on S^2 and L_z is on S^1. The eigenfunction is

$$\psi_{n\ell m}(\chi, \theta, \phi) = e^{in\chi} Y_{\ell m}(\theta, \phi) \ . \tag{5}$$

The transition from $S^3 \to S^2 \times S^1$ is not straightforward. One observes that the metric (10) can be written in an equivalent form (k=1)

$$d\ell^2(S^3) = a^2 \left[\frac{dr^2}{1-kr^2} + r^2(d\theta^2 + \sin^2\theta d\phi^2) \right] \tag{6}$$

where k = 1, 0, -1 denote the closed, flat and open topologies and $r = \sin\chi$, χ, $\sinh\chi$ respectively. At small angles, χ can be regarded as a radial coordinate r. For flat space the range of χ becomes $0 \le \chi < \infty$. It is in the limit when the angular momentum of S^2 is decoupled from the "radial" equation (3) governing $G_{n\ell}$ that G_{no} "reduces" to $e^{in\chi}$. In general relativity one encounters this transition in the joining of a Schwarzschild metric to a Friedmann metric (in the so-called "bag-of-gold" model). As we explained in Ref. 10 and earlier, one sees problems of this nature more frequently in perturbation treatment of quantum mechanical systems obeying different symmetries. Problems in curved spacetime is formally similar when the wave operator of quantum fields is treated as Hamiltonian describing an equivalent dynamical system (e.g. waves in mixmaster universe and motion of an asymmetric top, etc.).

Now the opposite effect of "closing-up" a dimension (in the finite size sense) we also see in a number of familiar examples: the five dimensional Kaluza-Klein (KK) theory $M^4 \times S^1$ where a periodic boundary condition is imposed on the fifth spatial dimension to account for the electromagnetic gauge potential. The scale generated has to be of the order of the Planck length to be compatible with the observed value of the fine structure constant. Another example is the finite temperature (FT) theory $R^3 \times S^1$, where a periodicity condition is imposed on the imaginary time dimension. The scale a_1 generated is the inverse-temperature β. We have discussed the infrared behavior of these systems from the viewpoint of finite size effect. The effective dimension depends crucially on the energy scale of observation and measurement (length scale ξ is to be compared with the geometric scales a for S^1 and ℓ for R^n). At low energy for KK theory or at high temperature for FT theory, ($a \ll \xi \ll L$) the system behaves like a N-1 dimensional system where N = 5 and 4 respectively. The finite size effect associated with S^1 becomes insignificant. However at Planck energy or low temperature ($\xi \ll a$, L) the full dimensionality becomes important. Note that it is the scale (finite size) but not the topology or geometry which ultimately determines the effective physical properties of the system.

III. FINITE TEMPERATURE THEORY, EVENT HORIZON AND DYNAMICAL FINITE SIZE EFFECT

In the two examples given above, the scale of S^1 is extrinsically and formally introduced by imposing the periodicity condition. Neither the compactness nor the periodicity conditions is necessary for the manifestation of the finite size effect (cf. Kaluza-Klein theories with non-compact internal dimensions). By contrast the effect we want to focus on here depends on a scale generated dynamically in time (H^{-1}) or space (λ^{-1}) from the exponential transformation of the scale factor (a

$\sim e^{Ht}$ in time or $a \sim e^{\lambda x}$ in space) describing the system. In the temporal case $a \sim e^{Ht}$, we call this dynamical finite size effect because the constraint (event horizon) is dynamically induced. For the spatial case $a \sim e^{\lambda x}$, it is a special class of scaling transformation. In both cases the constraint (or screening) is induced by the dynamics or the scaling from the equation of motion, but not explicitly imposed. We will sometimes refer to them as induced-constrained systems. To see the connection with the problem we posed earlier for the de Sitter universe, let us compare the symmetry behavior of the following cases:

1) the Einstein Universe (ultrastatic case)
2) finite temperature theory in Einstein universe
3) Robertson-Walker universe (dynamic case)
4) de Sitter universe (in what we would call "ultra-dynamic" case)

The metric of the spatially-closed RW universe is given by

$$ds^2 = dt^2 - a^2(t)\, d\Omega_3^2 \tag{7}$$

where $d\Omega_3^2$ is the interval on the unit three sphere. The Einstein Universe has a constant a. The de Sitter universe in the closed RW coordination has

$$a = H^{-1} \cosh Ht \tag{8}$$

At late times

$$a \sim H^{-1} e^{Ht} . \tag{9}$$

This is the behavior of de Sitter universe in the flat RW coordinatization. The fluctuation field operator A to linear order is given by

$$A = \Box + M_1^2 \tag{10}$$

where $M_1^2 = m^2 + (1-\nu)R/6 + \lambda\hat{\phi}^2/2$ for a $\lambda\phi^4$ theory.
Here R is the scalar 4-curvature, $\hat{\phi}$ is the background field and $\nu = 0$, 1 denote conformal and minimal coupling respectively. (See Ref. 10 for details). The wave function on a $R^1 \times S^3$ RW space can be Fourier expanded as

$$\Phi(t,x) = \Sigma\, \phi_{n\ell}(t)\, Y_{n\ell m}(\chi,\theta,\phi) \tag{11}$$

where $Y_{n\ell m}(\chi,\theta.\phi)$ is the hyperspherical harmonics on S^3. The amplitude functions $\phi_{n\ell}$ satisfy the wave equation

$$A\phi_{n\ell} = \left[\frac{d^2}{dt^2} + 3\left(\frac{\dot{a}}{a}\right)\frac{d}{dt} + \kappa_N^2 + M_1^2 \right] \phi_{n\ell}(t) = 0 \tag{12}$$

The spatial contribution to the wave operator comes through the Casimir operator L^2 on S^3 which has a discrete spectrum $n(n+2)$, $n=0, 1,2, \dots$. But the scale factor a in the physical wave numbers κ_n renders it continuous. For R^3 spatial section the wave number κ_n is of course continuous to begin with.

To simplify matters we can also work with the wave operator A' without the linear (damping) term. Define $f_n(t) = \phi_n(t)\, a^{3/2}$, the wave equation (10) can be written as

$$A'f_n = \left[\frac{d^2}{dt^2} + W_n^2(t) \right] f_n(t) = 0 \tag{13}$$

where $W_n^2(t) = \mu^2(t) + \omega_n^2(t)$, $\omega_n^2 = \kappa_n^2 + M_1^2$,

$$\mu^2(t) = -\gamma^2(t) - \dot{\gamma}(t) \text{ and } \gamma = \frac{3}{2}(\dot{a}/a). \tag{14}$$

Let us analyze the spectrum λ_n of the fluctuation field wave operator $A'\phi_n = \lambda_n \phi_n$. From (13) and (14)

$$\lambda_n = k_o^2 + \mu^2 + \kappa_n^2 + M_1^2 , \tag{15}$$

where k_o is the eigenvalue of $\partial/\partial t$. (Strictly speaking this is well-defined only in the time-independent limit, e.g., late time de Sitter. We use this expression mainly to illustrate the ideas.) With each of the terms in (15) we can associate a scale: ℓ_t, L, ℓ_s and ξ respectively. There are two geometric scales: a time scale ℓ_t ($\sim k_o^{-1}$) which is a continuum, a spatial scale ℓ_s ($\sim \kappa_n^{-1}$) which vanishes at late times; a dynamic scale L ($\sim \mu^{-1}$) and a physical scale $\xi(\sim M_{eff}^{-1})$. The physical scale is determined by

$$\xi^{-2} \equiv M_{eff}^2 \equiv \frac{\partial^2 V_{eff}}{\partial\phi^2}\Big|_{\hat{\phi}_{min}} = M_1^2 + \frac{\lambda}{2}(\langle\phi^2\rangle_o + \langle\phi^2\rangle_T) \text{ to linear order,} \tag{16}$$

where M_{eff}^2 includes the quantum and thermal contributions. It depends on curvature, field coupling [see (10)], frequency modes, and interaction, which runs with energy. The physics of the system is an interplay between ξ and the geometric and dynamic scales. We have disucssed the effect of geometric scales previously.[10] The present discussion focuses on the dynamic scale. Consider first the <u>static</u> cases:

1) For the Einstein universe ($S^3 \times R^1$) $\lambda_n = k_o^2 + \omega_n^2$ where $k_o = (-\infty, \infty)$. Since ω_n is discrete, the spectrum has a band structure.

2) For finite temperature theory in the Einstein universe ($S^3 \times S^1$), one imposes a periodicity condition on the imaginary time. Thus $\lambda_n = k_o^2 + \omega_n^2$, where $k_o = 2\pi n_o/\beta$, $n_o = 0, \pm 1, \pm 2, \ldots$. The scale in the (imaginary) time direction is supplied by the inverse temperature β. The spectrum of A′ becomes discrete. As $\beta \to \infty$ (low temperature) one recovers the continuum ($\Sigma_n \to \beta \int dk_o/2\pi$) limit $S^3 \times R^1$ as in case 1. As $\beta \to 0$ (high temperature) infrared dimensional reduction occurs (for modes $\omega_n < \beta^{-1}$). We note again that the division between cases 1 and 2, i.e., whether the time dimension is "open" or "closed" really depends on the physical scale ξ in question. (For example, high mode with wavelength $\omega_n^{-1} < \beta$ will never experience dimensional reduction even at "high" temperature).

Now for the <u>dynamic</u> cases, a new mass term μ^2 enters which depends on how fast the scale factor changes:

3) For the Robertson-Walker universes with power-law dependence $a \sim t^p$, $H = \frac{\dot{a}}{a} = \frac{p}{t}$, and $\mu^2 \propto t^{-2}$ which goes to zero at late times. However,

4) For the de Sitter universe with $a = e^{Ht}$, $H = \dot{a}/a =$ constant, a term $\mu^2 = -\frac{9H^2}{4}$ is generated by this particular class of dynamics. (It is of the same nature as the extrinsic curvature-induced mass term $\sim (1-\nu)R/6$ in M_1^2 For de Sitter $R = 12H^2$)

This dynamically generated scale provides the equivalent effect either as due to curvature (vacuum fluctuation) or finite temperature (thermal energy). Comparing λ_n here with case 2, one sees that if the constant term $\frac{3}{2}$ H in de Sitter is identified with the periodicity of the imaginary time $\frac{2\pi}{\beta}$ (n_o=1) in the finite temperature theory, then the "closing up" of the time dimension by the induced scale $L = H^{-1}$ due to the exponential expansion introduces a temperature $T = c\,\frac{H}{2\pi}$. (The coefficient $c = d/2$ depends on the spatial dimension d.) This temperature is related to the Hawking temperature and the scale is associated with the event horizon.[2,21] Note that the existence of the k_o^2 term in λ_n is irrelevant because again, as explained earlier, the relevance of a scale (geometric or dynamic) is measured by the degree of interference or overlap it has with the scale characteristic of the particular physical process in question. (E.g. this is why the importance of any physical quantity should always be measured against the renormalization scale.) Conversely, the physical meaning of a process can also be restricted by the relevant scale in question. An example is in the evolution of vacuum fluctuations in de Sitter universe. The horizon (H^{-1}) acts as a natural divider of the physically communicable perturbances from the non-communicable ones.

Physically, the rapid "stacking up" of wave modes in the ground state due to exponential red-shifting results at late times in the formation of an event horizon of infinite redshift. The event horizon formed in such ultra-dynamical conditions now plays the role of finite size and its consequence is an example of what we call dynamical finite size effect. Notice, however, the difference in how this concept is arrived at, as distinct from the geometric viewpoint. The exponential expansion in some dimension R^1 introduces a finite size with scale $L = H^{-1} = (\dot{a}/a)^{-1}$ which has a finite temperature interpretation: the temperature experienced by the observer is given by $T \simeq H/2\pi$. (Compare this with the relation $2\pi aT = 1$ for finite temperature theory in the static Einstein universe with radius a.) The more conventional and well-established way to understand these phenomena is based on the event horizon regarded as a geometric entity intrinsic to spacetime. Our viewpoint differs from this in two fundamental aspects:

1) The source of appearance of the event horizon is kinematical rather than geometrical.
2) Dynamical finite size effect on the symmetry behavior depends on the interplay of the correlation length and the induced

scale. The former changes with energy, curvature and the interaction of fields with geometry. The latter depends on the dynamics of the system.

In our view this effect results fundamentally from a particular kind of scaling transformation. In some cases it permits a geometric interpretation (like the generation or existence of event horizon) but it is not itself a geometric effect.

IV. THERMAL PARTICLE PRODUCTION AND HAWKING-UNRUH EFFECT

Another interesting manifestation of dynamical finite size effect is in the thermal nature of particle production in spacetimes undergoing exponential expansion. One often distinguishes particle production in general cosmological spacetimes from those in spacetimes with event horizons. The later class includes black holes, Rindler space, moving mirror, etc., where the celebrated Hawking effect was discovered.[3] The nature of the two processes are different. Loosely put: in general cosmological spacetimes[4] it is due to the backscattering of waves, whereas in black hole spacetimes the main cause is commonly attributed to the workings of an event horizon, and the spectrum is always nearly thermal. The later class also includes the de Sitter universe.[21] The way we propose to understand this process is, like before, to attribute the basic cause to the exponential red-shifting of waves as they emerge from the event horizon rather than the geometry of the event horizon *per se*. This is seen to be the case indeed for the uniformly-accelerated observer, the moving mirror in hyperbolic trajectory and the Schwarzschild black hole.[22] If we view this exponential redshift as the more fundamental cause than the geometry of spacetime, we may be able to gain a little more physical insight into this process and generalize the Hawking effect to other systems. This is what Unruh did in his accelerated detector[7] interpretation and the sonic black hole[23] analogy of Hawking effect - note that they need not have anything intrinsic to do with topology, geometry or curved spacetime. Taking this viewpoint, we can also understand why certain systems in some cosmological spacetimes whose geometry do not possess an event horizon also produce particles with a thermal spectrum. One may say that thermal particle production in de Sitter space is no surprise because it has an event horizon. The way Gibbons and Hawking[2] originally demonstrated the existence of Hawking effect in de Sitter space was by a geometric approach - indeed, by exploiting the formal similarity of de Sitter space in the static coordinatization with the Schwarzschild metric and identifying the event horizon. The thermal nature can also be deduced from the Robertson-Walker universe coordinatization, which has been discussed by a large number of authors. One needs to be careful in correctly interpreting the results using the appropriate vacuum states with different symmetry properties. (For example, the results obtained from the RW and the de Sitter-invariant vacuua are very different). An example which does not lend itself to a direct geometric interpretation of the thermal nature of particle production in cosmological spacetime is that of Parker's 1976 work.[24] There he considered a minimally-coupled scalar field in a spatially-flat RW universe with $a(\tau) \sim e^{\rho\tau}$ initially ($\rho =$ constant), then in a vacuum defined with respect to the τ time ($dt = Vd\tau$, where $V=a^3$) the particles produced have a thermal spectrum. The wave equation for massless minimal field has the form (for the kth mode)

$$\frac{d^2\phi_k}{d\tau^2} + a^4(\tau)\, k^2\, \phi_k(\tau) = 0 \tag{17}$$

Parker considered a statically-bounded situation with $a(\tau) \to a_{1,2}$ at $\tau = \mp\infty$ and calculated particle production by the method of mode decomposition and Bogolubov transformation. A phase difference in the large argument expansion of the Bessel functions between $\tau \to \pm\infty$ yields the ratio of the Bogolubov coefficients in the form

$$\left|\frac{\beta_k}{\alpha_k}\right| \simeq e^{-4\pi k\rho^{-1} a_1^2} \tag{18}$$

which accounts for the thermal spectrum. The exponential form in (18) is a common feature in almost all derivations of thermal spectrum via the Bogolubov transformation method. As explained by Parker the central cause lies in the initial exponential rise of the scale factor

$a \sim e^{\rho\tau}$. In our way of interpretation this class of dynamics generates a finite size with scale $L \simeq (\dot{a}/a)^{-1} = \rho^{-1}$ (here a dot denotes $d/d\tau$) (from Sec. II) and one should expect that the temperature of the spectrum is $T \simeq \rho^{-1}$ (from Sec. III). This is indeed what was found in Parker's calculation. We are using this example to point out the generality of this feature and the connection between the ultra-dynamic property of spacetime, the finite size effect and thermal particle production. Of course the thermal property is attached to the particular set of observers for a particular set of vacuum defined with respect to a particular set of time (τ time in this case). For example the same dynamics will not yield a thermal spectrum with respect to cosmic time t. Dynamical finite size effect would therefore depend on the dynamics of the system and the kinematics of the observer.

To see a clearer distinction between the geometric view (e.g. in terms of event horizons) and the kinematic view (e.g. in terms of finite size effect) and to appreciate the more fundamental character and the wider range of applicability of the kinematic approach we are proposing, we can examine cases where the spacetime has event horizon but does not manifest Hawking effect. The extreme Reissner-Nordstrom spacetime is such a case.[25] In our viewpoint this is because the outgoing waves are subjected to a power-law rather than an exponential red-shifting, which is instrumental to giving rise to a thermal radiation.[26] This explanation also renders the disparity between apparent and event horizons as the true cause of Hawking radiation[27] less relevant. We can also examine situations which are not amenable to exact descriptions (e.g. existence or non-existence of event horizon) but can only be treated in approximate terms. An example is that used by Brandenberger and Khan[28] to discuss the approach to the inflationary state. The scale factor obeys a form

$$a(t) = a_o \, e^{\alpha(t)}$$

where

$$\alpha(t) = \frac{H}{(\Gamma/2)} \left(1 - e^{-\Gamma t/2}\right) \tag{19}$$

which tends to the de Sitter state $\alpha = Ht$ at large times ($\Gamma \ll H$). Another example is an accelerated observer which becomes uniformly accelerating asymptotically. In both cases one sees a nearly-thermal spectrum when the system approaches the exponential form asymptotically. In the kinematic viewpoint the approximately thermal property can be explained easily. But in the geometric viewpoint, a geometric concept of "quasi-" or "almost-formed" event horizon to describe these conditions is not so easy to define. Similarly, even though the finite temperature formulation of black hole Hawking radiation has proven to be very useful, it remains confined to the conditions of equilibrium thermodynamics. Any study of perturbed and interacting quantum statistical systems like symmetry breaking and thermodynamic stability of black holes, and phase transitions and critical dynamics in the early universe, which involve dynamical and non-equilibrium conditions, the kinematical approach would be more adaptible. There may exist geometric descriptions of non-equilibrium thermodynamics of quantum gravitational systems, but the techniques are not as well-developed or wieldy as the quantum field formulation which dynamical finite size effect is based on.

What then are the advantages of viewing Hawking effect as a manifestation of dynamical finite size effect? We think that viewing thermal particle production as a result of a constraint induced by extreme red-shifting (due to exponential expansion of the background scale factor) has the advantage that it can provide a common conceptual basis for thermal particle production from spacetimes with event horizons and from certain types of (ultradynamic) cosmological spacetimes. It can also suggest many occurances of Hawking radiation in situations completely unrelated to gravity or curved spacetime. Unruh's sonic black hole is one such case we have mentioned. One can also entertain the thought that Hawking effect - now in its more general sense as we have explained - can also appear in the spatial dimensions. Consider wave propagation in an inhomogeneous media (with, say, an exponentially-varying index of refraction) satisfying the wave equation with the form

$$u''(x) + e^{\lambda x} u(x) = 0 \qquad (20)$$

It should experience a finite size in the spatial dimension at distance λ^{-1} Here, rather than a thermal spectrum one may get a thermal momentum distribution (from the spatial components of $\beta_\mu k^\mu$ where β_μ is the temperature 4-vector and k^μ is the momentum 4-vector). The implications are not clear to me now but are certainly intriguing and worth pursuing. Generalized Hawking-Unruh effect should appear in systems whose scale factors undergo an exponential scaling transformation. We think this should be a common characteristic for all such induced-constrained systems.

V. FINITE SIZE EFFECT AND COARSE-GRAINING

The exponential scaling transformation associated with dynamical finite size effect has another interesting consequence: The infinite redshifting of modes in any system subjected to these conditions creates a coarse-graining which can reduce the information content of a system and effectively transform a microscopic world to a macroscopic world. Two familiar examples are the black hole[1,2] and the inflationary universe.[9] In both cases one usually attributes these effects to the working of the event horizon. In the quantum black hole case this brings up also the issue of pure states going into mixed states as a consequence of Hawking radiation.[5,6] The success of the inflationary universe in resolving the flatness and homogeneity problems really lies in the ability of inflation to dilute away these problems - "shuffling" them, so to speak, to scales much larger than our observable universe, of the size of domain boundaries. One can regard these as consequences of the existence of event horizons in these spacetimes, and indeed so. But the emancipation of the meaning of event horizon from a geometric notion to a kinematic notion and viewing its consequences as dynamical finite size effect in an induced-constrained system make it easier for us to discuss the dynamical and statistical properties of the system under general physical conditions. Viewing it in terms of finite size effect is also more fruitful because it incorporates the interplay of the system (quantum fields) and its environment (geometry or boundary condition) as well as the dependence of the physical parameters (e.g. correlation length in a symmetry breaking problem or the interaction strength and nonadiabaticity parameter in particle creation problem) on the energy scale of the system or the curvature of the background spacetime.

For illustrative purpose here it would suffice to mention two classes of problems: one on the quantum dynamical, the other on the statistical aspect. In many problems one needs to introduce a scale to separate the background field from the fluctuation field, the system from the bath variables and to implement a two-time (slow versus fast) approximation.[29] The event horizon (or Hubble radius) is often used as such a scale. This is really a very coarse scale which can act only as an upper bound. It is not sensitive to the behavior of the system in different frequency ranges. It also cannot account for the result of the interaction of quantum fields with the background geometry (e.g., as embodied in the effective action or effective mass). A more accurate measure should depend on the modes, the dynamics (how fast it is being red-shifted away), and the interaction. An example is the separation of the quantum regime from the classical regime for each normal mode in a particle creation and backreaction problem, as exemplified in Hu and Parker's early work.[29] Nonlocal effects (in time) exist throughout the process. The backreaction of created particles can also cause the spacetime dynamics to deviate from the initial exponential behavior. The event horizon picture would be too simplistic to handle such situations. The study of infrared behavior and phase transition in these spacetimes are examples we have mentioned before. Another class of problem is the study of entropy generation and dissipative effects in quantum gravitational systems.[30] Henry Kandrup and I have recently applied a subdynamics analysis to the study of entropy generation from particle interaction in cosmological spacetimes.[31] But as I have discussed elsewhere before,[30] the fact that particle horizons exist in many cosmological solutions and the particles are created at spacelike separations make it possible to have entropy generation even with the creation of free field particles. Certain global structures of spacetime (geometric or topological) would act as agents for coarse-graining. In such cases, again the quantum field approach in terms of dynamical finite size effect would probably be more adaptible than the geometric or topological methods. This is so not only because the geometric constraints cannot quite differentiate the response of each mode, but also because it cannot take into account the interaction of the system and its constraints. (Note again the constraints are induced by DFSE, not imposed externally, as they vary with energy and dynamics.) More simply put, the distinction and difficulty lies in the fact that we don't have as complete and versatile a geometric theory of dynamics as a dynamical theory of geometry.

There are many other areas where exponentiation transformation occurs and can lead to interesting results, e.g. in infinitesimal generators and continuous groups, information theory[32], aspects of chaotic dynamics, etc.. We know in statistical physics that macroscopic quantities like entropy measure the logarithm of the number of accessible states of a system which changes exponentially fast in response to disturbances. Topological entropy of a dynamical system which measures how fast trajectories in phase space deviate are defined in terms of exponential mappings. In all cases this class of mappings transforms infinitesimal to finite quantities, microscopic to macroscopic variables. What we have done here is to connect a number of well-known processes and effects in gravity like Hawking radiation, inflationary universe, and view them as consequences of systems which can be understood in the common framework of exponential scale transformations.

VI. SUMMARY AND REMARKS

It is perhaps of some interest to review where we began and where we've come to in trying to understand the effect of geometry and topology on the symmetry and dynamics of quantum systems in curved space. We began by examining in detail some representative spacetimes and gradually come to understand that the key factor which determines the symmetry behavior (as in a phase transition problem) is an interplay of correlation length, which depends on energy and interaction, and the geometric scales. We showed that the properties of quantum fields in curved space can be understood in terms of a Hamiltonian system with constraints. The Hamiltonian corresponds to the wave operator, the constraints come from the compact or finite dimensions or from stipulated boundary conditions on the fields. We learned to deduce its symmetry behavior by analyzing its spectrum. The influence of constraints on the quantum system is what we have categorically called "finite size effect". One can recast the problem of finding the symmetry behavior of the subspaces of a space of higher symmetry to that of an equivalent Hamiltonian system which can be decomposed into separate parts with different lower symmetries. In so doing one can invoke the many well-developed techniques of separation of variables, mode decomposition and spectral analysis for such studies. We then generalized the notion of constraints from the extrinsically imposed ones (e.g. Casimir effect in a box, periodicity condition on $R^1 \rightarrow S^1$) to the dynamically induced ones - specifically for the class of exponentially expanding spacetimes. We found that this class of dynamics generates the same effects as those produced by an event horizon. We used these concepts to discuss the relation with finite temperature theory and Hawking radiation. We also pointed out the difference between the kinematic viewpoint and the geometric viewpoint associated with event horizons. In particular we used the concept of dynamical finite size effect to relate thermal particle production in certain cosmological spacetimes and spacetimes with event horizons. In this vein we also suggested that Hawking effect is by no means special to curved space but is common to quantum systems with kinematically-induced constraints. Finally we attributed the essense of these phenomena as the result of a system subjected to an exponentiation scaling transformation. This class of transformations also provide a natural measure of coarse-graining, which brings about information loss and macroscopic averaging. If there is a philosophical point underneath this sequential rendition of ideas, it is in trying to understand what is special about physics in curved spacetime and how gravity distinguishes itself from other fundamental forces. I hope our proposal to view the effect of gravity in certain quantum systems, as equivalent to the action of constraints - extrinsic or induced - could help to reduce the barrier between the geometric and the field-theoretical formulations of gravity, thereby helping us to better understand its more unique and distinctive features.

As explained in the beginning, for the sake of lucidity I have chosen to discuss only ideas in this talk. Many of these ideas are based on observations rather than proofs. Whatever little derivation seen here is also merely suggestive rather than substantiative. Every point made here has yet to be sharpened and claims to be scrutinized closely. Let me name a few technical problems which need be worked out for this purpose:

1) In the discussion of dynamically-induced symmetry breaking since the fluctuation operator spectrum is continuous, one cannot rely on isolating the zero mode or the lowest-lying modes as the dominant contribution to the infrared behavior. One may instead use the spectral density function to examine these contributions.
2) Although it is not absolutely necessary to understand the topological changes accompanying the different modes of symmetry breaking, (e.g. the large deformation limit of Taub universe as discussed in Ref. 10) it would be interesting to see the relation between topological changes, geometric deformation and symmetry transformations.
3) Define relation between real-time continuum in ultradynamic spacetime and the periodic imaginary-time in finite temperature theory.
4) Establish the exponential expansion of the scale factor as a sufficient condition leading to thermal particle creation. Examine with this alternative viewpoint processes in those special geometries where no Hawking radiation is predicted.
5) Analyze the nature of exponentiation scaling transformation in more general systems. Understand its implications on symmetry breaking, on particle creation (or quasiparticle formation as in Bogolubov transformations), on coarse-graining, and on information theory.

Acknowledgement Many of the ideas on dynamical finite size effect and phase transition discussed here were generated in discussions with Denjoe O'Connor during his last two years as a graduate student at Maryland. Chris Stephens extended some aspects of our work further in his Ph.D. thesis work, especially interesting is his devise of a proper-time integral method in detecting the existence of these effects in more general quantum systems and relating them to thermal particle production. Sukanya Sinha has helped to clarify and develop further our work. I thank them for enriching my understanding of these interesting phenomena. I would also like to thank the organizers of this conference for their hospitality and especially Dr. Gabor Kunstatter for his patience and enthusiasm. This work is supported in part by the National Science Foundation under Grant No. PHY 84-18199 and by the Department of Energy under Grant No. DE-AC02-76ER02220.

References

1. J. D. Bekenstein, Phys. Rev. D7, 2333 (1972).
2. S. W. Hawking, Nature (London) 248, 30 (1974); Comm. Math. Phys. 42, 199 (1975). G. Gibbons and S. W. Hawking, Phys. Rev. D15, 2738 (1977).
3. P.C.W. Davies and S. A. Fulling, Proc. Roy. Soc. Lond. A356, 237 (1977); W. G. Unruh, Phys. Rev. D14, 870 (1976); J. B. Hartle and S. W. Hawking, Phys. Rev. D13, 2188 (1976); G. W. Gibbons and M. J. Parry, Phys. Rev. Lett. 36, 985 (1976); P. Candelas and D. W. Sciama, Phys. Rev. Lett. 38, 1372 (1977). See also standard reviews or monographs, e.g. N. D. Birrell and P.C.W. Davies, Quantum fields in Curved Space (Cambridge University Press, Cambridge, 1982); B. L. Hu and L. Parker, Quantum Field theory in Curved Spacetime - Reprint Volume I (World Scientific Publishing Co., Singapore, 1988).
4. L. Parker, Phys. Rev. 183, 1057 (1969).
5. See, e.g. S. W. Hawking, Phys. Rev. D14, 2460 (1976); R. Penrose in General Relativity, An Einstein Centenary Survey (ed.) S. W. Hawking and W. Israel (Cambridge University Press, Cambridge 1979); R. M. Wald in Quantum Gravity 2 (ed.) C. J. Isham, R, Penrose and D. W. Sciama (Clarendon Press, Oxford 1981); D. N.

Page, Phys. Rev. Lett. 44, 301 (1980).

6. S. W. Hawking, Comm. Math. Phys. 87, 395 (1983); L. Alvarez-Gaumé and C. Gomez, ibid 89, 235 (1983); D. J. Gross, Nucl. Phys. B236, 349 (1984); T. Banks, L. Susskind and M. E. Peskin, ibid B244, 125 (1984); S. W. Hawking, ibid B244, 135 (1984)
7. W. G. Unruh, Phys. Rev. D14, 870 (1976).
8. D. W. Sciama, in Centenario di Einstein, Editrice Giunti Barbara-Universitaria (1979).
9. A. H. Guth, Phys. Rev. D32, 347 (1981); A. D. Linde, Phys. Lett. 108B, 389 (1982), A. Albrecht and P. J. Steinhardt, Phys. Rev. Lett. 48, 1220 (1982); A. A. Starobinsky, Phys. Lett. 91B, 99 (1980); S. W. Hawking and I. Moss, Phys. Lett. 110B, 35 (1982).
10. B. L. Hu and D. J. O'Connor, Phys. Rev. D36, 1701 (1987); B. L. Hu in Proc. Fourth Marcel Grossmann Meeting (ed.) R. Ruffini (North-Holland, Amsterdam, 1986); D. J. O'Connor, Ph.D. thesis, University of Maryland (1985).
11. D. J. O'Connor, B. L. Hu and T. C. Shen, Phys. Lett. 130B, 31 (1983).
12. T. C. Shen, B. L. Hu and D. J. O'Connor, Phys. Rev. D31, 2401 (1985).
13. B. L. Hu and D. J. O'Connor, Phys. Rev. Lett. 56, 1613 (1986).
14. B. L. Hu and D. J. O'Connor, Phys. Rev. D34, 2535 (1986).
15. C. Stephens, Ph.D. thesis, University of Maryland (1986).
16. H. B. G. Casimir, Proc. Kan. Ned. Acad. Wet., 51, 793 (1948).
17. L. H. Ford, Phys. Rev. D12, 2963 (1975). For a review of recent work on Kaluza-Klein theories, see, e.g. T. Appelquist, A. Chodos and P. O. Freund, Modern Kaluza-Klein Theory (Addison-Wesley, New York, 1987).
18. B. L. Hu and D. J. O'Connor, Phys. Rev. D30, 743 (1984).
19. B. L. Hu, D. J. O'Connor and S. Sinha (1988); B. L. Hu, D. J. O'Connor and A. Stylianopoulos (1988).
20. Symmetry behavior in de Sitter space has been discussed by many authors. See e.g., The Very Early Universe (ed.) G. Gibbons, S. W. Hawking and S. Siklos (Cambridge Univ. Press, Cambridge, 1983).

For an explanation of the effect of large redshift, see the contributions of Ford, Hu and Linde therein. See also A. Vilenkin and L. H. Ford, Phys. Rev. D26, 1231 (1982), A. D. Linde, Phys. Lett. B116, 335 (1982), A. A. Starobinsky, Phys. Lett. B117, 175 (1982).

21. For a recent discussion on thermal Green's function, event horizon and Hawking effect, see D. W. Sciama, P. Candelas and D. Deutsch, Adv. in Physics. 30, 327 (1981), S. A. Fulling and S. N. M. Ruijenaars, Phys. Rep. 152, 135 (1987) and references therein.
22. See, e.g., Sec. 4.4, 4.5, 8.1 in Birrell and Davies (Ref. 3).
23. W. G. Unruh, Phys. Rev. Lett. 46, 1351 (1981).
24. L. Parker, Nature (London) 261, 20 (1976).
25. I thank J. Hartle for asking this question.
26. B. L. Hu and W. G. Unruh, private communication.
27. See, e.g. J. York, in Quantum Theory of Gravity (ed.) S. M. Christensen (Hilger Bristol, London 1984); P. Hajicek, Phys. Rev. D36, 1065 (1987).
28. R. Brandenberger and R. Khan, Phys. Lett. 119B, 75 (1982).
29. See, e.g., B. L. Hu and L. Parker, Phys. Rev. D17, 933 (1978); R. Balian and M. Vénéroni, Ann. Phys. (N.Y.) 135, 270 (1981); E. Calzetta and B. L. Hu (1986), E. Calzetta (1987); A. A. Starobinsky in Field Theory, Quantum Gravity and Strings, eds. H. J. deVega and N. Sanchez (Springer-Verlag, Heidelberg 1986).
30. B. L. Hu, Phys. Lett. 90A, 375 (1982), 97A, 368 (1983), in Cosmology of the Early Universe ed. by L. Z. Fang and R. Ruffini (World Scientific Publ. Co., Singapore, 1984).
31. B. L. Hu and H. E. Kandrup, Phys. Rev. D35, 1776 (1987).
32. Understanding black hole thermodynamics in terms of information theory was attempted originally by J. D. Bekenstein, Phys. Rev. D12, 3077 (1975) and recently by W. H. Zurek, Phys. Lett. 77A, 399 (1980); and W. H. Zurek and K. S. Thorne, Phys. Rev. Lett. 54, 2171 (1986).

The Vilkovisky-DeWitt Effective Action Beyond One Loop

Anton REBHAN

Institut für Theoretische Physik
Technische Universität Wien
Karlsplatz 13, A–1040 Vienna, Austria

Abstract

The reparametrization invariant definition of the effective action in quantum field theories due to Vilkovisky and DeWitt, which in gauge theories achieves manifest gauge invariance and quantum gauge fixing independence even off mass shell, is reviewed and its perturbative expansion in primary Feynman diagrams is discussed. The relationship to the conventional formalism is considered, and it is checked that the S-matrix remains unchanged.

1. Introduction

The effective action [1,2,4] is the central object in covariantly quantized field theories. It is the quantum analogue of the classical action, providing effective field equations by a variation principle, and generating 1-particle-irreducible (1pi) vertex functions which replace the classical vertices in perturbation theory.

In the conventional approach it is introduced as the Legendre transform of a path integral generating connected Green functions with respect to external sources. The latter, however, are responsible for its off-shell dependence on field parametrization. In gauge theories this general weakness is responsible for the gauge dependence of the off-shell effective action, which even pertains to the gauge invariant effective action(s) constructed by means of background field gauges [7].

The manifestly reparametrization invariant "Vilkovisky-DeWitt effective action" (VDEA) [2,4,5,9,10,11] achieves both gauge invariance and quantum gauge fixing independence. It has already found some applications in cases where the conventional framework encounters seemingly unavoidable gauge dependences, as in self-consistent dimensional reduction in Kaluza-Klein theories [12], decoupling thresholds in grand unified theories [10,13], and quantum chromodynamics [14,15] (see also [16]).

Here we shall restrict ourselves to the formal aspects of the VDEA. After a recapitulation of the conventional approach in sect. 2, we discuss the reparametrization invariant perturbation theory basing on normal coordinates introduced by Honerkamp and Ecker [3] in sect. 3. This formalism, however, does not provide an effective action beyond 1 loop, but is achieved in the construction initiated by Vilkovisky [2] and completed by DeWitt [4]. This development is reviewed in sect. 4, where we make the resulting perturbation theory explicit. We also check that the corresponding S-matrix is the usual one. This is done first for non-gauge theories, and in sect. 5 for the more intricate case of gauge theories.

2. Conventional Definition of the Effective Action

The content of the covariantly quantized field theory associated with the classical action $S[\varphi]$ can be elegantly summarized by a path integral acting as generating functional for connected Green functions

$$e^{iW[J]} = \int \mathcal{D}\varphi\, \mu[\varphi] \exp i\{S[\varphi] + J_i\varphi^i\}, \quad \mathcal{D}\varphi = \prod_i d\varphi^i, \tag{2.1}$$

where J_i are external sources and $\mu[\varphi]$ is an appropriate measure functional. Here and in the following we employ the compact notation of DeWitt [1] in which the symbol φ^i comprises all physical fields, the index i encompassing all discrete and continuous (space-time) labels. Correspondingly, the summation convention is extended to include integration. Functional differentiation is abbreviated by a comma and subsequent indices if it refers to the first argument of a functional; other derivatives are given explicitly.

Conventionally, the effective action is defined by the Legendre transformation

$$\Gamma[\bar{\varphi}] = W[J] - J_i \bar{\varphi}^i, \tag{2.2}$$

where the dependence on external sources is replaced by the dependence on the "mean field" $\bar{\varphi}^i$,

$$\bar{\varphi}^i[J] = \frac{\delta W[J]}{\delta J_i}. \tag{2.3}$$

(2.2) and (2.3) imply

$$\Gamma_{,i}[\bar{\varphi}] = -J_i, \tag{2.4}$$

which are the "effective field equations", representing the quantum version of the classical action principle $S_{,i}[\varphi] = -J_i$. In perturbation theory $\Gamma[\bar{\varphi}]$ generates one-particle-irreducible vertex functions, the quantum analogue of the classical vertices $S_{,ijk\ldots}$.

The above construction can be summarized in one functional integro-differential equation for $\Gamma[\bar{\varphi}]$ [2]

$$e^{\frac{i}{\hbar}\Gamma[\bar{\varphi}]} = \int \mathcal{D}\varphi\, \mu[\varphi] \exp \frac{i}{\hbar} \left\{ S[\varphi] + \Gamma_{,i}[\bar{\varphi}](\bar{\varphi} - \varphi)^i \right\}. \tag{2.5}$$

Iteration of (2.5) with $\hbar$ as expansion parameter yields the perturbative expansion in primary Feynman diagrams:

$$\begin{aligned} \Gamma[\bar{\varphi}] =& S[\bar{\varphi}] + \hbar \left(-i \log \mu[\bar{\varphi}] + \frac{i}{2} \log \det (S_{,ij}[\bar{\varphi}]) \right) \\ &+ \hbar^2 \left(-\frac{1}{12} \ominus - \frac{1}{8} \bigcirc\!\!\bigcirc \right) + \ldots, \end{aligned} \tag{2.6}$$

where in the diagrams a full line represents the Feynman-Green function $G^{ij}[\bar{\varphi}]$ of the operator $S_{,ij}[\bar{\varphi}]$, and vertices are given by $S_{,ijk\ldots}[\bar{\varphi}]$.

Furthermore, time ordered expectation values of operators

$$\langle \Omega[\varphi] \rangle_J = \Omega \left[\frac{\delta}{\delta J} \right] W[J] \tag{2.7}$$

are obtained through

$$\langle \Omega[\varphi] \rangle_{-\Gamma_{,i}} = e^{-i\Gamma[\bar{\varphi}]} \int \mathcal{D}\varphi\, \mu[\varphi] \Omega[\varphi] \exp i \left\{ S[\varphi] + \Gamma_{,i}[\bar{\varphi}](\bar{\varphi} - \varphi)^i \right\} \tag{2.8}$$

without direct reference to external sources.

However, the introduction of external sources has left its imprints on the implicit definition (2.5) of the effective action in that it has singled out a certain parametrization φ^i of the field configuration space $\Phi = \{\varphi^i\}$. Had we started with another parametrization, e.g. $\chi = \varphi + \delta\varphi[\varphi]$, we would have obtained another effective action which in general is not just the old one in the parametrization $\bar{\chi} = \bar{\varphi} + \delta\varphi[\bar{\varphi}]$. Rather,

$$\delta\Gamma[\bar{\varphi}] = \Gamma_{,i}[\bar{\varphi}]\left(\delta\varphi^i[\bar{\varphi}] - \langle\delta\varphi^i[\varphi]\rangle\right) + O(\delta\varphi^2) \tag{2.9}$$

which vanishes only on-shell ($\Gamma_{,i}[\bar{\varphi}] = 0$) if $\delta\varphi[\varphi]$ is non-linear.

In addition to this off-shell parametrization dependence of the effective action we have effective field equations depending on the parametrization even *on* mass shell:

$$\delta\Gamma_{,i}[\bar{\varphi}]\Big|_{\Gamma_{,i}=0} = \Gamma_{,ij}[\bar{\varphi}]\left(\delta\varphi^j[\bar{\varphi}] - \langle\delta\varphi^j[\varphi]\rangle\right). \tag{2.9}$$

The same holds for any vertex function derived from the effective action. Only the S-matrix, which essentially is contained in the undifferentiated on-shell effective action, is reparametrization invariant. Frequently, however, the central objects are effective field equations (cf. quantum cosmology) or off-shell Green functions (cf. quantum chromodynamics) rather than the S-matrix, and the problem of parametrization dependence enters.

Parametrization dependence actually is the most general weakness of the usual definition of the effective action. In gauge theories, the nuisance of off-shell gauge dependence is just another manifestion: In order to regulate the path integral one is led to introduce gauge conditions which can be viewed as singling out a coordinate system within the gauge orbits on configuration space.

3. Normal Coordinate Expansions

If configuration space is naturally endowed with an (affine) connection $\Gamma^i_{jk}[\varphi]$, then one actually has a preferred coordinate system on Φ, viz. geodesic normal coordinates around some given point, i.e. configuration.

Geodesics on Φ are defined by

$$\frac{d^2\varphi^i(s)}{ds^2} + \Gamma^i_{jk}[\varphi(s)]\frac{d\varphi^j(s)}{ds}\frac{d\varphi^k(s)}{ds} = 0 \tag{3.1}$$

with s being an affine parameter. Then the two-point functional

$$\sigma^i[\varphi(s), \varphi(0)] = \frac{d\varphi^i(s)}{d\log s}, \tag{3.2}$$

which is tangent to the geodesic at $\varphi(s)$, provides a geometrically sensible generalization of the "coordinate difference" $\varphi - \varphi'$:

$$\sigma^i[\varphi, \varphi'] = (\varphi - \varphi')^i - \frac{1}{2}\Gamma^i_{jk}[\varphi](\varphi - \varphi')^j(\varphi - \varphi')^k + \ldots, \tag{3.3}$$

for it is a vector with respect to the first argument and scalar with respect to the second.

In the context of non-linear σ-models, where the target space metric defines a natural (Riemannian) connection on configuration space, Honerkamp and Ecker [3] have employed normal coordinate expansions around a classical solution of $S_{,i}[\varphi_{cl.}] = -J_i$ in the framework of the (old version of the) background field method introduced by DeWitt [1]:

$$e^{iW^H[\varphi_{cl.}]} = \int \mathcal{D}\varphi \, \exp i\left\{S[\varphi] + J_i[\varphi_{cl.}](\varphi^i_{cl.} - \sigma^i[\varphi_{cl.}, \varphi])\right\} \tag{3.4}$$

This expression* can be split into a classical factor generating external tree graphs and a functional containing loops,

$$e^{iW^H[\varphi_{cl.}]} = e^{iW_0[\varphi_{cl.}]}e^{iW_L[\varphi_{cl.}]} \tag{3.5}$$

$$e^{iW_0[\varphi_{cl.}]} = \exp i\left(S[\varphi_{cl.}] + J_i[\varphi_{cl.}]\varphi^i_{cl.}\right) \tag{3.6}$$

$$e^{iW_L[\varphi_{cl.}]} = \int \mathcal{D}\varphi \, \exp i\left(S[\varphi] - S[\varphi_{cl.}] + S_{,i}[\varphi_{cl.}]\sigma^i[\varphi_{cl.}, \varphi]\right). \tag{3.7}$$

Note that the integrand in the last equation is manifestly scalar with respect to both φ and $\varphi_{cl.}$, hence $W_L[\varphi_{cl.}]$ is a reparametrization invariant functional. It is only the "classical" functional $W_0[\varphi_{cl.}]$ which still depends on the parametrization.

---

* From now on we forgo the explicit path integration measure $\mu[\varphi]$.

The perturbative expansion using normal coordinates is easily derived from

$$e^{iW_L[\varphi_{cl.}]} = \int \mathcal{D}\sigma \left|\frac{\delta(\varphi)}{\delta(\sigma)}\right| \exp i \sum_{n=2}^{\infty} \frac{(-1)^n}{n!} S_{;i_1 \ldots i_n}[\varphi_{cl.}]\sigma^{i_1} \cdots \sigma^{i_n}. \tag{3.8}$$

At 1-loop order $W_0 + W_L = S + \frac{i}{2}\log\det(S_{;ij})$ generates 1pi vertex functions, hence represents a reparametrization invariant effective action. However, this does not hold for higher loop orders. There the 1pi pieces have to be extracted "by hand", and there is no straightforward way of defining a corresponding effective action.

4. The Vilkovisky-DeWitt Effective Action

Vilkovisky [2] has pointed out that what prevents the usual effective action from being a reparametrization invariant functional is the appearance of a "coordinate difference" $(\bar{\varphi} - \varphi)^i$ in the implicit equation (2.5) defining $\Gamma[\bar{\varphi}]$. Therefore he proposed to substitute the geometrically more meaningful two-point functional $\sigma^i[\bar{\varphi}, \varphi]$ for $(\bar{\varphi} - \varphi)^i$ in (2.5):

$$e^{i\Gamma^V[\bar{\varphi}]} = \int \mathcal{D}\varphi \exp i \left\{ S[\varphi] + \Gamma^V_{,i}[\bar{\varphi}]\sigma^i[\bar{\varphi}, \varphi] \right\}. \tag{4.1}$$

This is manifestly a reparametrization invariant functional. However, compared to the conventional definition, one important thing has changed. In (2.5) a variation of $\bar{\varphi}^i$ yields

$$\Gamma_{,i}[\bar{\varphi}] = \Gamma_{,i}[\bar{\varphi}] + \Gamma_{,ij}[\bar{\varphi}] \left\langle (\bar{\varphi} - \varphi)^i \right\rangle, \tag{4.2}$$

implying $\langle (\bar{\varphi} - \varphi)^i \rangle = 0$ if $\Gamma_{,ij}[\bar{\varphi}]$ is non-singular. Now from (4.1) we obtain

$$\Gamma_{,i}[\bar{\varphi}] = \Gamma_{,j} \left\langle \sigma^j_{;i}[\bar{\varphi}, \varphi] \right\rangle + \Gamma_{,ij}[\bar{\varphi}] \left\langle \sigma^j[\bar{\varphi}, \varphi] \right\rangle, \tag{4.3}$$

and $\bar{\varphi}$ is *not* defined by $\langle \sigma^i[\bar{\varphi}, \varphi] \rangle = 0$ as one might expect. Rather,

$$\begin{aligned} \left\langle \sigma^j[\bar{\varphi}, \varphi] \right\rangle &= (\Gamma_{,ij}[\bar{\varphi}])^{-1} \left\{ \delta^k_i - \left\langle \sigma^k_{;i}[\bar{\varphi}, \varphi] \right\rangle \right\} \Gamma_{,k}[\bar{\varphi}] \\ &= (\Gamma_{,ij}[\bar{\varphi}])^{-1} \left\{ -\frac{i\hbar}{3} R^k_{min}[\bar{\varphi}] G^{mn}[\bar{\varphi}] + O(\hbar^2) \right\} \Gamma_{,k}[\bar{\varphi}], \end{aligned} \tag{4.4}$$

where $R^k_{min}[\bar\varphi]$ is the curvature functional associated with $\Gamma^i_{jk}[\bar\varphi]$, and $G^{mn}[\bar\varphi]$ is the Green function of $S_{;mn}[\bar\varphi]$. Hence in general $\langle\sigma^i[\bar\varphi,\varphi]\rangle \neq 0$, which means that $\Gamma^V[\bar\varphi]$ does not generate 1pi vertex functions beyond 1 loop.

This can be readily seen by expanding and iterating (4.1):

$$e^{\frac{i}{\hbar}\Gamma^V[\bar\varphi]} = \int \mathcal{D}\sigma \left|\frac{\delta(\varphi)}{\delta(\sigma)}\right| \exp\frac{i}{\hbar}\left\{\sum_{n=0}^{\infty} S_{;(i_1\ldots i_n)}[\bar\varphi]\sigma^{i_1}\cdots\sigma^{i_n} + \Gamma^V_{,i}[\bar\varphi]\sigma^i\right\}. \tag{4.5}$$

Rescaling $\sigma^i \to \sqrt{\hbar}\sigma^i$ allows the usual expansion in Gaussian integrals. In order $\hbar^{-1}$ we have* $\Gamma^{(0)}[\bar\varphi] = S[\bar\varphi]$. With $\Gamma^{(0)}_{,i}[\bar\varphi] = S_{,i}[\bar\varphi] \equiv S_{;i}[\bar\varphi]$ we obtain a pure Gaussian integral in order $\hbar^0$ giving $\Gamma^{(1)}[\bar\varphi] = \frac{i}{2}\log\det S_{;ij}[\bar\varphi]$, which is just $W^{(1)}_L$ of Honerkamp. In the next step of iteration we have to plug in

$$\begin{aligned}\Gamma^{(0)+(1)}_{,i}[\bar\varphi] &= S_{,i}[\bar\varphi] - \frac{i}{2}(S_{;mn})_{,j}[\bar\varphi]G^{nm}[\bar\varphi] \\ &= S_{,i}[\bar\varphi] - \frac{i}{2}S_{;mni}[\bar\varphi]G^nm[\bar\varphi] - i\Gamma^m_{im}[\bar\varphi].\end{aligned} \tag{4.6}$$

The last term in (4.6) can be neglected when the connection functional is local. The second term on the r.h.s. of (4.6) - one would expect - ought to subtract 1-particle reducible parts from $\Gamma^{(2)}$ when one performs Wick contractions in $\langle S_{;(3)}\varphi^{(3)}\Gamma_{,i}\varphi^i\rangle$. However, $S_{;mni}$ is not symmetric in its indices (unless $R^i_{jkl} = 0$), whereas all vertices stemming from the covariant Taylor expansion of $S[\varphi]$ in (4.5) are symmetrized! Consequently, 1-particle-irreducibility is lost with the simple substitution $(\bar\varphi - \varphi)^i \to \sigma^i[\bar\varphi,\varphi]$ in (4.1).

Recently, DeWitt [4] has taken up Vilkovisky's proposal and has presented a modified construction. He sticks to the conventional definition in which he introduces normal coordinates around a base point φ^i_* which only in the very end is identified with the mean field $\bar\varphi^i$. As we shall see, this preserves 1-particle-irreducibility to all orders.

In normal coordinates $\phi^i \equiv -\sigma^i[\varphi_*,\varphi]$, the φ_*-dependent conventional definition reads

$$e^{i\hat\Gamma[\bar\phi;\varphi_*]} = \int \mathcal{D}\varphi \exp i\left\{S[\varphi] + (\bar\phi - \phi)^i\hat\Gamma_{,i}[\bar\phi;\varphi_*]\right\}. \tag{4.7}$$

---

* Instead of the superscript "V" we give the loop orders in brackets.

In terms of the original (arbitrary) parametrization φ the corresponding mean field $\bar{\varphi}$ is given implicitly by $\bar{\phi}^i = -\sigma^i[\varphi_*, \bar{\varphi}]$, and the dependence on the base point φ_* can be eliminated by identifying $\varphi_*^i = \bar{\varphi}^i$ in $\tilde{\Gamma}[\bar{\varphi}; \varphi_*] \equiv \hat{\Gamma}[\bar{\phi}[\varphi_*, \varphi]; \varphi_*]$:

$$\Gamma[\bar{\varphi}] := \tilde{\Gamma}[\bar{\varphi}; \bar{\varphi}]. \tag{4.8}$$

The "Vilkovisky-DeWitt effective action" (VDEA) $\Gamma[\bar{\varphi}]$ can be shown [4] to be given as solution of the following coupled pair of functional equations

$$e^{i\Gamma[\bar{\varphi}]} = \int \mathcal{D}\varphi \exp i \left\{ S[\varphi] + \Gamma_{,i}[\bar{\varphi}] C^{-1}{}^i_j[\bar{\varphi}] \sigma^j[\bar{\varphi}, \varphi] \right\} \tag{4.9}$$

$$C^i_j[\bar{\varphi}] = \left\langle \sigma^i_{;j}[\bar{\varphi}, \varphi] \right\rangle_{-\Gamma_{,i}}, \tag{4.10}$$

which can be solved by a coupled iteration using

$$\sigma^i_{;j}[\bar{\varphi}, \varphi] = \delta^i_j - \tfrac{1}{3} R^i_{kjl}[\bar{\varphi}] \sigma^k[\bar{\varphi}, \varphi] \sigma^l[\bar{\varphi}, \varphi] + O(\sigma^{(3)}). \tag{4.11}$$

One can easily check that now $\langle \sigma^i[\bar{\varphi}, \varphi] \rangle = 0$ is fulfilled. Indeed, looking again at the iteration performed above with Vilkovisky's original definition, we find that 1-particle-irreducibility is restored. There the modification of DeWitt amounts to

$$\begin{aligned} \Gamma^{(1)}_{,i}[\bar{\varphi}] \to & \Gamma^{(1)}_{,i}[\bar{\varphi}] + S_{,j}[\bar{\varphi}] C^{(1)-1}{}^j_i[\bar{\varphi}] \\ = & \Gamma^{(1)}_{,i}[\bar{\varphi}] - \tfrac{i}{3} S_{,j}[\bar{\varphi}] R^j_{kil}[\bar{\varphi}] G^{kl}[\bar{\varphi}] \end{aligned} \tag{4.12}$$

and thus to the desired symmetrization

$$S_{;mni} G^{nm} + \tfrac{2}{3} R^j_{nim} S_{,j} G^{nm} = S_{;(mni)} G^{nm}. \tag{4.13}$$

In general the iteration of (4.9) and (4.10) yielding the perturbative expansion seems to be rather complicated. It is, however, most easily derived from the auxiliary functional $\hat{\Gamma}[\bar{\phi}; \varphi_*]$ prior to the identification $\varphi_* = \bar{\varphi}$ [5]. Apart from a Jacobian, by expanding

$$e^{i\hat{\Gamma}[\bar{\phi};\varphi_*]} = \int \mathcal{D}\phi \left| \frac{\delta(\varphi)}{\delta(\phi)} \right| \exp i \left\{ \sum_{n=0}^{\infty} \tfrac{1}{n!} S_{;i_1 \ldots i_n}[\varphi_*] \phi^{i_1} \cdots \phi^{i_n} + (\bar{\phi} - \phi)^i \hat{\Gamma}_{,i}[\bar{\phi}; \varphi_*] \right\} \tag{4.14}$$

one arrives at primary Feynman diagrams which differ from the usual ones only in that n-vertices are given by

$$S_{;(i_1\dots i_n)}[\varphi_*] + \sum_{m=n+1}^{\infty} \frac{1}{(m-n)!} S_{;(i_{n+1}\dots i_m)} \bar{\phi}^{i_{n+1}} \dots \bar{\phi}^{i_m} \tag{4.15}$$

instead of $S_{,i_1\dots i_n}[\bar{\varphi}]$.

The VDEA finally is obtained by $\varphi_* = \bar{\varphi}$, that is

$$\bar{\phi}^i \Big|_{\varphi_* = \bar{\varphi}} = -\sigma^i[\bar{\varphi}, \bar{\varphi}] = 0,$$

hence*

$$\Gamma[\bar{\varphi}] = \hat{\Gamma}[0; \bar{\varphi}]. \tag{4.16}$$

We therefore arrive at the conventional series of primary Feynman diagrams with n-vertices $S_{;(i_1\dots i_n)}[\bar{\varphi}]$.

In contrast to the conventional approach we have to distinguish between internal vertices contained in the primary diagrams, and external ones which are produced by differentiating these primary diagrams. A vertex function $\Gamma_{,jkl\dots}$ will contain vertices $(S_{;(i_1\dots i_n)})_{,jkl\dots}$. In general even covariant differentiation does not help because of the non-commutativity of covariant derivatives. This situation is analogous to the one in the construction of a gauge invariant effective action by means of background field gauges [7,8].

One question to be answered is whether all this makes any difference when one is calculating a physical quantity. We shall restrict ourselves to the S-matrix and argue that it is not affected by the departure from the standard approach implied in taking (4.16) as effective action.

The identification $\varphi_* = \bar{\varphi}$ means that the new effective action is not given as Legendre transform of a generating functional of connected Green functions with a usual path integral representation. This is again analogous to the new

---

* This definition has been independently proposed in ref. 6, where the auxiliary functional $\hat{\Gamma}[\bar{\phi}; \varphi_*]$ has been found necessary in studying renormalization of $\Gamma[\bar{\varphi}]$.

formulation of the background field method where it has been proved that the S-matrix remains the same [8]. In this latter case, however, the modifications as compared to the standard approach are restricted to the gauge fixing sector, while (4.16) means a more radical change.

Prior to the identification $\varphi_* = \bar{\varphi}$, $\hat{\Gamma}[\bar{\phi}; \varphi_*]$ can be viewed as a conventional effective action in an unusual parametrization. Reparametrizations leave the S-matrix unchanged, when they are ultra-local with respect to space-time (i.e. involving no space-time derivatives) [1]. $\varphi_* = \bar{\varphi}$, however, introduces additional vertices stemming from differentiations with respect to the second argument of the functional $\tilde{\Gamma}[\bar{\varphi}; \varphi_*]$ (cf. (4.8)).

These additional vertices can be studied by considering an interpolating functional $\tilde{\Gamma}[\bar{\varphi}; \varphi_*(t)]$ with $\varphi_*(0) = \varphi_*$ and $\varphi_*(1) = \bar{\varphi}$, and its Legendre transform $\tilde{W}[J; \varphi_*]$. It can be shown [5] that all additional contributions are contained in

$$\frac{\delta \tilde{W}[J; \varphi_*]}{\delta \varphi_*^k} = J_i D^{-1}{}^i{}_j[\varphi_*, \bar{\varphi}] \left(C^j_k[\varphi_*, \bar{\varphi}] - \left\langle C^j_k[\varphi_*, \varphi] \right\rangle \right) \Big|_{\bar{\varphi} = \frac{\delta \tilde{W}}{\delta J}} \tag{4.17}$$

with

$$C^j_k[\varphi, \varphi'] = \sigma^j_{;k}[\varphi, \varphi'], \quad D^i_j[\varphi, \varphi'] = -\frac{\delta \sigma^i[\varphi, \varphi']}{\delta \varphi'^j}. \tag{4.18}$$

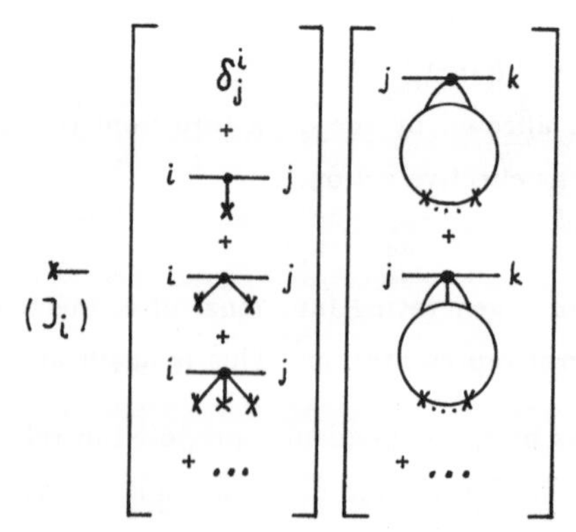

Fig. 1: Diagrammatic representation of the r.h.s. of eq. (4.17).

The diagrammatic expansion of (4.17) is given in fig. 1, where [symbol] represents some functional consisting of Γ^i_{jk} and its functional derivatives. The line k corresponds to an inner line of a connected Green function, whereas i is the index of one of the external lines to be amputated on-shell in the construction of the S-matrix. If Γ^i_{jk} is local with respect to space-time, the on-shell momentum of the source J_i in fig. 1 gets split by local vertices so that the amputation operation at J_i encounters no on-shell propagator, thus annihilating the whole expression. Usually, e.g. in non-linear σ-models, the connection functional Γ^i_{jk} is even ultra-local, so for those applications, the above argumentation is sufficient to show S-matrix equivalence. In the case of gauge theories, which we shall deal with presently, we shall have to take a closer look at (4.17).

5. The Vilkovisky-DeWitt Effective Action for Gauge Theories

We have already mentioned that the gauge dependences of the conventional off-shell effective action in gauge theories are but one manifestation of its parametrization dependence. Indeed, here the VDEA comes into its own, providing for the first time an off-shell gauge invariant and quantum gauge fixing independent effective action.

Gauge theories are characterized by an infinite set of Noether identities

$$S_{,i}[\varphi]D^i_\alpha[\varphi] \equiv 0, \tag{5.1}$$

with $D^i_\alpha[\varphi]$ being the generators of gauge transformations which we require to form a closed* gauge algebra

$$D^i_{\alpha,j}[\varphi]D^j_\beta[\varphi] - D^i_{\beta,j}[\varphi]D^j_\alpha[\varphi] = D^i_\gamma[\varphi]c^\gamma_{\alpha\beta}[\varphi], \tag{5.2}$$

where $c^\gamma_{\alpha\beta}[\varphi]$ are the structure functions. It will *not* be necessary to restrict ourselves to the case of inhomogeneously linear gauge generators.

---

* An "open" gauge algebra would contain further terms proportional to the field equations $S_{,i}[\varphi]$.

In order to remove the redundancy of the path integration caused by the existence of gauge orbits where the action is stationary, one has to introduce gauge fixing terms and a Faddeev-Popov determinant. In effect, one substitutes

$$S[\varphi] \to S[\varphi] + \frac{1}{2}\eta_{\alpha\beta}F^{\alpha}[\varphi]F^{\beta}[\varphi] - i\log\det Q^{\alpha}_{\beta}[\varphi], \tag{5.3}$$

where $Q^{\alpha}_{\beta}[\varphi] = F^{\alpha}_{,i}[\varphi]D^{i}_{\beta}[\varphi]$ is the Faddeev-Popov operator associated with the gauge condition $F^{\alpha}[\varphi]$.

Because of the source term, the conventional path integral (2.1) becomes dependent on the gauge conditions F^{α}. This quantum gauge fixing dependence persists even for the gauge invariant effective action(s) of ref. 7.

The VDEA provides a unique construction, if there is a unique connection $\Gamma^{i}_{jk}[\varphi]$ on configuration space. Such a connection has to conform to the gauge structure on Φ which is reflected in a parametrization $\varphi^{i} \leftrightarrow (\mathcal{I}^{A}, \mathcal{F}^{\alpha})$ where the $\mathcal{I}^{A}$ label the gauge orbits, $\mathcal{I}^{A}_{,i}[\varphi]D^{i}_{\alpha}[\varphi] = 0$, and the $\mathcal{F}^{\alpha}$ provide a coordinate system within the gauge orbits.

Geodesics on Φ in this parametrization are given by

$$\ddot{\mathcal{I}}^{A} + \Gamma^{A}_{BC}\dot{\mathcal{I}}^{B}\mathcal{I}^{C} + 2\Gamma^{A}_{B\gamma}\dot{\mathcal{I}}^{B}\dot{\mathcal{F}}^{\gamma} + \Gamma^{A}_{\beta\gamma}\dot{\mathcal{F}}^{\beta}\dot{\mathcal{F}}^{\gamma} = 0, \tag{5.4}$$

$$\ddot{\mathcal{F}}^{\alpha} + \Gamma^{\alpha}_{BC}\dot{\mathcal{I}}^{B}\mathcal{I}^{C} + 2\Gamma^{\alpha}_{B\gamma}\dot{\mathcal{I}}^{B}\dot{\mathcal{F}}^{\gamma} + \Gamma^{\alpha}_{\beta\gamma}\dot{\mathcal{F}}^{\beta}\dot{\mathcal{F}}^{\gamma} = 0, \tag{5.5}$$

where a dot means differentiation with respect to the affine parameter. The requirement [4] that geodesics on Φ project onto geodesics on the orbit space Φ/G leads to

$$\Gamma^{A}_{BC} = \Gamma^{A}_{BC}[\mathcal{I}], \quad \Gamma^{A}_{B\gamma} = 0 = \Gamma^{A}_{\beta\gamma}, \tag{5.6}$$

so that the component $\sigma^{A}[\bar{\varphi}, \varphi]$ only depends on the invariants $\bar{\mathcal{I}}$,$\mathcal{I}$.

Given a metric $\gamma_{ij}[\varphi]$ on Φ whose Killing vectors are the gauge generators $D^{i}_{\alpha}[\varphi]$, (5.6) is met by [2,4,9]

$$\begin{aligned}\Gamma^{i}_{mn}[\varphi] = \left\{ {i \atop mn} \right\}(\gamma) - 2D^{i}_{\alpha.(m}\gamma_{n)k}D^{k}_{\beta}N^{-1\alpha\beta} \\ + N^{-1\alpha\delta}D^{j}_{\delta}\gamma_{jm}N^{-1\beta\gamma}D^{l}_{\gamma}\gamma_{ln}D^{k}_{(\beta}D^{i}_{\alpha).k},\end{aligned} \tag{5.7}$$

where $\left\{ {i \atop mn} \right\}(\gamma)$ is the Christoffel symbol associated with $\gamma_{ij}[\varphi]$ and a dot with subsequent indices abbreviates the corresponding covariant differentiation. $N_{\alpha\beta} \equiv$

$D^i_\alpha \gamma_{ij} D^j_\beta$ in general is a differential operator, hence the full connection (5.7) is non-local with respect to space-time. A starting metric γ_{ij} can be uniquely singled out in the examples of Yang-Mills and Einstein gravity theories [2], so that (5.7) is unique in these cases.

The connection (5.7) gives rise to a 2-point functional $\sigma^i[\bar{\varphi}, \varphi]$ which fulfills

$$\sigma^i_{;k}[\bar{\varphi}, \varphi] D^k_\alpha[\bar{\varphi}] \propto D^i_\beta[\bar{\varphi}], \tag{5.8}$$

$$\frac{\delta \sigma^i[\bar{\varphi}, \varphi]}{\delta \varphi^k} D^k_\alpha[\varphi] \propto D^i_\beta[\bar{\varphi}]. \tag{5.9}$$

A gauge transformation in the source term of the VDEA (4.9) produces the term

$$\Gamma_{,i}[\bar{\varphi}] C^{-1}{}^i_j[\bar{\varphi}] \frac{\delta \sigma^j[\bar{\varphi}, \varphi]}{\delta \varphi^k} D^k_\alpha[\varphi] \tag{5.10}$$

which by virtue of (5.8) and (5.9) is in fact proportional to $\Gamma_{,i}[\bar{\varphi}] D^i_\beta[\bar{\varphi}]$. This shows that the VDEA has a chance of being a gauge invariant and gauge fixing independent functional. Indeed, complete gauge invariance and gauge independence has been proved [2,4,10] for

$$\begin{aligned} e^{i\Gamma[\bar{\varphi}]} =& (\det \eta[\bar{\varphi}])^{\frac{1}{2}} \int \mathcal{D}\varphi \, \det Q[\varphi; \bar{\varphi}] \exp i \Big\{ S[\varphi] \\ &+ \frac{1}{2} \eta_{\alpha\beta}[\bar{\varphi}] F^\alpha F^\beta[\varphi; \bar{\varphi}] + \Gamma_{,i}[\bar{\varphi}] C^{-1}{}^i_j[\bar{\varphi}] \sigma^j[\bar{\varphi}, \varphi] \Big\}, \end{aligned} \tag{5.11}$$

provided the gauge conditions satisfy $F^\alpha[\bar{\varphi}; \bar{\varphi}] = 0$.

To simplify the perturbative expansion of (5.11) one would like to go back to the auxiliary functional $\hat{\Gamma}[\bar{\phi}; \varphi_*]$. Indeed, it can be shown [5] that an appropriately gauge fixed functional $\hat{\Gamma}$ reproduces (5.11), despite the fact that in (5.11) one had to introduce a mean field dependent gauge breaking term. The simplest perturbative expansion is obtained from

$$\begin{aligned} e^{i\hat{\Gamma}[\bar{\phi}; \varphi_*]} =& (\det \eta[\varphi_*])^{\frac{1}{2}} \int \mathcal{D}\varphi \, \det Q[\varphi; \varphi_*] \exp i \Big\{ S[\varphi] \\ &+ \frac{1}{2} \eta_{\alpha\beta}[\varphi_*] F^\alpha_i[\varphi_*] F^\beta_j[\varphi_*] (\phi - \bar{\phi})^i (\phi - \bar{\phi})^j + (\bar{\phi} - \phi)^i \hat{\Gamma}_{,i}[\bar{\phi}; \varphi_*] \Big\}. \end{aligned} \tag{5.12}$$

At $\varphi_* = \bar{\varphi}$, i.e. $\bar{\phi} = 0$, apart from a Jacobian contribution the Feynman rules are similar to those of a conventional gauge theory with non-linear gauge fixing. The vertices between internal $\bar{\varphi}$-lines are given by $S_{;(i_1 \ldots i_n)}[\bar{\varphi}]$ for $n \geq 3$, while the kinetic kernel is $S_{;ij}[\bar{\varphi}] + \eta_{\alpha\beta} F_i^\alpha[\bar{\varphi}] F_j^\beta[\bar{\varphi}]$. The Faddeev-Popov kernel is $F_i^\alpha[\bar{\varphi}] D_\beta^i[\bar{\varphi}]$ as usual, however the vertices of the Faddeev-Popov ghosts are to be derived from the covariant expansion of

$$Q_\beta^\alpha[\varphi;\bar{\varphi}] = F_i^\alpha[\bar{\varphi}] \frac{\delta(-\sigma^i[\bar{\varphi},\varphi])}{\delta\varphi^j} D_\beta^j[\varphi]. \tag{5.13}$$

Again, vertices between external and internal lines of a vertex function are different, as discussed in the previous section.

Starting from 2 loops additional graphs [5] emerge from the Jacobian

$$\mathcal{D}\varphi = \mathcal{D}\phi \ \det \frac{\delta\varphi}{\delta\phi} \tag{5.14}$$

which now cannot be neglected due to the non-locality of $\Gamma^i_{jk}[\varphi]$. For example, at 2-loop order (5.14) gives rise to the additional contribution

$$\left(\tfrac{1}{3} R^m_{ijm}[\bar{\varphi}] + \frac{1}{2} \Gamma^m_{ij,m}[\bar{\varphi}] - \frac{1}{2} \Gamma^m_{in}[\bar{\varphi}] \Gamma^n_{jm}[\bar{\varphi}] \right) G^{ij}[\bar{\varphi}] \tag{5.15}$$

to the effective action.

The non-locality of the connection (5.7) also makes it necessary to reconsider the S-matrix equivalence argument of the previous section. First, the change of parametrization $\varphi \to \sigma$ is non-local this time. However, if the starting metric $\gamma_{ij}[\varphi]$ is ultra-local (containing only undifferentiated δ-functions), then all non-localities occur only along gauge orbits [2] and are harmless due to the gauge independence of the S-matrix. As to the additional vertices introduced by the identification $\varphi_* = \bar{\varphi}$, we have seen that in the case of non-gauge theories they do not contribute to the S-matrix because of the locality of the vertices in fig. 1. Now they are non-local. However, the one explicit external source J_i is connected to an upper index of $\Gamma^i_{jk,\ldots}$ which either comes from $\left\{ {i \atop jk} \right\}_{,\ldots}$ or from $D^i_{\alpha.j} \equiv D^i_{\alpha,j} + \left\{ {i \atop jm} \right\} D^m_\alpha$ and its derivatives. If $\left\{ {i \atop jk} \right\}$ is a local functional, then again the on-shell momentum of J_i is distributed among at least 2 internal lines as is necessary in our argumentation. This is guaranteed if $\gamma_{ij}[\varphi]$ is ultra-local as we have already required. An ultra-local starting metric γ_{ij} indeed exists for Yang-Mills theories and Einstein gravity, and our argument applies. In the application to superspace supergravity, though, Fradkin and Tseytlin [9] have found it necessary to depart from this requirement.

References

1. B.S. DeWitt, *Dynamical Theory of Groups and Fields*, Gordon and Breach, N.Y. 1965; Phys. Rev. 162, 1195 (1967); "Spacetime Approach to Quantum Field Theory", in *Relativity, Groups and Topology* II, Les Houches 1983, eds. B.S. DeWitt and R. Stora, North Holland, Amsterdam 1984.

2. G.A. Vilkovisky, "The Gospel According to DeWitt", in *Quantum Theory of Gravity, Essays in Honor of the 60th Birthday of Bryce S. DeWitt*, ed. S.M. Christensen, A. Hilger, Bristol 1984; Nucl. Phys. B234, 125 (1984); A.O. Barvinsky and G.A. Vilkovisky, Phys. Lett. 131B, 313 (1983); Phys. Rep. 119, 1 (1985).

3. J. Honerkamp, Nucl. Phys. B36, 130 (1972);
G. Ecker and J. Honerkamp, Phys. Lett. 42B, 253 (1972);
J. Honerkamp, F. Krause and M. Scheunert, Nucl. Phys. B69, 618 (1974)
L. Alvarez-Gaumé, D.Z. Freedman and S. Mukhi, Ann. of Phys. 134, 85 (1981).

4. B.S. DeWitt, "The Effective Action", in E.S. Fradkin's 60th Birthday Festschrift, A. Hilger, to appear.

5. A. Rebhan, "Feynman Rules and S-Matrix Equivalence of the Vilkovisky-DeWitt Effective Action", Techn. Univ. Wien preprint, April 1987.

6. P.S. Howe, G. Papadopoulos and K.S. Stelle, "The Background Field Method and the Non-Linear σ-Model", preprint CERN-TH. 4744/87.

7. B.S. DeWitt, "A Gauge Invariant Effective Action", in *Quantum Gravity* II, eds. C.J. Isham, R. Penrose and D.W. Sciama, Oxford Univ. Press, Oxford 1981;
D.G. Boulware, Phys. Rev. D23, 389 (1981);
L.F. Abbott, Nucl. Phys. B185, 189 (1981);
D.M. Capper and A. McLean, Nucl. Phys. B203, 413 (1982);
C.F. Hart, Phys. Rev. D28, 1993 (1983).

8. L.F. Abbott, M.T. Grisaru and R.K. Schaefer, Nucl. Phys. B229, 372 (1983);
A. Rebhan and G. Wirthumer, Z. Phys. C28, 269 (1985).

9. E.S. Fradkin and A.A. Tseytlin, Nucl. Phys. B234, 509 (1984).

10. A. Rebhan, Nucl. Phys. B288, 832 (1987).

11. G. Kunstatter, "Vilkovisky's Unique Effective Action", to appear in *Superfield Theories*, ed. H.C. Lee, Plenum, N.Y. 1987;
D.J. Toms, "The Effective Action", preprint NCL 87 TP9, to appear in the proceedings of the 2nd Canadian Conference on General Relativity and Relativistic Astrophysics, May 1987;
C.P. Burgess and G. Kunstatter, "On the Physical Interpretation of the Vilkovisky-DeWitt Effective Action", Univ. of Winnipeg preprint, June 1987.

12. G. Kunstatter and H.P. Leivo, Phys. Lett. 166B, 321 (1986); Nucl. Phys. B279, 641 (1987);
S.R. Huggins, G. Kunstatter, H.P. Leivo and D.J. Toms, Phys. Rev. Lett. 58, 296 (1987); "The Vilkovisky-DeWitt Effective Action for Quantum Gravity", Univ. of Newcastle preprint 1987.

13. M. Kreuzer and A. Rebhan, paper in preparation.

14. A. Rebhan, Z. Phys. C30, 309 (1986).

15. T.H. Hansson and I. Zahed, Phys. Rev. Lett. 58, 2397 (1987).

16. Y.-C. Kao, J. Koller and H. Yamagishi, Phys. Rev. Lett. 58, 1077 (1987).

Physical Interpretation and Uniqueness of the Vilkovisky-DeWitt Effective Action

C.P. Burgess
Physics Department, McGill University
3600 University St., Montreal P.Q.
Canada H3A 2T8

and

G. Kunstatter
Physics Department, The University of Winnipeg
515 Portage Ave., Winnipeg, Manitoba
Canada, R3B 2E9

ABSTRACT The covariant background field method is used to generate an infinite class of parametrization invariant effective actions, which contains as a special case the Vilkovisky-DeWitt effective action. The invariant actions are shown to share those features at the root of the utility of the standard effective action. In particular, it is shown that they have a physical interpretation as the minimum energy in states that are subject to a reparametrization invariant constraint. Certain "Nielsen-type" identities are used to prove that all members of this class of invariant actions predict the same vacuum energy. It is also pointed out that when quantizing on curved configuration spaces, the concept of a mean field is not uniquely defined.

1. Introduction

The effective action as usually defined[1] has three fundamental properties which make it an important tool in quantum field theory:

(a) Derivatives of the effective action with respect to the fields generate one particle irreducible(1PI) correlation functions for the field operator. In other words, the full connected Green's functions can be constructed via tree graphs from these 1PI correlation functions. This property makes the effective action useful for analyzing the divergences of a theory[2].

(b) It has a physical interpretation related to the minimum energy of the source free Hamiltonian under the constraint that the expectation value of the field equal a particular value. This justifies the minimization of the effective action as a criterion for identifying the ground states of a system, and makes the effective action useful in analyzing symmetry properties of the vacuum[3].

(c) It can be calculated perturbatively by summing the one particle irreducible graphs of an appropriately defined field variable, using Feynman rules obtained from the classical action. It is important to note that this property is *a priori* distinct from (a) above.

Despite these many virtues, the usual effective action formalism has a very serious shortcoming, which becomes apparent when considering either sigma models or gauge theories. In particular, the ordinary effective action is parametrization dependent: it does not transform as a scalar under non-linear field redefinitions. That is, if ϕ^i denotes the complete set of classical fields describing a particular theory[4], then under a non-linear field redefinition $\phi'^i = \phi'^i(\phi)$ the new effective action $\Gamma'(\phi')$ does not equal $\Gamma(\phi)$, even if the classical action $I(\phi)$ is invariant. As noted by Vilkovisky[5] this lack of parametrization invariance is responsible for the gauge parameter dependence of the ordinary effective action for gauge theories[6]. In the case of sigma models, the absence of a preferred(i.e. cartesian) coordinate system for the target space results in an ambiguity in the off-shell effective action.

The purpose of this talk is to describe a new formalism due originally to Vilkovisky[5] and modified by DeWitt[11], which eliminates the problem of parametrization and gauge parameter dependence by considering the geometrical structure of the configuration space for the field theory. Since the Vilkovisky-DeWitt formalism has been discussed in general by A. Rebhan in a previous talk, the present discussion will focus on the question of whether the Vilkovisky-DeWitt formalism yields an effective action possessing properties (a)-(c) above. The Vilkovisky-DeWitt effective action does indeed possess the last two properties. However, if one requires the effective action to generate 1PI correlation functions, then one is forced to consider not a unique effective action but an infinite class of invariant effective actions, all possessing properties (a) - (c). This is nonetheless a considerable improvement over the standard formalism, because the parametrization invariant effective actions have a physical interpretation in terms of the expectation values of well defined operators, even in the case of curved configuration spaces and gauge theories.

The outline of this talk is as follows. In section 2, the ordinary background field effective action[6,7] will be briefly presented, with particular emphasis on its derivation via the Legendre transform, and the nature of its dependence on the "background field". Section 3 contains an introduction to the covariant background field method[7,8,12], and shows that this formalism leads to an infinite family of physically equivalent effective actions, of which the Vilkovisky-DeWitt effective action is but a single representative. (For simplicity, it is assumed that the Lagrangian possesses no gauge invariance. Most of the results should nonetheless be valid for gauge theories as well, providing the configuration space is taken to be the space of connections modulo the group of gauge transformations.) Section 4 contains the proof that all members of the family of invariant actions have the physical interpretation as the minimum energy of the system under a covariant constraint, and in section 5 it is shown that, as expected, they all yield the same physical predictions for the vacuum state. An interesting fact brought out by this analysis is that in general the concept of a mean field is not uniquely defined when quantizing on a curved configuration space[22]. Finally, section 6 contains conclusions and prospects for future work.

2. The Ordinary Background Field Method

One possible way of defining the ordinary background field effective action[6,7], is to couple the source in the generating functional not to the quantum field ϕ, but to the fluctuations of the quantum field about an arbitrary but fixed background field, ϕ_*. (This definition is most suitable for comparison with the covariant background field method, to be described in the following section). In particular consider the vacuum-vacuum transition amplitude in the presence of an arbitrary source $J_i(x)$:

$$\exp iW[J;\phi_*] = \int \mathcal{D}\phi \exp i\left(S[\phi] + J_i\sigma^i(\phi_*,\phi)\right) \tag{2.1}$$

where $\sigma^i(\phi_*,\phi) \equiv \phi^i - \phi^i_*$ is the quantum fluctuation and $\mathcal{D}\phi$ denotes an invariant measure on the space of fields. Standard arguments[11] can be used to show that W defined in Eq.(2.1) is the generator for connected Green functions for the operator $\sigma^i(\phi_*,\phi)$. The effective action is the Legendre transform of the generating functional $W[J;\phi_*]$:

$$\hat{\Gamma}[v^i;\phi_*] = W[J;\phi_*] - v^i J_i \tag{2.2}$$

where J is understood to be a functional of v, given implicitly by:

$$\begin{aligned} v^i &\equiv \frac{\delta W[J;\phi_*]}{\delta J_i} \\ &= \langle \sigma^i(\phi_*,\phi)\rangle_J \\ &= \langle \phi^i \rangle_J - \phi^i_* . \end{aligned} \tag{2.3}$$

Note that functional differentiation when applied to a functional of two arguments denotes partial functional differentiation; i.e. the other argument is to be held fixed. For example, $\frac{\delta W[J;\phi_*]}{\delta J_i}$ is defined by

$$\begin{aligned} \delta W[J;\phi_*] &= W[J+\delta J;\phi_*] - W[J;\phi_*] \\ &= \frac{\delta W[J;\phi_*]}{\delta J_i}\delta J_i \end{aligned}$$

In the above, $\langle \quad \rangle_J$ denotes the T^*-ordered ground state expectation value in the presence of the source J. In particular:

$$\langle f[\phi]\rangle_J = \exp\left(-iW[J;\phi_*]\right)\int \mathcal{D}\phi f[\phi] \exp i\left(S[\phi] + J_i\sigma^i(\phi_*,\phi)\right) \tag{2.4}$$

Using the definitions (2.2) and (2.3) it follows that

$$\frac{\delta\hat{\Gamma}[v^i;\phi_*]}{\delta v^i} = -J_i . \tag{2.5}$$

The content of Eqs(2.2)-(2.5) can be summarized in the single functional differentional equation for the effective action:

$$\exp i\hat{\Gamma}[v^i;\phi_*] = \int \mathcal{D}\phi \exp i\left[S[\phi] - \frac{\delta\hat{\Gamma}}{\delta v^i}\left(\sigma^i(\phi_*,\phi) - v^i\right)\right] \tag{2.6}$$

Substituting for v^i from Eq.(2.3) gives the effective action as a functional of the expectation value, $\overline{\phi} \equiv \langle\phi\rangle_J$ of the field ϕ:

$$\exp i\Gamma[\overline{\phi};\phi_*] = \int \mathcal{D}\phi \exp i\left[S[\phi] - \frac{\delta\Gamma[\overline{\phi};\phi_*]}{\delta\overline{\phi}^i}(\phi^i - \overline{\phi}^i)\right]. \tag{2.7}$$

By differentiating Eq.(2.7) with respect to ϕ^i_* one derives the important identity:

$$\frac{\delta\Gamma[\overline{\phi};\phi_*]}{\delta\phi^i_*} = 0 \qquad (2.8),$$

which states that the effective action is independent of the arbitrarily chosen background field ϕ_*.

The problem with the ordinary background field method arises due to the fact that $\sigma^i(\phi_*,\phi)$ defined above is the difference between the coordinates of two distinct points in field space, and does not transform covariantly under field reparametrizations. Consequently, neither $W[J;\phi_*]$ nor $\Gamma[\overline{\phi};\phi_*]$ is a scalar function of the configuration space coordinates. This results in the off-shell parametrization and gauge dependence of the ordinary effective action. In the following section, the cure for this non-invariance, as proposed by Vilkovisky[5] and subsequently modified by DeWitt[12] is presented.(It has also been considered in various forms in the context of non-linear sigma models[7,8,13].)

3. Covariant Background Field Method

In order to define an invariant effective action, it is necessary to choose a more suitable definition for the operator $\sigma^i(\phi_*,\phi)$. The most natural and simple choice, which also has the property of reducing to the standard expression in the case of vector spaces, is to define $\sigma^i(\phi_*,\phi)$ to be the Gaussian normal coordinates of the point ϕ with origin at ϕ_*. Since Gaussian normal coordinates on field space are discussed in detail in Rebhan's contribution to these proceedings[14], it is sufficient here to simply mention that $\sigma^i(\phi_*,\phi)$ are the components of the suitably normalized tangent vector at the point ϕ_* to the geodesic starting at ϕ and ending at ϕ_*. When the configuration space is a vector space, the Gaussian normal coordinates reduce to Cartesian coordinates with ϕ_* as origin. Note that $\sigma^i(\phi_*,\phi)$ is defined to be minus that of Rebhan[14], so that when the configuration space is flat, it is equal to the usual quantum fluctuation about the background field ϕ_*.

The components of $\sigma^i(\phi_*,\phi)$ transform as vectors under coordinate transformations of the point ϕ_*, but they are scalars under coordinate transformations of the point ϕ, so that the source term in the definition of the generating functional $W[J;\phi_*]$ will be a scalar under field redefinitions, as long as the source J is taken to lie in the co-tangent space of ϕ_*. An important question concerns the choice of connection on the field space. For non-linear sigma models[7,8,13], the target space is endowed with a metric, whose Christoffel symbol provides a natural choice of connection. In the case of gauge theories, Vilkovisky has shown[5] that there is a suitable connection on the space of gauge potentials modulo the group of gauge transformations which does yield a gauge and parametrization independent effective action. This has been verified in several explicit calculations[14-16]. With the above choice for $\sigma^i(\phi_*,\phi)$ in Eq.(2.1), Equations (2.2) through (2.6) follow exactly as before, except of course for the last line of Eq.(2.3). As in the case of the ordinary effective action, it is straightforward to show that $\hat{\Gamma}$ generates one particle irreducible correlation functions for the operator $\sigma^i(\phi_*,\phi)$. Moreover $\hat{\Gamma}$ can be constructed perturbatively by summing one-particle irreducible graphs for the field $\sigma^i(\phi_*,\phi) - v^i$. In the next section we will verify that it has the usual physical interpretation as well.

The argument v^i of $\hat{\Gamma}$ has the physical interpretation as the vacuum expection value of the operator $\sigma^i(\phi_*,\phi)$ in the presence of the source J. It is possible to define an effective action whose argument is not the expectation value of a normal coordinate such as v^i, but the "mean field" $\overline{\phi}$, but here one encounters the first significant departure from the usual formalism. First one has to define the concept of a mean field. The most natural definition is to define $\overline{\phi}$ to be the point in field space whose normal coordinate with respect to ϕ_* is v^i. That is:

$$\begin{aligned}\sigma^i(\phi_*,\overline{\phi}) &= v^i \\ &= \langle\sigma^i(\phi_*,\phi)\rangle_J\end{aligned} \tag{3.1}$$

The equation for the resulting effective action is

$$e^{i\Gamma[\overline{\phi};\phi_*]} = \int \mathcal{D}\phi e^{i\left[S[\phi] - \frac{\delta\Gamma[\overline{\phi};\phi_*]}{\delta\overline{\phi}^j}(D^{-1})^j_i\left(\sigma^i(\phi_*,\phi) - \sigma^i(\phi_*,\overline{\phi})\right)\right]} \tag{3.2}$$

where

$$D^i_j[\overline{\phi};\phi_*] = \frac{\delta\sigma^i(\phi_*,\overline{\phi})}{\delta\overline{\phi}^j} \tag{3.3}$$

It is clear that $\Gamma[\overline{\phi};\phi_*]$ defined above is a scalar functional of both $\overline{\phi}$ and ϕ_*. It is also true that the 1PI correlation functions of $\sigma^i(\phi_*,\phi)$ are proportional to the partial derivatives of $\Gamma[\overline{\phi};\phi_*]$ with respect to $v^i = \sigma^i(\phi_*,\overline{\phi})$.

Until now the reference point ϕ_* has been left arbitrary. By differentiating Eq.(3.2) with respect to ϕ_* it is possible to derive the following identity[12]:

$$\frac{\delta\Gamma[\overline{\phi};\phi_*]}{\delta\phi^i_*} + E^j_i[\overline{\phi};\phi_*]\frac{\delta\Gamma[\overline{\phi};\phi_*]}{\delta\overline{\phi}^j} = 0, \tag{3.4}$$

where

$$E_i^j[\overline{\phi};\phi_*] \equiv (D^{-1})^j_k \left(\langle\sigma^k_{;i}(\phi_*,\phi)\rangle_J - (\langle\sigma^k(\phi_*,\phi)\rangle_J)_{;i}\right) \tag{3.5}$$

and the semi-colon denotes covariant differentiation with respect to ϕ_*. Thus the dependence of $\Gamma[\overline{\phi};\phi_*]$ on ϕ_* is *not* trivial in general. In order to remove this unphysical dependence on the background field, DeWitt[12] makes the choice $\phi_* = \overline{\phi}$, which leads to the so-called Vilkovisky-DeWitt effective action (this definition for the effective action was also considered in refs. 7,8), whose defining functional differential equation is:

$$\begin{aligned} \exp i\Gamma_{VD}[\overline{\phi}] &= \lim_{\phi_* \to \overline{\phi}} \exp i\Gamma[\overline{\phi};\phi_*] \\ &= \int \mathcal{D}\phi \exp i\left[S[\phi] + \frac{\delta\Gamma_{VD}[\overline{\phi}]}{\delta\overline{\phi}^j}(C^{-1})^j{}_i\sigma^i(\overline{\phi},\phi)\right] \end{aligned} \tag{3.6}$$

The appearance of

$$C^i_j \equiv \langle\sigma^i_{;j}(\overline{\phi},\phi)\rangle_J \tag{3.7},$$

in Eq.(3.6) is due to the fact that the limit $\phi_* \to \overline{\phi}$ must be taken *after* $\hat{\Gamma}$ has been evaluated, so that:

$$\frac{\delta\Gamma_{VD}[\overline{\phi}]}{\delta\overline{\phi}^i} = \lim_{\phi_* \to \overline{\phi}} \left(\frac{\delta\Gamma[\overline{\phi};\phi_*]}{\delta\overline{\phi}^i} + \frac{\delta\Gamma[\overline{\phi};\phi_*]}{\delta\phi_*^i}\right) \tag{3.8}$$

Eq.(3.6) then follows from Eqs(3.4), (3.5) and the fact that

$$\lim_{\phi_* \to \overline{\phi}} \sigma^i_{;j}(\phi_*,\overline{\phi}) = -\delta^i_j. \tag{3.9}$$

A general argument, presented in ref. 9, can be used to show that both $\Gamma[\overline{\phi};\phi_*]$ and $\Gamma_{VD}[\overline{\phi}]$ can be evaluated semi-classically by summing only one particle irreducible graphs using Feynman rules derived from the classical action. The only complication is that in the former case, the expansion of the classical action must be done in powers of the operator $\psi^i = \sigma^i(\phi_*,\phi) - \sigma^i(\phi_*,\overline{\phi})$ while in the case of the $\Gamma_{VD}[\overline{\phi}]$, $\sigma^i(\overline{\phi},\phi)$ is the relevant operator. However, although derivatives of $\Gamma[\overline{\phi};\phi_*]$ with respect to v^i generate 1PI correlation functions for the operator $\sigma^i(\phi_*,\phi)$, $\Gamma_{VD}[\overline{\phi}]$ cannot be used to generate 1PI correlation functions for any known operator[17]. It was shown in ref. 7, in fact, that one must consider $\Gamma[\overline{\phi};\phi_*]$ and not $\Gamma_{VD}[\overline{\phi}]$ in order to correctly analyze the divergences of non-linear sigma models. In order to see that $\Gamma_{VD}[\overline{\phi}]$ does not generate 1PI correlation functions, consider a general Legendre transform of $\Gamma_{VD}[\overline{\phi}]$[18]:

$$W_{VD}[\tilde{J}] = \Gamma_{VD}[\overline{\phi}] + \tilde{J}_i u^i \tag{3.10}$$

where u^i is an arbitrary covariant function of $\overline{\phi}$ (i.e.vector field on the configuration space), and $\tilde{J}_i$ is defined by

$$\tilde{J}_i = -\frac{\delta \Gamma_{VD}[\overline{\phi}]}{\delta u^i}. \tag{3.11}$$

It is straightforward to show that

$$\exp iW_{VD}[\tilde{J}] = \int \mathcal{D}\phi \exp i\left(S[\phi] + \tilde{J}_i F^i[\overline{\phi}, \phi]\right) \qquad (3.12),$$

where

$$F^i[\overline{\phi}, \phi] = -\left(\frac{Du^i}{D\overline{\phi}^j}(C^{-1})^j_k \sigma^k(\overline{\phi}, \phi) - u^i(\overline{\phi})\right) \tag{3.13}$$

Here D denotes covariant differentiation with respect to $\overline{\phi}$. In Eq.(3.12), $\overline{\phi}$ is a function of $\tilde{J}$ given implicitly by

$$\langle \sigma^i(\overline{\phi}, \phi) \rangle_{\tilde{J}} = 0 \tag{3.14}$$

Consequently, $F^i[\overline{\phi}, \phi]$ is a function of $\tilde{J}$. Since the coupling of the field to source in $W[\tilde{J}]$ is non-linear, the n-th derivative of W with respect to $\tilde{J}$ will not in general equal the n-point correlation function of F^i.

4.. The Energy Interpretation

The family of parametrization invariant effective actions $\Gamma[\overline{\phi}; \phi_*]$ defined above admit a simple energy interpretation[9] that is a natural, covariant extension of that of the ordinary effective action[10]. The standard effective action, when evaluated at a particular time-independent function $\overline{\phi}(x)$, is proportional to the minimum expectation value of the system Hamiltonian, H, in that part of the Hilbert space of normalized states, $|\psi\rangle$, satisfying the constraint $\langle\psi|\phi|\psi\rangle = \overline{\phi}$. The corresponding property of $\Gamma_{VD}[\overline{\phi}]$ comes from a similar interpretation of $\Gamma[\overline{\phi}; \phi_*]$. In particular, if an external source J is adiabatically turned on at time $t = -T/2$ and off at $t = T/2$, then

$$\lim_{T\to\infty} \frac{-\Gamma[\overline{\phi}; \phi_*]}{T} = min_{\overline{\phi}}\{\langle\psi|H|\psi\rangle\}, \tag{4.1}$$

where H is the Hamiltonian operator derived from $S[\phi]$, and $min_{\overline{\phi}}$ denotes minimization of the expectation value for all normalized states $|\psi\rangle$ in the Hilbert space such that

$$\langle\psi|\sigma^i(\phi_*,\phi)|\psi\rangle = \sigma^i(\phi_*,\overline{\phi}), \tag{4.2}$$

for time independent $\sigma^i(\phi_*,\overline{\phi})$. Note that $\langle\psi|\ |\psi\rangle$ denotes the usual inner product on the Hilbert space, which formally corresponds to $\langle\ \rangle_J$ when $J = 0$.

The first step in the proof is to establish, via standard arguments[11] that in the limit $T \to \infty$, the in-out vacuum transition amplitude

$$\begin{aligned} \exp iW[J;\phi_*] &= {}_J\langle\Omega_{out}|\Omega_{in}\rangle_J \\ &\to \exp\left(-iE_0[J;\phi_*]T\right) \end{aligned} \tag{4.3}$$

In the above, $E_0[J;\phi_*]$ is the lowest eigenvalue of the modified Hamiltonian

$$\hat{H}[J;\phi_*] = \left(H - \frac{J_i\sigma^i(\phi_*,\phi)}{T}\right), \tag{4.4}$$

with corresponding ground state $|\psi_0[J;\phi_*]\rangle$. In order for $\hat{H}$ to be well-defined as a Hamiltonian, it is necessary that $\sigma^i(\phi_*,\phi)$ not contain time derivatives of the field ϕ.

Step 2 in the argument is to use the method of Lagrange multipliers to minimize $\langle\psi|H|\psi\rangle$ subject to the constraints $\langle\psi|\psi\rangle = 1$ and $\langle\psi|\sigma^i(\phi_*,\phi)|\psi\rangle = \sigma^i(\phi_*,\overline{\phi})$. Demanding that the variation of

$$\langle\psi|H|\psi\rangle + \alpha(\langle\psi|\psi\rangle - 1) + \frac{\beta_i}{T}\left(\langle\psi|\sigma^i(\phi_*,\phi)|\psi\rangle - \sigma^i(\phi_*,\overline{\phi})\right)$$

vanish for arbitrary variations of the state $|\psi\rangle$ and lagrange multipliers α and β_i, respectively, yields the three conditions;

$$H|\psi\rangle + \alpha|\psi\rangle + \frac{\beta_i\sigma^i(\phi_*,\phi)}{T}|\psi\rangle = 0 \tag{4.5a}$$

$$\langle\psi|\psi\rangle = 1 \tag{4.5b}$$

$$\langle\psi|\sigma^i(\phi_*,\phi)|\psi\rangle = \sigma^i(\phi_*,\overline{\phi}) \tag{4.5c}$$

Finally, one notes that the following constitutes a solution to the variational problem of Eqs(4.5a-c):

$$\beta_i = -J_i[\overline{\phi};\phi_*] \tag{4.6a}$$

$$\alpha = -E_0[J_i[\overline{\phi};\phi_*];\phi_*] \tag{4.6b}$$

$$|\psi\rangle = |\psi_0[J;\phi_*]\rangle, \tag{4.6c}$$

where $J_i[\overline{\phi};\phi_*]$ is determined by solving

$$\frac{\delta W[J;\phi_*]}{\delta J_i} = \sigma^i(\phi_*,\overline{\phi}). \tag{4.7}$$

The fact that Eqs(4.6a)-(4.6c) do indeed solve Eqs.(4.5a)-(4.5c) can be verified by noting that Eq.(4.5a) follows from (4.4) above, Eq.(4.5b) is obvious and Eq.(4.5c) is a direct consequence of the definition of $\sigma^i(\phi_*,\overline{\phi})$ as the ground state expectation value of the operator $\sigma^i(\phi_*,\phi)$ in the presence of the source J. Given the solution (4.6a-c), it follows automatically via (4.5a), that

$$\begin{aligned} min_{\overline{\phi}}\{\langle\psi|H|\psi\rangle\} &= -\alpha - \frac{\beta_i\sigma^i(\phi_*,\overline{\phi})}{T} \\ &= \lim_{T\to\infty}\left(-\frac{W[J;\phi_*]}{T} + \frac{J_i\sigma^i(\phi_*,\overline{\phi})}{T}\right). \end{aligned} \tag{4.8}$$

This completes the proof of (4.1).

We have shown that $\Gamma[\overline{\phi};\phi_*]$ is proportional to the minimum of the energy of the source-free system, subject to a covariant, ϕ_*-dependent constraint. It is therefore clear that the Vilkoviky-DeWitt effective action, which chooses the particular reference point $\phi_* = \overline{\phi}$, also possesses the same physical interpretation. The corresponding constraint in the case of $\Gamma_{VD}[\overline{\phi}]$, however, is that $\langle\psi|\sigma^i(\overline{\phi},\phi)|\psi\rangle = 0$. An immediate consequence of this result is that for both $\Gamma[\overline{\phi};\phi_*]$ and $\Gamma_{VD}[\overline{\phi}]$, minimizing the effective action over the set of all possible constraints (i.e. minimizing with respect to $\overline{\phi}$) should yield the ground state energy of the source free system. Moreover, the value of $\sigma^i(\phi_*,\overline{\phi})$ which minimizes $\Gamma[\overline{\phi};\phi_*]$ gives the ground state expectation value of the operator $\sigma^i(\phi_*,\phi)$. In other words, if $\overline{\phi}_0$ gives the location of the absolute minimum of $\Gamma[\overline{\phi};\phi_*]$ for a particular ϕ_*, then

$$\frac{-\Gamma[\overline{\phi}_0;\phi_*]}{T} = E_0 \equiv \langle\psi_0|H|\psi_0\rangle \tag{4.9}$$

$$\sigma^i(\phi_*,\overline{\phi}_0) = \langle\psi_0|\sigma^i(\phi_*,\phi)|\psi_0\rangle, \tag{4.10}$$

where $|\psi_0\rangle$ is the ground state of the source free system. Eq.(4.9) implies that $\Gamma[\overline{\phi}_0;\phi_*]$ is independent of ϕ_*:

$$\frac{d\Gamma[\overline{\phi}_0;\phi_*]}{d\phi_*^i} = \frac{\delta\Gamma[\overline{\phi}_0;\phi_*]}{\delta\overline{\phi}_0^j}\frac{d\overline{\phi}_0^j}{d\phi_*^i} + \frac{\delta\Gamma[\overline{\phi}_0;\phi_*]}{\delta\phi_*^i}$$
$$= 0, \tag{4.11}$$

where d denotes a total derivative. In the following section we will prove that Eq.(4.11) does indeed follow from the defining equation (3.2) of $\Gamma[\overline{\phi};\phi_*]$.

5. Nielsen-Type Identities for $\Gamma[\overline{\phi};\phi_*]$

In the mid-seventies, the issue of the gauge parameter dependence of the effective potential, V, for Yang-Mills theories was resolved by the work of Nielsen[19,20], who showed that the existence of identities of the form

$$\frac{\partial V[\phi;\alpha]}{\partial\alpha} + C(\phi;\alpha)\frac{\partial V[\phi;\alpha]}{\partial\phi} = 0, \tag{5.1}$$

implied that physical quantities derived from the effective potential V would be independent of the gauge parameter α. In effect, Eq.(5.1) states that although V contains explicit gauge parameter dependence, any shift in the gauge parameter can be compensated by an appropriate shift in the field ϕ. As we shall see, the gauge parameter dependence addressed by Nielsen is very similar to the issue of the ϕ_*-dependence of the covariant effective action $\Gamma[\overline{\phi};\phi_*]$.

In the previous section it was shown that

$$\frac{\delta\Gamma[\overline{\phi};\phi_*]}{\delta\phi_*^i} + E_i^j[\overline{\phi};\phi_*]\frac{\delta\Gamma[\overline{\phi};\phi_*]}{\delta\overline{\phi}^j} = 0, \tag{5.2}$$

where $E_i^j[\overline{\phi};\phi_*]$ was defined in Eq.(3.5). This implies that $\Gamma[\overline{\phi};\phi_*]$ is constant along characteristic surfaces defined locally by

$$d\overline{\phi}^j - E_i^j[\overline{\phi};\phi_*]d\phi_*^i = 0. \tag{5.3}$$

By differentiating Eq.(5.2) with respect to $\overline{\phi}^j$, keeping in mind that the lower index of $E_i^j[\overline{\phi};\phi_*]$ is a vector index at ϕ_* and a scalar at $\overline{\phi}$, while the upper index, j, is the opposite, one derives the following identity:

$$\frac{\delta^2\Gamma[\overline{\phi};\phi_*]}{\delta\overline{\phi}^k\delta\phi_*^i} + E_i^j[\overline{\phi};\phi_*]\frac{\mathrm{D}^2\Gamma[\overline{\phi};\phi_*]}{\mathrm{D}\overline{\phi}^k\mathrm{D}\overline{\phi}^j} = -\frac{\mathrm{D}E_i^j[\overline{\phi};\phi_*]}{\mathrm{D}\overline{\phi}^k}\frac{\delta\Gamma[\overline{\phi};\phi_*]}{\delta\overline{\phi}^j} \tag{5.4}$$

Interchanging the order of differentiation in the terms on the left hand side of Eq(5.4) yields:

$$\left[\frac{\delta}{\delta\phi_*^i} + E_i^j[\overline{\phi};\phi_*]\frac{\mathrm{D}}{\mathrm{D}\overline{\phi}^j}\right]\frac{\delta\Gamma[\overline{\phi};\phi_*]}{\delta\overline{\phi}^k}$$

$$= -\left(\frac{\mathrm{D}E_i^j[\overline{\phi};\phi_*]}{\mathrm{D}\overline{\phi}^k} + \mathrm{T}_{ik}^j\right)\frac{\delta\Gamma[\overline{\phi};\phi_*]}{\delta\overline{\phi}^j}, \tag{5.5}$$

where for the sake of generality, Eq.(5.5) allows for the possibility of non-zero torsion, T_{ik}^j, on the configuration space[21]. Eq.(5.5) states that the covariant directional derivative of the vector $\frac{\delta\Gamma[\overline{\phi};\phi_*]}{\delta\overline{\phi}}$ along the characteristic surfaces defined by (5.3) is proportional to the vector itself. Thus, if $\frac{\delta\Gamma[\overline{\phi};\phi_*]}{\delta\overline{\phi}^j} = 0$ at one point on a characteristic surface, then it is covariantly constant, and hence zero at every point on that surface. This result implies that under an infinitesmal change in the reference point ϕ_*, the solution, $\overline{\phi}_0$, to the quantum corrected equations of motion moves along the appropriate characteristic surface, so that:

$$\frac{d\overline{\phi}_0^j}{d\phi_*^i} = \mathrm{E}_i^j[\overline{\phi}_0;\phi_*]. \tag{5.6}$$

Eq(4.11) therefore follows immediately as a consequence of (5.2) and (5.6), and as expected, the functional equation for $\Gamma[\overline{\phi};\phi_*]$ is consistent with the energy interpretation.

One interesting consequence of Eq(5.6) is that unless $E_i^j[\overline{\phi};\phi_*]$ vanishes when $J = 0$, the vacuum mean field $\overline{\phi}_0$ is not uniquely defined, even up to diffeomorphisms: It depends explicitly on ϕ_*. As a consistency check, the same result will now be derived using an operator formalism. The energy interpretation of the previous section implies that

$$\sigma^i(\phi_*,\overline{\phi}_0(\phi_*)) = \langle\psi_0|\sigma^i(\phi_*,\phi)|\psi_0\rangle. \tag{5.7}$$

Under an infinitesmal change in reference point $\phi_* \to \phi_*' = \phi_* + d\phi_*$, the expectation value of $\sigma^i(\phi_*,\phi)$ goes to

$$\sigma^i(\phi_*';\overline{\phi}_0') = \langle\psi_0|\sigma^i(\phi_*,\phi) + \sigma^i_{;j}(\phi_*,\phi)d\phi_*^j|\psi_0\rangle, \tag{5.8}$$

whereas the normal coordinates of the original mean field $\overline{\phi}_0$ go to

$$\sigma^i(\phi'_*;\overline{\phi}_0) = \sigma^i(\phi_*,\overline{\phi}_0) + \sigma^i_{;j}(\phi_*,\overline{\phi}_0)d\phi^j_*. \tag{5.9}$$

Thus for a given, unique ground state $|\psi_0\rangle$, the shift in the mean field as defined by Eq.(3.2) is

$$d\overline{\phi}^i_0 = \frac{d\overline{\phi}_0}{d\sigma^i(\phi'_*,\overline{\phi}'_0)} d\sigma^i, \tag{5.10}$$

where

$$\begin{aligned} d\sigma^i &= \sigma^i(\phi'_*,\overline{\phi}'_0) - \sigma^i(\phi'_*,\overline{\phi}_0) \\ &= \langle\psi_0|\sigma^i_{;j}(\phi_*,\phi)|\psi_0\rangle - \sigma^i_{;j}(\phi_*,\overline{\phi}_0) \end{aligned} \tag{5.11}$$

Eq.(5.10) therefore agrees precisely with Eq.(5.6), as claimed.

6. Conclusions

In summary, a natural class of covariant effective actions exists which includes as a limiting case that of Vilkovisky and DeWitt. These actions share the key features which are central to the utility of the conventional effective action with the additional property of covariance with respect to field redefinitions. Specifically, these effective actions generate the one-particle irreducible correlation functions of the covariant field variable $\sigma^i(\phi_*,\phi)$, they have a simple perturbative expansion, and they can be interpreted as the minimum energy subject to the constraint that the expectation value of $\sigma^i(\phi_*,\phi)$ is fixed equal to a given function. Due to the noncommutativity of the limiting process with functional differentiation, the Vilkovisky-DeWitt action does not generate 1PI correlation functions.

One might ask at this point whether much has been gained over the conventional formalism. After all, if the effective action is just used to generate propagators for the computation of some physical quantity, then, since physical quantities must be parametrization independent, the parametrization-dependence of the intermediate steps should not matter. Indeed, the value of a physical quantity does not depend on the formalism used to compute it. For these purposes the only advantage offered by the covariant effective actions defined above is in the utility of organizing the calculation in a manner in which parametrization-independence is manifest at every step.

The central difference between the conventional and the covariant effective action lies in the fact that the covariant actions, when evaluated at any point in field space, are themselves physical quantities. (A property not shared by the conventional effective action since it is parametrization dependent.) The value of the covariant effective action as a physical quantity is made most clear by its energy interpretation. The minimum energy subject to a covariant constraint is, of course, a physical thing.

Apart from its conceptual advantage, this has practical implications. For some applications it is the effective action itself that is of interest in a problem. Its finite-temperature generalization is the system's Helmholtz free energy, for example, and so its derivatives with respect to field variables and temperature give the thermodynamic response functions. For these purposes, the covariant quantity must be used. (In this regard, it should be noted that when the derivatives of the free energy with respect to the field variables are of interest, the Vilkovisky-DeWitt action is just as inappropriate as the conventional effective action.) This may in fact be the source of the conflicting results which emerge when the conventional formalism is used to analyze the stability of plasma oscillations in finite temperature quantum chromodynamics[24]. This issue is currently under investigation.

Acknowledgements

The authors would like to thank P. Caldas, C.J. Isham, H.C. Lee, H.P. Leivo, C. Ordonez, A. Rebhan, D.J. Toms and Zhonguan Zhu for helpful converstations. This work has been supported in part by the Natural Sciences and Engineering Research Council of Canada.

References

1) J. Schwinger, lectures given at the Institute for Advanced Study, Princeton, 1954 (unpublished). B.S. DeWitt, in *Relativity Groups and Topology*, Eds. B.S. DeWitt and C. DeWitt, Gordon and Breach, 1964; G. Jona-Lasinio, Nuovo Cimento **34**, 1790 (1964).
2) See for example, R. Becchi, in the proceedings of the Les Houches summer school, *Relativity, Groups and Topology*, edited by B.S. DeWitt, (North Holland, 1983) and references therein.
3) G. Jona-Lasinio, in 1) above; S. Coleman and E. Weinberg, Phys. Rev. **D7**, 1887 (1973); R. Jackiw, Phys. Rev. **D9**, 1686 (1974).

4) We use DeWitt's condensed index notation in which the single index i represents all internal and spacetime indices on the fields, as well as the coordinates of the point at which the field is to be evaluated. Repeated indices should be summed, including integration over spacetime points with appropriate measure. For example, in electromagnetism

$$J_i \phi^i \sim \int d^4 x \sqrt{-g} J_\mu(x) A^\mu(x).$$

5) G.A. Vilkovisky, in *Quantum Theory of Gravity*, edited by S. Christensen, (Adam Hilger, Bristol, 1984); G.A. Vilkovisky, Nucl. Phys. **B234**, 125 (1984); A.O. Barvinsky and G.A. Vilkovisky, Phys. Lett. **131B**, 313 (1983).
6) This problem is distinct from the question of gauge invariance of the effective action, which was solved by the introduction of the "background-gauge invariant effective action". See B.S. DeWitt, *Dynamical Theory of Groups and Fields*, (Gordon Breach, New York, 1965); G. 't Hooft, in *Proc. 12th Winter School in Theoretical Physics, Karpacz*, Acta. Univ. Wrats. no. 38, (1975);, L.F. Abbot, Nucl. Phys. **B185**, 189 (1981); B.G. Boulware, Phys. Rev. **D23**, 389, (1981); B.S. DeWitt, "A Gauge Invariant Effective Action", in *Quantum Gravity II*, eds. C.J. Isham, R. Penrose and D.W. Sciama, (Oxford Univ. Press, Oxford 1981).
7) P.S. Howe, G. Papadopoulos, and K.S. Stelle, Nucl. Phys. **B296**, 26 (1988).
8) C.M. Hull, in *Super Field Theories*, Edited by H.C. Lee *et al*, (Plenum, New York, 1987).
9) C.P. Burgess and G. Kunstatter, J. Mod. Phys. A (Letts) (to appear).
10) S. Coleman, *Laws of Hadronic Matter*, edited by A. Zichichi(Periodici Scientific, Milan, 1975); T.D. Lee and G.C. Wick, Phys. Rev. **D9**, 2291 (1974).
11) See for example, D.J. Amit, **Field Theory Renormalization Group, and Critical Phenomena**, (McGraw Hill, New York, 1978).
12) B.S. DeWitt, "The Effective Action", in *E.S. Fradkin's 60th Birthday Festschrift*, A. Hilger, to appear.
13) J. Honerkamp, Nucl. Phys. **B36**, 130 (1972); G. Ecker and J. HonerKamp, Phys. Lett. **42B**, 253(1972); D. Friedan, Phys. Rev. Lett. **45**, 1057 (1980); Ann. Phys. **163**, 318 (1985); L. Alvarez-Gaume, D.Z. Freedman and S. Mukhi, Ann. Phys. (N.Y.) **134**, 85 (1981); D. Boulware and L. Brown, Ann.

Phys. (N.Y.) **138**), 392 (1982); E. Braaten, T.L. Curtwright and C.K. zachos, Nucl. Phys. **B260**, 630 (1985); S. Mukho, Nucl. Phys. **264**, 640 (1986).

14) A. Rebhan, these proceedings; Nucl. Phys. **B288**, 832 (1987); "Feynman Rules and S-Matrix Equivalence of the Vilkovisky-DeWitt Effective Action", Tehn. Univ. Wien preprint, April 1987.

15) H.P. Leivo, these proceedings; PhD. Thesis, University of Toronto, 1987.

16) E.S. Fradkin and A.A. Tseytlin, Nucl. Phys. **B234**, 509 (1984); G. Kunstatter, in *Super Field Theories*, ed H.C. Lee (World Scientific, Singapore 1987); S.R. Huggins, G. Kunstatter, H.P. Leivo and D.J. Toms, Phys. Rev. Lett. **58**, 196 (1987); "The Vilkovisky-DeWitt Effective Action for Quantum Gravity", University of Winnipeg preprint, 1987; D.J. Toms, "The Effective Action", Proceedings of the 2nd Canadian Conference on General Relativity and Relativistic Astrophysics, (to appear).

17) P. Ellicott and D.J. Toms, Univ. of Newcastle preprint 1987.

18) G. K. is grateful to Zhongyuan Zhu for valuable discussions on this point.

19) N.K. Nielsen, Nucl. Phys. **B101**, 173 (1975).

20) For more recent discussions of the Nielsen identities, see I.J.R. Aitchison and C.M. Fraser, Ann. Phys.(NY) **156**, 1 (1984). N.K. Nielsen, University of Odense preprint (1986); D. Johnson, L.P.T. Univ. of Paris preprint (1986).

21) C. Hull has considered the effects of introducing torsion into the configuration space connection for non-linear sigma models. See ref. 8 above.

22) We are grateful to C.J. Isham for clarifying this point. See C.J. Isham, these proceedings; and in *Relativity, Groups and Topology II. Proceedings, 1983 Les Houches Summer School*, eds. B.S. DeWitt and R. Stora (North-Holland, Amsterdam 1984).

23) S. Randjbar-Daemi and M.H. Sarmadi, Phys. Lett. **151B**, 343 (1985); G. Kunstatter and H.P. Leivo, Phys. Lett. **166B**, 321 (1986); Nucl. Phys. **B279**, 641 (1987); Phys. Lett. **183B**, 75-80 (1987).

24) K. Kajante and J. Kapusta, Ann. Phys. **160**, 477 (1985); U. Heinz, K. Kajantie and T. Toimela, Phys. Lett. **183B**, 96 (1987); T.H. Hansson and I. Zahed, Phys. Rev. Lett **58**, 2397 (1987); Nucl. Phys. **B292**, 725 (1987). H.-T. Elze, U. Heinz, K. Kajantie and T. Toimela, "High Temperature Gluon Matter in the Background Gauge", R.I.T.P. Helsinki preprint, 1987.

The Fate of Bosonic Superconducting Cosmic Strings

R. MACKENZIE

Department of Physics
Ohio State University
Columbus, OH 43210

ABSTRACT

The energetics of bosonic superconducting cosmic strings is explored in detail. This is used to investigate an important aspect of the string's life cycle: whether or not so-called cosmic springs form. It is found that springs could form in a larger region of parameter space than previously observed. As cosmic springs give rise to cosmological problems, this presents a stronger restriction on the allowed parameters of the theory.

INTRODUCTION

It is widely believed that the symmetry group we observe in low-energy particle interactions, SU(3)xU(1), is smaller than the underlying symmetry group of fundamental physics, and that spontaneous symmetry breaking is the mechanism by which this reduction of the symmetry is achieved. Grand unified theories postulate that at high energy, or temperature, the full underlying symmetry would be manifest, but that as the temperature is decreased a number of phase transitions occur which break the symmetry in stages down to SU(3)xU(1).

This is important in cosmology, as according to the big bang theory the universe was at some early time an extremely hot, dense 'fireball', and it has been expanding and cooling ever since. Thus, the sequence of symmetry breakings inferred from particle physics actually occurred in the evolution of the universe.

One of the important implications of this is that various topological objects (domain walls, cosmic strings, monopoles) may have been formed as the universe cooled. Particle physics considerations indicate that monopoles must have been produced (if SU(3)xU(1) arose from a semisimple group), whereas for domain walls and cosmic strings a weaker statement is true: realistic theories which predict domain walls and/or cosmic strings can be constructed, but the existence of these objects is not a necessity.

While monopoles and domain walls are believed to cause cosmological problems (unless they are produced before a period of inflation), cosmic strings may have been created often enough to have important consequences, without coming to dominate the universe.

Strings coexist with less exotic matter in the universe; the matter sees strings as regions of higher energy density and is thus attracted gravitationally to strings. This is a possible mechanism for the formation of structure in the universe.[1] The time scale of string evaporation is critical to this scenario. This can be seen at an intuitive level as follows. If evaporation occurs too quickly, the strings disappear before any appreciable matter density fluctuations arise. If it occurs too slowly, they give rise to density fluctuations which are too large. In an intermediate region the strings survive long enough to cause matter to accrete around them; even after the strings decay these perturbations continue to grow, forming galaxies and galactic clusters.

The possibility that cosmic strings may be superconducting and carry very large electromagnetic currents was realized by Witten.[2] This leads immediately to a number of questions: If superconducting strings exist, would they attain large currents? Would such currents lead to new possibilities of observation of cosmic strings? How does the dynamics of superconducting strings differ from that of ordinary cosmic strings? What would be their importance on the formation of structure in the universe?

The issue I would like to address here concerns the final stages of the superconducting string's life cycle. Superconducting strings, like ordinary cosmic strings, oscillate and radiate energy. As will be shown below, the current carried by the string increases as it shrinks. There are two reasons why the current cannot increase indefinitely. First, there is a critical current above which the string goes 'normal', ie, non-superconducting. In this process, to quote Witten, "sparks fly": the string emits high-energy gamma rays and the current momentarily decreases. (The nature of this process will be made more explicit below, after I have explained the mechanism of superconducting strings.) Second, as a loop shrinks and its current increases, a point is reached where the string's energy is minimized; further shrinking of the loop would increase its energy. Thus, a classically stable configuration, a cosmic spring [3] exists.

The formation of springs causes cosmological problems regarding nucleosynthesis,[3] so a crucial question is whether the critical current or spring current is lower. This is because the lower of the two currents is reached first in the string's evolution, so if the spring current is the lower of the two strings form springs before going critical; the critical current is never reached and an unacceptable cosmology results.

As was emphasized by Witten, one theoretically disappointing feature of superconducting strings is that there is no powerful topological reason why a string should be superconducting, as there is for the string's existence; even given the appropriate fields and interactions, a theory will only have superconducting strings if the parameters of the theory lie in a certain range. There is thus a micro-physical restriction on the parameters. The cosmic spring idea presents another, cosmological restriction.[4]

In this talk I will re-examine the evaluation of the critical and spring currents. The approach will be to study the energetics of static superconducting string configurations and to determine the critical and spring currents from this. As a result, the spring current will be seen to be lower than previously obtained, giving a stronger cosmological restriction on the parameters of a theory with superconducting strings.

The outline of the talk is as follows. I will first explain the mechanism of bosonic superconducting strings as devised by Witten, and illustrate that these strings carry long-lived currents.

(This will essentially be a rehash of Witten's original treatment, as well as of other rehashes of it.) Next, I will write down the energy of a static string configuration. From this the spring current and critical current can be obtained. I will argue that a contribution to the energy which is neglected in previous treatments, the energy in the electromagnetic field, should be included in calculating the spring current but excluded in calculating the critical current. With this the spring current is lower by a large factor than the previous result, giving a stronger cosmological restriction on the parameters.

THE MECHANISM OF BOSONIC SUPERCONDUCTING STRINGS

Following Witten's treatment, I will use the simplest toy model to illustrate the mechanism of bosonic superconducting strings; according to Witten, realistic grand unified theories exist which have the necessary features.

Ordinary cosmic strings arise in the Abelian Higgs model, with spontaneously broken $U(\tilde{1})$ symmetry. (Note that this $U(\tilde{1})$ symmetry is not electromagnetism!) The Lagrangian is

$$L = -\frac{1}{4}R_{\mu\nu}^2 + |D_\mu\phi|^2 - \frac{\lambda}{4}(|\phi|^2 - \mu^2)^2$$

With ϕ a complex field with $U(\tilde{1})$ charge g, the vacua lie on a circle, $\phi = \mu e^{i\theta}$. If we were in 2+1 dimensions, this theory would contain solitons (vortices), where around a large circle at spatial infinity σ changes by 2π. Such a field configuration is, in polar coordinates (ρ,φ),

$$\phi = \mu\hat{\phi}_o(\rho)e^{i\varphi} \qquad R_\varphi = \frac{1}{-g\rho}\hat{R}_o(\rho)$$

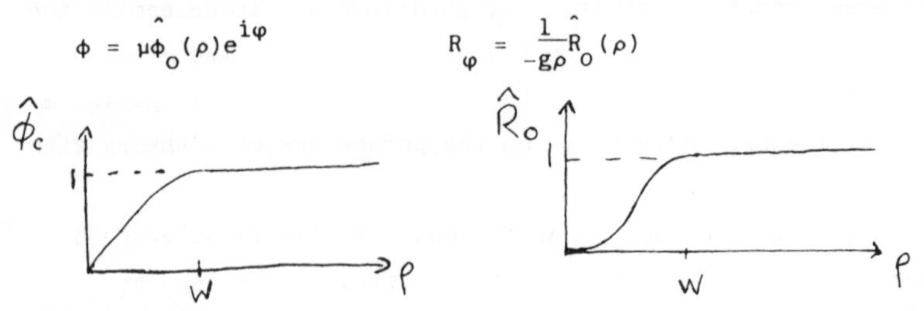

where ϕ and R approach one as $\rho\to\infty$, and approach zero linearly and quadratically, respectively, as $\rho\to 0$. The energy of this configuration is $\sim\mu^2$ and its width is $w\approx(\lambda\mu^2)^{-1/2}$.

In three dimensions there are no true solitons, although it is easy to imagine a field configuration which is a one-dimensional line-like extension of the two-dimensional soliton, i.e., a configuration whose cross-section looks like the two-dimensional soliton.

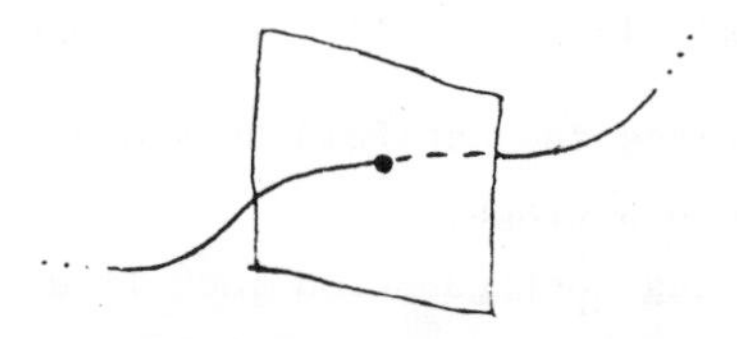

Such a configuration is a cosmic string; its energy <u>per unit length</u> is μ^2.

In spite of the topological nature of the cross-section of the string, there is no topological barrier preventing a closed string from decaying; in any large plane the winding number is zero.

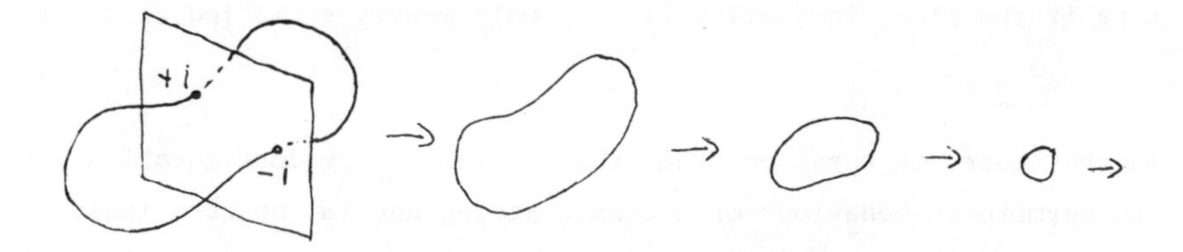

To make the string superconducting, we enlarge the theory to include another scalar field σ charged under a second U(1) gauge group, which will be identified with electromagnetism. The Lagrangian is

$$L = -\tfrac{1}{4}R_{\mu\nu}^2 + |D_\mu\phi|^2 - \tfrac{1}{4}F_{\mu\nu}^2 + |D_\mu\sigma|^2 - V(\phi)$$

$$V(\phi) = \frac{\lambda}{4}(\phi^2-\mu^2)^2 + \frac{\bar{\lambda}}{4}|\sigma|^4 - m^2|\sigma|^2 + f|\phi|^2|\sigma|^2$$

The σ-independent piece of the potential wants to break $U(\tilde{1})$, while the ϕ-independent piece wants to break U(1). However, the last term, which links ϕ and σ, inhibits the simultaneous breaking of both symmetries. If, for example, ϕ attains a vev, the last term acts as a $|\sigma|^2$ term with a positive coefficient, reducing the tendency for σ to attain a vev.

It turns out that if the parameters satisfy the restrictions

$$f\mu^2 > m^2 \tag{1a}$$

$$\lambda\mu^4 > m^4/\bar{\lambda} \tag{1b}$$

then the vacuum is $(|\phi|, |\sigma|) = (\mu, 0)$: electromagnetism is preserved but $U(\tilde{1})$ is broken. Since the manifold of vacua is still a circle, once again we get cosmic strings.

What does the string configuration look like? We must find the minimum-energy field configuration with fields of the appropriate asymptotic form for large ρ:

$$\phi(\rho) \rightarrow \mu e^{i\varphi}$$

$$R_\varphi(\rho) \rightarrow -1/g\rho$$

$$\sigma(\rho) \rightarrow 0$$

Naively, we might guess $\sigma = 0$ everywhere, that it is just along for the ride. As was shown by Witten, however, σ can be more interesting if the first inequality (1a) is only <u>weakly</u> satisfied, i.e., if

$$f\mu^2 \gtrsim m^2$$

In that case, it turns out that the minimum-energy configuration with the asymptotic behaviour of a cosmic string has $|\sigma|$ nonzero inside the core of the string; i.e., electromagnetism is broken there. The exact form of the magnitude of this configuration, denoted $\sigma_o(\rho)$, requires a detailed calculation; it turns out that σ_o is of order $m\bar{\lambda}^{-½}$, and σ_o makes a contribution to the energy per unit length of order $-m^4w^2/\bar{\lambda}$.

LONG-LIVED CURRENTS

The important feature of these strings is that they can carry long-lived electromagnetic currents. This can be demonstrated in two steps. First, I will use a topological argument to show that certain field configurations where the phase of σ varies along the length of the string are long-lived, and second, I will show that a space-dependent phase results in an electromagnetic current.

To illustrate the first step, consider a closed loop of string. σ attains an expectation value in the core of the string, so it essentially lives on S^1, the closed loop of string. Furthermore, as outlined above the magnitude is determined by minimizing the energy, but the phase is undetermined, so it takes values in S^1. So the string defines a map from $S^1 \rightarrow S^1$, the phase of σ as a function of the position along the string. Topologically nontrivial configurations are possible where the phase changes by an integer multiple N of 2π along the string. So there are two levels of topological nontriviality at work: one giving rise to the original cosmic string, and the other involving the global structure of a closed loop of string.

There is an energy barrier separating configurations with different winding numbers. Classically, therefore, N is conserved. However, this is not true quantum mechanically, since the energy barrier is not infinite: the winding number can change if $|\sigma|$ goes temporarily to zero at some point along the string, making the winding number ill-defined. These tunnelling events are responsible for the critical current, as will be explained below.

To illustrate the second step, note that the current density derived from the Lagrangian is

$$\vec{j} = -\frac{\partial L}{\partial \vec{A}} = -2e\sigma_o(\rho)^2(\vec{\nabla}\theta + e\vec{A})$$

where θ is the phase of σ, which depends upon the coordinate along the string. The equation of motion of the gauge field is, in the Coulomb gauge,

$$\nabla^2\vec{A} = -\vec{j}$$

We must solve these two equations for the integrated current flowing along the string which results due to the space-dependence of θ. In the approximation that the string is very thin compared to its radius of curvature, these equations can be solved for the integrated current J and the tangential component of the gauge field inside the string A; the result is

$$A = \frac{J \log R/w}{2\pi}$$

$$J = -\frac{2Ke^2}{1+\frac{Ke^2}{\pi}\log\frac{R}{w}}\frac{N}{R} \approx -\frac{2\pi}{e \log\frac{R}{w}}\frac{N}{R}$$

where R is the size of the loop, w is the width of the string, and K is a dimensionless quantity representing, in a sense, the 'strength' of the expectation value of σ:

$$K = \int d^2p\, \sigma_o^{\,2} \approx \frac{w^2m^2}{\bar{\lambda}}$$

Topological arguments told us that states with a space-dependent phase are long-lived; we have now learned that these states carry electromagnetic currents. Thus our task has been accomplished.

ENERGETICS, CRITICAL CURRENT AND SPRING CURRENT

We now have an idea of what the fields and current look like; the next step is to estimate the energy of a static string of length $2\pi R$ and winding number N. To do this, the energy can be divided up into four pieces:

$$E = E_{R,\phi} + E_{\sigma_o} + E_\theta + E_F .$$

The first part is the energy in the (R,ϕ) sector, i.e., the energy of the ordinary cosmic string. This is

$$E_{R,\phi} = 2\pi R\mu^2 .$$

E_{σ_o} is the energy from the magnitude of σ, part of which comes from the gradient energy and part of which comes from the potential energy. Its value was given above,

$$E_{\sigma_o} = 2\pi R(-\frac{m^4w^2}{\bar{\lambda}}) = -2\pi RKm^2 .$$

E_θ is the kinetic energy for the phase of σ:

$$E_\theta = \int d^3x\sigma_o^2(\theta'+eA)^2 = \ldots = 2\pi R\, J^2/4Ke^2 .$$

Finally, E_F is the energy in the electromagnetic field:

$$E_F = \int d^3xF_{ij}^2 = \int d^3x\vec{j}\cdot\vec{A} = 2\pi R\frac{\log R/w}{2\pi}J^2$$

Combining these, we have

$$E = 2\pi R\{\mu^2 - Km^2 + \frac{J^2}{4Ke^2} + \frac{\log R/w}{2\pi} J^2\} \tag{2}$$

We are now in a position to calculate the critical current and spring current.

To understand the nature of the critical and spring currents, suppose a string exists and that it is carrying a current. The string will oscillate, radiating energy and therefore shrinking, which increases the gradient along the string of the phase, θ. This causes the current to increase, and also the last two terms in the energy increase. This increase in energy has two important effects.

First, it was observed above that the energy barrier between topological sectors of different winding numbers is finite. This means that quantum tunnelling events which decrease the winding number are in principle possible. These events are, however, exceedingly rare, for essentially the same reason that, for example, instanton-mediated decay of the proton is unimportant: when one computes the Euclidian classical action for a "bounce" event between configurations with different winding numbers it is of the form $1/\lambda$ where λ generically represents a 4-scalar field coupling constant. This means the amplitude for the tunnelling event has a factor $e^{-1/\lambda}$ and in the semiclassical limit, where all dimensionless coupling constants go to zero (and outside of which the entire classical discussion above is of questionable validity), this factor is exceedingly small.

This argument no longer applies, however, when the current increases to a certain value: the energetics must be reexamined since the energy in the current makes a significant contribution. Intuitively, the energy required to set $|\sigma|$ to zero inside the string can be supplied by the energy due to the current. The critical current, then, is calculated by finding the value of the current at which the energy which arises due to it, the current, equals the energy required to set $|\sigma|$ to zero.

A subtle point arises here: naively we might think the critical current is calculated by setting the sum of the last three terms of (2) to zero. Since the electromagnetic energy E_F is much greater than E_θ, due to the large logarithm, it might seem like the third term can be neglected. However, the degree of localization of energy on the string is important. The energy stored in σ_o, E_{σ_o}, is exponentially localized inside the core of the string, as is E_θ. However, the last term in (2), E_F, is the electromagnetic energy, which is spread out over all space. Thus, intuitively, the energy in the electromagnetic field is not available to set $|\sigma|$ to zero, and hence the critical current should be calculated by equating the second and third terms in (2); the result is

$$J_c^2 = 4K^2e^2m^2$$

The second effect of the increase in energy due to the current is that there is a stable, static configuration (actually, a family of such configurations) for a string of a given winding number. If we make the R-dependence of J explicit, the energy becomes

$$E = 2\pi R\{\mu^2 - Km^2 + \frac{\pi^2}{Ke^4\log\frac{2R}{w}}\frac{N^2}{R^2} + \frac{2\pi}{e^2\log\frac{R}{w}}\frac{N^2}{R^2}\}$$

This time the degree of localization of the various contributions to the energy is not important, and we may ignore the third term relative to the fourth. The first two terms increase with R while the last term diverges as R→0; therefore the energy is minimized for a certain radius; setting $\partial E/\partial R$ to zero gives us the spring current:

$$J_S{}^2 \cong \frac{2\pi(\mu^2-Km^2)}{\log\frac{R}{w}}$$

where log(R/w) has been assumed large and relatively slowly-varying as a function of R.

The constraints given above ensure that the quantity μ^2-Km^2, which appears both in J_S^2 and in E, is positive.

The cosmological restriction on the parameters of the theory arises from the requirement that the spring current must be greater than the critical current. We are now in a position to quantify this.

$$\frac{J_S^2}{J_C^2} = \frac{\pi(\mu^2 - Km^2)}{2K^2e^2m^2 \log R/w} > 1$$

This is quite a strong restriction on the parameters, since the argument of the logarithm is the ratio of a cosmological distance to a particle physics distance; Witten suggests $R/w >> e^{100}$.

An alternative way of displaying this result which exhibits several aspects in the following:

$$\frac{\bar{\lambda}}{e^2}\left(\frac{m_\phi}{m}\right)^2\left(\frac{\bar{\lambda}}{\lambda}\left(\frac{m_\phi}{m}\right)^4 - 1\right) \gtrsim \log\frac{R}{w} >> 1$$

ignoring numerical factors and defining $m_\phi^2 = \lambda\mu^2$. The last factor on the left is greater than zero due to the constraint (1b). The semi-classical limit has λ, $\bar{\lambda}$, $e^2 \to 0$ in fixed ratios, with the mass parameters fixed, so this restriction is independent of the semi-classical limit.

If we assume the dimensionless coupling constants are similar, as are the masses, the restriction is not satisfied. The restriction would be satisfied if, for example, $\bar{\lambda} >> \lambda$, e^2 or $m_\phi^2 >> m^2$. While there is nothing a priori which says this cannot be true, it begs the question: why are the mass scales (or coupling constants) so different?

SUMMARY

The critical current and spring current for a bosonic superconducting string were calculated by studying the energetics of static string configurations. It was found that the requirement that the spring current be higher than the critical current presented a strong restriction on the parameters of the theory.

ACKNOWLEDGEMENTS

I wish to thank Robert Brandenberger for discussions and encouragement.

REFERENCES

1. A. Vilenkin, Phys. Rep. 121 (1985) 263, and references therein.
2. E. Witten, Nucl. Phys. B249 (1985) 557.
3. E. Copeland, M. Hindmarsh and N. Turok, Phys. Rev. Lett. 58 (1957) 1910.
4. R. MacKenzie, DAMTP preprint 87-13 (To appear, Phys. Lett. B).

Are 2-branes better than 1?

K.S. Stelle
The Blackett Laboratory
Imperial College
London SW7, England

and

P.K. Townsend
D.A.M.T.P.
University of Cambridge
Cambridge CB3 9EW, England

ABSTRACT

Recent progress in the theory of supermembranes and their yet higher dimensional generalizations, super p-branes, is reviewed. On the one hand, some remarkable properties of the 11-dimensional supermembrane, in particular, have led to speculations that it might be a consistent quantum mechanical extension of 11-dimensional supergravity. On the other hand, a classification of super p-branes has led to an understanding of how they are realized as supersymmetric "extended solitons".

1. - CLASSICAL SUPERMEMBRANES

There is a natural generalization of the action

$$I = m \int dt (\dot{X}^2)^{1/2} \tag{1.1}$$

for a relativistic point particle to an extended object of dimension p. If ξ^i are coordinates on the $(p+1)$–dimensional world volume swept out by the object in the course of its time evolution, and if $X^m(\xi)$ gives the configuration of this world volume in a d-dimensional spacetime with metric $g_{mn}(X)$, then the generalization of (1.1) is

$$I = T \int d^{p+1}\xi \det{}^{1/2} \left(\partial_i X^m \partial_j X^n g_{mn}(X) \right) \tag{1.2}$$

where T is the tension, with units of $(\text{mass})^{p+1}$. For p=1, this is just the string action of Nambu and Goto, but it was originally proposed by Dirac for a membrane [1], i.e. p=2. In fact, there has been quite a lot of work done on membranes since then, but the advent of the supermembrane [2,3] has raised many new and interesting issues and has led to some striking connections with, for example, 11-dimensional supergravity, singletons, and supersymmetric solitons, to name a few. This review is an attempt to persuade the reader that there is much to be gained from the wider perspective of the p-brane panorama wherein the string is but a 1-brane (in other words, a thinly-veiled attempt at "brane-washing").

Strings come in two varieties: "elementary" and "cosmic". The quantized excitations of an elementary string are considered as elementary particles, whereas a cosmic string is generally treated classically, quantum fluctuations being assumed negligible. One might see a cosmic string through a telescope, but not an elementary string. Nevertheless, the *same* action, the Nambu-Goto action, can be used to describe both. This action is appropriate for a "structureless" string; no information on how the string was formed is encoded in it. This is to be expected for an elementary string, for which no such information is required, and for a cosmic string in the long-wavelength limit, for which such information is lost. The same considerations apply to p-branes; the action (1.2) can be studied from the point of view of "elementary p-branes" or "cosmic p-branes". We shall be considering both points of view in what follows, but in either case our first task is to supersymmetrize the action (1.2).

To this end, it is convenient to rewrite (1.2) in the Howe-Tucker form [4] ($\gamma = |\det \gamma_{ij}|$)

$$I = \frac{T}{2} \int d^{p+1}\xi \, \{\sqrt{\gamma}\gamma^{ij}\partial_i X^m \partial_j X^n g_{mn} - (p-1)\sqrt{\gamma}\} \tag{1.3}$$

where γ_{ij} is an *independent* (dimensionless) world-volume metric. The X field equation that follows from (1.3) is

$$\gamma^{ij}\left(\partial_i\partial_j X^m - \left\{ {k \atop ij} \right\}(\gamma)\ \partial_k X^m + \partial_i X^p \partial_j X^q \left\{ {m \atop pq} \right\}(g)\right) = 0 \tag{1.4}$$

(the "brane-wave equation"), where $\left\{ {k \atop ij} \right\}$ and $\left\{ {m \atop pq} \right\}$ are the Christoffel connections for the world volume metric $\gamma_{ij}(\xi)$ and the spacetime metric $g_{mn}(X)$ respectively. The γ field equation is

$$\gamma_{ij} = \partial_i X^m \partial_j X^n g_{mn}(X) \ , \tag{1.5}$$

which states that γ_{ij} is equal to the metric induced from g_{mn} by the embedding $X^m(\xi)$. Inserting (1.5) into (1.4), one obtains the equation that follows from the action (1.2), thereby demonstrating the *classical* equivalence of (1.3) and (1.4).

Observe that if p=1 the (p+1)–dimensional "cosmological term" in (3) vanishes and the action reduces to the Brink-DiVecchia-Howe form of the Nambu-Goto string action. For p=0, one can rescale the action and the metric so as to remove the cosmological term, and in this case one obtains the action for the massless point particle, which is inequivalent to (1.1). For $p \geq 2$ the cosmological term is essential; without it the γ equation would impose the constraint that the induced metric $\partial_i X^m \partial_j X^n g_{mn}$ vanish, and one cannot make sense of this for $p \geq 2$.

We should also specify the boundary conditions in (1.3). The boundary conditions for an open membrane have been discussed by Collins and Tucker [5]. They are rather complicated and we shall suppose that the p-brane is *closed* for $p \geq 2$. In any case it turns out that supersymmetry and Lorentz invariance require this [6].

To supersymmetrize the action (1.3), one can proceed in one of two ways. Considering (1.3) as a (p+1)–dimensional field theory, one can extend it to a locally supersymmetric field theory by introducing world-volume spinor superpartners for X^m and γ_{ij}. This involves the techniques of (p+1)–dimensional supergravity and would lead to a "spinning" p-brane action (i.e. an analogue of the spinning particle action, which describes a spin $\frac{1}{2}$ particle). Alternatively, one can extend the notion of a p-brane moving through spacetime to one moving through a superspace with coordinates Z^M. This leads to the super p-brane.

For the string, these two options lead to the Lorentz covariant Ramond-Neveu-Schwarz and Green-Schwarz formulations, respectively, of the superstring. Remarkably, these two formulations are *equivalent* descriptions of the same theory (in the critical dimension d=10 and after the GSO projections). For $p \geq 2$ there is a problem with the spinning p-brane, however. The supersymmetric extension of the cosmological term involves an auxiliary field S in a term of the form $\sqrt{\gamma}S$. In supergravity theories, there is also an S^2 term, so that elimination of S yields the usual cosmological term, but S^2 comes with the graviton kinetic term and one *cannot* have one without the other. If one includes a kinetic term for γ_{ij}, then it no longer appears as an auxiliary variable and the action cannot then be considered as an extension of (2).

For this reason, we pass now to the super p-brane. Here one starts from an N-extended superspace in d dimensions with coordinates $Z^M = (X^m, \Theta^\mu)$ (we suppress internal symmetry indices on Θ for simplicity; in any case, only N=1 will ultimately be relevant). In place of the spacetime metric $g_{mn}(X)$ we now have the superspace vielbein $E_M{}^A$ relating the Lorentz covariant one-form basis $\{E^A; A = (a, \alpha)\}$ to the coordinate 1-form basis $\{dZ^M\}$,

$$E^A = dZ^M E_M{}^A \quad . \tag{1.6}$$

Given a map $Z^M(\xi)$ from superspace to the world volume, we can pull back the forms E^A on superspace to the world volume,

$$^*E^A = d\xi^i E_i{}^A \qquad E_i{}^A = \partial_i Z^M(\xi)\; E_M^A(Z(\xi)) \ . \tag{1.7}$$

The action (3) is now readily generalized to

$$I_{kin} + I_{cosm} = \frac{T}{2}\int d^{p+1}\xi \left\{ \sqrt{\gamma}\gamma^{ij} E_i{}^a E_j{}^b \eta_{ab} - (p-1)\sqrt{\gamma} \right\} \ , \tag{1.8}$$

but for the string, at least, we know that an additional term of Wess-Zumino form is possible and, in fact, is very important [7,8]. Hughes, Liu and Polchinski [2] showed how this could be generalized to $p > 1$, and they wrote down an action for a super 3-brane in d=6 flat superspace. In [3] this was further generalised to curved superspace, and an 11-dimensional supermembrane action was proposed. We shall comment further on the significance of the Wess-Zumino term later, when we consider "cosmic" p-branes; for the moment we shall just explain the construction. The starting point is a *closed* (p+2)–form $H = dB$ in superspace. The pull-back of the (p+1)–form

$$B = \frac{1}{(p+1)!} E^{A_1} \ldots E^{A_{p+1}} B_{A_{p+1}\ldots A_1} \tag{1.9}$$

is used to construct the WZ term $\int {}^*B$ which is then added to (1.8) to yield the action (for T=1)

$$\begin{aligned} I = \int d^{p+1}\xi \Big\{ & \frac{1}{2}\sqrt{\gamma}\gamma^{ij} E_i^a E_j^b \eta_{ab} \; - \frac{(p-1)}{2}\sqrt{\gamma} \\ & + \frac{1}{(p+1)!}\epsilon^{i_1\ldots i_{p+1}} E_{i_1}^{A_1} \ldots E_{i_{p+1}}^{A_{p+1}} B_{A_{p+1}\ldots A_1} \Big\} \ . \end{aligned} \tag{1.10}$$

The important feature of the WZ term in this action is that it allows the possibility of a certain "fermionic gauge invariance", or "κ-invariance". Defining

$$\delta Z^A \equiv \delta Z^M E_M{}^A \ , \tag{1.11}$$

the infinitesimal κ-transformation can be written as

$$\delta Z^a = 0 \quad \delta Z^\alpha = (1+\Gamma)^\alpha{}_\beta \kappa^\beta(\xi) \ , \tag{1.12}$$

where κ is an *anticommuting spacetime spinor* parameter, and

$$\Gamma = \frac{(-1)^{\frac{(p+1)(p-2)}{4}}}{(p+1)!\sqrt{\gamma}} \epsilon^{i_1\ldots i_{p+1}} E_{i_1}^{a_1} \ldots E_{i_{p+1}}^{a_{p+1}} \left(\Gamma_{a_1\ldots a_{p+1}}\right) \ , \tag{1.13}$$

where $\Gamma_{a_1\ldots a_n}$ is the antisymmetrized product of n Γ-matrices*,e.g. $\Gamma_{ab} = \frac{1}{2}(\Gamma_a\Gamma_b - \Gamma_b\Gamma_a)$. If γ_{ij} is kept as an *independent* field, then we should specify its κ-variation too, but it is convenient for present purposes to take γ_{ij} to be given by (1.5). In this case one has

$$\Gamma^2 = 1 \quad , \tag{1.14}$$

so that the $(1 + \Gamma)$ factor in (12) is a projection operator (×2). One can show that the action (1.10) is κ-invariant provided [3] (i) that the (p+2)–form $H = [(p+2)!]^{-1}E^{A_{p+1}}\ldots E^{A_1}H_{A_1\ldots A_{p+1}}$ satisfies

$$H_{\alpha a_{p+1}\ldots a_1} = \frac{(-1)^{\frac{(p+1)(p-2)}{4}}}{p!}(\Gamma_{a_1\ldots a_{p+1}}\Lambda)_\alpha \qquad H_{\alpha\beta a_p\ldots a_1} = \frac{(-1)^{p+\frac{(p+1)(p-2)}{4}}}{2p!}(\Gamma_{a_1\ldots a_p})_{\alpha\beta}$$

$$H_{\alpha\beta\gamma A_{p-1}\ldots A_1} = 0 \quad , \tag{1.15}$$

for an arbitrary spinor Λ_α, (ii) that the superspace torsion tensor $T_{AB}{}^C$ satisfies

$$T_{\alpha\beta}{}^a = (\gamma^a)_{\alpha\beta} \qquad \eta_{c(a}T_{b)\alpha}{}^c = \eta_{ab}\Lambda_\alpha \tag{1.16}$$

and (iii) that H is indeed *closed*.

In *flat* superspace, H and T do satisfy (1.15) and (1.16) (with $\Lambda = 0$), so the only non-trivial constraint is that H be closed. Since

$$E^A = \{(dX^m - i\overline{\Theta}\gamma d\Theta)\delta^a_m, d\Theta^\mu\delta^\alpha_\mu\} \tag{1.17}$$

in flat superspace, it is clear that the closure of H requires that

$$(d\overline{\Theta}\Gamma_a d\Theta)(d\overline{\Theta}\Gamma^{ab_1\ldots b_{p-1}}d\Theta) = 0 \quad . \tag{1.18}$$

Consistency with (1.15) requires that $(\Gamma^{ab_1\ldots b_{p-1}})_{\alpha\beta}$ be symmetric in $(\alpha\beta)$, so that the second factor in (1.18) does not vanish by itself (note that $d\Theta$ is *commuting*). Equation (1.18) is equivalent to the requirement that the Γ-matrices satisfy a certain identity. In the simplest cases, where we have a single Majorana, or Majorana-Weyl, spinor Θ, this identity is

$$(\Gamma_a P)_{(\alpha\beta}(\Gamma^{ab_1\ldots b_{p-1}}P)_{\gamma\delta)} = 0 \quad , \tag{1.19}$$

where P is a chirality projection operator if Θ is Majorana-Weyl, and is the unit matrix otherwise. For p=1, this identity is equivalent to the (apparently weaker) one

$$(\Gamma_a P)_{\alpha(\beta}(\Gamma^a P)_{\gamma\delta)} = 0 \quad , \tag{1.20}$$

---

* Spinor indices are raised and lowered with the charge conjugation matrix $C_{\alpha\beta}$; thus $(\gamma^a)_{\alpha\beta} = (\gamma^a)_\alpha{}^\gamma C_{\gamma\beta}$.

which is satisfied for d=3,4&10. An analogous identity is satisfied for d=6 (where Θ is $SU(2)$-Majorana-Weyl) and this accounts for the existence of the superstring action in d=3,4,6&10 [7].

A complete analysis of the constraint (1.16) [9] leads to the conclusion that the super p-brane action (1.10) exists only for a very limited set of values of (p,d,N). In fact, for $p \geq 2$, only N=1 is possible, and in all cases $p \leq 5$ (we assume that $p > d-1$). The allowed cases for $p \geq 2$ are shown by the ticks in the table. The dots refer to supersymmetric particle theories, and will be discussed in section three.

Table

Super p-branes in d spacetime dimensions

d \ p	0	1	2	3	4	5
11			√			
10		√				√
9	•				√	
8				√		
7			√			
6		√		√		
5	•		√			
4		√	√			
3	•	√				
2	•					
		N=1 N=2	N=1			

For p=1, both N=1 and N=2 are possible, corresponding to the heterotic and to the type II (or type I open) superstrings, respectively. For $p > 1$, only N=1 is possible, a fact that leads to the conclusion that super p-branes for $p \geq 2$ must be closed. Recall that the open superstring with N=1 supersymmetry actually requires an N=2 superspace action, but the two Θ's are identified at the ends of the string. Similar considerations apply to open supermembranes, but now we have only *one* Θ, which is *irreducible* under the Lorentz group. For this reason, open super p-brane boundary conditions are not compatible with Lorentz invariance, as we have mentioned earlier.

Observe that the table of allowed super p-branes includes the six-dimensional 3-brane of [2] and the 11-dimensional membrane of [3], as it should. A remarkable feature of both these models, and in fact of *all* (κ-invariant) super p-branes, is that when considered as (p+1)–dimensional field theories, the boson and "fermion" (i.e. Θ) degrees of freedom match. For example, for the d=11 supermembrane, we have 11 boson fields X^m, but 3 are unphysical as a result of the three-dimensional world-sheet diffeomorphism invariance. We therefore have a net total of 8 physical bosons. Similarly, κ-invariance allows us to fix a gauge in which

$$\Theta = \begin{pmatrix} S \\ 0 \end{pmatrix} , \tag{1.21}$$

where S is a 16-component spinor of $SO(9)$. Eight components of S are the momenta conjugate to the other eight, so that we have a net total of eight fermions. One can easily work out the values of P and d for which this match is possible. Note that the counting of degrees of freedom for ($p \geq 2$)–branes is straightforward. For the superstring case, $p = 1$, the counting is complicated by the need to separate the left- and right-moving modes. One finds that the bose-fermi match is possible for precisely those combinations given in the table, which was derived by requiring the closure of H, ostensibly a very different requirement. Thus, we have the result that *the (κ-invariant) super p-brane action is possible if and only if the world-volume boson and fermion degrees of freedom match* [9].

This is evidence that, as for the superstring, spacetime supersymmetry and κ-invariance imply *world-volume supersymmetry*. This is perhaps surprising in view of the apparent absence of a "spinning p-brane" action, but recall that the latter requires *local* N=1 world-volume supersymmetry, whereas we have here an extended rigid world-volume supersymmetry. It is perhaps also surprising in view of the fact that although Θ is a *spacetime* spinor, it is a *world-volume scalar*, and world-volume supersymmetry requires world-volume fermions. Here, however, one must remember that, as for any other gauge theory, properties of the physically significant degrees of freedom can be determined unambiguously only after gauge fixing. Locally, at least, one can always fix the world-volume reparameterizations by the choice $X^i = \xi^i$, $i = 0, 1, 2, \ldots, p$. In order to maintain this gauge, a Lorentz transformation must now be accompanied by a compensating reparameterization with a specific field-dependent parameter η.

To determine the compensating reparameterization, we write $(X^i, X^{i'})$ for X^m ($i' = p{+}1, \ldots, d{-}1$) and $(L^i{}_j, L^i{}_{j'}, L^{i'}{}_{j'})$ for the (constant) Lorentz transformation parameter $L^m{}_n$. The combined Lorentz and reparameterization transformation of X^i is therefore

$$\delta X^i = \eta^j(\xi)\partial_j X^i + L^i{}_j X^j + L^i{}_{j'} X^{j'} \quad . \tag{1.22}$$

Requiring that δX^i vanish when $X^i = \xi^i$ fixes η^i to be

$$\eta^i = -L^i{}_j\xi^j - L^i{}_{j'}X^{j'} \ . \tag{1.23}$$

The combined transformation of $X^{i'}$ in this gauge is therefore

$$\delta X^{i'} = -L^k{}_j\xi^j\partial_k X^{i'} + L^{i'}{}_{j'}X^{j'} - L^k{}_{j'}X^{j'}\partial_k X^{i'} \ . \tag{1.24}$$

The transformations of the coset $SO(d-1,1)/\left(SO(p,1)\times SO(d-p-1)\right)$ are now *non-linear*, so that the original Lorentz group has effectively been spontaneously broken to $SO(p,1)\times SO(d-p-1)$. The latter factor is simply a linearly-realized internal symmetry, whereas the SO(p,1) transformation of $X^{i'}$ is that of a $(p+1)$–dimensional world-volume scalar. Similarly, using (1.23), the $SO(p,1)$ transformation of Θ is found to be

$$\delta\Theta = -L^i{}_j\xi^j\partial_i\Theta + \frac{1}{4}L_{ij}\Gamma^{ij}\Theta \ . \tag{1.25}$$

The matrices Γ^{ij} provide a (reducible) spinor representation of the $SO(p,1)$ algebra, so that Θ can now be identified as a *world-volume* spinor [6].

The κ-gauge has still to be fixed, of course. After fixing it, the space-time supersymmetry will have to be accompanied by a specific compensating κ-transformation, and the combined transformation can be interpreted as that of a world-volume supersymmetry. This interpretation is easily checked for small fluctuations about solutions of the brane-wave equations that maintain the symmetry; we shall call such solutions "supersymmetric". For *arbitrary* fluctuations we can invoke the Haag-Łopuszanski-Sohnius theorem, which essentially tells us that the only symmetry of an *interacting* field theory that can interchange bosons and fermions is supersymmetry. A direct proof would be desirable, of course, but the fully non-linear transformations are quite complicated.

Another striking feature of the results of the table is that all super p-branes lie on one of four sequences connected to the superstrings in d=3,4,6&10. In fact, this is no accident. In ref. [10], it was shown how a process of "double-dimensional reduction" allows one to reduce the action of the 11-dimensional supermembrane to the type IIa 10-dimensional superstring action. Applying this to the other super p-branes of the table w e see that all of them (except the other types of superstring) are obtainable from the "top-brane" by double-dimensional reduction.

Of special interest is the 11-dimensional supermembrane. It is a remarkable fact that when Λ_α=0 the superspace constraints (1.15, 1.16), are just the field equations of 11-dimensional supergravity in the form given by Cremmer and Ferrara and by Brink and Howe [11]. Moreover, it was shown in [10] that the equations for $\Lambda \neq 0$ are equivalent to those for $\Lambda = 0$. We conclude, therefore, that *the κ-invariance of the supermembrane implies the 11-dimensional supergravity field equations.*

One solution to the 11-dimensional field equations is flat space. The action for the supermembrane in flat superspace is, including a conventional normalization factor,

$$I = \frac{-1}{4\pi^2} \int d^3\xi \left[\frac{1}{2}\sqrt{-\gamma}\ \gamma^{ij}\Pi_i^a \Pi_{ja} - \frac{1}{2}\sqrt{-\gamma} + \varepsilon^{ijk}\Pi_i^A \Pi_j^B \Pi_k^C B_{CBA} \right] , \tag{1.22}$$

where $\Pi_i^A = \partial_i Z^M E_M^A = (\Pi_i^a, \Pi_i^\alpha)$, with

$$\Pi_i^a = \partial_i X^a - i\overline{\Theta}\Gamma^a \partial_i \Theta \ , \tag{1.23}$$

$$\Pi_i^\alpha = \partial_i \Theta^\alpha . \tag{1.24}$$

Θ^α is a 32-component Majorana spinor, $a = 0, 1, \cdots, 10$, and we have set the membrane tension T equal to unity. The super 3-form B_{CBA} is found by solving (1.15) with $\Lambda = 0$. For 11-dimensional spacetime with $p = 2$, this requires that all components of H vanish except for $H_{ab\alpha\beta} = -\frac{1}{3}(\Gamma_{ab})_{\alpha\beta}$.

After solving for B_{CBA}, the action may be written as

$$\begin{aligned} I = \frac{-1}{8\pi^2} \int d^3\xi\ [\sqrt{-\gamma}\ \gamma^{ij}\Pi_i^a \Pi_{ja} - \sqrt{-\gamma} + \varepsilon^{ijk}\Pi_i^a \Pi_j^b \overline{\Theta}\Gamma_{ab}\partial_k \Theta \\ - i\varepsilon^{ijk}\Pi_i^a \overline{\Theta}\Gamma_{ab}\partial_j\Theta\overline{\Theta}\Gamma^b\partial_k\Theta \\ - \frac{1}{3}\ \varepsilon^{ijk}\overline{\Theta}\Gamma_{ab}\partial_i\Theta\overline{\Theta}\Gamma^a\partial_j\Theta\overline{\Theta}\Gamma^b\partial_k\Theta] . \end{aligned} \tag{1.25}$$

The action (1.25) is invariant under the rigid spacetime supersymmetry transformation

$$\begin{aligned} \delta\Theta &= \varepsilon \\ \delta X^a &= i\ \overline{\varepsilon}\ \Gamma^a \Theta \ , \end{aligned} \tag{1.26}$$

and the κ-transformations

$$\begin{aligned} \delta\Theta &= (1 + \Gamma)\kappa \\ \delta X^a &= i\overline{\Theta}\Gamma^a \delta\Theta \ , \end{aligned} \tag{1.27}$$

where κ is a world-volume scalar, spacetime Majorana spinor function of τ, σ and ρ, and

$$\Gamma = \frac{i}{6\sqrt{-\gamma}}\ \varepsilon^{ijk}\Pi_i^a \Pi_j^b \Pi_k^c \Gamma_{abc} \ . \tag{1.28}$$

For the action (1.25), κ-invariance requires that γ_{ij} also transform. The variation of γ_{ij} is complicated, and may be found in [3]. In practice, it is more convenient to exhibit the κ-invariance by going on "half-shell", i.e. substituting the algebraic equation of motion for γ_{ij},

$$\gamma_{ij} = \Pi_i^a \Pi_{ja} \ , \tag{1.29}$$

into the action (1.25) and working in 1.5 order formalism.

Varying (1.26) with respect to X^a and Θ yields

$$\partial_i(\sqrt{-\gamma}\ \gamma^{ij}\Pi_j^a) + \varepsilon^{ijk}\Pi_i^b \partial_j \overline{\Theta}\Gamma^a{}_b \partial_k \Theta = 0 \ , \tag{1.30}$$

$$(1-\Gamma)\gamma^{ij}\Pi_i^a \Gamma_a \partial_j \Theta = 0 \ , \tag{1.31}$$

where we have used the identity

$$\Gamma\Pi_k^a\Gamma_a = \Pi_k^a\Gamma_a\Gamma = \frac{-i}{2\sqrt{-\gamma}}\ \gamma_{k\ell}\varepsilon^{ij\ell}\Pi_i^a\Pi_j^b\Gamma_{ab} \ . \tag{1.32}$$

The κ-invariance may be used to gauge away 16 of the 32 components of Θ. A convenient gauge to choose is

$$\Gamma\Theta = -\Theta \ . \tag{1.33}$$

For the bosonic gauge choices, we work in an analogue [12] of the light-cone gauge for the string. We first use the three general coordinate invariances to set

$$X^+ = p^+\tau \ , \tag{1.34}$$

$$\gamma_{0\bar{\imath}} = 0 \quad , \quad (\bar{\imath} = 1,2) \tag{1.35}$$

where $X^\pm = \frac{1}{\sqrt{2}}(X^0 \pm X^{10})$.

Although these conditions fix the gauge-freedoms that are functions of all three coordinates (τ, σ, ρ), there remains a residual τ-independent reparametrization invariance of (1.34) and (1.35),

$$\delta\xi^0 = 0, \quad \delta\xi^{\bar{\imath}} = f^{\bar{\imath}}(\sigma, \rho) \ . \tag{1.36}$$

Consequently, at a fixed time τ^*, one may impose further gauge conditions. Thus, we use one of the two gauge freedoms (1.36) to impose

$$\gamma_{00} = -\overline{\gamma} \tag{1.37}$$

at $\tau = \tau^*$, where $\overline{\gamma} \equiv \det\ \gamma_{\bar{\imath}\bar{\jmath}}$. Eq. (1.37) is in fact maintained for all τ, by virtue of the $a = +$ component of the X field equation (1.30) and by the Θ field equation

(1.3) This may be seen by substituting (1.34) and (1.35) into (1.30, 1.31), to obtain

$$\frac{\partial}{\partial \tau}\left(\frac{\overline{\gamma}}{\gamma_{00}}\right)^{\frac{1}{2}} = 0 \ . \tag{1.38}$$

There still remains one reparametrization freedom in (1.38) that leaves the gauge condition (1.37) invariant, namely diffeomorphisms of the form (1.36) for which $\partial_{\bar{\imath}}(\delta\xi^{\bar{\imath}}) = 0$, i.e. [12]

$$\delta\xi^0 = 0, \quad \delta\xi^{\bar{\imath}} = \varepsilon^{\bar{\imath}\bar{\jmath}}\partial_{\bar{\jmath}} f(\sigma,\rho) \ , \tag{1.39}$$

where $\varepsilon^{12} = -\varepsilon^{21} = 1$. The Jacobian of this transformation is unity. It is the generalization for membranes of the residual freedom in the case of the string to perform a shift of σ by a constant. For the time being, we shall leave this residual gauge freedom unfixed.

2. – SEMICLASSICAL QUANTIZATION

Owing to the intrinsically non-linear structure of the membrane theory, we are forced to adopt a semiclassical approach to quantization. This involves choosing a stable classical solution and quantizing the fluctuations about it. From this procedure, we may investigate the relative contributions of the bosons and fermions to the vacuum energy in the topological sector of the theory containing the classical solution. This requires calculating the invariant $(\text{mass})^2 = -P^a P_a$ of the supermembrane, which can be obtained from the canonical momentum [13] $K_a = \partial L/\partial \dot{X}^a$,

$$K^a = -\frac{1}{4\pi^2}\sqrt{-\gamma}\,\gamma^{0i}\Pi_i^a - \frac{1}{4\pi^2}\,\varepsilon^{0ij}(\Pi_i^b + \frac{i}{2}\,\overline{\Theta}\Gamma^b\partial_i\Theta)\overline{\Theta}\Gamma^a{}_b\partial_j\Theta \ . \tag{2.1}$$

The total conserved momentum is then given by

$$P^a = \int d\sigma d\rho K^a \ . \tag{2.2}$$

In order to simplify the problem as much as possible, in reference [13] the supermembrane was taken to propagate in a spacetime with topology $R^9 \times S^1 \times S^1$. With such a topology, there is a classical solution with a toroidal membrane stretched around the $S^1 \times S^1$. Choosing the $S^1 \times S^1$ to be in the X^1 and X^2 directions, the classical solution takes the form of a purely bosonic background, with

$$X^1 = \ell_1 R_1 \sigma \quad , \quad X^2 = \ell_2 R_2 \rho \quad , \quad X^I = 0, \quad I = 3,\cdots,9 \ , \tag{2.3}$$

$$\Theta = 0 \ , \tag{2.4}$$

where $0 \le \sigma \le 2\pi$, $0 \le \rho \le 2\pi$, R_1 and R_2 are the radii of the two circles, and ℓ_1 and ℓ_2 are integers characterizing the winding numbers of the membrane around the two circles. The most general flat 2-torus is characterized by 3 parameters, but for simplicity we take the two circles to be orthogonal.

In the background (2.3, 2.4), the world-volume metric is flat,

$$\gamma_{ij} = diag(-(\ell_1\ell_2 R_1 R_2)^2, (\ell_1 R_1)^2, (\ell_2 R_2)^2) \quad , \tag{2.5}$$

and solving for X^- we find

$$X^- = \frac{1}{2p^+}(\ell_2\ell_2 R_1 R_2)^2 \tau \quad . \tag{2.6}$$

The classical (mass)^2 of the solution is then given by inserting (2.3 - 2.6) into (2.1, 2.2), with the result

$$(\mathrm{mass})^2 = (\ell_1\ell_2 R_1 R_2)^2 \quad . \tag{2.7}$$

For simplicity, we shall set $\ell_1 = \ell_2 = R_1 = R_2 = 1$ for the rest of the calculations in this section, restoring them in the final results. Thus $X^+ = p^+\tau, X^- = \tau/(2p^+), X^1 = \sigma, X^2 = \rho$ and $\gamma_{ij} = diag(-1,1,1)$ in the background.

We now consider fluctuations $\underline{Z}$ of the transverse coordinate around the classical solution, and write $\underline{X} = \underline{X}_{classical} + \underline{Z}$. Thus,

$$\begin{aligned} X^1 &= \sigma + Z^1 \quad , \\ X^2 &= \rho + Z^2 \quad , \\ X^I &= Z^I \quad . \end{aligned} \tag{2.8}$$

The fermions, being zero in the background, are pure fluctuations Θ. Substituting (2.8) into (1.30) and (1.31), and keeping only the terms of linear order in $\underline{Z}$ and Θ, we find, in the gauges chosen in section 1,

$$\ddot{Z}^1 = \partial_\sigma\partial_\sigma Z^1 + \partial_\sigma\partial_\rho Z^2 \quad , \tag{2.9}$$

$$\ddot{Z}^2 = \partial_\rho\partial_\rho Z^2 + \partial_\sigma\partial_\rho Z^1 \quad , \tag{2.10}$$

$$\ddot{Z}^I = \partial_\sigma\partial_\sigma Z^I + \partial_\rho\partial_\rho Z^I \quad , \tag{2.11}$$

$$\dot{\Theta} = i\Gamma_2\partial_\sigma\Theta - i\Gamma_1\partial_\rho\Theta \quad . \tag{2.12}$$

At this stage, we must fix the remaining gauge invariance of Eq.(1.39). The gauge choice $\gamma_{0\bar{\imath}} = 0$ yields an equation that can be solved for $\partial_{\bar{\imath}} X^-$. Taking the curl, we find an integrability condition for this equation. Upon linearization about our background, this condition is

$$\partial_\rho \dot{Z}^1 = \partial_\sigma \dot{Z}^2 \quad . \tag{2.13}$$

This can be integrated to give $\partial_\rho Z^1 = \partial_\sigma Z^2 + h(\sigma,\rho)$, which leads us to exploit the residual gauge symmetry (1.39) to set the undetermined function $h(\sigma,\rho)$ to zero, thus obtaining for all τ

$$\partial_\rho Z^1 = \partial_\sigma Z^2 \ . \tag{2.14}$$

This allows us to cast (2.9) and (2.10) into the form of standard wave equations,

$$\ddot{Z}^1 = \partial_\sigma \partial_\sigma Z^1 + \partial_\rho \partial_\rho Z^1 \ , \tag{2.15}$$

$$\ddot{Z}^2 = \partial_\sigma \partial_\sigma Z^2 + \partial_\rho \partial_\rho Z^2 \ . \tag{2.16}$$

The relation (2.14) reduces the number of independent functions of τ, σ and ρ by one, yielding 8 bosonic degrees of freedom. Since the fermion equation of motion (2.12) is of first order, the 16 κ-gauge-fixed components of Θ satisfying (1.33) also give rise to 8 fermionic degrees of freedom.

For the eleven-dimensional Γ-matrices, we use a representation appropriate to the 11=9+2 split,

$$\begin{aligned}
\Gamma^I &= \gamma^I \otimes \sigma_3 \otimes \sigma_3, \quad I = 3, \cdots 9 \ , \\
\Gamma^\pm &= 1 \otimes \sigma_\pm \otimes 1, \quad \sigma_\pm = \frac{1}{\sqrt{2}}(\sigma_1 \pm i\sigma_2), \\
\Gamma^{\bar{\imath}} &= 1 \otimes \sigma_3 \otimes i\sigma_{\bar{\imath}}, \quad \bar{\imath} = 1, 2 \ .
\end{aligned} \tag{2.17}$$

Using this representation, we can express Θ in the κ-gauge (1.33) as

$$\Theta = (16\sqrt{2}\, p^+)^{-\frac{1}{4}} \begin{pmatrix} \chi \\ -i\chi^* \\ -\sqrt{2}p^+\chi \\ -i\sqrt{2}p^+\chi^* \end{pmatrix} \ , \tag{2.18}$$

where χ is a complex 8-component spinor of $SO(7) \times U(1) \subset SO(9)$. Substituting (2.18) into (2.17) casts the fermion equation into the form

$$\dot{\chi} = \partial_\sigma \chi^* - i\partial_\rho \chi^* \ . \tag{2.19}$$

Substituting (2.8) and (2.18) into the gauge conditions $\gamma_{oo} = -\overline{\gamma}$ and $\gamma_{o\bar{\imath}} = 0$, and keeping terms to quadratic order in fluctuations, one finds

$$\begin{aligned}
\dot{X}^- = \frac{1}{2p^+} \Big[& 1 + \underline{\dot{Z}}^2 + (\partial_\sigma \underline{Z})^2 + (\partial_\rho \underline{Z})^2 + G(\underline{Z}) \\
& + \frac{i}{2} \left[(\chi^\dagger \partial_\sigma \chi^* + \chi^T \partial_\sigma \chi) - i(\chi^\dagger \partial_\rho \chi^* - \chi^T \partial_\rho \chi) \right] \Big] \ ,
\end{aligned} \tag{2.20}$$

$$\partial_\sigma X^- = \frac{1}{p^+}\left[\dot{Z}^1 + \partial_\sigma \underline{Z}\cdot\dot{\underline{Z}} + \frac{i}{8}\left[(\chi^T\dot{\chi} + \chi^\dagger\dot{\chi}^*) + (\chi^\dagger\partial_\sigma\chi + \chi^T\partial_\sigma\chi^*)\right]\right] , \quad (2.21)$$

$$\partial_\rho X^- = \frac{1}{p^+}\left[\dot{Z}^2 + \partial_\rho \underline{Z}\cdot\dot{\underline{Z}} + \frac{i}{8}\left[i(\chi^T\dot{\chi} - \chi^\dagger\dot{\chi}^*) + (\chi^\dagger\partial_\rho\chi^* + \chi^T\partial_\rho\chi^*)\right]\right] , \quad (2.22)$$

where

$$G(\underline{Z}) = 2(\partial_\sigma Z^1 + \partial_\rho Z^2) + 4(\partial_\sigma Z^1\partial_\rho Z^2 - \partial_\rho Z^1\partial_\sigma Z^2) . \quad (2.23)$$

When we come to construct the mass formula, G($\underline{Z}$) will drop out in the $\int d\sigma d\rho$ integral.

As a preliminary to quantization, we write the general solutions of (2.15), (2.16), (2.17) and (2.18) as

$$\underline{Z} = \underline{z}_o + \underline{p}\tau + \frac{1}{\sqrt{2}}\sum_{m^2+n^2\neq 0}\frac{1}{\omega_{mn}}e^{i(m\sigma+n\rho)}[\underline{\alpha}^\dagger_{mn}e^{i\omega_{mn}\tau} + \underline{\alpha}_{-m-n}e^{-i\omega_{mn}\tau}] , \quad (2.24)$$

$$\chi = \sqrt{2}S_{oo} + \sum_{m^2+n^2\neq 0} e^{i(m\sigma+n\rho)}[\frac{m-in}{\omega_{mn}}S^\dagger_{mn}e^{i\omega_{mn}\tau} + S_{-m-n}e^{-i\omega_{mn}\tau}] , \quad (2.25)$$

and

$$\omega_{mn} = (m^2+n^2)^{\frac{1}{2}} , \quad (2.26)$$

where the Fourier coefficients have been chosen in (2.24) so that $\underline{Z}$ is real.

Proceeding to the semiclassical quantization, we must distinguish between the unconstrained variables Z^I ($I = 3,\cdots,9$) and the two variables Z^1 and Z^2, which satisfy the constraint (2.13) and the gauge condition (2.14). For the unconstrained variables, we have the canonical commutation relation $[K^I, Z^J] = -i\delta^{IJ}\delta(\sigma-\sigma')\delta(\rho-\rho')$, which implies that

$$[\dot{Z}^I, Z^J] = -(2\pi)^2 i\delta^{IJ}\delta(\sigma-\sigma')\delta(\rho-\rho') . \quad (2.27)$$

For the constrained variables Z^1 and Z^2, we must proceed differently, owing to the need to incorporate (2.13) and (2.14). For this, we adopt Dirac's procedure for quantizing constrained variables [14]. Eqs. (2.13) and (2.14) constitute a non-commuting (i.e. second class) system of constraints. The canonical momenta conjugate to Z^1 and Z^2 are $k^1 = (2\pi)^{-2}\dot{Z}^1$, $k^2 = (2\pi)^{-2}\dot{Z}^2$. Denoting the constraints (2.13) and (2.14) by $\phi_s = 0, s = 1,2$, with

$$\phi_1 = \partial_\rho k^1 - \partial_\sigma k^2 ,$$

$$\phi_2 = \partial_\rho Z^1 - \partial_\sigma Z^2 , \quad (2.28)$$

we have the Poisson bracket algebra

$$[\phi_1(\sigma,\rho),\phi_2(\sigma',\rho')]_P = \nabla^2\delta(\sigma-\sigma')\delta(\rho-\rho') \equiv C_{12} \ , \tag{2.29}$$

where $\nabla^2 = \partial_\sigma^2 + \partial_\rho^2$.

The Dirac brackets are now defined by

$$[A,B]_D = [A,B]_P - [A,\phi_s]_P (C^{-1})^{st}[\phi_t,B]_P \ , \tag{2.30}$$

for arbitrary operators A and B. The Dirac brackets have the property that ϕ_1 and ϕ_2 now commute with each other and with the Hamiltonian. The Dirac brackets of the conjugate pairs (Z^1,k^1) and (Z^2,k^2) are:

$$\begin{aligned}
[k^1(\sigma,\rho),Z^1(\sigma',\rho')]_D &= -\left(1-\frac{\partial_\rho^2}{\nabla^2}\right)\delta(\sigma-\sigma')\delta(\rho-\rho') \ , \\
[k^2(\sigma,\rho),Z^2(\sigma',\rho')]_D &= -\left(1-\frac{\partial_\sigma^2}{\nabla^2}\right)\delta(\sigma-\sigma')\delta(\rho-\rho') \ .
\end{aligned} \tag{2.31}$$

We now pass from the Dirac brackets to the quantum commutators

$$\begin{aligned}
[\dot{Z}^1(\sigma,\rho),Z^1(\sigma',\rho')] &= -(2\pi)^2 i\left(1-\frac{\partial_\rho^2}{\nabla^2}\right)\delta(\sigma-\sigma')\delta(\rho-\rho') \ , \\
[\dot{Z}^2(\sigma,\rho),Z^2(\sigma',\rho')] &= -(2\pi)^2 i\left(1-\frac{\partial_\sigma^2}{\nabla^2}\right)\delta(\sigma-\sigma')\delta(\rho-\rho') \ .
\end{aligned} \tag{2.32}$$

For the fermions, one could proceed as above using the Dirac formalism for the eleven-dimensional spinor satisfying the Majorana and κ-gauge conditions. In practice, we find it more convenient to work with the unconstrained spinor χ. Substituting (2.18) into the action (1.22), one can quantize χ canonically, yielding

$$\{\chi^{*A},\chi^B\} = 2(2\pi)^2\delta^{AB}\delta(\sigma-\sigma')\delta(\rho-\rho') \ , \tag{2.33}$$

where $A,B = 1,\cdots 8$ are SO(7) spinor indices.

Substituting (2.24) and (2.25) into (2.27), (2.32) and (2.33), we find the following commutation relations for the α and S oscillators:

$$[\alpha^1_{mn},\alpha^{1\dagger}_{m'n'}] = \frac{m^2}{\omega_{mn}}\delta_{mm'}\delta_{nn'} \tag{2.34}$$

$$[\alpha^2_{mn},\alpha^{2\dagger}_{m'n'}] = \frac{n^2}{\omega_{mn}}\delta_{mm'}\delta_{nn'} \tag{2.35}$$

$$[\alpha^I_{mn},\alpha^{J\dagger}_{m'n'}] = \omega_{mn}\delta^{IJ}\delta_{mm'}\delta_{nn'} \tag{2.36}$$

$$\{S^A_{mn},S^{B\dagger}_{m'n'}\} = \delta^{AB}\delta_{mm'}\delta_{nn'} \tag{2.37}$$

$$[p^1,z^1_0] = [p^2,z^2_0] = -i \quad , \quad [p^I,z^J_0] = -i\delta^{IJ}, \tag{2.38}$$

$$\{S^A_{00},S^{B\dagger}_{00}\} = \delta^{AB}, \tag{2.39}$$

with all other independent commutators vanishing. Note that the constraint (2.14) implies that

$$n\alpha^1_{mn} = m\alpha^2_{mn} \ , \tag{2.40}$$

and that this is consistent with (2.34) and (2.35). It also implies that $[\alpha^1_{mn}, \alpha^{2\dagger}_{m'n'}] = \frac{mn}{\omega_{mn}}\delta_{mm'}\delta_{nn'}$, etc.

Substituting (2.24) and (2.25) into (2.20), and then the result into (2.1) and (2.2), we find

$$\begin{aligned} P^- = \frac{1}{2p^+}[1 + \underline{p}^2 + &\sum_{m^2+n^2\neq 0} (\underline{\alpha}_{mn}\cdot\underline{\alpha}^\dagger_{mn} + \underline{\alpha}^\dagger_{mn}\cdot\underline{\alpha}_{mn}) \\ &+ \sum_{m^2+n^2\neq 0} \omega_{mn}\,(-S^A_{mn}S^{A\dagger}_{mn} + S^{A\dagger}_{mn}S^A_{mn})] \ . \end{aligned} \tag{2.41}$$

Using the fact that $P^+ = p^+$ and $\underline{P} = \underline{p}$, we therefore have the eleven-dimensional mass formula

$$(\text{mass})^2 = (\ell_1\ell_2R_1R_2)^2 + H \ , \tag{2.42}$$

where

$$H = 2\sum_{m^2+n^2\neq 0} (\underline{\alpha}^\dagger_{mn}\cdot\underline{\alpha}_{mn} + \omega_{mn}S^{A\dagger}_{mn}S^A_{mn}) \ . \tag{2.43}$$

In obtaining the result (2.42) from (2.41), we have restored the winding numbers ℓ_1 and ℓ_2 and radii R_1 and R_2 of the two circles, which we had set equal to unity earlier; with these restored, the frequencies are

$$\omega_{mn} = [(m\ell_2R_2)^2 + (n\ell_1R_1)^2]^{\frac{1}{2}} \ . \tag{2.44}$$

The key result is that the mass formula (2.42, 2.43) is correct as it stands, without a vacuum energy term. This is because the bosonic contribution, which takes the form of an Epstein zeta function,

$$\Delta_B(\text{mass})^2 = \sum_{m^2+n^2\neq 0} [(m\ell_2R_2)^2 + (n\ell_1R_1)^2]^{\frac{1}{2}} \ , \tag{2.45}$$

coming from normal-ordering (2.41), is cancelled mode-by-mode by an equal but opposite fermionic contribution. Note that $\alpha^\dagger_{mn}$ and $S^\dagger_{mn}$ are creation operators for all m and n, while α_{mn} and S_{mn} are annihilation operators. The difference between this convention and the usual string convention stems from the form of the oscillator expansions (2.24) and (2.25).

The cancellation of the vacuum energy contributions (2.44) between bosons and fermions depends crucially on the κ-symmetry (1.27), which caused the number of fermionic physical degrees of freedom to be halved, (eq.(2.18)).

The cancellation that we have observed between semiclassical boson and fermion contributions to the vacuum energy gives support to our expectation that this cancellation will take place also in the exact theory. Although the supermembrane has only a Green-Schwarz formulation, the combination of spacetime supersymmetry and the fermionic gauge invariance appears to have the same consequence for the cancellation of the vacuum energy correction as would a world-volume supersymmetry.

3. – RECENT RESULTS

Quantum Consistency

In computing the total energy of any solution of the membrane equations, one should include both the classical energy and the zero point energy of the quantum fluctuations. For any extended object, for which there are an infinity of possible fluctuation modes, one would expect the zero point energy to be infinite and require renormalization. For *bosonic* extended objects, this is in fact just what happens. For the bosonic string, the renormalization should be consistent with conformal invariance and this can be accomplished using ζ-function regularization/renormalization. It is one of the "miracles " of string theory that the renormalized zero-point energy is not only *negative*, but is just such as to be cancelled (in 26 dimensions and for the open string) by the *positive* classical contribution of a rotating string with *integral* angular momontum (1, in fact, in units of $\hbar$). The integral values of angular momentum are precisely those allowed by the quantum theory, of course, so this cancellation means that the lowest lying spin excitations of the bosonic string are *massless*. This fact is an essential ingredient in the argument that 26-dimensional Yang-Mills theory is the "field theory limit" of the open bosonic string.

Could a similar miracle occur for membranes? This question was addressed by Kikkawa and Yamasaki [15]. As an analogue of the rotating string, they considered a solution in which the membrane rotates simultaneously in two orthogonal planes; this is possible in dimensions $d \geq 5$. They computed the energy of the zero-point fluctuations about this solution. As expected, it is infinite, but ζ-function regularization can again be used to arrive at a renormalized, finite result (although the status of ζ-function regularization is less clear than in string theory). The result is *not* such that it can be cancelled by a classical contribution for any value of the angular momentum allowed by quantum mechanics. Therefore, Kikkawa and Yamasaki concluded that, unlike the string, the spectrum of the bosonic membrane does not include massless particles.

Leaving aside a number of objections that one could make to any statements about the value of the vacuum energy in a non-renormalizable three-dimensional theory on the basis of a semiclassical calculation, the question of the possibility of massless states must be completely reconsidered when one includes supersymmetry. Provided that we consider fluctuations about a supersymmetric solution, supersymmetry "on the brane" will ensure the *complete cancellation* of the quantum correction to the classical energy. As we have explained previously, supersymmetry "on the brane" is a consequence of both (i) the rigid spacetime supersymmetry and (ii) the local κ-invariance. As illustrated by the foregoing semiclassical quantization of the toroidal membrane, this leads to a set of linear equations for the small fluctuations that are those of a three-dimensional supersymmetric field theory. In principle, one should consider the three-dimensional *interacting* field theory of fluctuations about the classical solution. Again, supersymmetry will continue to guarantee the cancellation of the quantum corrections to the vacuum energy provided that one can show that the symmetry in question really is *supersymmetry*. As we have explained above, the arguments for this, while fairly convincing, are still indirect.

Given that the quantum correction to the classical energy of a supermembrane vanishes, one clearly cannot expect a *cancellation* between classical and quantum energies. *Massless* states can only be associated with configurations of *zero* classical energy, i.e. collapsed membranes. This also applies to the superstring, so let us first consider how the massless spectrum of the type IIA d=10 superstring may be deduced. The string equations may be solved by setting

$$X^\mu(0,\tau) = q^\mu + p^\mu \tau \quad , \tag{3.1}$$

where p and q are constant and p^2=0, which represents a string collapsed to a point. Upon quantization, we replace the constants q^μ, p^μ by canonically conjugate operators satisfying $[q^\mu, p^\nu] = i\hbar\eta^{\mu\nu}$. The spinors Θ^α of the type IIA GS superstring can be reduced to the $SO(8)$ spinor S_a and $SO(8)$ conjugate spinor $S_{\dot{a}}$ by a gauge choice. The equations for Θ^α may be solved by setting

$$S_a(\sigma,\tau) = \Psi_a \ , \qquad S_{\dot{a}}(\sigma,\tau) = \Psi_{\dot{a}} \ , \tag{3.2}$$

where Ψ_a and $\Psi_{\dot{a}}$ are anticommuting constants; after quantization, these anticommuting constants are replaced by fermion operators obeying the canonical commutation relations

$$\{\Psi_a, \Psi_b\} = \hbar\delta_{ab} \ , \ \ \{\Psi_{\dot{a}}, \Psi_{\dot{b}}\} = \hbar\delta_{\dot{a}\dot{b}} \ , \ \ \{\Psi_a, \Psi_{\dot{a}}\} = 0 \ . \tag{3.3}$$

The (Ψ_a,$\Psi_{\dot{a}}$) are therefore generators of the Clifford algebra of $SO(8+8)$, for which the unique faithful irrep. is 2^8-dimensional. Thus, the degeneracy of the massless states is 2^8, consisting of 128 bosons and 128 fermions. The only d=10 left-right symmetric irreducible supermultiplet with this degeneracy is the massless supermultiplet of type IIA supergravity, so we conclude that type IIA supergravity is the field theory limit of the type IIA superstring.

Precisely the same argument may be applied to the 11-dimensional supermembrane collapsed to a point, except that with the choice of the $SO(8)$-covariant light-cone gauge, the two $SO(8)$ spinors S_a, $S_{\dot{a}}$ appear as a single 16-component spinor of $SO(9)$, but which again generates the Clifford algebra of $SO(16)$. The degeneracy is therefore again 128+128, but now each 128 will come as a sum of irreps of $SO(9)$, the little group for massless particles in 11 dimensions. In fact, one can show [16] that the $SO(9)$ irreps are *84* $\oplus$ *44* for the bosons and *128* for the fermions, which is just the $SO(9)$ content of d=11 supergravity. Once again, we see that the d=11 supermembrane is closely related to d=11 supergravity.

A natural question is whether a similar relation holds for the other super p-branes. It was shown in [16] that *only* in the case of the d=10 superstring and the d=11 supermembrane does the massless spectrum coincide with that of a supergravity model. Supergravity is also, for $p \geq 2$, the only kind of spacetime background that we know how to incorporate in a super p-brane action. It is not known how to couple a super p-brane to background Yang-Mills, for example; it may be that this is impossible. Thus, only for the d=10 superstring and the d=11 supermembrane is there a coincidence between the massless spectrum of the quantized extended object and the background to which it can couple. This might be taken as evidence that the "critical" dimension for supermembranes is 11 [16].

The nature of the spectrum of the d=11 supermembrane has been clarified to some extent by the above analysis. One point that is clear is that topologically distinct classical solutions will lead to distinct sectors of the full spectrum and one should presumably include all of these. In particular, massless particles are to be expected only in a sector containing collapsed membranes. Since a topological sector containing collapsed membranes does not have a static classically stable solution about which one can linearize, one might then resort to a semiclassical calculation of the zero point energies using *dynamically* stabilized classical solutions. This was done by Kikkawa and Yamasaki [15] for the bosonic membrane, as discussed above. A similar calculation has been performed recently [17] for the closed supermembrane, with the result that the frequencies of the boson and fermion fluctuations about a *given* dynamical classical solution are not equal. One must, however, take into account the degeneracy of the original classical solution, since it breaks all the world-volume supersymmetries. Since this has not been taken into account, it is still too early to draw conclusions about the massless states from this calculation.

The individual topological sectors of the spectrum presumably interact when one goes beyond the semiclassical approximation. This would be analogous to the interaction of open and closed string sectors in string field theory. As in that case, the interactions could change the spectrum drastically. A more pressing problem is that, whereas a string can collapse only to a point, a membrane can collapse to a *line*; the general classical solution of this type can be found in [6]. Any such line can be continuously deformed into another one, so that there would appear to be a *continuum* of massless states. In terms of the field theory limit, we would have to describe these by fields of 11-dimensional supergravity that depend on an additional continuous parameter. It is almost as if we are dealing with a kind of 12-dimensional supergravity (but with 11-dimensional covariance). Speculations about a possible 12-dimensional formulation of 11-dimensional supergravity have been voiced in the past, and perhaps now is the time to reexamine them. In any case, this problem seems to us to present a serious difficulty of interpretation for the 11-dimensional supermembrane.

One question concerning the quantum consistency of membranes that is often raised is the lack of conformal invariance. However, three-dimensional reparameterization invariance is *sufficient* to remove all unphysical degrees of freedom in the classical bosonic membrane. The relevant question, therefore, is whether *this* symmetry is preserved by quantum corrections. Conformal invariance is simply not relevant to the question of ghosts in membrane theories. In fact, there is no reason to believe that there are anomalies in reparameterization invariance in three dimensions. For the super p-brane, we have to consider whether the κ-invariance could be anomalous. As for string theory, one way to address this question is to fix gauges so as to eliminate all the unphysical variables. In doing so, we lose manifest Lorentz invariance, however, and any anomaly can and must show up as a Lorentz anomaly. A simple way to check whether Lorentz anomalies show up has been discussed by Bars [18]. Consider first the d=10 superstring: the *massive* states at each mass level are found in irreps of $SO(8)$ in the light-cone gauge-fixed theory, but Lorentz covariance requires that they be reassembled into irreps of $SO(9)$, the "little group" for massive particled in d=10. It was a well-known fact that they can be so reassembled, but it was apparently not appreciated that the GS superstrings in d=3,4&6 *fail* this test.

To apply this idea to super p-branes with $p > 1$, Bars then considered the first massive excitations about a p-torus solution of the type considered in the last section in the context of semiclassical quantization. He asked whether these excitations are of the correct multiplicities to fill out representations of the little group for massive particles in d dimensions and showed that *all* super p-branes with $p \geq 2$ *fail* this test with the exception of the d=11 supermembrane. In the latter case, he also showed that the multiplicities of the massive states also correspond to those of massive d=11 supermultiplets. When this analysis is applied to the other p-branes, one does not find multiplicities corresponding to the little group for massive particles in d dimensions. Unfortunately, this analysis does not in itself establish the presence of Lorentz anomalies in theories other than the 2-brane in d=11. This is because a theory compactified on a p-torus has only $SO(d-p-1,1)$ Lorentz symmetry, for which the massive representation little group is $SO(d-p-1)$, and *this* group is not broken by the quantization procedure. Nevertheless, Bars' result certainly indicates that the supermembrane theory in d=11 is special.

One might consider applying the above type of argument to a membrane compactified on a circle, as in the double dimensional reduction [10] discussed in the first section. Although the tower of "massive strings" may prove to have very complicated interactions, the "massless" string at the lowest level should already reveal whether Lorentz invariance is preserved at the quantum level or not. The results of [18] for strings indicate that only in the case of the 11-dimensional supermembrane, which gives the type IIA superstring in d=10, will the massive states of this superstring fill out the $SO(9)$ little group representations required for the d=10 Lorentz symmetry that should be unbroken for this compactified membrane theory. It would certainly be worthwhile trying to formulate such an argument in more detail. In any case, it would appear that, as a consequence of the results of [10] and [18], the only two supersymmetric quantum theories of extended objects that have a chance of being consistent are the d=10 superstring and the d=11 supermembrane.

There is more to the question of quantum consistency than anomalies, however. An immediate problem to be faced by the 11-dimensional supermembrane theory is the non-renormalizability of three-dimensional non-linear σ-models. Here, however, one must carefully consider the physical significance of renormalizability, or lack of it. Consider the closed bosonic string theory. Each mode of the string is associated with a σ-model operator, those of (naïve) dimension two being associated with the massless modes and those of dimension greater than two with the massive modes (we ignore the tachyon). If we were to include any of the operators of dimension greater than two in the σ-model action, it would be non-renormalizable and we should have to consider them all. Technically, this is a non-renormalizable theory, but all this means is that the conditions for σ-model conformal invariance are now an *infinite* component equation for *all* the modes of the string, which is physically reasonable. *The fact that the dimension two operators alone yield a renormalizable σ-model is equivalent to the statement that this infinite component equation can be consistently truncated to the sector involving the massless modes alone.* This would imply that the massive modes cannot decay into the massless ones. Of course, this conclusion could be changed by the presence of non-perturbative contributions to the β-functions that might couple the massive modes to appropriate currents built from the massless modes. From this point of view, renormalizability is not crucial for the consistency of the theory, but is an indication of whether one can (in some appproximation) obtain an effective action for the massless modes alone.

In the case of membranes, the issue of conformal invariance does not arise, but one can still require finiteness of the three-dimensional σ-model. This would again yield an infinite-component equation for *all* the modes of the membrane. The absence of a renormalizable sector means that it is *not possible* to consistently truncate this equation in any approximation. This makes construction of a low energy effective theory problematic, but may not pose a fundamental problem of consistency.

Extended Solitons

It is easy to interpret a p-brane moving in spacetime, but the physical meaning of a super p-brane moving in superspace is not so clear. Investigations into "elementary" p-branes give no insight into this question, but from the "cosmic" p-brane point of view, the answer is easily found. Consider first the bosonic case, and suppose that we have found a solution of a d-dimensional field theory with a topological defect of dimension p. At long wavelengths, this defect will appear as a structureless p-dimensional object of negligible width, a p-brane. At any point on the $(p+1)$–dimensional world volume, translational invariance is spontaneously broken in directions *transverse* to the world volume. As a consequence, there will be $(d\text{-}p\text{-}1)$ Goldstone modes, or collective coordinates, $\mathbf{x}_T(\xi)$ associated to the world volume, describing the transverse fluctuations of the p-brane. At sufficiently long wavelengths, these are the only fluctuations of relevance, so the motion of the object should be describable in this limit by a $(p+1)$–dimensional action for the $(d\text{-}p\text{-}1)$–vector $\mathbf{x}_T(\xi)$. In order to make spacetime Lorentz invariance manifest, we can augment $\mathbf{x}_T$ with $(p+1)$ unphysical variables, replacing it by $x^m(\xi)$, provided we build in a gauge invariance that will ensure that these unphysical variables have no physical effect. Clearly, the gauge invariance in question is reparameterization invariance. By this argument, one is led to (1.2) as essentially the only possible Lorentz covariant action describing the p-brane in the long wavelength limit.

The above argument is a generalization of that proposed for the vortex in the 4-dimensional Abelian Higgs model by Nielsen and Olesen. There is a supersymmetric version of this model, and an interesting question is what is the low-energy effective action for the vortex in this case. This was investigated by Hughes and Polchinski [19], who argued that the long-wavelength action must be the Green-Schwarz superstring action. The crucial point is that the vortex solution has the property that it breaks only *half* of the spacetime supersymmetry. If supersymmetry were completely broken, there would be as many Goldstone fermion modes on the world volume as there were components of the original supersymmetry generator Q_α. If only *half* of the supersymmetry is broken, we have only half as many Goldstone fermion modes, but to achieve a *covariant* action for this half, we must augment them with the remaining unphysical half. Again, this requires the introduction of a gauge symmetry (κ-symmetry) to ensure that the unphysical variables have no physical effect. The Green-Schwarz action is the unique action (without higher derivative terms) having this gauge invariance. The same considerations apply to any p-dimensional extended object solution of a supersymmetric field theory. The Nielsen-Olesen solution of the Abelian Higgs model, for example, represents a three-dimensional extended object if spacetime is six-dimensional. The effective action is therefore that of a super 3-brane [2].

We have now traced the need for a κ-invariant action to the fact that certain extended object solutions of supersymmetric field theories break only half the supersymmetry, but is this a *general* property of these solutions? To answer this question we need to understand better *why* only half the supersymmetry is broken. We can do this by considering the nature of the effective action for solitons, i.e. 0-branes, in supersymmetric field theories [20]. As first shown by Witten and Olive [21], the topological charge T of the soliton appears in the supersymmetry algebra as a central charge. As a consequence, its mass M is subject to the Bogomol'nyi bound

$$M \geq |T| \tag{3.4}$$

and configurations for which this bound is saturated, i.e. $M = |T|$, (which are necessarily *solutions* of the classical field equations [22]) are rather special. It is a well-known fact that representations of the supersymmetry algebra with central charge T are qualitatively different when $M = T$; "multiplet shortening" occurs because there are (non-Hermitian) linear combinations of *half* the supersymmetry generators that annihilate the "Clifford vacuum". Applied to a soliton with $M = |T|$, this means that linear combinations of half the supersymmetry generators will annihilate the soliton state, i.e. half the supersymmetry will be preserved and half broken. We need only half as many fermionic collective coordinates as there are components of the supersymmetry charge. The effective action must be that of a κ-invariant massive superparticle, i.e. the super p-brane action of (1.10) for p=0. In the classification of super p-branes, the p=0 case was excluded, but it is easy to see that the action must exist for d=2,3,5&9, at least, because it can be found by double dimensional reduction of the superstring action in d=3,4,6&10. This fact is illustrated by the dots in the $p = 0$ column of the table of allowed (p,d) values given in section one (although there are other possibilities for p=0 if $N > 1$).

Given a solution in a (flat space) d-dimensional field theory we can interpret it as a string in a $(d+1)$–dimensional field theory, a membrane in a $(d+2)$–dimensional field theory, etc.. We can call this process "double dimensional oxidation" because it is the converse of the process of double dimensional reduction mentioned earlier. The Bogomol'nyi bound now takes the form

$$\frac{\text{Mass}}{\text{unit } p\text{–vol.}} \geq |T| \tag{3.5}$$

for the p-brane tension. Solutions for which this bound is saturated must be described by κ-invariant actions, so it remains to understand *why* the bound should be saturated. Of course, we have only to understand this for solitons because whatever we learn there can immediately be applied to p-branes by "oxidation".

In fact, it is well appreciated (although apparently not proved) that solitons of supersymmetric field theories must saturate the Bogomol'nyi bound because they would otherwise be unstable. They will radiate away mass in the form of "mesons" of the underlying field theory until a static configuration of minimum energy is reached. This is necessarily one for which $M = |T|$. We have seen, therefore, that from the perspective of the "cosmic p-brane" the ultimate reason for the particular super p-brane action (1.10) is one of stability.

An obvious question now is whether *all* the super p-branes of the table can be "realised" as extended object solutions of supersymmetric field theories. It is not difficult to see that all those of the **R**, **C**, **H** sequences can be interpreted this way [20]. One has only to think of the instanton solutions in the d=1,2&4 dimensional supersymmetric quantum mechanics, supersymmetric sigma-model, and supersymmetric Yang-Mills theory, respectively. These solutions may be considered as solitons in d=2,3&5, strings in d=3,4&6, etc.. In this way we rediscover the **R**, **C**, **H** sequences. It is interesting to try to understand from this point of view why these sequences terminate where they do. The multiple well potential of supersymmetric quantum mechanics can be "oxidised" as far as a d=4 field theory (of the Wess-Zumino supermultiplet), but not beyond d=4. The supersymmetric sigma-model can be "oxidised" as far as d=6, but the model doesn't exist for $d > 6$. Similarly, the supersymmetric Yang-Mills theory exists as far as d=10 but not beyond. The agreement with the analysis of [9] is satisfying inasmuch as no appeal to the details of an underlying field theory was made there.

Can the last, octonionic, sequence also be understood in this way? As of this writing, no answer to this question, as to many others, is known. We hope only to have persuaded the reader that these are questions worth asking.

REFERENCES

[1] P.A.M. Dirac, *Proc. Roy. Soc.* **A268** (1962) 57.

[2] J. Hughes, J. Liu, and J. Polchinski, *Phys.Lett.* **180B** (1986) 370.

[3] E. Bergshoeff, E. Sezgin and P.K. Townsend, *Phys.Lett.* **189B** (1987) 75.

[4] P.S. Howe and R.W. Tucker, *J. Phys.* **A10** (1977) L155.

[5] P.A. Collins and R.W. Tucker, *Nucl. Phys.* **B112** (1976) 150.

[6] E. Bergshoeff, E. Sezgin and P.K. Townsend, *Properties of the 11-dimensional supermembrane*, preprint IC/87/255 (1987).

[7] M.B. Green and J.H. Schwarz, *Nucl. Phys.* **B243** (1984) 285.

[8] M. Henneaux and L. Mezincescu, *Phys. Lett.* **152B** (1985) 340.

[9] A. Achucarro, J.M. Evans, P.K. Townsend and D.L. Wiltshire, *Phys. Lett.* **198B** (1987) 441.

[10] M.J. Duff, P.S. Howe, T. Inami and K.S. Stelle, *Phys. Lett.* **191B** (1987) 70.

[11] E. Cremmer and S. Ferrara, *Phys. Lett.* **91B** (1980), 61.
L. Brink and P.S. Howe, *Phys Lett.* **91B** (1960), 384.

[12] J. Hoppe, Aachen preprint PITHA 86/24; and Ph.D. Thesis, MIT (1982).

[13] M.J. Duff, T. Inami, C.N. Pope, E. Sezgin and K.S. Stelle, *Nucl. Phys.* **B** (in press).

[14] P.A.M. Dirac *Lectures on Quantum Mechanics*, Belfer Graduate School of Science, Yeshiva University, New York (1964).

[15] K. Kikkawa and M. Yamasaki, *Prog.Theor.Phys.* **76** (1986) 1379.

[16] I. Bars, C.N. Pope and E. Sezgin, *Phys. Lett.* **198B** (1987) 455.

[17] L. Mezincescu, R.I. Nepomechie and P. van Nieuwenhuizen, *Do supermembranes contain massless particles?*, preprint UTMG-139, ITP-SB-87-43 (1987).

[18] I. Bars, *First massive level and anomalies in the supermembrane*, preprint USC-87/HEP06.

[19] J. Hughes and J. Polchinski, *Nucl. Phys.* **B278** (1986) 147.

[20] P.K. Townsend, *Supersymmetric extended solitons*, University of Cambridge D.A.M.T.P. preprint (1987).

[21] E. Witten and D. Olive, *Phys. Lett.* **78B** (1978) 97.

[22] W. Boucher, *Phys. Lett.* **132B** (1983) 88.

BOSONIC STRING THEORY AT FINITE TEMPERATURE*

Yvan Leblanc

Center for Theoretical Physics
Laboratory for Nuclear Science
and Department of Physics
Massachusetts Institute of Technology
Cambridge, MA 02139 U.S.A.

ABSTRACT

Most studies of the thermodynamics of strings have been carried out for an ideal gas of string excitations. In this talk we shall describe a real-time finite temperature formalism for interacting string field theory, which is a generalization of the thermofield dynamics formalism of ordinary field theory. We also show how to compute finite temperature amplitudes in the context of the first quantized open bosonic string theory. Finally we comment on interesting thermodynamical problems which may be addressed within the present real-time formalism.

I. INTRODUCTION

To this day, superstring theories[1)] (especially the heterotic strings[2)]) are the best candidates for the unification of all known forces in nature and have received enormous attention, mostly since the discovery of anomaly and infinity cancellations for some superstring models by Green and Schwarz[3)] a few years ago. Superstring theories are the only known grand-unified models which yield a renormalizable quantum theory of gravity. Consequently, much interest has been

* This work was supported in part by funds provided by the U. S. Department of Energy (D.O.E.) under contract #DE-AC02-76ER03069, and by the Natural Sciences and Engineering Research Council of Canada.

devoted to the study of compactification, superstring cosmology and the finite temperature aspects of superstring theories at high temperature and high energy density.[4-7]

Parallel to these investigations, endeavor has also been brought to a deeper understanding of the mathematical structure of string theories in general, including the quantization problem as well as the perturbative and the non-perturbative aspects. These include Polyakov's path integral quantization over random surfaces of the two-dimensional linear σ-model,[8] geometric quantization approaches on appropriate manifolds,[9] second quantization of interacting string theories,[10-20] instantons on the world-sheet and symmetry breaking.[20] For a review on recent developments in superstring theory, see Ref. [20].

Also, recent years have seen the birth of superstring phenomenology,[21] which studies the implicationsof superstring models for particle physics at present day accelerators energy scales. Given the high degree of uniqueness of string theories (they have very few free parameters), such studies are essential to determine the most viable candidates as well as the most realistic compactification schemes.

Perhaps as equally important to the determination of realistic theories is the study of string cosmology. Modern scenarios for the evolution and the expansion of the universe may have to be revised or modified in view of the predictions of string theories. Of course, because the early universe was in a state of very high temperature and density (and most likely not in thermal equilibrium), it is necessary to understand better string thermodynamics. This brings us to the main subject of this presentation.

Most recent studies of string thermodynamics rely on the ideal gas picture of string excitations. For dynamical problems, such as the expansion (inflation) of the universe, a real-time formalism for real (interacting) gas is needed as well as a formalism treating processes departing from thermodynamical equilibrium. In the following sections we present such a formalism for equilibrium processes in the context of the interacting open bosonic string field theory and leave the non-equilibrium problem open to further studies. Our formalism is an extension of the so-called thermo-field dynamics (TFD) formalism to string systems.[5,22–31]

In the following, we choose natural units and use the signature $(-++\cdots)$ for the metric tensor.

II. THERMO-FIELD DYNAMICS

In ordinary field theory, the main advantage of TFD is to formally replace the usual statistical average of a dynamical variable A by a suitable vacuum expectation value[22], that is,

$$\langle A\rangle \equiv \mathrm{tr}\left(A\, e^{-\beta H}\right)/\,\mathrm{tr}\left(e^{-\beta H}\right) = \langle\theta(\beta)\,|A|\,\theta(\beta)\rangle \quad . \tag{2.1}$$

In order to find a thermal vacuum $|\theta(\beta)\rangle$ satisfying Eq. (2.1), it has been realized that one must first implement an effective doubling of the physical degrees of freedom of the zero temperature theory. These new unphysical dynamical variables are usually denoted by a tilde (˜).

The entire structure of TFD can be extracted from the following list of axioms:[5,30]

<u>Axiom 1</u>: Dynamical variables belonging to different subspaces are independent,

$$\left[A(t),\tilde{B}(t)\right]_{\mp} = 0 \quad , \tag{2.2}$$

<u>Axiom 2</u>: The tilde conjugation (mapping between the two subspaces) is defined by the following tilde conjugation rules,

$$\text{(a)}\quad (AB)^{\sim} = \tilde{A}\tilde{B} \quad , \tag{2.3}$$

$$\text{(b)}\quad (C_1 A + C_2 B)^{\sim} = C_1^{*}\tilde{A} + C_2^{*}\tilde{B} \quad , \tag{2.4}$$

$$\text{(c)}\quad \left(\tilde{A}\right)^{\dagger} = \left(A^{\dagger}\right)^{\sim} \quad , \tag{2.5}$$

where C_1, C_3 are c-numbers.

Axiom 3: The thermal vacuum is invariant under tilde conjugation,

$$|\widetilde{\theta(\beta)}\rangle = |\theta(\beta)\rangle \quad . \tag{2.6}$$

Axiom 4: The thermal vacuum satisfies the thermal state conditions,

$$A(t,\vec{x})|\theta(\beta)\rangle = \sigma \tilde{A}^\dagger \left(t - i\frac{\beta}{2}, \vec{x}\right)|\theta(\beta)\rangle \quad , \tag{2.7}$$

$$\langle\theta(\beta)|A(t,\vec{x}) = \langle\theta(\beta)|\tilde{A}^\dagger \left(t + i\frac{\beta}{2}, \vec{x}\right)\sigma^* \quad , \tag{2.8}$$

where $|\sigma| = 1$.

Axiom 5: The double tilde conjugation is defined as,

$$\tilde{\tilde{A}} = \sigma A \quad . \tag{2.9}$$

The thermal state conditions (2.7) – (2.8) are equivalent to the Kubo–Martin–Schwinger (KMS) condition of the axiomatic C^*-algebra approach to statistical mechanics.

Introducing the thermal doublet notation,

$$A^\alpha = \begin{cases} A & ; \quad \alpha = 1 \ , \\ \tilde{A}^\dagger & ; \quad \alpha = 2 \ , \end{cases} \tag{2.10}$$

the real-time finite temperature formalism constructed from the above axioms accommodates a perturbation theory very similar to the zero temperature case. The Gell–Mann–Low formula now reads as [23]

$$\begin{aligned} &\langle\theta(\beta)\,|T\phi_1^{\alpha_1}(x_1)\cdots\phi_n^{\alpha_n}(x_n)|\,\theta(\beta)\rangle \\ &= \frac{\left\langle\theta;\beta\left|T\phi_1^{\alpha_1}(x_1)\cdots\psi_n^{\alpha_n}(x_n)\exp\left[i\int_{-\infty}^{\infty}d^4x\hat{\mathcal{L}}_I(x)\right]\right|\theta;\beta\right\rangle}{\left\langle\theta;\beta\left|T\,\exp\left[i\int_{-\infty}^{\infty}d^4x\,\hat{\mathcal{L}}_I(x)\right]\right|\theta;\beta\right\rangle} \quad , \end{aligned} \tag{2.11}$$

where the right-hand side is in the interaction representation. Also one has

$$\hat{\mathcal{L}} = \mathcal{L} - \tilde{\mathcal{L}} \quad , \tag{2.12}$$

where

$$\mathcal{L} = \mathcal{L}_0 + \mathcal{L}_I \; . \tag{2.13}$$

Note that we restricted ourselves to the boson case only.

The $\hat{\mathcal{L}}_I$-term in (2.11) originates from the fact that the generator of time translations is the total Hamiltonian,

$$\hat{H} = H - \tilde{H} \; . \tag{2.14}$$

The Gell–Mann–Low formula supplemented by Wick's theorem yields the finite temperature Feynman rules of the theory. These are very similar to the rules at zero temperature except for the fact that the vertices now carry thermal indices which should be summed over and that propagators are now thermal propagators,

$$\begin{aligned} \Delta_0^{\alpha\beta}(x,y) &= i \langle \theta;\beta \,| T\phi^\alpha(x)\phi^\beta(y) |\, \theta;\beta \rangle \\ &= \frac{1}{(2\pi)^4} \int d^4p\, e^{ipx}\, \Delta_0^{\alpha\beta}(p) \;, \end{aligned} \tag{2.15}$$

where

$$\Delta_0^{\alpha\beta}(p) = \left[U_B(|p_0|) \frac{\tau}{p^2 + M^2 - i\tau\delta} U_B(|p_0|) \right]^{\alpha\beta} \; . \tag{2.16}$$

The Bogoliubov transformation matrix $U_B^{\alpha\beta}(\omega)$ is obtained explicitly as

$$U_B^{\alpha\beta}(\omega) = \frac{1}{\sqrt{e^{\beta\omega} - 1}} \begin{bmatrix} e^{\frac{\beta\omega}{2}} & 1 \\ 1 & e^{\frac{\beta\omega}{2}} \end{bmatrix} \;, \tag{2.17}$$

and

$$\tau \equiv \begin{bmatrix} 1 & 0 \\ 0 & -1 \end{bmatrix} \; . \tag{2.18}$$

The matrix $U_B^{\alpha\beta}(\omega)$ arises as a consequence of normal ordering with respect to the thermal vacuum as well as the thermal state condition (2.7) – (2.8).

The above finite temperature Feynman rules also arise from the path-integral formalism. In such a case one has the following generating functional,[27)]

$$\begin{aligned} \hat{Z}[J] \propto \int \left[\prod_{\alpha,x} d\phi^\alpha(x) \right] \exp i \int_{-\infty}^{\infty} d^4x \\ \times \left\{ \frac{1}{2}\phi^\alpha(x) \Delta_0^{-1^{\alpha\beta}}(-i\partial)\phi^\beta(x) + \mathcal{L}_I(\phi^1) - \mathcal{L}_I^*(\phi^2) + J^\alpha(x)\tau^{\alpha\beta}\phi^\beta(x) \right\} \; . \end{aligned} \tag{2.19}$$

Performing Gaussian integrals, one obtains the familiar form,

$$\hat{Z}[J] = \hat{N}^{-1} \exp i \int d^4x \left\{ \mathcal{L}_I \left[\frac{-i\delta}{\delta J} \right] - \mathcal{L}_I^* \left[\frac{i\delta}{\delta \tilde{J}} \right] \right\}$$
$$\times \exp i \int d^4x \, d^4y \left[\frac{1}{2} J(x) \tau \Delta_0(x-y) \tau J(y) \right]^{\alpha\alpha} \quad . \tag{2.20}$$

It is important to realize that, in the path-integral formalism, the theory is non-trivial with respect to the thermal degrees of freedom because of the Feynman convergence factor in the expression for causal propagator (2.16). Inversion of the propagator while leaving the $i\delta$-term finite but small yields,

$$\Delta_0^{-1^{\alpha\beta}}(p) = \left(p^2 + M^2\right) \tau^{\alpha\beta} - i\delta \left[\tau U_B(|p_0|) U_B(|p_0|) \tau\right]^{\alpha\beta} \quad . \tag{2.21}$$

It is the boundary term in (2.21) which mixes both sectors ϕ_1 and ϕ_2 in Eq. (2.19) and therefore renders the theory non-trivial with respect to temperature. Such a fact will be remembered when constructing a quantum field theory of strings at finite temperature in Section IV by use of the path-integral method.

III. A BRIEF OVERVIEW OF STRING FIELD THEORY

When one considers the problem of string interactions, it may be advantageous to have a second quantized formalism for strings. Of course such a quantum field theory should reproduce the scattering amplitudes obtained in the context of the first quantized theory.

Another advantage, which is closer to our considerations, is the close resemblance with ordinary field theory, a feature which will help us formulate the real-time finite temperature formalism for strings in the next section.

Historically, canonical quantization of string field theory was carried out in the light-cone gauge.[10, 11] More recently, covariant formulations have been obtained in the path-integral representation.[12–20]

In general, a string field Φ (here we restrict ourselves to open bosonic strings) is a functional of string coordinates $X_\mu(\sigma)$ and of Faddeev–Popov ghosts and anti-ghosts $c(\sigma)$ and $b(\sigma)$,

$$\Phi = \Phi\left[X_\mu(\sigma), c(\sigma), b(\sigma)\right] \quad . \tag{3.1}$$

The latter coordinates and ghosts are first quantized operators of a string theory with reparametrization (gauge) invariance of the world sheet. In such a theory, constraints must be imposed on the physical sector of the Hilbert spaces $\mathcal{F}$ of the first quantized operators to ensure unitarity. The latter space is the Fock space spanned by the oscillator modes of the coordinates and ghosts,

$$\begin{aligned} [\alpha_n^\mu, \alpha_m^\nu] &= n\, g^{\mu\nu} \delta_{m+n,0} \ ; \qquad & [x^\mu, p^\nu] &= i\, g^{\mu\nu} \ ; \\ \{c_n, c_m\} &= \{b_n, b_m\} = 0 \ ; \qquad & \{c_n, b_m\} &= \delta_{n+m,0} \ . \end{aligned} \tag{3.2}$$

Since the Fock space of oscillator modes has indefinite metric, a subsidiary condition is imposed to leave the physical subspace free from the negative norm states,

$$Q|\chi_{\text{phys.}}\rangle = 0 \ , \tag{3.3}$$

where Q is the first quantized Becchi–Rouet–Stora–Tyutin (BRST) charge associated with the reparametrization (conformal) invariance of the world sheet. The BRST operator is known to the mathematicians as the operator which computes the cohomology of a Lie (gauge) algebra. In string systems, the Lie algebra is the Virasoro algebra. Equation (3.3) is the statement that physical states belong to a given cohomology class of the Lie algebra up to equivalence (gauge transformation), that is,

$$|\chi_{\text{phys.}}\rangle = |\chi'_{\text{phys.}}\rangle + Q|\lambda\rangle \ , \tag{3.4}$$

for some state $|\lambda\rangle$.

It can be shown that the nilpotency of the BRST charge is equivalent to the Virasoro algebra with vanishing anomaly. Such a case occurs only for $D = 26$ and $\alpha(0) = 1$ ($\alpha(0)$ is the intercept of a Regge trajectory) for the bosonic string. The equivalence classes of solutions of (3.3) with definite ghost number N form the Nth cohomology group of the Virasoro algebra V with given representation $R : H^N(V; R)$. For the bosonic string, it can be shown that physical states belong to cohomology classes of ghost number $N = -\frac{1}{2}$.

It is useful to note that the field functional Φ can be viewed as a Dirac bracket,

$$\Phi = \langle X_\mu(\sigma), c(\sigma), b(\sigma)|\Phi\rangle \quad , \tag{3.5}$$

where the ket $|\Phi\rangle$ can be expanded as,

$$|\Phi\rangle = \sum_s |s\rangle \Phi_s \quad , \tag{3.6}$$

in which $|s\rangle$ denotes a state of the string. The coefficients Φ_s completely characterize the state of the string. BRST-invariant actions for string fields have also been constructed.[16-19] As in ordinary gauge theories, quantization is carried out in a particular gauge. Without going into the details of the gauge fixing procedure, here we simply give an example of a BRST-invariant gauge fixed action which has the following form (we now use ϕ_s instead of Φ_s to emphasize that the action is gauge fixed),

$$S[\phi_s] = S_0[\phi_s] + S_I[\phi_s] \quad , \tag{3.7}$$

where,

$$S_0[\phi_s] = \frac{1}{2} \sum_{s_1 s_2} \Delta^{-1}_{s_1 s_2} \phi_{s_2} \phi_{s_1} \quad , \tag{3.8}$$

and,

$$S_I[\phi_s] = \frac{g}{3} \sum_{s_1 s_2 s_3} V_{s_1 s_2 s_3} \phi_{s_3} \phi_{s_2} \phi_{s_1} \quad . \tag{3.9}$$

The action described by (3.7) – (3.9) is Witten's gauge fixed action.[17-19] Other actions have also been considered[16] which include a quartic interaction in addition to the cubic term. Also, the vertex in (3.9) is in Witten's configuration.

String field correlation functions can now be obtained from the following path-integral formula,

$$\langle\langle \phi_{s_1} \cdots \phi_{s_n} \rangle\rangle \equiv \frac{\int [\prod_s d\phi_s] \, (\phi_{s_1} \cdots \phi_{s_n}) \exp iS[\phi_s]}{\int [\prod_s d\phi_s] \exp iS[\phi_s]} \quad , \tag{3.10}$$

which can also be obtained in terms of a generating functional,

$$\langle\langle \phi_{s_1} \cdots \phi_{s_n} \rangle\rangle = \frac{i^{-n} \delta^n}{\delta J_{s_n} \cdots \delta J_{s_1}} Z[J] \bigg|_{J_s = 0} \quad , \tag{3.11}$$

where,

$$Z[J] = N^{-1} \exp iS_I \left[\frac{-i\delta}{\delta J_s} \right] Z_0[J] \quad , \tag{3.12}$$

in which,

$$Z_0[J] = \exp \frac{i}{2} \sum_{s_1 s_2} J_{s_2} \Delta_{s_2 s_1} J_{s_1} \quad . \tag{3.13}$$

Note that in Eq. (3.10) we assumed that the functional integral measure is BRST-invariant.

A useful expression for the free string propagator $\Delta_{s_1 s_2}$ is the following,

$$\Delta_{s_1 s_2} = \alpha' \left\langle z_1 \left| \frac{1}{L_0 - \alpha(0) - i\alpha'\delta} \right| z_2 \right\rangle (2\pi)^D \delta(k_1 + k_2) \quad , \tag{3.14}$$

where,

$$L_0 - \alpha(0) \equiv \alpha' \left[k_1^2 + M^2 \right] \quad , \tag{3.15}$$

in which,

$$M^2 = \frac{1}{\alpha'} \left[\sum_{n=1}^{\infty} \alpha_{-n} \cdot \alpha_n + \sum_{n=1}^{\infty} n \left(c_{-n} b_n + b_{-n} c_n \right) \right] - \frac{\alpha(0)}{\alpha'} \quad . \tag{3.16}$$

Here, α' is the Regge slope. The states $|z\rangle$ in (3.14) are related to $|s\rangle$ by the following relations,

$$\phi[s] = \phi_s = \langle s|\phi\rangle = \langle z|\phi(k)\rangle = \phi_z(k) \quad . \tag{3.17}$$

Therefore,

$$\Delta_{s_1 s_2} = \langle\langle \phi_{s_1} \phi_{s_2} \rangle\rangle = \langle z_1 \left| \langle\langle |\phi(k_1)\rangle \, \langle \phi(k_2)| \rangle\rangle \right| z_2 \rangle \quad , \tag{3.18}$$

where,

$$\langle\langle |\phi(k_1)\rangle \, \langle\phi(k_2)| \rangle\rangle = \alpha' \Delta(k_1) (2\pi)^D \delta(k_1 + k_2) \quad , \tag{3.19}$$

in which,

$$\Delta(k) \equiv \frac{1}{L_0 - \alpha(0) - i\alpha'\delta} \tag{3.20}$$

is the propagator of the first quantized operator formalism.

IV. FINITE TEMPERATURE STRING FIELD THEORY

In order to see more clearly how to generalize the finite temperature field theory described in Section II to the case of string field theory, let us first consider the expression for the generating functional of the free string field theory,

$$Z_0[J] = N^{-1} \int \left[\prod_s d\phi_s\right] \exp\left\{iS_0[\phi_s] + iJ_s\phi_s\right\} \quad , \tag{4.1}$$

If one recalls the decomposition (3.17) for the string zero-mode, the functional measure in (4.1) can be rewritten as,

$$\prod_s d\phi_s = \prod_z \prod_k d\phi_z(k) \quad . \tag{4.2}$$

Therefore we may write,

$$Z_0[J] = \prod_z Z_0[J;z] \quad , \tag{4.3}$$

where,

$$Z_0[J,z] \equiv N_z^{-1} \int \prod_k d\phi_z(k) \times \exp\left\{\frac{i}{2}\sum_k \phi_z(k)\Delta^{-1}(k;z)\phi_z(-k) + i\sum_k J_z(k)\phi_z(k)\right\} \quad . \tag{4.4}$$

The above consideration clearly indicates that the free string field generating functional can be expressed as an infinite product of generating functional of ordinary free fields $\phi_z(k)$ where the masses are labelled by z. The statistical mechanics for such a case has been described in Section II. We now have a clear prescription. We first perform an effective doubling of the string field degrees of freedom by introducing unphysical fields $\tilde{\phi}_s$ and currents $\tilde{J}_s$ into the theory.[5] The total action now reads as,

$$\hat{S}_J \equiv S[\phi_s] - S^*\left[\tilde{\phi}_s\right] + J_s\phi_s - \tilde{J}_s\tilde{\phi}_s \quad . \tag{4.5}$$

Introducing the thermal doublet notation,

$$\phi_s^\alpha = \begin{bmatrix} \phi_s \\ \tilde{\phi}_s \end{bmatrix} \quad ; \quad J_s^\alpha = \begin{bmatrix} J_s \\ \tilde{J}_s \end{bmatrix} \quad , \tag{4.6}$$

one then has,

$$\langle\langle|\phi^{\alpha}(k_1)\rangle\,\langle\phi^{\beta}(k_2)|\rangle\rangle = \alpha'\,(2\pi)^D\,\delta(k_1+k_2)\Delta^{\alpha\beta}(k_1)\ , \tag{4.7}$$

where,

$$\Delta^{\alpha\beta}(k) \equiv \left[U_B(|k_0|)\frac{\tau}{L_0-\alpha(0)-i\alpha'\tau\delta}U_B(|k_0|)\right]^{\alpha\beta}\ . \tag{4.8}$$

Finite temperature string field theory correlation functions are now expressed as,

$$\langle\langle\phi_{s_1}^{\alpha_1}\cdots\phi_{s_n}^{\alpha_n}\rangle\rangle = \frac{i^{-n}\epsilon^{\alpha_1}\cdots\epsilon^{\alpha_n}\delta^n}{\delta J_{s_n}^{\alpha_n}\cdots\delta J_{s_1}^{\alpha_1}}\hat{Z}[J]\bigg|_{J=0}\ , \tag{4.9}$$

where

$$\hat{Z}[J] = \hat{N}^{-1}\exp\left\{iS_I\left[\frac{-i\delta}{\delta J_s}\right] - iS_I^*\left[\frac{i\delta}{\delta\tilde{J}_s}\right]\right\}\hat{Z}_0[J]\ , \tag{4.10}$$

in which,

$$\hat{Z}_0[J] = \exp\frac{i}{2}\sum_{s_1 s_2}[J_{s_2}\tau\Delta_{s_2 s_1}\tau J_{s_1}]^{\alpha\alpha}\ . \tag{4.11}$$

The above generating functional can be obtained from the following path integral formula with proper boundary term inducing the finite temperature mixing of ϕ_s and $\tilde{\phi}_s$:

$$\hat{Z}[J] = \hat{N}^{-1}\int\left[\prod_{\alpha=1}^{2}\prod_{s}d\phi_s^{\alpha}\right]\exp\left\{i\hat{S}_J[\phi] + \text{boundary term}\right\}\ . \tag{4.12}$$

Equations (4.9) $\sim$ (4.11) yield the finite temperature Feynman rules of our bosonic string field theory.

As a practical example of the above rules, we can compute the finite temperature expressions for the four-tachyon scattering amplitudes in the tree and one-loop approximations. Here we perform such computations making use of the operator formalism of the first quantized bosonic string (BRST formalism).

In the operator approach of the BRST covariant first quantized formalism, propagators and vertices are now operators acting on the Fock space $\mathcal{F}$ of the string modes. Correspondence with the rules of string field theory suggests the following,

$$V(k) \longrightarrow V^{\alpha\beta\gamma}(k) = \begin{cases} V(k) \ ; & \alpha = \beta = \gamma = 1 \ , \\ -V^*(k) \ ; & \alpha = \beta = \gamma = 2 \ , \\ 0 \ ; & \text{otherwise} \ , \end{cases} \tag{4.13}$$

and,

$$\Delta(p) \longrightarrow \Delta^{\alpha\beta}(p) = \left[U_B(|p_0|) \frac{\tau}{L_0 - 1 - i\alpha'\delta\tau} U_B(|p_0|) \right]^{\alpha\beta} \ , \tag{4.14}$$

in which we set the intercept $\alpha(0)$ to 1 and where we choose the convention $\alpha' = \frac{1}{2}$. The four-tachyon amplitude in the tree approximation now reads as,

$$A_4^{\text{F.T.}}(s,t) = g^2 \left\langle 0; k_1 \left| V^{111}(k_2) \Delta^{11} V^{111}(k_3) \right| 0; k_4 \right\rangle \ , \tag{4.15}$$

where the tachyon vertex is given as,

$$V(k) =: e^{ik\cdot X(1)} : \ , \tag{4.16}$$

in which,

$$X^\mu(z) = x^\mu - ip^\mu \ln z + i \sum_{n=1}^{\infty} \frac{1}{n} \left[\alpha_n^\mu z^{-n} - \alpha_{-n}^\mu z^n \right] \ . \tag{4.17}$$

Note that we have defined the Mandelstam variables (s,t) as,

$$s \equiv -(k_1 + k_2)^2 \ , \tag{4.18}$$

$$t \equiv -(k_2 + k_3)^2 \ . \tag{4.19}$$

The following parametrized form for the propagator is also useful,

$$\Delta^{11}(p) = \int_0^1 dx \left[\frac{e^{\beta|p_0|}}{(e^{\beta|p_0|} - 1)} x^{L_0 - 2 - i\alpha'\delta} - \frac{1}{(e^{\beta|p_0|} - 1)} x^{L_0 - 2 + i\alpha'\delta} \right] \ . \tag{4.20}$$

Insertion of the above propagator and vertex operators into Eq. (4.15) yields the following,

$$A_4^{\text{F.T.}}(s,t) = g^2 \int_0^1 dx \left\{ \left[\frac{e^{\beta|k_{0_1} + k_{0_2}|}}{(e^{\beta|k_{0_1} + k_{0_2}|} - 1)} x^{-\alpha(s) - i\alpha'\delta - 1} \right.\right. \\ \left.\left. - \frac{1}{(e^{\beta|k_{0_1} + k_{0_2}|} - 1)} x^{-\alpha(s) + i\alpha'\delta - 1} \right] (1 - x)^{-\alpha(t) - 1} \right\} \ , \tag{4.21}$$

where use has been made of coherent states techniques and where $\alpha(s) = \alpha(0) + \alpha' s$.

Recalling that,

$$B(a,b) = \int_0^1 dx\, x^{a-1} (1-x)^{b-1} \quad , \tag{4.22}$$

where $B(a,b)$ is the Euler beta function, as well as,

$$2\pi i \delta(y) = \frac{1}{y - i\delta} - \frac{1}{y + i\delta} \quad , \tag{4.23}$$

one obtains finally,

$$\begin{aligned} A_4^{\text{F.T.}}(s,t) = g^2 \Bigg[& B(-\alpha(s), -\alpha(t)) + \frac{2\pi i}{\left(e^{\beta |k_{0_1} + k_{0_2}|} - 1\right)} \\ & \times \sum_{n=0}^{\infty} \frac{1}{n!} (\alpha(t)+1)(\alpha(t)+2) \cdots (\alpha(t)+n)\, \delta(\alpha(s) - n) \Bigg] \quad , \end{aligned} \tag{4.24}$$

a result which has also been obtained in the context of string field theory.[5] Next we compute the planar one-loop correction to (4.24). The one-loop diagram is obtained by sewing together two external legs of the tree amplitude. One has,

$$B_4^{\text{F.T.}}(k_1 k_2 k_3 k_4) = g^4 \int d^D p \, \text{Tr} \left[\Delta^{11} V(k_1) \Delta^{11} V(k_2) \Delta^{11}(k_3) \Delta^{11} V(k_4) \right] \quad . \tag{4.25}$$

Making use of the following formula for traces,

$$\text{Tr}(A) = \int \frac{dz\, dz^*}{\pi} e^{-|z|^2} \langle z | A | z \rangle \quad , \tag{4.26}$$

where $|z\rangle$ is a coherent state, we have,

$$\begin{aligned} B_4^{\text{F.T.}}(k_1 k_2 k_3 k_4) = g^4 & \int d^D p \int_0^1 \frac{dx_1 dx_2 dx_3 dx_4}{w^2} \\ \times & \frac{\text{tr}_{\,\text{gh}} \left(w^{n[c_{-n} b_n + b_{-n} c_n]} \right)}{\left(1 - e^{-\beta |p_{0_1}|}\right)\left(1 - e^{-\beta |p_{0_2}|}\right)\left(1 - e^{-\beta |p_{0_3}|}\right)\left(1 - e^{-\beta |p_{0_4}|}\right)} \\ \times & \prod_{n=1}^{\infty} \prod_{\mu=0}^{D-1} \int \frac{dz\, dz^*}{\pi} e^{-|z|^2} \left\langle z \left| \prod_{r=1}^{4} \exp\left[\frac{k_{r\mu} \alpha^{\mu}_{-n}}{n} \rho_r^n \right] \exp\left[\frac{-k_{r\mu} \alpha^{\mu}_{n}}{n} \rho_r^{-n} \right] \right| w^n z \right\rangle \end{aligned}$$

$$\times \left[x_1^{\sigma_1} x_2^{\sigma_2} x_3^{\sigma_3} x_4^{\sigma_3} + \left(\sum_{i=1}^{4} e^{-\beta|p_{0_i}|} x_1^{\sigma_1} \cdots x_i^{\sigma_i^*} \cdots x_4^{\sigma_4} \right) \right.$$

$$+ \left(\sum_{i<j} e^{-\beta|p_{0_i}|} x_1^{\sigma_1} \cdots x_i^{\sigma_i^*} \cdots x_j^{\sigma_j^*} \cdots x_4^{\sigma_4} \right)$$

$$+ \left(\sum_{\substack{i=1 \\ i \neq j,k,\ell}}^{4} e^{-\beta|p_{0_j}|} e^{-\beta|p_{0_k}|} e^{-\beta|p_{0_\ell}|} x_1^{\sigma_1^*} \cdots x_i^{\sigma_i} \cdots x_4^{\sigma_4^*} \right)$$

$$\left. + e^{-\beta|p_{0_1}|} e^{-\beta|p_{0_2}|} e^{-\beta|p_{0_3}|} e^{-\beta|p_{0_4}|} x_1^{\sigma_1^*} x_2^{\sigma_2^*} x_3^{\sigma_3^*} x_4^{\sigma_4^*} \right] , \tag{4.27}$$

where we defined,

$$\rho_r \equiv x_1 \cdots x_r \quad ; \qquad w \equiv \rho_4 \quad ;$$

and

$$\sigma_i \equiv \alpha' p_i^2 - i\alpha' \delta_i \quad , \tag{4.28}$$

as well as,

$$p_i^\mu \equiv p^\mu - (k_1 + k_2 + \cdots + k_{i-1})^\mu \quad . \tag{4.29}$$

Again making use of coherent states techniques and the formula for complex Gaussian integration,

$$\int \frac{dz\, dz^*}{\pi} e^{-c|z|^2} e^{(az+bz^*)} = \frac{1}{c} e^{\frac{ab}{c}} \quad , \tag{4.30}$$

together with the trace over ghost modes,

$$tr_{\text{gh}} \left(w^{n[c_{-n}b_n + b_{-n}c_n]} \right) = \prod_{n=1}^{\infty} (1 - w^n)^2 \equiv [f(w)]^2 \quad , \tag{4.31}$$

one obtains finally,

$$B_4^{\text{F.T.}}(k_1 k_2 k_3 k_4) = g^4 \int d^D p \int_0^1 \frac{dx_1 dx_2 dx_3 dx_4}{w^2} [f(w)]^{2-D}$$

$$\times \exp\left[\sum_{r<s} k_r \cdot k_s \ln\left(\psi'_{rs}\right) \right]$$

$$\times \left(1 - e^{-\beta|p_{0_1}|}\right)^{-1} \left(1 - e^{-\beta|p_{0_2}|}\right)^{-1} \left(1 - e^{-\beta|p_{0_3}|}\right)^{-1} \left(1 - e^{-\beta|p_{0_4}|}\right)^{-1}$$

$$
\begin{aligned}
\times \Bigg\{ & x_1^{\sigma_1} x_2^{\sigma_2} x_3^{\sigma_3} x_4^{\sigma_4} + \left(\sum_{i=1}^{4} e^{-\beta|p_{0_i}|} x_1^{\sigma_1} \cdots x_i^{\sigma_i^*} \cdots x_4^{\sigma_4} \right) \\
+ & \left(\sum_{i<j} e^{-\beta|p_{0_i}|} e^{-\beta|p_{0_j}|} x_1^{\sigma_1} \cdots x_i^{\sigma_i^*} \cdots x_j^{\sigma_j^*} \cdots x_4^{\sigma_4} \right) \\
+ & \left(\sum_{\substack{i=1 \\ i\neq j,\,k,\,\ell}}^{4} e^{-\beta|p_{0_j}|} e^{-\beta|p_{0_k}|} e^{-\beta|p_{0_\ell}|} x_1^{\sigma_1^*} \cdots x_i^{\sigma_i} \cdots x_4^{\sigma_4^*} \right) \\
+ & e^{-\beta|p_{0_1}|} e^{-\beta|p_{0_2}|} e^{-\beta|p_{0_3}|} e^{-\beta|p_{0_4}|} x_1^{\sigma_1^*} x_2^{\sigma_2^*} x_3^{\sigma_3^*} x_4^{\sigma_4^*} \Bigg\} \quad , \qquad (4.32)
\end{aligned}
$$

where,

$$
\ln(\psi'_{rs}) \equiv - \sum_{n=1}^{\infty} \frac{\left[c_{sr}^n + (w/c_{sr})^n - 2w^n \right]}{n(1-w^n)} \quad , \qquad (4.33)
$$

and,

$$
c_{sr} \equiv \rho_s/\rho_r = x_{r+1} \cdots x_s \quad . \qquad (4.34)
$$

At zero temperature, only the $x_1^{\sigma_1} x_2^{\sigma_2} x_3^{\sigma_3} x_4^{\sigma_4}$-term survives and the momentum integration can be performed explicitly. A more detailed analysis of the above results is beyond the scope of this presentation. Our purpose here was to show how temperature comes in through explicit computations.

V. THERMODYNAMICAL PROBLEMS IN STRING SYSTEMS

The finite temperature formalism developed in the previous section enables us to address in more generality (that is, for interacting systems) some physical issues previously discussed in the ideal string gas picture, exclusively. Among such issues, the problem of string cosmology and its relation to the maximum temperature may be the most important one.

If is a general feature of string (dual) systems that the degeneracy of states with mass M increases exponentially for large M. Consequently, the density of states in mass space is of generic form,

$$
\rho(M) = C M^{-a} \exp(bM) \quad ; \qquad M \to \infty \quad , \qquad (5.1)
$$

with model-dependent constants a and b.

The canonical partition function for an ideal gas of string excitations can then be written as,

$$\ln Z \simeq \frac{V}{(2\pi)^{D-1}} \int_{\eta}^{\infty} dM\, \rho(M) \int d^{D-1}k \ln \left[1 - \exp\left(-\beta\sqrt{k^2 + M^2}\right)\right]^{-1} , \tag{5.2}$$

where η is an infrared cut-off below which the density (5.1) is no longer valid.

Inserting Eq. (5.1) into (5.2) at high M, expanding the logarithm in the partition function and integrating over space momenta, one shows that,

$$\ln Z \sim F\left[\frac{TT_0}{T_0 - T}\right] , \tag{5.3}$$

in which F diverges for $T > T_0$ and where $T_0 = \frac{1}{b}$. Therefore, T_0 plays the role of a maximum (Hagedorn) temperature.

Early universe scenarios in the context of the above exponentially rising mass density were first considered by Huang and Weinberg[32] in 1970 and more recently by E. Alvarez[4] for superstring models, as well as others.[7]

For some string models, such as the closed superstring and the heterotic string, energy fluctuations becomes so large that the canonical ensemble can no longer describe the thermodynamics of the system. One then has to re-analyze the problem making use of the microcanonical ensemble, as has been done by Bowick and Wijewardhana.[7] It has been shown for the latter model that an ideal gas of purely massive string excitations has negative microcanonical specific heat. The thermodynamics is therefore similar to that of a black hole. Equilibrium between massless and massive modes for the heterotic string can then be achieved above the Hagedorn temperature provided that the energy of higher string modes is greater than a certain fraction of the total energy. Consequently, there is no true maximum temperature, at least in the ideal string gas picture.

Another approach, followed by Gleiser and Taylor,[7] is to assume that, at high temperature, only string modes with long enough lifetime contribute to the equilibrium partition function, thereby inducing an effective cut-off in the integration over the mass density and avoiding the maximum temperature problem. A resultant cosmology has been discussed by the latter authors who found a short-lived inflationary era.

Finally, perhaps not completely unrelated to the problem of string cosmology, is the possibility of a deconfining phase transition[33,34] as one reaches the Hagedorn temperature.

Simple considerations for an ideal string gas lead to the following expression for the effective string tension at finite temperature,

$$\sigma(T) = \frac{1}{2\pi\alpha'}\sqrt{1 - \frac{T^2}{T_0^2}} \quad . \tag{5.4}$$

At the Hagedorn temperature T_0, the string tension vanishes,

$$\sigma(T_0) = 0 \quad , \tag{5.5}$$

signaling a deconfining phase transition. For a more detailed discussion of this problem and its relation to the stabilization of the tachyon mode, see Ref. [34].

VI. CONCLUSIONS

In this talk, we outlined the general construction of a real-time finite temperature formalism for interacting string field theory in the context of the open bosonic string. We then showed how the finite temperature Feynman rules carry over to the first quantized (BRST-invariant) operator formalism and obtained expressions for finite temperature four-tachyon scattering amplitudes in the tree and one-loop (planar) approximation. We also briefly discussed interesting thermodynamical problems for which the above formalism may help improve the understanding of string thermodynamics by taking into account string interactions.

Finally, in view of the recent progress in the generalization of the Thermo-Field Dynamics formalism to physical situations departing from thermodynamical equilibrium in the context of ordinary field theory,[35−40] it seems very likely that similar developments can also be achieved for string field theory. This is of importance in particular for superstring cosmology where one would like to have a formalism treating the case of a non-adiabatic expansion of the universe.

REFERENCES

1. See for example, Green, M. B., Schwarz, J. H. and Witten, E., *Superstring Theory*, Vols. I-II (Cambridge University Press, Cambridge, UK, 1986).

2. Gross, D. J., Harvey, J. A., Martinec, E. and Rohm, R., *Nucl. Phys.* **B256**, 253 (1985); **B267**, 75 (1986).

3. Green, M. B. and Schwarz, J. H., *Phys. Lett.* **149B**, 117 (1984); **151B**, 21 (1985).

4. Alvarez, E., *Phys. Rev.* **D31**, 418 (1985); *Nucl. Phys.* **B269**, 596 (1986).

5. Leblanc, Y., "String Field Theory at Finite Temperature," MIT preprint CTP #1435 (1987).

6. Enqvist, K., Mohanty, S., and Nanopoulos, D. V., "Quantum Cosomology of Superstrings", University of Wisconsin-Madison preprint (1987).

7. Bowick, M. J. and Wijewardhana, L. C. R., *Phys. Rev. Lett.* **54**, 2485 (1985); Gleiser, M. and Taylor, J. G., *Phys. Lett.* **164B**, 36 (1985); Tye, S. H. H., *Phys. Lett.* **158B**, 388 (1985); Sundborg, B., *Nucl. Phys.* **B254**, 583 (1985).

8. Polyakov, A. M., *Phys. Lett.* **103B**, 207 (1981); *Phys. Lett.* **103B**, 211 (1981).

9. See for example, Bowick, M. J. and Rajeev, S. G., "String Theory as the Kähler Geometry of Loop Space," *Phys. Rev. Lett.* **58**, 535 (1987); "The Holomorphic Geometry of Closed Bosonic String Theory and $\mathcal{D}iff\ S^1/S^1$," MIT preprint CTP #1450 (1987); and references cited therein. See also, Pilch, K. and Warner, N. P., "Holomorphic Structure of Superstring Vacua," MIT preprint CTP #1457 (1987).

10. Kaku, M. and Kikkawa, K., *Phys. Rev.* **D10**, 1110 (1974); *Phys. Rev.* **D10**, 1823 (1974).

11. Cremmer, E. and Gervais, J. L., *Nucl. Phys.* **B76**, 209 (1974); *Nucl. Phys.* **B90**, 410 (1975).

12. Green, M. B. and Schwarz, J. H., *Nucl. Phys.* **B218**, 43 (1983); *Nucl. Phys.* **B243**, 475 (1984).

13. Thorn, C. B., *Nucl. Phys.* **B263**, 493 (1986).

14. Siegel, W., *Phys. Lett.* **149B**, 162 (1984); *Phys. Lett.* **151B**, 391 (1985); *Phys. Lett.* **151B**, 396 (1983).

15. Siegel, W. and Zwiebach, B., *Nucl. Phys.* **B263**, 105 (1986).

16. Hata, H., Itoh, K., Kugo, T., Kunitomo, H. and Ogawa, K., *Phys. Lett.* **172B**, 186 (1986); *Phys. Lett.* **172B**, 195 (1986); *Phys. Rev.* **D34**, 2360 (1986); *Phys. Rev.* **D35**, 1318 (1987); *Phys. Rev.* **D35**, 1356 (1987).

17. Witten, E., *Nucl. Phys.* **B268**, 253 (1986).

18. Giddings, S., Martinec, E. and Witten, E., *Phys. Lett.* **176B**, 362 (1986).

19. Thorn, C. B., "Perturbation Theory for Quantized String Fields," Institute for Advanced Study preprint, Princeton, NJ (1986).

20. Schwarz, J. H., "Review of Recent Developments in Superstring Theory," CalTech preprint (1987).

21. See Ref. [20] and references cited therein.

22. Takahaski, Y., and Umezawa, H., *Collective Phenomena* **2**, 55 (1975).

23. Matsumoto, H. *Fortsch. der Physik* **25**, 1 (1977).

24. Umezawa, H., Matsumoto, H. and Tachiki, M., *Thermo-Field Dynamics and Condensed States* (North-Holland, Amsterdam, 1982).

25. Matsumoto, H., Ojima, I. and Umezawa, H., *Ann. of Phys.* **152**, 348 (1984).

26. Matsumoto, H., Nakano, Y. and Umezawa H., *Phys. Rev.* **D29**, 1116 (1984).

27. Semenoff, G. and Umezawa, H., *Nucl. Phys.* **B220** [FS8], 196 (1983).

28. Matsumoto, H., Nakano, Y., Umezawa H., Mancini, F. and Marinaro, M., *Prog. Theor. Phys.* **70**, 599 (1983).

29. Matsumoto, H., Nakano, Y. and Umezawa, H., *J. Math. Phys.* **25**, 3076 (1984).

30. Matsumoto, H., in *Progress in Quantum Field Theory*, Ezawa, H. and Kamefuchi, S., eds. (North-Holland, Amsterdam, 1985).

31. For a review on Field Theory at finite temperature see: Landsman, N. P. and van Weert, Ch., *Phys. Rep.* **145**, 141 (1987).

32. Huang, K. and Weinberg, S., *Phys. Rev. Lett.* **25**, 895 (1970).

33. Pisarski, R. D. and Alvarez, O., *Phys. Rev.* **D26**, 3735 (1982).

34. Olesen, P., *Phys. Lett.* **160B**, 144 (1985); *Phys. Lett.* **160B**, 408 (1985); *Phys. Lett.* **168B**, 220 (1986); *Nucl. Phys.* **B267**, 539 (1986).

35. Arimitsu, T and Umezawa, H., *Prog. Theor. Phys.* **74**, 429 (1985); *Prog. Theor. Phys.* **77**, No. 1 (1987).

36. Arimitsu, T. and Umezawa, H., *J. Phys. Soc.* (Japan) **55**, **1475** (1986).

37. Umezawa, H., Yamanaka, Y., Hardman, I. and Arimitsu, T., "The Thermally Dissipative Free Field and Canonical Formalism," University of Alberta preprint (1986).

38. Hardman, I., Umezawa, H. and Y. Yamanaka, "Time-Dependent Canonical Formalism of Thermally Dissipative Fields and Renormalization Scheme", University of Alberta preprint (1987).

39. Umezawa, H. and Yamanaka, Y., "Time-Dependent Non-Equilibrium Thermo-Field Dynamics and Self-Consistent Renormalization." University of Alberta preprint (1987).

40. Umezawa, H. and Yamanaka, Y., "Time-Dependent Non-Equilibrium Thermo-Field Dynamics of Type–II Fields," University of Alberta preprint (1987).

ANOMALIES IN D=2 FIELD THEORIES WITH (1,0) WORLD-SHEET SUPERSYMMETRY

Yong-Shi Wu
Department of Physics, University of Utah
Salt Lake City, Utah 84112, U.S.A.

ABSTRACT

The first half of the paper is devoted to a brief but comprehensive review which highlights anomalies in gauge theories and sigma models, with emphasis on the cohomological point of view and methods. In the second half we present a new analysis of possible world-sheet supersymmetry in D=2 (1,0) models anomaly. Our results show no indication for such anomaly and we argue that there should be a generalization of (family) index theorem in the supersymmetrical cases. Critical dimensions of (1,0) super-chiral models (with group manifolds as target spaces) are also discussed.

INTRODUCTION

In recent years the resurgence of string theories[1] has aroused interests in D=2 field theories, especially in those which are conformally and/or supersymmetrically invariant. This is because the first quantization of a string is essentially a quantized nonlinear sigma model[2] on the world-sheet (swept by the string) with space-time as target space, either flat or compactified. On the other hand, the low-energy effective theory of strings can be formulated in terms of[3] and extracted from[4] the quantum nonlinear sigma model on the world-sheet. From the point of view of string theories, two-dimensionality is more fundamental. It is generally believed that space-time physics has its origin in D=2 world-sheet physics, although the connection between them is only partially understood. We have the feeling that the proper formulation of second quantized strings probably needs more and deeper understanding of D=2 field theories, specially of those on arbitrary Riemann (or super-Reimann) surfaces.

An important link between world-sheet physics and space-time physics has been given by anomalies in D=2 quantum field theories, which describe the propagation of a string in certain space-time backgrounds. For example, the well-known critical dimension and modular invariance of a string theory[1] are both actually anomaly-free conditions for the D=2 field theory. The discovery of any new anomaly would impose extra constraints on string model construction and would provide new insight into space-time physics.

Since supersymmetry turns out to be an important ingredient of construction of realistic string models, it is interesting to see whether or not there is an anomaly for it. Although in all presently known models there is no anomaly associated with supersymmetry, a deep understanding of this fact is in demand, because none of the usual regularizations respects supersymmetry. Different approaches to this problem have been suggested and worked out for theories in higher dimensions,[5] where a direct comparison[6] of the results is often very hard to make. In D=2, the situation is much simpler so that a comparison of different approaches[7] is possible.

Here I will report on a recent study in collaboration with Z. Wang[8] on (1,0) world-sheet supersymmetry. In addition to checking the consistency of various approaches, the main motivation behind our work is to gain some insights into the problem of whether there is a generalization of family index theorem in the space of super-connections. It is well-known that the validity of usual cohomology argument for chiral gauge anomaly is assured by the family index theorem. If in supersymmetric theories a generalized cohomology argument can also reproduce the anomalies obtained by perturbative calculations or other means, there should be a sort of generalized family index theorem which validates the generalized cohomology argument. I will also represent a recent result obtained by F. Yu and myself[9] on critical dimensions of (1,0) super-chiral models, which reflect the separate cancellation of super-conformal anomalies in left and right moving sectors (for a closed string).

BRIEF REVIEW OF ANOMALIES

If a symmetry in a classical theory is violated merely by quantization (or by quantum effects), then we say the quantized theory has an anomaly for that symmetry. One indication for the possible existence of an anomaly is that no regularization scheme respects all symmetries in the corresponding classical theory.

Although it was thought originally that the occurrence of anomaly is abnormal, now we know its occurrence is quite often and common for many systems.

Anomalies can occur either for a global (non-gauge) symmetry or for a local (or gauge) symmetry. (The Noether current associated with a global symmetry does not couple to a gauge field, but that associated with a local symmetry normally does). In four dimensions, the well-known axial (or $U_A(1)$) anomaly[10] and trace anomaly[11] belong to the former case. Also, the breaking of scale invariance in perturbation theory, in the form of anomalous dimensions and Callan-Symanzik equations[12], can be interpreted as anomaly for scale symmetry[13] in classical massless theories. On the other hand, chiral gauge anomalies[14] and gravitational anomalies[15] in D=2n and D=4k+2, respectively, are well-known examples of anomalous gauge symmetries. The conformal anomaly in D=2, in contrast to that in D=4, has many features which are shared by gauge anomalies.

According to the manner an anomaly arises, we can have either a local anomaly or a global anomaly. For a local anomaly, the anomaly occurs for a continuous symmetry under infinitesimal transformations and, therefore, appears as an extra term in the would-be conservation equation for the corresponding Noether current. For a global anomaly, it occurs for a discrete symmetry or for a continuous symmetry under large transformations (i.e., those which are not connected to the identity transformation). Sometimes a local (global) anomaly is also referred to as a (non-) perturbative anomaly, because it can(not) be obtained by perturbative calculations. Well-known examples for global anomalies include Witten's SU(2) anomaly in D=4[16], Redlich's parity anomaly in D=3[17] and modular non-invariance in D=2[18], which can be viewed as a sort of global gravitational anomaly on world-sheet.

Normally we can always compute anomalies by means of field-theoretical methods such as direct diagrammatic calculations[10], or Fujikawa's path integral method[19], the heat kernal method[20] and so on. The amazing thing discovered in recent years is that there is a very deep and fundamental relationship between certain anomalies and topology (index theorem, family index theorem and cohomology). Therefore, such anomalies as gauge anomalies and gravitational anomalies, either local or global ones, and D=2 conformal anomalies can be derived or inferred from topological methods.[21,22]

However, after obtaining a non-vanishing anomaly for a certain symmetry, we still have to check whether it can be removed by adding local counterterms in the Lagrangian or not. Only when it is not removable, the symmetry under consideration is anomalous. However, when the anomaly is removable by local counterterms, the symmetry is not really anomalous. We need to check whether or not these local counterterms violate other classical symmetries in the theory. If some other is violated that symmetry becomes anomalous. In other words, anomaly may be shifted from one symmetry to another by adding local counterterms in the Lagrangian. This fact is very important for understanding why one can derive the same critical dimension for a string theory from the anomaly-free conditions for different symmetries. In conformity with this, an anomalous theory should be defined as a theory of which <u>not all</u> classical symmetries can survive quantization; that which symmetry becomes anomalous may depend on the regularization and/or the local counterterms. A well-known example is three-dimensional gauge theories with fermions in which the global gauge anomaly may be shifted to parity anomaly. A detailed analysis of the relationship between anomalies and regularizations in such theories has been presented in [23].

Undoubtedly anomalies play a very important role in physics, since it represents the third way to break a symmetry, in addition to the well-known explicit breaking and spontaneous (or dynamical) symmetry breaking. Normally the anomaly for a global symmetry is not harmful and has physical consequences. For example, the axial anomaly gives us an interpretation of $\pi^\circ \to 2\gamma$ which was theoretically used to determine the number of colors for quark (=3)[24]. The scale anomaly gives rise to the Callan-Symanzik equations which determine the short-distance behavior of the quantum field theory under consideration[12]. However, the anomaly for local gauge symmetry is usually considered as so harmful as to completely ruin the internal consistency of the quantized theory[25]. Therefore, the anomaly-free conditions for local gauge symmetries are used as essential constraints for consistency on the model building. This has been conventional practice in model building, both in (electroweak and grand) unified theories and in string theories. Nevertheless, here I would like to emphasize that theoretically in principle it is still an open question whether an anomalous gauge theory can be consistently quantized or not[26].

For readers who are not familiar with anomalies, the following two sections serve as a concrete presentation of above general and abstract statements. We will be restricted ourselves to gauge anomalies for local symmetries.

GAUGE ANOMALIES AND COHOMOLOGICAL METHOD

Consider a Yang-Mill theory with fermions

$$L = L_{YM} + i\ \Psi(x)\gamma^{\mu}(\partial_{\mu} - i\ A^{a}_{\mu}\ \lambda^{a})\Psi(x) \qquad (1)$$

where $\Psi(x)$ belongs to an irreducible spinor representation in D-dimensional Euclidean space-time and a representation of Lie group G whose generators are denoted by λ^a. Note that when D=even, $\Psi(x)$ must be Weyl fermions. First integrate fermions with fixed gauge background and write the partition function as

$$Z = \int DA\ \exp\ \{-\ S_{YM}[A] + i\ W_{eff}[A]\} \qquad (2)$$

$$Z_{eff}[A] \equiv \exp\{i W_{eff}[A]\} = \int D\Psi D\bar{\Psi} \exp\{- i\int d^Dx \bar{\Psi} \gamma^\mu D_\mu \Psi\} \quad (3)$$

Now whether the path integral quantization preserves gauge invariance depends on whether

$$Z_{eff}[A^g] = Z_{eff}[A] \qquad \text{for } A^g \equiv g^{-1}Ag + g^{-1}dg \quad (4)$$

(Here A^g represents the gauge-transformed of A by g(x)). Or

$$\Delta W_{eff}[A] \equiv W_{eff}[A^g] - W_{eff}[A] \overset{?}{=} 0 \ (\text{mod } 2\pi) \quad (5)$$

Here Δ can be understood as the coboundary operator in a gauge orbit.

If $\Delta W_{eff}[A] = 0$, then there is no gauge anomaly. When

$$0 \neq \Delta W_{eff}[A] = \Delta f_{loc}[A] \equiv f_{loc}[A^g] - f_{loc}[A] \quad (6)$$

where $f_{loc}[A]$ is a local functional of A in the following sense:

$$f_{loc}[A] = \int d^Dx\ f(A, \partial A, ---) \quad (7)$$

then the anomaly is removable, since $L \to L - if \equiv L'$ will lead to $W_{eff}[A] \to W'_{eff}[A] = W_{eff}[A] - f_{loc}[A]$ which satisfies $\Delta W'_{eff}[A] = 0$. So only when

$$\Delta W_{eff}[A] \neq \Delta f_{loc}[A] \quad (8)$$

we have a genuine gauge anomaly in the quantized Yang-Mills theory. In the language of cohomology, eq. (8) implies that $\Delta W_{eff}[A]$ is a 1-cocycle which is nontrivial in local cohomology on a gauge orbit.

For <u>local</u> anomaly, let us consider infinitesimal variation $g'(x) = g(x) + \delta g(x)$. This leads to, on a gauge orbit,

$$\delta(A^g) = -dv - [A^g, v], \qquad \delta v = -v^2 \quad (9)$$

for $v \equiv g(x)^{-1}dg(x)$. This can be identified with the well-known BRST transformation. Taking $g(x) = 1$, we have

$$\delta W_{eff}[A] \equiv A[A;v] = \int d^Dx\ v^a(x) G^a(x) \quad (10)$$

where $v^a(x)\lambda^a = \delta g(x)$. Here $A[A;v]$ or $G^a(x)$ is called the local anomaly. The latter appears in the would-be conservation equation as follows:

$$D_\mu J^{a\mu}(x) \equiv D_\mu \delta W_{eff}[A]/\delta A^a_\mu(x) = - G^a(x) \quad (11)$$

For $Z_{eff}[A]$ to be single-valued on a gauge orbit we must have

$$\delta A[A;v] = 0 \qquad (\text{since } \delta^2 = 0) \tag{12}$$

This leads to the Wess-Zumino consistency condition for a local anomaly[27]

$$X^a(x)G^b(y) - X^b(y)G^a(x) = f^{abc}G^c(x)\delta^{(D)}(x-y) \tag{13}$$

where $X^a(x) \equiv D_\mu[\delta/\delta A^a_\mu(x)]$. In cohomology, eq. (12) is just the 1-cocycle condition for the local anomaly $A[A;v]$.

To consider the <u>global</u> anomaly, we need to consider a large gauge transformation $g(x)$ which exists only when the gauge orbit $OG = \{g(x): S^D \to G\}$ has more than one connected component. (Here we assume that $g(x) \to g_o \epsilon G$ as $|x| \to \infty$). This in turn requires the homotopy group $\pi_o(OG) = \pi_D(G) \neq 0$. Thus, the global gauge anomaly occurs when

$W[A^g] \neq W[A] + 2n\pi$. Well-known examples include

$$D{=}4,\ G = SU(2) : \pi_4(SU(2)) = Z_2 \tag{14}$$

$$D{=}3,\ G = \text{simple}: \pi_3(G) = Z \tag{14'}$$

In a gravitational theory, the local gauge symmetry may consist of either general coordinate transformations or local Lorentz frame rotations or both, depending on whether we use the metric $g_{\mu\nu}$ or the frame $e^a{}_\mu$ or both as basic variables. Thus, a gravitational theory may have global[28] or local anomaly for either of these symmetries. It can be shown that only in D=4k+2 there may be local gravitational anomaly[15], and that it can be shifted between general coordinate symmetry and local Lorentz symmetry[29].

The cohomological point of view not only provides us with an understanding of the topological meaning of anomalies, but also indicates a possible systematic procedure for evaluating anomalies. Consider, e.g., D=2n gauge theory. Our goal is to construct a nontrivial 1-cocycle in the space of potentials A. Let us start with the following quantity in

$2n+2 \equiv D+2$ dimensions[21]:

$$\Omega_{2n+2}(F) = P(F^{n+1}) \equiv P(F,\text{---},F) \quad (n+1) \text{ entries} \tag{15}$$

where $F \equiv F^a\lambda^a = dA + A^2$ in terms of differential forms; P is a symmetric invariant polynomial. It is easy to verify that

$$d\Omega_{2n+2}(F) = 0 \rightarrow \Omega_{2n+2}(F) = d\omega^{\circ}_{2n+2}(A) \tag{16}$$

On the other hand, the gauge invariance of Ω_{2n+2} leads to $\delta\Omega_{2n+2}(F) = 0$. Applying δ to eq. (16) and using $d\delta + \delta d = 0$, we have

$$d(\delta\omega^{\circ}_{2n+1}) = 0 \rightarrow \delta\omega^{1}_{2n+1} = -d\omega^{1}_{2n}(A;v) \tag{17}$$

Since $\delta^2 = 0$, applying δ again leads to

$$d(\delta\omega^{1}_{2n}) = 0 \rightarrow \delta\omega^{1}_{2n} = -d\omega^{2}_{2n-1}(A;v) \tag{18}$$

Integrating over D=2n space-time M, from Stokes' theorem we have

$$\delta\int_M \omega^{1}_{2n}(A;v) = 0 \tag{19}$$

Thus, $\int_M\omega^{1}_{2n}(A;v)$ is a 1-cocycle in the space of A. The cohomological argument identifies the D=2n gauge anomaly to be

$$\delta W_{eff}[A] \equiv A[A;v] = \int_M\omega^{1}_{2n}(A;v) \tag{20}$$

However, the cohomological argument does not tell us which P_{n+1} we should start with. (Normally there are more than one symmetric invariant polynomials P_{n+1} in D=2n with $n > 2$). So we need the following rule to supplement above procedure: P_{n+1} should be that given by the Atiyah-Singer index theorem for the Dirac operator in (D+2) dimensions. (This rule is also true for the gravitational case and for gauge-gravitational mixed anomalies)[15]. Originally the rule is obtained empirically to reproduce diagrammatic results. Now it can be derived from the family index theorem in the space of A in D=2n[30].

One of the motivations for studying anomalies in supersymmetric theories is to see whether a generalized cohomological argument exists and works in supersymmetric cases. If it really exists and works, reproducing the results of field-theoretical calculations, then there should be a similar generalized (family) index theorem working in the supersymmetric cases. Mathematically, this is an open question. But some clues may be provided by certain calculations in quantum field theory!

We will be concentrated to D=2 cases which are not only very simple but also related to superstring models.

ANOMALIES IN NONLINEAR SIGMA MODELS

The first quantization of a string model is actually a quantized D=2 nonlinear sigma model on the world-sheet with space-time as target space. The simplest model is described by scalar fields $X^{\mu}(\tau,\sigma)$ on the world sheet whose points are labelled by (τ,σ) with X^{μ} to be coordinates of the target manifold N and by the world-sheet metric $g_{mn}(\tau,\sigma)$. The typical action reads

$$S = \frac{1}{2\pi}\int d\tau d\sigma \sqrt{-g}\ (\frac{1}{2}\ g^{mn}\partial_m X^{\nu}\partial_n X^{\mu} G_{\mu\nu}(X) + \ldots)$$

where $\partial_m = (\partial_{\tau},\partial_{U})$, $G_{\mu\nu}(X)$ is the metric of the target space (or space-time). From this example it is easy to see that such a sigma model has generally two kinds of classical symmetries: i.e.

(1) World-sheet symmetries, such as reparametrization invariance $(\tau,\sigma) \to (\tau',\sigma')$ and conformal symmetry for g_{mn};

(2) Space-time (or target space) symmetries, such as isometry of the space-time background $G_{\mu\nu}(X)$; for flat space-time background it is the Poincare or Lorentz symmetry for X^{μ}.

If there are gauge fields on the world sheet or gauge background in the space-time, one will have also gauge invariance on world-sheet or in space-time, respectively.

Each symmetry mentioned above may become anomalous after quantization. The world-sheet anomalies are similar to the anomalies discussed in preceding sections. The space-time (or target space) anomalies are special to sigma models and, therefore, are usually referred to as sigma-model anomalies[31]. Conceptually the distinction of these two kinds of anomalies is important, although in most papers the distinction is not explicitly mentioned.

For superstrings, there are two different approaches[1]. In the Neveu-Schwartz-Ramond approach, one has explicit world-sheet supersymmetry and super-Weyl invariance at the classical level. The space-time supersymmetry in this model is derived from the GSO projection[32] or, equivalently, from the summation over all possible spin structures on the world sheet[33]. In the Green-Schwartz approach, one has explicit space-time supersymmetry without mentioning any world-sheet supersymmetry. In the following we will be concentrated on the first (world-sheet supersymmetry) approach. We will consider only closed strings. The D=2 world-sheet supersymmetries can be classified by two integers (p.q.), since the left and right moving sectors are independent to each other. The simplest one is the (1,0) supersymmetry[34]; higher supersymmetries can be obtained by imposing certain restrictions on the space-time backgrounds in the action[35]. The world-sheet gravitational, Lorentz and super-Weyl anomalies in the (1,0) models have been discussed in the literature[36,37], but no one has explicitly considered anomalies associated with the (1,0) world-sheet supersymmetry. This is the subject of the next section.

ABSENCE OF SUSY ANOMALIES IN GLOBAL (1,0) SUSY GAUGE THEORIES

We adopt the notations for the (1,0) flat superspace used by Ovrut and his collaborators[38]. (See also ref. [39]). The superspace coordinates are $Z^M = (X^{\pm}, \theta)$ where θ is a Weyl-Majorana spinor in D=2. The (1,0) supercharge, $Q = i(\frac{\partial}{\partial\theta} - i\theta\frac{\partial}{\partial x^+})$, satisfying $\{Q,Q\} = 2i\frac{\partial}{\partial x^+}$. The super-covariant derivatives are given by $D_M = (D_{\pm}, D_\theta)$: $D_{\pm} = \frac{\partial}{\partial x^+}$, $D_\theta = \frac{\partial}{\partial\theta} + i\theta\frac{\partial}{\partial x^+}$. In the world-sheet superspace, one can define scalar superfield

$\Phi(X^M,\theta) = \phi(x) + \theta\lambda(x)$ spinor superfield $\Psi(X^M,\theta) = \psi(x) + \theta F(x)$ and the potential super 1-form. $A = e^A A_A^a (iT^a) \equiv e^A A_A(x,\theta)$ (21)

where $e^A \equiv dz^M e_M^A$ are frame super 1-forms. With appropriate constraints imposed and the Bianchi identities solved, only the fields χ_θ^a, $V_\pm^a$, χ_-^a defined as follows are independent components:

$$A_\theta^a = i(\chi_\theta^a + \theta V_+^a),\ A_-^a = V_-^a + i\theta(\chi_-^a + D_-\chi_\theta^a) \tag{22}$$

SUSY transformation is given by $\delta_S(\epsilon) = i\epsilon Q$ and the generalized gauge transformation for A_A reads

$$\delta_G(\Lambda)A_A = - D_A\Lambda + [A_A,\Lambda] \tag{23}$$

where $\Lambda(X^\mu,\theta)$ is a scalar superfield taking values in Lie group G. The gauge and SUSY invariant action for this D=2 field theory is given by

$$S = \int d^2x d\theta\ \{-\mathrm{Tr}F_{-\theta}D_\theta F_{-\theta} + D_\theta\Phi^a D_-\Phi^a - \Psi^a(D_\theta)_{ab}\Psi^b\} \tag{24}$$

where $F_{-\theta}^a = D_-A_\theta^a - D_\theta A_-^a + [A_-,A_\theta]^a$ and $D_A = D_A - A_A$. First integrating over Φ^a and Ψ^a, one obtains the effective action

$$W_{eff}[A] = \int D\Phi^a D\Psi^a \exp\{i\int d^2x d\theta [D_\theta\Phi^a D_-\Phi^a - \Psi^a(D_\theta)_{ab}\Psi^b]\} \tag{25}$$

The gauge and SUSY anomalies are respectively

$$A_G[A;\Lambda] = \delta_G(\Lambda)W_{eff}[A],\ A_S[A;\epsilon] = \delta_S(\epsilon)W_{eff}[A] \tag{26}$$

In ref. [37] the gauge anomaly is obtained by a generalized cohomological argument as follows

$$A_G[A;\Lambda] = \mathrm{Tr}\int d^2x d\theta\ [\Lambda(D_-A_\theta - D_\theta A_-)] \tag{27}$$

without mentioning $A_S[A;\epsilon]$.

In order to prove the absence of SUSY anomaly in this model we use a technique which we invented before[6], i.e. study the anomalies in the Wess-Zumino gauge and compare them with the above $A_G[A;\Lambda]$. In this model the Wess-Zumino gauge is given by χ_θ=0. So the physical components are $(V^a_\pm, \chi^a_-)$.

The usual gauge transformation in the W-Z gauge is

$$\delta_g(\alpha)V^a_\pm = -D_\pm\alpha^a + [V_\pm,\alpha]^a, \ \delta_g(\alpha)\chi^a_- = [\chi_-,\alpha^a] \tag{28}$$

where the parameter $\alpha(x) = \alpha^a(x)\lambda^a$ depends on x only. The SUSY transformation in the W-Z gauge is given by

$$\delta_s(\epsilon)V_+ = 0, \ \delta_s(\epsilon)V_- = 0, \ \delta_s(\epsilon)\chi_- = -\epsilon F_{+-} \tag{29}$$

The commutation relations among these transformations read

$$\begin{cases} [\delta_s(\rho), \ \delta_s(\epsilon)] = \delta_g(2i\epsilon\rho V_+) + 2i\epsilon\rho D_+ \\ [\delta_g(\alpha), \ \delta_g(\beta)] = \delta_g([\alpha,\beta]), \ [\delta_g(\alpha), \ \delta_s(\epsilon)] = 0 \end{cases} \tag{30}$$

Therefore the consistency conditions for anomalies in W-Z gauge are given by

$$\begin{cases} \delta_s(\rho)A_s[\tilde{A};\epsilon] - \delta_s(\epsilon)A_s[\tilde{A};\rho] = A_g[\tilde{A};2i\epsilon\rho V_+] \\ \delta_g(\beta)A_g[\tilde{A};\alpha] - \delta_g(\alpha)A_g[\tilde{A};\beta] = A_g[\tilde{A};[\alpha,\beta]) \\ \delta_g(\alpha)A_s[\tilde{A};\epsilon] - \delta_s(\epsilon)A_g[\tilde{A};\alpha] = 0 \end{cases} \tag{31}$$

where $\tilde{A} = A\big|_{\chi_\theta = 0}$ is the superconnection in W-Z gauge, and

$$A_g(\tilde{A};\alpha) = \delta_g(\alpha)W_{eff}[\tilde{A}], \ A_s(\tilde{A};\epsilon) = \delta_s(\epsilon)W_{eff}[\tilde{A}] \tag{32}$$

To find the solution to this set of consistency conditions we note that one must have the usual gauge anomaly

$$A_g(\tilde{A};\alpha) = \int d^2x \ \mathrm{Tr}[\alpha(\partial_-V_+ - \partial_+V_-)] \tag{33}$$

since the gauge transformation $\delta_g(\alpha)$ is the same as usual. Given this, by trial and error we find

$$A_s[\tilde{A};\epsilon] = \int d^2x \ \mathrm{Tr}(\epsilon V_+\chi_-) \tag{34}$$

It is straight forward to check that the above A_s together with A_g satisfy the consistency conditions (31).

Now we come to the key point of our proof. It is well-known that the SUSY transformation (29) preserving the W-Z gauge is actually the genuine SUSY transformation ($\delta_S(\epsilon) = i\epsilon Q$) followed by a compensating generalized gauge transformation (23) with $\Lambda = -i\theta\epsilon V_+$:

$$\delta_s(\epsilon) = \delta_S(\epsilon) + \delta_G(\Lambda = -i\theta\epsilon V_+) \tag{35}$$

(The second term is needed since $\delta_S(\epsilon)$ does not generally preserve the W-Z gauge). Applying this equation to $W_{eff}[\tilde{A}]$ we obtain

$$A_s[\tilde{A};\epsilon] = A_S[\tilde{A};\epsilon] + A_G[\tilde{A};\Lambda = -i\theta\epsilon V_+] \tag{36}$$

It is easy to verify from eqs.(27), (33) and (34) that

$$A_G[\tilde{A};\Lambda = \alpha] = A_g[\tilde{A};\alpha] \tag{37}$$

$$A_G[\tilde{A};\Lambda = -i\theta\epsilon V_+] = A_s[\tilde{A};\epsilon]$$

Therefore,

$$A_S[\tilde{A};\epsilon] \equiv \delta_S(\epsilon)W_{eff}[\tilde{A}] = 0 \tag{38}$$

This means that for A in the Wess-Zumino gauge, there is no genuine supersymmetry anomaly. Since the Wess-Zumino gauge can be reached by an appropriate generalized gauge transformation (23) and the latter gives rise to an anomaly $A_G[A;\Lambda]$ which is SUSY invariant, one can conclude that $A_S[A;\epsilon] = 0$ for a generic A. Namely there is no supersymmetry anomaly at all in this model. A similar result has been obtained for the (1,0) locally supersymmetric (supergravity) case that if ordinary gravitational anomaly is cancelled, then there is no local supersymmetry anomaly either[40]. Because of the space limit, we are not able to present it here.

CRITICAL DIMENSIONS OF (1,0) SUPER-CHIRAL MODELS

Here (1,0) superchiral models refers to sigma models with group manifolds (plus some flat or toroidal space) as target spaces and with (1,0) world-sheet supersymmetry.

Superconformal anomaly cancellation conditions in (1,0) supersymmetric models with flat or toroidally compactified space-time have been discussed by Gates in this summer institute[41]. Here I am reporting the result of a joint work with Feng Yu on the compactification on group manifolds[9].

The critical dimensions of chiral models with (0,0) or (1,1) world-sheet supersymmetry have been discussed in the literature[42]. The results are respectively

$$d+n = 26 - \frac{d_G}{1+C_A/2|k|} \equiv D(k,G) \tag{39}$$

for bosonic strings in space-time $M^d \times T^n \times G$ and

$$d+n = 10 - \frac{2}{3}\frac{d_G}{1+C_A/2|k|} - \frac{1}{3} d_G \equiv D'(k;G) \tag{40}$$

for (1,1) superstrings in space-time $M^d \times T^n \times G$. Here

d_G = dim. of G, k = integer coefficient in WZW term

C_A = Casimir eigenvalue of the adjoint representation of G.

It is well-known that for the theory to be conformal anomaly free one has to add the Wess-Zumino-Witten term to the Lagrangian which represents the effect of a two-rank manifold antisymmetric tensor background or a torsion on the group.

For the heterotic (1,0) superstrings, the left and right movers can live in different manifolds. We assume that

left (bosonic) movers live in $M^d \times T^{n_L} \times G$

right (spinning) movers live in $M^d \times T^{n_R} \times G$

This can be achieved by having scalar superfields with $M^d \times G$ as target space and N_L leftons with a torus T^{n_L} as target and N_R rightons with T^{n_R} as target. For the theory to be superconformal anomaly free, we also need to add a Wess-Zumino-Witten term for scalar superfields living on G. The integer coefficient of this term (in appropriate unit) is still denoted by k. We have shown that the critical dimension of the heterotic (1,0) super-chiral model is determined by

$$d+n_L = D(k,G), \quad d+n_R = D'(k,G) \tag{41}$$

where D(k,G) and D'(k,G) are given respectively by the right side of eqs. (39) and (40). Namely the left-movers have to satisfy the constraint for (0,0) bosonic strings and the right-movers have to satisfy the constraint for (1,1) spinning strings. In turn, this implies that superconformal anomalies should be cancelled within the left-moving and right-moving sectors separately. Since here the left (right) moving sector is just half of the usual (0,0) bosonic ((1,1) spinning) models, we can infer in retrospect that in the usual (0,0) bosonic and (1,1) spinning chiral models, (super)conformal anomalies are also cancelled in either left- or right-moving sector.

Since now the level k and Lie group G have to satisfy both equations (39) and (40), the constraints on k and G become more restrictive than before. For simple Lie group G, the only solutions are

$$\begin{array}{llll} k=1 & G=SU(3) & d+n_L=24 & d+n_R=6 \\ k=1 & SU(4) & 23 & 3 \\ k=2 & SU(5) & 22 & 4 \\ k=5 & SU(3) & 21 & 4 \\ k=7 & SU(5) & 19 & 2 \end{array} \qquad (42)$$

From the above mentioned separate cancellation of superconformal anomalies within the left- or right-moving sector, we also see that we can take the left half of a (1,0) super-chiral model labelled by (k,G) and the right half of another (1,0) super-chiral model labelled by (k',G') to form a more heterotic chiral model which is still superconformal anomaly free. In other words we anticipate the existence of the model in which

left bosonic movers live in $M^d \times T^{n_L} \times G$

right spinning movers live in $M^d \times T^{n_R} \times G'$

(The d dimensional Minkowski space-time is assumed to be common for both left and right movers). And we anticipate that the critical dimension for the model is given by

$$d + n_L = D(k,G), \quad d + n_R = D'(k',G'). \qquad (43)$$

Moreover, when these conditions are satisfied, each half of the model (the left or right moving sector) itself is actually (super)conformal anomaly free. To construct models which are equivalent to either half of the above model, we have to construct and quantize (1,0) heterotic models in which either the leftons or rightons live in curved space-time. There are some technical problems, but they can be overcome. The work is in progress.

Acknowledgements. I am grateful to the organizers of the Summer Institute for the warm hospitality and the wonderful atmosphere. I acknowledge the enjoyable collaboration with Zi Wang and Feng Yu and also the hospitality of Aspen Center for Physics where the paper is written.

REFERENCES

1. M.B. Green, J.H. Schwartz and E. Witten, Superstring Theory, vol. I and II, Cambridge University Press, 1987; and references therein.
2. S. Deser and B. Zumino, Phys. Lett. 65B (1976) 369; L. Brink, P. DiVecchia and P. Howe, Phys. Lett. 65B (1976) 471.
3. E.S. Fradkin and A.A. Tseytlin, Phys. Lett. 158B (1985) 316.
4. C.G. Callan, D. Friedan, E.J. Martinec and M.J. Perry, Nucl. Phys. B262 (1985) 593; A. Sen, in Unified String Theories, ed. by M. Green and D. Gross, World Scientific, Singapore, 1986; and references therein.
5. See, e.g., N.K. Nielson, Nucl. Phys. B244 (1984) 499; O. Piquet and K. Sibold, Nucl. Phys. B247 (1984) 484; G. Girardi, R. Grimm and R. Stora, Phys. Lett. 156B (1985) 203; L. Borona, P. Pasti and M. Tonin, Phys. Lett. 156B (1985) 341; Nucl. Phys. B252 (1985) 458; H. Itoyama, V.P. Nair and H.C. Ren, Nucl. Phys. B262 (1985) 317; R. Garries, M. Scholl and J. Wess, Z. Phys. C28 (1985) 623.
6. Z. Wang and Y.S. Wu, Phys. Lett. 164B (1985) 305.
7. D.S. Hwang, Nucl. Phys. B267 (1986) 349; Y. Tanii, Nucl. Phys. B259 (1985) 677; Phys. Lett. 165B (1985) 275; Nucl. Phys. B289 (1987) 187.
8. Z. Wang and Y.S. Wu, Utah preprint, 1987.
9. F. Yu and Y.S. Wu, Utah preprint, 1987.
10. S. Adler, Phys. Rev. 177 (1969) 2426; J. Bell and R. Jackiw, Nuovo Cimento 60A (1969) 47.
11. S.L. Adler, J.C. Collins and A. Duncan, Phys. Rev. D15 (1977) 1712, L.S. Brown, Phys. Rev. D15 (1977) 1469; L.S. Brown asnd J.P. Cassidy, Phys. Rev. D15 (1977) 2810; J.C. Collins, A. Duncan and S.D. Joglekar, Phys. Rev. D16 (1977) 438.
12. C.G. Callan, Phys. Rev. D2 (1970) 1541; K. Symanzik, Comm. Math. Phys. 23 (1971) 491.
13. K.G. Wilson, Phys. Rev. 179 (1969) 1499; Phys. Rev. D2 (1970) 1478.
14. W.A. Bardeen, Phys. Rev. 184 (1969) 1848; P.H. Frampton and T.W. Kephart, Phys. Rev. Lett. 50 (1983) 1343, 1347.
15. L. Alvarez-Gaumé and E. Witten, Nucl. Phys. B234 (1983) 269.

16. E. Witten, Phys. Lett. 117B (1982) 324.
17. A. Niemi and G. W. Semenoff, Phys. Rev. Lett. 51, (1983) 2077; A.N. Redlich, Phys. Rev. Lett. 52 (1984) 1.
18. D.J. Gross, J.A. Harvey, E. Martinec and R. Rohm, Phys. Rev. Lett. 54 (1985) 502; Nucl. Phys. B256 (1985) 253; B267 (1986) 75.
19. K. Fujikawa, Phys. Rev. Lett. 42 (1979) 1195; Phys. Rev. D21 (1980) 2848; D29 (1984) 285.
20. J. Schwinger, Phys. Rev. 82 (1984) 664; B.S. DeWitt, Dynamical Theory of Groups and Fields, Gordon and Beach, 1965; S. M. Christensen and M.J. Duff, Nucl. Phys. B154 (1979) 301.
21. J. Dixson (unpublished); R. Stora, In Progress in Gauge Theory, Ed. G.T. Hooft et al. (Plenum Press, N.Y., 1984); B. Zumino, in Relativity, Group and Topology II, eds. B.S. DeWitt and R. Stora (North Holland, 1984); B. Zumino, Y.S. Wu and A. Zee, Nucl. Phys. B239 (1984) 477; L. Baulieu, Nucl. Phys B241 (1984) 557. L.D. Faddeev, Phys. Lett. 145B (1984) 81.
22. L. Alvarez-Gaumé and P. Ginsparg, Nucl. Phys. B243 (1984) 449; M.F. Atiyah and I.M. Singer, Proc. Nat. Acad. Sci. USA 81 (1984) 2597; O. Alvarez, I.M. Singer and B. Zumino, Comm. Math. Phys. 96 (1984) 409; J. Lott, ibid. 93 (1984) 533; E. Witten, Comm. Math. Phys. 100 (1985) 197; O. Alvarez, Nucl. Phys.
23. A.J. Niemi, G.W. Semenoff and Y.S. Wu, Nucl. Phys. B276 (1986) 173.
24. S.Adler, Phys. Rev. 177 (1969) 2426.
25. D.J. Gross and R. Jackiw, Phys. Rev. D6 (1972) 477.
26. L.D. Faddeev and S. Shatashvili, Theor. Math. Phys. 60 (1984) 770; R. Jackiw and R. Rajaraman, Phys. Rev. Lett. 54 (1985) 1219.
27. J. Wess and B. Zumino, Phys. Lett. 37B (1971) 95.
28. E. Witten, in ref. [22].
29. W. Bardeen and B. Zumino, Nucl. Phys. B244 (1984) 421.
30. See ref. [22].

31. G. Moore and P. Nelson, Phys. Rev. Lett. 53 (1984) 1519; Comm. Math. Phys. 100 (1985) 83; L. Alvarez-Gaumé and P. Ginsparg, Nucl. Phys. B262 (1985) 439; J. Bagger, D. Nemechansky and S. Yankielowicz, Nucl. Phys. B262 (1985) 478; E. Cohen and E. Gomez, Nucl. Phys. B254 (1985) 235; P. DiVecchia, S. Ferrara and L. Girardello., Phys. Lett. 151B (1985) 199.
32. F. Gliozzi, J. Scherk and D. Olive, Nucl. Phys. B122 (1977) 253.
33. N. Seiberg and E. Witten, Nucl. Phys. B276 (1986) 272.
34. M. Sakamoto, Phys. Lett. 151B (1985) 115; C. Hull and E. Witten, Phys. Lett. 160B 91985) 398.
35. C. Hull and E. Witten, in ref. [34]; C. Hull, Phys. Lett. 178B (1986) 357.
36. Ref. [18] and S. Gates, M. Grisaru, L. Mezincescu and P. Townsend, Nucl. Phys. B286 (1987) 1.
37. J. Louis and B.A. Ovrut, University of Pennsylvania Report, UPR-0322T; R. Garreis, J. Louis and B. Ovrut, UPR-0332T and UPR-0337T.
38. M. Evans and B. Ovrut, Phys. Lett. 174B (1986) 63; 175B (1986) 145; 184B (1987) 153; 186B (1987) 134; M. Evans, J. Louis and B. Ovrut, Phys. Rev. D35 (1987) 3045.
39. R. Brooks, F. Muhammad and S. Gates, Nucl. Phys. B268 (1986) 599; G. Moore and P. Nelson, Nucl. Phys. B274 (1986) 509.
40. Z. Wang and Y.S. Wu, in preparation.
41. S.J. Gates, in this Proceedings.
42. D. Nemenschansky and S. Yankielowicz, Phys. Rev. Lett. 54 (1985) 620; S. Jain, R. Shankar and S. Wadia, Phys. Rev. D32 (1985) 2713; E. Bergshoeff, S. Randjbar-Daemi, A. Salam, H. Sarmadi and E. Sezgin, Nucl. Phys. B269 (1986) 77.

Current Algebra for Chiral G x G Theory with Wess-Zumino Term*

A.J. MACFARLANE

Department of Applied Mathematics and Theoretical Physics
Cambridge University, Silver Street,
Cambridge CB3 9EW, U.K.

ABSTRACT

We study chiral G x G field theory with Wess-Zumino term. Using canonical formalism in terms of Goldstone fields, we derive current algebras by systematic methods, valid both in four and two dimensional Minkowski space. We display the anomalous structures that arise, discuss briefly and contrast their mathematical interpretation.

1. THEORY WITHOUT WESS-ZUMINO TERM

We begin by discussing field theory with chiral G x G invariance[1)] in (3 + 1) dimensional Minkowski space M_4 in the absence of Wess-Zumino terms. We add these later, and consider the (1+1) case later. Although almost all our results are valid for any compact semi-simple G, we work with G = SU(n), because the Wess-Zumino term for G in M_4 (but not M_2) involves the totally symmetric tensor d_{abc} and vanishes unless this tensor is non-trivial, as for SU(n), n > 2.

Write $U \varepsilon G$ in the form $U = \exp \frac{1}{2} i \lambda \cdot \phi$, where the λ_a are Gell-Mann matrices of SU(n) and the ϕ^a are Goldstone fields. An action invariant under G x G, i.e. under $U \to G_L U G_R^{-1}$, where G_L and G_R are SU(n) transformations, is

$$S_o = \frac{1}{f^2} \int_M d^4x \, \mathrm{Tr} \, \partial^\mu U^\dagger \, \partial^\nu U \, \eta_{\mu\nu}$$

$$= - \frac{1}{f^2} \int_M d^4x \, \mathrm{Tr} \, (U^{-1} \partial^\mu U) \, (U^{-1} \partial^\nu U) \eta_{\mu\nu} \tag{1a}$$

$$= -\frac{1}{f^2}\int_M d^4x \, \mathrm{Tr}\, (\partial^\mu U\, U^{-1})\, (\partial^\nu U\, U^{-1})\eta_{\mu\nu} \tag{1b}$$

Set $f = 1$, $\partial_a = \partial/\partial\phi^a$, and write (1a) as

$$S_o = \frac{1}{2}\int d^4x \, \partial^\mu\phi^a \, \partial^\nu\phi^b \, \eta_{\mu\nu} \, g_{ab} \, , \tag{2}$$

$$g_{ab} = - 2 \, \mathrm{Tr}\, (U^{-1}\partial_a U)\, (U^{-1}\partial_b U) \, . \tag{3}$$

Since $U^{-1}\partial_b U$ is traceless, it is natural to define a matrix R^{-1} by

$$- i \, U^{-1}\partial_b U = \frac{1}{2}\lambda_c \, R^{-1}{}_{cb} \, . \tag{4}$$

From (2) to (4), we see that the metric of the theory is given by

$$g_{ab} = R^{-1}{}_{ca} \, R^{-1}{}_{cb} \, . \tag{5}$$

Eq. (5) shows R^{-1} plays the role of a (right) vielbein. It also (see section 5, and De Witt[2)]) plays a key role in the Lie group theory of G.

From (1b) and $- i(\partial_b U)U^{-1} = \frac{1}{2}\lambda_c L^{-1}{}_{cb}$, we can alternatively write the theory in terms of the left vielbein L^{-1}. Clearly $g_{ab} = L^{-1}{}_{ca}L^{-1}{}_{cb}$ yields the same metric as (5), a fact which arises naturally also in the group theory.

Here we work usually with G_R invariance. Analogous statements for G_L invariance arise similarly from a starting point that brackets factors of the same traces in an evident alternative fashion that exhibits the G_L invariance.

One can calculate R^{-1} and L^{-1} explicitly from their definitions, and check trivially they give rise to the same metric.

For small ε, $G_R = \exp\, (-\frac{1}{2}i\varepsilon\cdot\lambda)$ gives

$$\delta_R U = U\,\frac{1}{2}\, i\, \varepsilon \cdot \lambda = (\partial_b U)\,\delta_R\phi^b \, . \tag{6}$$

This defines $\delta_R\phi^b$, and allows explicit calculation of it as

$\delta_R \phi^b = R_{bc}\,\varepsilon_c$. Now Noether's procedure and $\varepsilon_a j_{\mu Ra} = \delta_R \phi^b \delta L_o/\delta\partial^\mu \phi^b$ yield the Noether current

$$j_{\mu Ra} = g_{bc}\,\partial_\mu \phi^c\, R_{ba} = R^{-1}_{ac}\,\partial_\mu \phi^c\ . \tag{7}$$

This identification is true provided that the transformation in question leaves L_o (and not just S_o) invariant, as can be explicitly verified here.

To set up the canonical formalism for (2), we define $\Pi_a = \partial L_o/\partial \dot{\phi}^a$, and impose

$$[\phi^a(t,\underline{x})\ ,\ \Pi_b(t,\underline{y})] = i\delta^a_b\,\delta(\underline{x} - \underline{y})\ . \tag{8}$$

As all our calculations are classical, one should understand Poisson brackets in place of [,]/i always. Eqs. (7) and (8) easily yield

$$[j_{oRa}\ ,\ j_{oRb}] = if_{abc}\, j_{oRc}\, \delta(\underline{x} - \underline{y})\ ,\ \text{etc,} \tag{9}$$

provided that R obeys

$$R^c{}_a\, R^d{}_{b,c} - (a \leftrightarrow b) = -\, R^d{}_e\, f_{eab}\ . \tag{10}$$

While it is easy to prove (10) directly from the definition (4), the result arises naturally in relation to the Lie algebra of G (see section 5). There are similar results for $j_{\mu La}$ and $[j_{\mu La}\ ,\ j_{\nu Rb}] = 0$.

2. THEORY WITH WESS-ZUMINO TERM

We extend the action S_o of (1) to include a Wess-Zumino term S_1 and study field theory in M_4 with action

$$S = S_o + S_1$$

$$S_1 = \frac{iK}{5!} \int_D d^5x\ \mathrm{Tr}\ \varepsilon_{\lambda\mu\nu\rho\sigma}\, W^\lambda W^\mu W^\nu W^\rho W^\sigma \tag{11}$$

where $K = N/30\pi^2$, N integral , D is a five dimensional region with M_4 as its boundary $\partial D = M_4$. Of course the relevance of N integral is to the quantum theory, where it assures the single valuedness of the quantal generating functional of the Green's functions of the theory. Further

$$W^\mu = U^{-1}\partial^\mu U = U^{-1}\partial_a U\, \partial^\mu\phi^a \tag{12}$$

$$= \frac{1}{2}\, i\, \lambda_c\, R^{-1}_{ca}\, \partial^\mu\phi^a \, .$$

We note that W^μ given by (12) makes the G_R invariance of S_1 obvious, but is not the Noether current for the G_R invariance of S, although it is a natural variable for use in discussion of the structure of the theory. Again we remark that rewriting S_1 in terms of $(\partial^\mu U)U^{-1}$, instead of W^μ, allows the parallel discussion of G_L invariance. Whenever the parallel treatments give rise to apparently different expressions for S_1, there are identities, with a natural interpretation within the group theory, which ensure that they agree.

Our work requires a nice selection of manipulations of S_1[4]. All involve the calculation of a suitable variation δS_1 of S_1, and all are facilitated by systematic use of the Lemma: variations δU of U induce variations δW of W^μ, as defined by (12), of the form

$$\delta W^\mu = \partial^\mu(U^{-1}\delta U) - [U^{-1}\delta U \, , W^\mu] \, .$$

<u>Explicit Minkowski space form for S_1.</u>

The lemma yields

$$\delta \,\mathrm{Tr}\, W^\lambda W^\mu W^\nu\, W^\rho W^\sigma \varepsilon_{\lambda\mu\nu\rho\sigma}$$

$$= 5\partial^\lambda (\varepsilon_{\lambda\mu\nu\rho\sigma}\, \mathrm{Tr}\, U^{-1}\delta U\, W^\mu W^\nu W^\rho W^\sigma) \, ,$$

and hence

$$\delta S_1 = \frac{iK}{4!} \int_D d^5x\, \partial^\lambda\, (\varepsilon_{\lambda\mu\nu\rho\sigma}\, \mathrm{Tr}\, U^{-1}\, \delta U\, W^\mu W^\nu W^\rho W^\sigma) \tag{13}$$

$$= \frac{iK}{4!} \int_{M=\partial D} d^4x\, \varepsilon_{\mu\nu\rho\sigma}\, \mathrm{Tr}\, U^{-1}\, \delta U\, W^\mu W^\nu W^\rho W^\sigma \, .$$

To apply this here, replace $U(\phi)$ in the definition of S_1 by $U(t\phi) = \exp \frac{1}{2}\, it\lambda\cdot\phi$, so that

$$S_1(t = 1) = S_1 \quad , \quad S_1(t = 0) = 0 \; .$$

Then use of $\delta = \varepsilon \frac{\partial}{\partial t}$ for small ε gives

$$\frac{\partial S_1(t)}{\partial t} = \frac{iK}{4!} \int_M d^4x \; \varepsilon_{\mu\nu\rho\sigma} \; \mathrm{Tr}\; U^{-1}\dot{U} \; W^\mu W^\nu W^\rho W^\sigma \; , \tag{14}$$

and $S_1 = \int_0^1 dt \, \frac{\partial S_1(t)}{\partial t}$. Using

$$W^\mu = U^{-1}\partial^\mu U = U^{-1} \; \partial_a U \; \partial^\mu \phi^a \equiv W_a \; \partial^\mu \phi^a \; , \text{ Eq. (14)}$$

yields

$$S_1 = - \frac{K}{4!} \int_M d^4x \; h_{abcd} \; \varepsilon_{\mu\nu\rho\sigma} \; \partial^\mu\phi^a \partial^\nu\phi^b \partial^\rho\phi^c \partial^\sigma\phi^d \tag{15}$$

where the totally antisymmetric fourth rank tensor function of ϕ is given by

$$h_{abcd} = \int_0^1 dt \; \mathrm{Tr} \; \frac{1}{2} \; \lambda\cdot\phi \; W_{[a} \; W_b \; W_c \; W_{d]} \; . \tag{16}$$

It is implied that in (14), (16) the argument of U is $t\phi$ rather than ϕ.

The curl of h_{abcd} .

It has not so far been possible to derive a concise closed form explicit expression for h_{abcd}, although one can do the corresponding job for the M_2 case. However, the only information about h_{abcd} needed for later work is the important identity[4)]

$$5i \; h_{[abcd,e]} = \mathrm{Tr} \; W_{[a} \; W_b \; W_c \; W_d \; W_{e]} \; , \tag{17}$$

where $W_a = U^{-1}\partial_a U = \frac{1}{2} \; i\lambda_g R^{-1}_{ga}$. One proves this directly by calculation δS_1 from (15) and comparing with (13).

The Euler Lagrange Equation

Writing S_o as $S_o = - \int_M d^4x \; \mathrm{Tr} \; W^\mu W^\nu \eta_{\mu\nu}$, we use the lemma to develop

$$\delta \; \mathrm{Tr} \; W^\mu W^\nu \eta_{\mu\nu} = 2 \; \mathrm{Tr} \; \eta_{\mu\nu} \; W^\nu \partial^\mu (U^{-1}\delta U)$$

$$\equiv - 2 \; \mathrm{Tr} \; U^{-1}\delta U \; \partial_\mu W^\mu \tag{18}$$

upon integration by parts for δS_o . Now (18) and (13) combine to give

$$\delta S = \int d^4x \ \mathrm{Tr}\ U^{-1}\delta U \ (2\ \partial_\mu W^\mu + \frac{iK}{4!}\ \varepsilon_{\mu\nu\rho\sigma}\ W^\mu W^\nu W^\rho W^\sigma) \ .$$

Requiring that S be stationary for arbitrary variations δU of U gives the equation of motion

$$\partial_\mu W^\mu + \frac{iK}{48}\ \varepsilon_{\mu\nu\rho\sigma}\ W^\mu W^\nu W^\rho W^\sigma = 0 \ . \tag{19}$$

To present this as a current conservation law, we note $\partial^\mu W^\nu - \partial^\mu W^\nu + [W^\mu, W^\nu] = 0$, so that

$$\varepsilon_{\mu\nu\rho\sigma}\ \partial^\mu\ (W^\nu W^\rho W^\sigma) = -\ \varepsilon_{\mu\nu\rho\sigma}\ W^\mu W^\nu W^\rho W^\sigma \ .$$

So (19) becomes

$$0 = \partial^\mu J_\mu \ ,$$

$$J_\mu = W_\mu - \frac{iK}{48}\ \varepsilon_{\mu\nu\rho\sigma}\ W^\nu W^\rho W^\sigma \ . \tag{20}$$

Set $J_\mu = \frac{1}{2}\ i\ \lambda_a\ J_{\mu a}$, $W_\mu = \frac{1}{2}\ i\lambda_a\ W_{\mu a}$, so that $W_{\mu a} = R^{-1}_{ab}\ \partial_\mu\ \phi^b$ equals what would be the Noether current in the absence of S_1. Then, after some Gell-Mann matrix work, we get

$$J_{\mu a} = W_{\mu a} - (K/192)\ \varepsilon_{\mu\nu\rho\sigma} d_{abg} f_{cdg}\ W^\nu_b W^\rho_c W^\sigma_d \ . \tag{21}$$

Noether's Theorem

The action $S = S_o + S_1$ is invariant under an infinitesimal transformation of G_R, but

$$\delta L = \delta L_o + \delta L_1 = \delta L_1$$

is a total divergence which we write as $\partial^\mu K_{\mu a}\varepsilon_a$. Hence Noether's theorem gives rise to

$$0 = \partial^\mu\ J_{\mu a}\ \varepsilon_a \ ,$$

$$J_{\mu a}\varepsilon_a = \delta_R\phi^a\ \delta L/\delta\ \partial^\mu\phi^a - K_{\mu a}\varepsilon_a \ ,$$

$$J_{\mu a} = R_{ea}\,(g_{eb}\partial_\mu\phi^b - \varepsilon_{\mu\nu\rho\sigma}h_{ebcd}\partial^\nu\phi^b\partial^\rho\phi^c\partial^\sigma\phi^d/6) - K_{\mu a}\ . \qquad (22)$$

Here the first term comes from L_o and equals the first term of (21). It is easiest just to verify that (22) agrees with (21). Thus one determines $K_{\mu a}$ explicitly by identifying (22) with (21), computes its divergence and establishes that this is exactly what arises from the calculation of $\partial^\mu K_{\mu a}$ from δL_1 for an infinitesimal transformation of G_R. The curl identity (17) is required for the completion of the task.

3. CURRENT ALGEBRA

Addition of S_1 to S_o has already been seen to complicate the discussion of Noether's theorem. It also complicates the canonical formalism via $\Pi_a = \partial L/\partial\dot\phi^a$, for now we have

$$\Pi_a = g_{ab}\,\dot\phi^b - Kh_{abcd}\,\varepsilon_{ijk}\,\partial^i\phi^b\partial^j\phi^c\,\partial^k\phi^d/6\ . \qquad (23)$$

As an intermediate stage on the path to the Noether current algebra for $S = S_o + S_1$, we calculate the Poisson brackets of $W^\mu_a = R^{-1}_{ab}\,\partial^\mu\phi_b$. Using

$$W^o_e = \pi_a R_{ae} + KR_{ae}h_{abcd}\,\varepsilon_{ijk}\,\partial^i\phi^b\partial^j\phi^c\partial^k\phi^d/6$$

to eliminate $\dot\phi$ in favour of π , canonical calculations give rise (!) to

$$[W^o_a, W^o_b] = if_{abc}\,\delta(\underline{x} - \underline{y})\,W^o_c$$

$$+\ i\delta(\underline{x} - \underline{y})\,K\varepsilon_{ijk}\,f_{bcp}\,f_{deq}\,d_{pqa}\,W^i_cW^j_dW^k_e/96\ , \qquad (24)$$

$$[W^o_a,\ W^k_b] = i\,f_{abc}\,\delta(\underline{x} - \underline{y})\,W^k_c - i\delta_{ab}\,\partial/\partial y^k\,\delta(\underline{x} - \underline{y})\ ,$$

$$[W^k_a,\ W^\ell_b] = 0$$

Eq. (24) has been derived previously by Rajeev[5] and by Wudka[6] but not using the canonical formalism based on the Goldstone fields. It has also been discussed by Fujiwara[7].

Next, we turn to the algebra of the Noether currents J_{oa} of $S = S_o + S_1$, given explicitly by (22). Eqs. (22) and (24) lead directly to

$$[J_{oa}, J_{ob}] = i f_{abc} J_{oc} \delta(\underline{x} - \underline{y}) + i(K/24) d_{abg} \varepsilon_{ijk} \partial_i W^j_g)(\underline{x}) \partial/\partial y^k \delta(\underline{x} - \underline{y}) . \tag{25}$$

Apart from routine and tedious calculation, the key ingredients of the proofs of (24) and (25) are

a) the Lie algebra result (10),

b) the ff and df Jacobi identities of SU(n),

c) the curl identity for h, (17),

d) the distributional identity

$$a(x)b(y) \partial/\partial y \delta(x - y) = - a(y)b(y) \partial/\partial x \delta(x - y) + a(x)b(y)\delta(x - y) .$$

The latter has to be used with care, as for example in showing that derivatives of delta function terms cancel out of the right side of (24). Once this cancellation is achieved, collection of terms involving derivatives of h enables use of (17) to complete proof of (24).

The result (25) seems to be new. Fujiwara[7)] has discussed possible anomalous additions to current alegbraic results by use of the methods of differential geometry as advocated by Zumino[8)] and Stora[9)]. He found that anomalous structures like those exhibited here in (24) and (25) are indeed allowed. He did not however correlate the latter allowed anomaly type to any specifically defined current or consider canonical methods.

Faddeev and Shatashvili[10)] have discovered previously (and discussed extensively from the standpoint of cohomology) anomalous structures exactly like that of (25). In the Hamiltonian formulation of the theory of SU(n) gauge fields coupled to Weyl fermions in the gauge $A_{oa} = 0$, one knows that the analogues of Gauss' Law enter as first-class constraints $G_a = 0$, with Poisson brackets

$$\{ G_a , G_b \} = f_{abc} G_c \delta(\underline{x} - \underline{y}) .$$

Upon quantization this algebra was shown[10] to develop an anomalous term exactly of the type found in (25).

4. CASE OF (1 + 1) DIMENSIONAL THEORY

In M_2, the Wess-Zumino term[11] is

$$S_1 = (N/24\pi) \int d^3x \, \varepsilon_{\rho\mu\nu} \, \mathrm{Tr} \, W^\rho W^\mu W^\nu , \tag{26}$$

$$= (N/8\pi) \int d^2x \, h_{ab} \, \varepsilon_{\mu\nu} \, \partial^\mu \phi^a \, \partial^\nu \phi^b , \tag{27}$$

where $h_{ab} = \int_o^1 dt \, \mathrm{Tr} \, \frac{1}{2} \, i\lambda\cdot\phi \, W_{[a} W_{b]}$. Everything is much easier than in the M_4 case. As (26) is quadratic in field derivatives, both the Noether current and Π_a are linear in them. Also the formula for h_{ab} can be easily calculated in closed form[11], although one does the current algebra without use of this, employing only the identity, analagous to (17), for $h_{[ab,c]}$.

One proceeds for (26) just as in section 2, doing the (here simpler) analogues of the same four calculations. Then, as in section 3, one gets, for the Noether currents, an algebra

$$[j_{oa} , j_{ob}] = i \, f_{abc} \, j_{oc} \, \delta(x - y)$$

$$+ \frac{N}{2\pi} \, i\delta_{ab} \, \partial/\partial x \, \delta(x - y) \tag{28}$$

of the familiar Kac-Moody type[12], involving an anomalous term that is independent of the variables of the theory.

5. SOME LIE GROUP THEORY

Let G be a Lie group, with elements a,b,c specified by r real parameters with multiplication law

$$c = b\cdot a \quad , \quad c^\rho = \phi^\rho \, (b,a) \quad , \quad 1 < \rho < r .$$

Parameters are chosen so that e = 0 for those of the identity. We view G as a manifold with points a,b,c. Then, for fixed a, the functions ϕ define the (group G_R of) right translations $\phi : M \to M$, or $s \to t = s \cdot a$, with $t^\rho = \phi^\rho(s,a)$.

To obtain the Lie PDEs of G_R, we compare

$t = s \cdot a$ and $t + dt = s \cdot (a + da)$,

First, we employ $t + dt = t \cdot \delta a$ to define δa, so that

$$dt^\rho = \partial\phi^\rho(t,u)/\partial u^\sigma \big|_{u=e} \, \delta a^\sigma . \tag{29}$$

Also $S \cdot (a + da) = t + dt = t \cdot \delta a = s \cdot a \cdot \delta a$, gives $a \cdot \delta a = a + da$, and hence

$$da^\sigma = \partial\phi^\sigma(a,u)/\partial u^\tau \big|_{u=e} \, \delta a^\tau . \tag{30}$$

Eqs. (29), (30) motivate introduction of a matrix, which we write in suggestive notation, as

$$R^\rho{}_\sigma(\cdot) = \partial\phi^\rho(\cdot,u)/\partial u^\sigma \big|_{u=e} . \tag{31}$$

Then (29), (30) read $da = R(a)\delta a$, $dt = R(t)\delta a$.
Thus the Lie PDEs are $dt = R(t)\, R(a)^{-1} da$,

or $$\partial t^\rho/\partial a^\sigma = R(t)^\rho{}_\tau \, R(a)^{-1\tau}{}_\sigma . \tag{32}$$

By hypothesis, these have solutions $t^\rho = \phi^\rho(s,a)$ in which now the s^σ feature as constants of integration. So they are integrable. Thus

$$\frac{\partial^2 t^\rho}{\partial a^\sigma \partial a^\tau} = \frac{\partial^2 t^\rho}{\partial a^\tau \partial a^\sigma} \tag{33}$$

Eq. (33) simplifies to

$$(R^\nu_\beta \, R^\tau_{\alpha,\nu} - R^\nu_\alpha \, R^\tau_{\beta,\nu})(t) = - f^\lambda_{\beta\alpha} \, R^\tau_\lambda \; (t) \tag{34}$$

where

$$f^\lambda_{\beta\alpha} = [\, R^{-1\lambda}{}_\rho \, R^\rho_{\beta,\sigma} \, R^\sigma_\alpha - (\alpha \leftrightarrow \beta)] \; (a) .$$

These must be constants, independent of a, as we have essentially done a separation of variables, so we may evaluate the $f^{\lambda}_{\beta\alpha}$ at a = e , using R(e) = 1 . Then

$$f^{\lambda}_{\beta\alpha} = (R^{\lambda}_{\beta,\alpha} - R^{\lambda}_{\alpha,\beta}) \tag{35}$$

$$= \left(\frac{\partial}{\partial v^{\alpha}} \frac{\partial}{\partial v^{\beta}} - \frac{\partial}{\partial v^{\beta}} \frac{\partial}{\partial u^{\alpha}} \right) \phi^{\lambda}(v,u)\Big|_{u=v=e} .$$

The f are just the structure constants of G, determined using (35) by the group multiplication law of G. Similar results hold for $L^{\rho}_{\ \sigma}(\cdot) = \partial\phi^{\rho}(v,\cdot)/\partial v^{\sigma}\big|_{v=e}$, and G_L.

Define generators of right and left translations by

$$R_{\sigma} = -\, i\, R^{\rho}_{\sigma}\, \partial/\partial t^{\rho} \ , \ L_{\sigma} = +\, iL^{\rho}_{\sigma}\, \partial/\partial t^{\rho} \ .$$

They satisfy the Lie algebra results

$$[\, R_{\beta} \ , R_{\alpha}\,] = i\, f^{\gamma}_{\beta\alpha}\, R_{\gamma} \ ,$$

$$[\, L_{\beta} \ , L_{\alpha}\,] = i\, f^{\gamma}_{\beta\alpha}\, L_{\gamma} \ , \tag{36}$$

$$[\, R_{\beta} \ , L_{\alpha}\,] = 0 \ .$$

Eq. (36) displays the typical chiral G x G structure, or that of the principal σ-model of G. One can also view the theory as having the coset space structure G_N/G_D , where G_N = G x G and G_D is the 'diagonal' subspace with generators V = L + R . In the chiral G x G view, V = L + R and A = L - R corresponding to vector and axial charges.

Next, we consider the adjoint action, for fixed b, of G, via

$$a \to a' = bab^{-1} \ .$$

Evaluating $\phi(a',b) = \phi(b,a)$ for a,a′ close to e, yields

$$a'^{\rho} = (L^{-1}R)^{\rho}_{\ \tau}$$

so that $D = L^{-1}R$ defines the adjoint representation of G.

By a smooth change of coordinates on M, we can pass from arbitrary to canonical coordinates. The latter correspond to the exponential parametrization of G, and we use latin letters for them. Just as the definition of U in section 1 is the exponential version of the defining representation of SU(n), so also is that of the adjoint representation given by

$$D(a) = \exp iF\cdot a \ , \ (F_a)_{bc} = - if_{abc} \ , \ F_a^\dagger = F_a \ . \tag{37}$$

Further, it is clear that the result (10) is just (34) in canonical coordinates, so that (31) for canonical coordinates is the group theoretical definition of the (right) vielbein of section one. Now since $D = L^{-1}R$, as given by (37), is unitary and real and hence orthogonal, we have

$$L^{-1} \, {}^{T}L^{-1} = R^{-1} \, {}^{T}R^{-1} \ .$$

This expresses the identity of the right and left vielbein expressions for the metric g of (5).

Finally the group multiplication law $D(b)D(a) = D(c)$ in canonical coordinates yields

$$c_k = b_k + a_k - \frac{1}{2} b_i \, a_j \, f_{ijk} \cdots ,$$

and is consistent with (36), as expected.

6. COCYCLES AND GROUP EXTENSIONS

In a quantum mechanical representation of a group G on a Hilbert space with physical states corresponding to rays, one meets representations $g \to U(g)$, which are representations 'only' up to a phase

$$U(g_2) \, U(g_1) = \exp i \, \omega \, (g_2,g_1) \, U(g_2 g_1) \ . \tag{38}$$

Here $\omega(g_2,g_1)$ is a real valued phase defined on G x G. If (38) provides an associative composition law for G, ω is called a two cocycle. It is trivial if removable by smooth redefinition of the phase of U(g). We deal with non-trivial cocycles.

Given G and such ω, we define the central extension $\hat{G}$ of G by providing the pairs (Θ,g), where Θ is a phase, with the composition law

$$(\Theta_2,g_2)\ (\Theta_1,g_1) = (\Theta_2 + \Theta_1 + \omega(g_2,g_1)\ ,\ g_2g_1)\ .$$

Using the notations $u\cdot v = u_a v_a$, $(u\wedge v)_a = f_{abc}U_b v_c$, and defining $U[\varepsilon] = \exp i \int dx\ j_o(x)\cdot\varepsilon(x)$, we turn first to the theory with action S_o defined in M_2. It is easy to use the BCH theorem and the current algebra of section 1 (the same for M_2 as M_4) to derive the group multiplication law of G as

$$U\ [\varepsilon_2]\ U\ [\varepsilon_1] = U\ [\varepsilon_3]\ ,\ \varepsilon_3 = \varepsilon_2 + \varepsilon_1 - \frac{1}{2}\,\varepsilon_2\wedge\varepsilon_1 + \ldots$$

Using the action $S = S_o + S_1$, it follows that the theory provides a representation of the central extension of G, via

$$\Delta\ [\Theta,\varepsilon] = e^{i\Theta}U\ [\varepsilon]\ ,$$

as the current algebra (28) yields

$$\Delta\ [\Theta_2,\varepsilon_2]\ \Delta\ [\Theta_1,\varepsilon_1] = \Delta\ [\Theta_3,\varepsilon_3]\ ,$$

with same result for ε_3 as before, and

$$\Theta_3 = \Theta_2 + \Theta_1 + \omega\ [\varepsilon_2,\varepsilon_1]$$

$$\omega\ [\varepsilon_2,\varepsilon_1] = N/(4\pi) \int dx(\varepsilon_2'\cdot\varepsilon_1 - \varepsilon_1'\cdot\varepsilon_2). \qquad (39)$$

Thus the Wess-Zumino term provides the central extension of the (locally realized) group.

In M_4, the situation is different. In place of $\omega\ [\varepsilon_2,\varepsilon_1]$ given by (39), we find using (25) the quantity[10)]

$$K/24 \int d^3x\ (\partial_i\varepsilon_{2a})\ (\partial_j\varepsilon_{1b})\varepsilon_{ijk}d_{abc}\ W_c^k\ .$$

In view of the dependence here upon W_c^k , we are not dealing with a central extension of G, but rather with a projective representation of it[10)].

I thank Anne Davis and John Gracey for their collaboration in the work discussed in this paper and elsewhere[13).

REFERENCES

1. S. Coleman, J. Wess and B. Zumino, Phys. Rev. 177,2239 (1968).
2. B.S. De Witt, Dynamical Theory of Groups and Fields, Gordon and Breach, New York, 1965.
3. J. Wess and B. Zumino, Phys. Lett. B37, 95 (1971).
4. E. Braaten, T.L. Curtwright and C.K. Zachos, Nucl. Phys. B260, 630 (1985).
5. S. Rajeev, Phys. Rev. D29, 2944 (1984).
6. J. Wudka, Canonical Quantization of the Skyrme Model, MIT preprint, #1185 (1984).
7. T. Fujiwara, Phys. Lett. B152, 103 (1985).
8. B. Zumino, Chiral Anomalies and Differential Geometry, Les Houches, 1983.
9. R. Stora, Algebra, Structure and Topological Origin of Anomalies, Cargèse, 1983.
10. L.D. Faddeev and S.L. Shatashvili, Theor. Math. Phys. 60, 206 (1986); L.D. Faddeev, Phys. Lett. B145, 81 (1984).
11. A. D'Adda, A.C. Davis and P. di Vecchia, Phys. Lett. B121, 335 (1985).
12. E. Witten, Commun. Math. Phys. 92, 455 (1984); B.A. Baaquie, Phys. Lett. B177, 310 (1986); P. Goddard and D.I. Olive, Intern. J. Mod. Phys. A1, 303 (1986).
13. A.C. Davis, J.A. Gracey and A.J. Macfarlane, Phys. Lett. B194, 415 (1987).

BRS ANOMALY

FOR STRING THEORIES

Takayuki Matsuki*

Dept. of Physics, Simon Fraser Univ.
Burnaby, B. C. , Canada V5A 1S6

Abstract

The anomaly due to breakdown of the Becchi-Rouet-Stora (BRS) invariance, by fixing both general coordinate and local conformal invariances, is calculated for a bosonic string theory by the path integral method. A critical dimension is derived by using a common conformal unit for fields and demanding a vanishing BRS anomaly. Relations are discussed with other two derivations of critical dimensions for relativistic string theories, i.e., methods of a local conformal anomaly and a nilpotency of the BRS charge.

*After September, 1987, the address is Physics Dept., Oakland Univ., Rochester, MI 48063, U.S.A.

I. Introduction

In deriving a critical dimension of relativistic string theories [1,2], local conformal invariance is forced on the first-quantized theory. That is, the conformal anomaly vanishes at a critical dimension, D = 26 for a bosonic string case. In the paper [2], the Becchi-Rouet-Stora (BRS) quantization [3] is adopted to derive a critical dimension. At first the gauge of general coordinate invariance is fixed and the corresponding Faddeev-Popov (FP) ghost terms are introduced. This Lagrangian is invariant under the global BRS transformation by construction and the functional measure of each field is fixed by demanding invariance under this BRS transformation. However there remains a local conformal invariance and the conformal anomaly is calculated via a path integral method, which vanishes at a critical dimension.

In the paper [4], although the Lagrangian is canonically quantized by using the BRS invariance, it looks that there is no local conformal invariance from the outset. Hence there is no conformal anomaly, instead a critical dimension is derived from a nilpotency of the BRS charge.

Comparing these derivations of critical dimensions, we can not yet judge which is the first principle to derive a critical dimension, vanishing of the local conformal anomaly or a nilpotency. With the BRS invariance, the Slavnov-Taylor identities can be derived, which may be utilized to prove unitarity and/or renormalizability. Hence to have a consistent quantum theory, the BRS anomaly, if there is, should vanish. In this paper, fixing both general coordinate and local conformal invariances of the bosonic string theory,

$$L_0 = -\frac{1}{2}\sqrt{g}\, g^{ab}\, \partial_a X\, \partial_b X \ , \tag{1.1}$$

we will calculate the BRS anomaly and show it vanishes at a critical dimension under certain circumstances. That is, a vanishing BRS anomaly is a necessary and sufficient condition to determine a critical dimension. Existence of the BRS anomaly is naturally expected from the following observation. The BRS transformations for matter fields are obtained by replacing a gauge function with the FP ghost multiplied by a constant Grassmann parameter. Hence if there is a "local" anomaly due to a local invariance, there must be a corresponding global BRS anomaly due to the matter fields after the Lagrangian is quantized a la BRS. Since we already know that there is a local conformal anomaly calculated in [2], there must be at least a corresponding BRS anomaly of that kind.

To calculate the BRS anomaly, the gauge condition for a general coordinate invariance in [2] will be adopted to show that the functional measure can not be fixed uniquely even if requiring a vanishing BRS anomaly at a critical dimension. Next it will be shown that when the gauge in [4] is taken, their Lagrangian corresponds to the one in which the local conformal invariance is already fixed. In this case there is no BRS anomaly hence they need to calculate a square of the BRS charge to show a nilpotency of this operator at a critical dimension. In Sect. II we will give BRS transformations for all fields, after introducing gauge fixing adopted in [2] and the FP ghost terms so that they satisfy a nilpotency. In Sect. III we will give the measure and calculate the BRS anomaly, which determines a critical dimension. In the same section we will also describe how to relate our work to that in [2].

II. BRS Quantization

We will introduce one more variable, ρ, as follows in terms of which conformal weights of all other fields are measured:

$$L_0 = E'\rho\,(1 + \rho^{-2} \det \tilde{g}_{ab})\ , \tag{2.1}$$

where E' is an auxiliary field and with $g = -\det g_{ab}$,

$$\tilde{g}_{ab} = g^{-1/4}\, g_{ab} = \begin{bmatrix} A^1 + A^2 & A^0 \\ A^0 & A^1 - A^2 \end{bmatrix} . \tag{2.2}$$

Using another auxiliary fields, B_0'', B_1'' and B, the gauge fixing condition for general coordinate invariance adopted in [2] together with the one for local conformal invariance is realized by

$$L_{GF} = \rho^{-1} B_a'' A^a + B(\rho - 1) = \rho^{-1} (B_0'' A^0 + B_1'' A^1) + B(\rho - 1) . \tag{2.3}$$

Here we have assumed that the field ρ has no singularity. In the case ρ has non-trivial topological properties, setting $\rho = 1$ generally changes the topological structure of the theory. To calculate the FP ghost terms, we need the BRS transformation of fields, ϕ_i, which is defined as $\lambda\delta\phi_i$ with a Grassmann constant λ. The transformations for matter fields are obtained by replacing gauge functions with the FP ghosts C^a and C_W and those for FP ghosts are obtained so that they satisfy a nilpotency, i.e., $\delta^2 = 0$.

$$\delta\rho = (C^a\partial_a + \frac{1}{2}\partial_a C^a + C_W)\, \rho \ , \tag{2.4a}$$

$$\delta\tilde{X} = \delta(\rho^\alpha X) = (C^a\partial_a + \frac{\alpha}{2}\partial_a C^a + \alpha C_W)\, \tilde{X} \ , \tag{2.4b}$$

$$\delta E' = (C^a\partial_a + \frac{1}{2}\partial_a C^a - C_W)\, E' \ , \tag{2.4c}$$

$$\begin{aligned} \delta A^0 &= (C^a\partial_a + \frac{1}{2}\partial_a C^a + C_W)\, A^0 \ , \\ &\quad + (\partial_1 C^0 + \partial_0 C^1)\, A^1 + (\partial_1 C^0 - \partial_0 C^1)\, A^2 \ , \end{aligned} \tag{2.4d}$$

$$\delta A^1 = (C^a\partial_a + \frac{1}{2}\partial_a C^a + C_W)\, A^1 \ ,$$

$$+ (\partial_1 C^0 + \partial_0 C^1) A^0 + (\partial_0 C^0 - \partial_1 C^1) A^2 , \tag{2.4e}$$

$$\delta A^2 = (C^a \partial_a + \frac{1}{2} \partial_a C^a + C_W) A^2 ,$$

$$+ (\partial_0 C^0 - \partial_1 C^1) A^1 - (\partial_1 C^0 - \partial_0 C^1) A^0 , \tag{2.4f}$$

$$\delta(d\tilde{C}^a) = \delta(\rho^{2\beta} dC^a) = (C^b \partial_b d\tilde{C}^a - \partial_b C^a d\tilde{C}^b + 2\partial_b C^b d\tilde{C}^a)$$

$$+ 2(\beta - 1) \partial_b C^b d\tilde{C}^a + 2\beta C_W d\tilde{C}^a , \tag{2.4g}$$

$$\delta\bar{\bar{C}}_a = \delta(\rho^{-\gamma} \bar{C}_a) = i \rho^{-\gamma} B_a'' - \gamma C_W \bar{\bar{C}}_a , \tag{2.4h}$$

$$\delta C_W = C^a \partial_a C_W , \tag{2.4i}$$

$$\delta \bar{C}_W = i B , \tag{2.4j}$$

$$\delta B_a'' = 0 , \tag{2.4k}$$

$$\delta B = 0 , \tag{2.4l}$$

where a conformal unit is taken differently for $\tilde{X}_\mu$, $\tilde{C}^a$, and $\bar{\bar{C}}_a$, α, β, and γ, are constants, and C^a, and $\bar{C}_a$, and C_W and $\bar{C}_W$ are the FP ghosts for general coordinate and local conformal (Weyl) invariances, respectively. Note that ρ behaves like $g^{1/4}$. Here the BRS transformation for the measure $d\tilde{C}^a$ is given instead of $\tilde{C}^a$ itself since the anomaly comes from the measure. The FP ghost terms can be easily found from the following observation that the sum of L_{GF} and L_{FP} is invariant under the BRS transformation and that it satisfies a nilpotency:

$$L_{GF} + L_{FP} = - i \,\delta[\bar{C}_a \rho^{-1} A^a + \bar{C}_W (\rho - 1)] . \tag{2.5}$$

Subtracting L_{GF} from the above equation, we obtain

$$L_{FP} = i \bar{C}_W (C^a \partial_a + \frac{1}{2} \partial_a C^a + C_W) \rho$$

$$+ i \bar{C}_0 [C^a \partial_a (\rho^{-1} A^0) + (\partial_1 C^0 + \partial_0 C^1) \rho^{-1} A^1 + (\partial_1 C^0 - \partial_0 C^1) \rho^{-1} A^2]$$

$$+ i \bar{C}_1 [C^a \partial_a (\rho^{-1} A^1) + (\partial_1 C^0 + \partial_0 C^1) \rho^{-1} A^0 + (\partial_0 C^0 - \partial_1 C^1) \rho^{-1} A^2] \ , \quad (2.6)$$

The total Lagrangian can be largely simplified by shifting auxiliary fields as follows:

$$E' = E - \frac{1}{2(\rho + A^2)} \eta^{ab} \partial_a (\rho^{-\alpha} \tilde{X}) \partial_b (\rho^{-\alpha} \tilde{X}) + \frac{1}{\rho + A^2} i \bar{C} i \partial\!\!\!/ C \ , \quad (2.7a)$$

$$B_0'' = B_0' + \partial_0 (\rho^{-\alpha} \tilde{X}) \partial_1 (\rho^{-\alpha} \tilde{X}) + E' A^0$$
$$+ i\partial_a (\bar{C}_0 C^a) - i \bar{C}_1 (\partial_1 C^0 + \partial_0 C^1) \ , \quad (2.7b)$$

$$B_1'' = B_1' - \frac{1}{2} [\partial_0 (\rho^{-\alpha} \tilde{X}) \partial_0 (\rho^{-\alpha} \tilde{X}) + \partial_1 (\rho^{-\alpha} \tilde{X}) \partial_1 (\rho^{-\alpha} \tilde{X})]$$
$$- E' A^1 + i\partial_a (\bar{C}_1 C^a) - i \bar{C}_0 (\partial_1 C^0 + \partial_0 C^1) \ , \quad (2.7c)$$

whose Jacobian is one. The total Lagrangian is given by

$$L = L_0 + L_C + L_{GF} + L_{FP} = -\frac{1}{2} \eta^{ab} \partial_a (\rho^{-\alpha} \tilde{X}) \partial_b (\rho^{-\alpha} \tilde{X}) + E\rho^{-1} [\rho^{-2} - (A^2)^2]$$
$$+ B_a A^a + B(\rho - 1) + i (\tilde{\bar{C}} \rho^{\gamma}) i \partial\!\!\!/ (\rho^{-2\beta} \tilde{C}) + i \bar{C}_W \tilde{C}_W \ , \quad (2.8)$$

where

$$\tilde{C}_W = \delta\rho = (C_W + C^a \partial_a + \frac{1}{2} \partial_a C^a) \rho \ ,$$
$$B_a = \rho^{-1} B_a' \quad \text{and} \quad i \partial\!\!\!/ = -i \sigma_2 \partial_0 + \sigma_3 \partial_1 \quad (2.9)$$

with the Pauli matrices σ_i. In Eq.(2.8), we have neglected a total divergence, $\partial_a(C^a f(x))$, which vanishes if the boundary condition on the ghost field is met:

$$C^1 = 0 \ , \quad \text{at} \quad \sigma = 0, \ \pi \ . \quad (2.10)$$

III. BRS Anomaly

Now we have to determine the functional measure appropriately, in which all the quantum effects are included and from which the BRS anomaly can be calculated via a path integral method. It is determined so that the BRS anomaly does not arise as far as possible. Most general form is given as $d\rho d\mu$ with

$$d\mu = (d\tilde{X}\, d\tilde{C}^a\, d\tilde{\tilde{C}}_a)\, (dE\, dB_a\, dA^i)\, (dB\, d\bar{C}_W)\, d\tilde{C}_W \ , \tag{3.1}$$

where brackets are inserted within which there occurs a cancellation of the BRS anomaly except for the first bracket. In the first bracket, there is the BRS anomaly, vanishing of which determines the critical dimension.

Calculation of the BRS anomaly for fields, $\tilde{X}_\mu$, $\tilde{C}^a$, and $\tilde{\tilde{C}}_a$, is given as follows. If the field ϕ BRS-transforms as

$$\delta\phi = (C^a\partial_a + a_1\, \partial_a C^a + a_2\, C_W)\, \phi \ , \tag{3.2}$$

then the contribution to the BRS anomaly from this field is given by, up to a total divergence,

$$\begin{aligned} A^\phi &= -i \sum_n \phi_n\, (C^a\partial_a + a_1\, \partial_a C^a + a_2\, C_W)\, \phi_n \\ &= -i\, [\, (a_1 - \tfrac{1}{2})\, \partial_a C^a + a_2\, C_W\,] \sum_n {\phi_n}^2 \end{aligned} \tag{3.3}$$

This argument holds for $\tilde{X}$ and $\tilde{\tilde{C}}_a$ since from Eqs.(2.4h), (2.7b) and (2.7c)

$$\delta\tilde{\tilde{C}}_a = (C^a\partial_a + \frac{2-\gamma}{2}\partial_a C^a - \gamma C_W)\tilde{\tilde{C}}_a + ..., \tag{2.4h'}$$

where only the terms proportional to $\tilde{\tilde{C}}_a$ are written. The $d\tilde{C}^a$ behaves like a contravariant vector and the terms in the first bracket of (2.4g) do not contribute by the same reason as the terms, $C^a\partial_a + \frac{1}{2}\partial_a C^a$, do not. The regularization of $\Sigma\, \phi_n\, \phi_n$ is given in [2] by

$$\lim_{M\to\infty} \sum_n \phi_n \exp(-\lambda_n^2 / M^2)\, \phi_n$$
$$= \lim_{M\to\infty} \mathrm{Tr} \int \frac{d^2k}{(2\pi)^2} e^{-ikx} \exp(-H^\phi / M^2)\, e^{ikx} \quad , \tag{3.4}$$

where the kinetic operator is given by

$$H^\phi = -\rho^{b_1} \partial\!\!\!/ \rho^{b_2} \bar{\partial}\!\!\!/ \rho^{b_1} \quad , \tag{3.5}$$

with ($b_1 = -\alpha$, $b_2 = 0$), ($b_1 = -2\beta$, $b_2 = 2\gamma$) and ($b_1 = \gamma$, $b_2 = -4\beta$) for $\phi = \tilde{X}_\mu$, $\tilde{C}^a$, and $\tilde{\tilde{C}}_a$, respectively, and

$$H^\phi \phi_n = \lambda_n^2 \phi_n \ , \quad \int d^2x\, \phi_m(x)^\dagger \phi_n(x) = \delta_{mn} \ . \tag{3.6}$$

We have neglected the divergent terms in Eq.(3.4) proportional to M^2 which may be regularized to be zero if we adopt the other regularization.

Noting that there is no Tr for the field $\tilde{X}$, we have the total BRS anomaly :

$$A_{BRS} = A^{\tilde{X}} + A^{\tilde{C}} + A^{\tilde{\tilde{C}}} = -\frac{1}{12\pi}[\,\frac{1}{2}\{D\alpha(\alpha - 1) - 8(\beta + \gamma)(\beta - 1) - 2(4\beta + \gamma)(\gamma - 1)\}$$

$$\cdot\,\partial_a C^a + \{\, D\alpha^2 - 8\beta(\beta + \gamma) - 2\gamma(4\beta + \gamma)\,\}\, C_W\,]\,\partial^2 \ln \rho\ . \tag{3.7}$$

The anomaly equation, after calculating the BRS current from the Lagrangian (2.8), is given by

$$\int d\rho d\mu\ e^{iS}\,(\,\partial_a J^a_{BRS} - A_{BRS}\,) = 0\ , \tag{3.8}$$

where

$$S = \int d^2x\, L\ , \tag{3.9}$$

$$J^a_{BRS} = -\,C^b\,\partial_b\,(\rho^{-\alpha}\,\tilde{X}\,)\,\partial^a\,(\rho^{-\alpha}\,\tilde{X}\,) + (\tilde{\tilde{C}}\rho^{\gamma})\,\gamma^a\,C^b\,\partial_b\,(\rho^{-2\beta}\,\tilde{C}\,)\ . \tag{3.10}$$

Now the condition that the anomaly vanish should be satisfied to have a unitary and renormalizable theory, which gives two equations from (3.7) as

$$D\alpha^2 - 8\beta(\beta + \gamma) - 2\gamma(4\beta + \gamma) = 0\ ,\ D\alpha - 16\beta - 10\gamma = 0\ . \tag{3.11}$$

Requiring ratios of α, β and γ are real, we get the condition on D as

$$D < 0 \ \text{ or } \ D \geq 26\ . \tag{3.12}$$

We have assumed that α, β and γ are different from each other. This is possible because the Lagrangian (2.8) seems to be essentially free theory and hence there is no relation among conformal weights, α, β and γ, for the fields, $\tilde{X}_\mu$, $\tilde{C}^a$, and $\tilde{\tilde{C}}_a$. One freedom, say α, is due to the scale invariance, however, there still remain two degrees of freedom, β and γ. This freedom can not be fixed even by taking into account the divergent terms proportional to ρ since those can be always subtracted by an appropriate counterterm.

The constraint would be given by the BRS symmetry as will be discussed below. The Lagrangian (2.8) is derived so that it is invariant under the BRS transformation at the tree level. At the one-loop level the anomaly arises in the BRS conservation equation as we have shown in this section. Due to the nilpotency of the BRS transformation, however, the BRS transformation of this anomaly should vanish since

$$\int d^2x\, \delta A_{BRS} = \delta \int d^2x\, \partial_a J^a_{BRS} = \delta^2 S = 0 \tag{3.13}$$

up to a total divergence. This equation is nothing but the Wess-Zumino consistency condition.[6] When we calculate δA_{BRS} from Eq.(3.7) without imposing the anomaly-vanishing condition given by Eq.(3.11), however, it does not vanish. This is due to the special regularization we have adopted, which is invariant under both general coordinate and local conformal transformations. This form of the anomaly, to be precise the coefficient of the FP ghosts in Eq.(3.7), is called a covariant form in [7]. We would like to claim that this freedom would be fixed by demanding the Wess-Zumino consistency condition on A_{BRS} [6], i.e., $\delta A_{BRS} = 0$ since this condition is nonlinear. To achieve this we must find the regularization which preserves the BRS invariances.[1] The relation of our results with the standard conformal anomaly in two dimensions should be mentioned here. That anomaly is given by $\sqrt{g}R$ up to a constant and if it is multiplied by C_W, it becomes BRS-invariant. In other words, the term $C_W\sqrt{g}R$ satisfies Eq.(3.13). Our result given by Eq.(3.7) is different from this expression and it does not satisfy Eq.(3.13).

Even from the form Eq.(3.7), we can derive the critical dimension as follows. We can determine the functional measure so that anomaly cancellation occurs as far as possible. This is achieved by demanding that at first there is no BRS anomaly due to the general coordinate invariance, which corresponds to take $\alpha = \beta = \gamma = 1$ as in [2]. However we might also be able to demand that at first there is no BRS anomaly due to the local conformal invariance. Unfortunately in this case the BRS anomaly vanishes since this case corresponds to take $\alpha = \beta = \gamma = 0$. As easily seen from (3.7) the anomaly vanishes. However there may be an intermediate case between these two extreme cases and is given by $\alpha = \beta = \gamma$. In this case the anomaly is given by

$$A_{BRS} = \alpha\,(\,\alpha\, C_W + \frac{\alpha - 1}{2}\partial_a C^a\,)\frac{26 - D}{12\pi}\partial^2 \ln\rho\ , \tag{3.14}$$

which vanishes at $D = 26$. This expression still has one parameter, α, which can take any value. If you take $\alpha = 0$, A_{BRS} vanishes identically and it can not determine the critical dimension. In this case we may need to calculate the BRS charge squared to obtain the critical dimension a la Ref.[4]. The parameter α would be fixed if the Wess-Zumino consistency condition [6] is imposed on A_{BRS} as mentioned above. $\alpha = 1$ is

---

[1]Work in pregress. The BRS invariant form of the anomaly corresponding to Eq.(3.7), up to a constant, is found to be $\delta(\ln\rho)\,\partial^2\ln\rho$, which is obviously different from $C_W\sqrt{g}R$.

the case adopted in [2], where there is no BRS anomaly due to the general coordinate invariance and $\alpha = 0$ is the case where there is no BRS anomaly due to the local conformal invariance. In other words, Eq.(3.14) includes two indistinguishable covariant anomalies due to breakdown of the general coordinate and local conformal invariances although the associated FP ghosts are different. The parameter α is related to a ratio of two anomalies included in Eq.(3.14).

Now the relation between [2] and ours is given as follows. In [2] the local conformal gauge is not fixed, the auxiliary field E in (2.8) is integrated, and $\alpha = 1$ is adopted. Since there is no BRS anomaly in [2], the local conformal anomaly is calculated, which is nothing but the coefficient of C_W in Eq.(3.14) for $\alpha = 1$.

Acknowledgements

The author would like to thank K.S. Viswanathan for discussions. This work has been partly supported by the Natural Science and Engineering Research Council of Canada.

References

[1] A. Polyakov, Phys. Lett. 103B (1981) 207.

[2] K. Fujikawa, Phys. Rev. D25 (1982) 2584.

[3] C. Becchi, A. Rouet, and R. Stora, Ann. Phys. (N.Y.) 98 (1976) 287.

[4] M. Kato and K. Ogawa, Nucl. Phys. B212 (1983) 443.

[5] K. Fujikawa, Nucl. Phys. B226 (1983) 437; K. Fujikawa and O. Yasuda, Nucl. Phys. B245 (1984) 436.

[6] J. Wess and B. Zumino, Phys. Lett. 37B (1971) 95.

[7] W. Bardeen and B. Zumino, Nucl. Phys. B244 (1984) 421.

CONFORMAL FIELD THEORY IN TWO DIMENSIONS AND CRITICAL PHENOMENA

H.C. Lee and Z.Y. Zhu*

Theoretical Physics Branch
Chalk River Nuclear Laboratories
Atomic Energy of Canada Limited
Chalk River, Ontario, Canada K0J 1J0

and

*Institute of Theoretical Physics
Academia Sinica, Beijing, China

TABLE OF CONTENT

I. CONFORMAL INVARIANCE AND CRITICAL BEHAVIOR

In 1984 Belavin, Polyakov and Zamolodchikov (BPZ) published two theoretical papers[1] that led to remarkable advances in our understanding of the properties of two dimensional systems at critical point. In their program, the rigorous languages of field theory and representation theory of algebra were used to show that conformal invariance is the key to understanding critical behaviour, and that in two dimensions the invariance yields a far richer theory than in other dimensions. The most imporant results of the program are:

i) Two dimensional systems at critical point are described by theories characterized by a quantity c known as the conformal anomaly. In each theory there is a set of "primary" scaling fields each with a specific scaling dimension h.

ii) When $c < 1$, each theory is labelled by an integer $m \geq 3$, and all scaling dimensions in the theory are rational numbers[2].

iii) All fields other than primary fields can be generated from the latter by infinitisimal conformal transformation.

iv) All correlation functions are reducible to correlation functions of primary fields, which satisfy differential equations rendering the functions calculable.

In a parallel development, conformal invariance was directly exploited successfully by using conformal mapping to extract properties of an infinite system from those of finite samples[3].

This paper represents the authors effort to understand the body of work briefly described above. A number of reviews on the subject exist[4-6], including a recent one by Cardy[4], which gives a particularly comprehensive physical description of the topic. Here we have striven to make the field theoretical aspects of the BPZ theory more accessible than in the original papers.

First a few words on why one may expect conformal invariance to have anything to do with critical behavior. A characteristic property of a statistical system at critical point is that the correlation length becomes infinite, and fluctuations at all length scales take place[7]. Clearly the system must then be invariant under uniform changes of length scales, since it does not admit any fixed scale. The realization of the importance of scale invariance[8] around the mid 1960's contributed greatly to the understanding of critical behavior and, a few years later, led to the development of the renormalization group technique[9] as a powerful computational tool. In 1970, Polyakov[10] first pointed out that critical systems should have conformal invariance. A rigorous definition of conformal transformation will be given in the next section. Roughly, it includes the Lorentz transformations - translation and rotation - and scale transformations in which the scale is a slowly varying local function. Thus, at any given point and within a sufficiently small region, a conformal transformation looks like a scale transformation. Therefore, one can expect a system with only short-range interaction to have conformal invariance at critical point. Statistical systems with nearest or next-to-nearest neighbor interactions - the Ising model is an example - satisfy this requirement. The group of conformal transformations is special in two dimensions: it is infinite dimensional there but is finite in all other dimensions. In two dimensions the group is isomorphic to the group of analytic functions, and has an associated infinite dimensional algebra - the Virasoro algebra[11]. It is the infinite dimensionality of this infrastructure that gives essence to the completeness of the BPZ formalism.

We give a few examples of conformal transformations on the complex z-plane:

i) $z \to w = (L/\alpha)\ell nz$, $0 < \alpha \leq 2\pi$. Transforms a wedge of open angle α on the z-plane to a semi-infinite strip of width L on the w-plane. If $\alpha = 2\pi$, the two edges of the strip are identified, in which case the strip is equivalent to the surface of a cylinder; otherwise the two edges are free.

ii) $z \to w = z^{\beta/\alpha}$. Transforms a wedge of open angle α on the z-plane to a wedge of open angle β on the w-plane. If $\alpha = 2\pi$, the w-plane wedge is equivalent to a conical surface.

iii) $z \to w = e^{-2\pi z/\ell}$. Transforms a strip of width ℓ and length L, with the edges identified, on the z-plane to a unit disc with a hole of radius $e^{-2\pi L/\ell}$ on the w-plane. The hole disappears in the limit of $L/\ell \to \infty$.

iv) $z \to w = (z-1)/(z+1)$. Special case of Möbius or projective transformation, transforms the whole z-plane to the whole w-plane, as follows: real axis in z to real axis in w; $z = \infty$ to $w = +1$; $z = -1$ to $w = \infty$; unit circle in z to imaginary axis in w and vice versa; interior (exterior) unit circle in z to left (right) half-plane in w and vice versa. This transformation can be decomposed into the product of translation, inversion, translation. Note that $z \to -1/z$ is conformal.

A key property of a system with conformal invariance is the existence of scaling fields $\phi(z)$ that transform as $\phi(z) \to (dz/dw)^{-h}\phi(w)$ under $z \to w$, where h is the scaling dimension. This gives the important relation

$$\langle\prod_i \phi_i(z_i)\rangle = \Big[\prod_i (dz_i/dw_i)^{-h_i}\Big] \langle\prod_i \phi_i(w_i)\rangle \tag{0}$$

for the correlation function. For simple scale transformation $z \to \lambda z$, (0) above reduces to the usual form. We will return to (0) several times.

The development of the rest of this article follows closely the outline given in the table of content.

2. CONFORMAL INVARIANCE AND TRACELESS STRESS TENSOR

Consider an action

$$I = \int d^D x \mathscr{L}\big(\phi_i(x), \phi_{i,\mu}(x)\big) \tag{1}$$

where the lagrangian density is a function of fields $\phi_i(x)$ and their derivatives

$$\phi_{i,\mu}(x) = \frac{\partial}{\partial x^\mu}\phi_i(x) = \partial_\mu \phi_i(x) \tag{2}$$

but is not an explicit function of the coordinates x^μ. Under a coordinate change

$$x^\mu \to x'^\mu = x^\mu + \delta x^\mu \tag{3}$$

and a form change of the field

$$\phi_i(x) \to \phi'_i(x) = \phi_i(x) + \bar{\delta}\phi_i(x) \tag{4}$$

the total field change is

$$\delta\phi_i(x) = \phi'_i(x') - \phi_i(x) = \bar{\delta}\phi_i(x) + \phi_{i,\mu}(x)\ \delta x^\mu \tag{5}$$

the lagrangian is changed by

$$\delta\mathcal{L} = \frac{\delta\mathcal{L}}{\delta\phi_i}\ \bar{\delta}\phi_i + \frac{\delta\mathcal{L}}{\delta\phi_{i,\mu}}\ \bar{\delta}\phi_{i,\mu} \tag{6}$$

and the integration measure by

$$\delta(d^D x) = d^D x\ \partial_\mu(\delta x^\mu) \tag{7}$$

The variation in the action is then

$$\begin{aligned}\delta I &= \int d^D x\{\partial_\mu(\mathcal{L}\ \delta x^\mu) + \frac{\delta\mathcal{L}}{\delta\phi_i}\ \bar{\delta}\phi_i + \frac{\delta\mathcal{L}}{\delta\phi_{i,\mu}}\ \bar{\delta}\phi_{i,\mu}\}\\ &= \int d^D x\ \partial_\mu\{\frac{\delta\mathcal{L}}{\delta\phi_{i,\mu}}\ \delta\phi_i + (g^{\mu\nu} - \frac{\delta\mathcal{L}}{\delta\phi_{i,\mu}}\ \phi^\nu_{i,}\ )\delta x_\nu\}\\ &\quad + \int d^D x\ (\frac{\delta\mathcal{L}}{\delta\phi_i} - \partial_\mu\ \frac{\delta\mathcal{L}}{\delta\phi_{i,\mu}})\ \bar{\delta}\phi_i\end{aligned} \tag{8}$$

The last term in (8) vanishes by the equation of motion. The remaining two terms separates δI into a part $\delta_\phi I$ coming from the field change $\delta\phi_i$ and a part $\delta_x I$ from the coordinate change δx^μ. The quantity

$$T^{\mu\nu} \equiv -g^{\mu\nu} + \frac{\delta\mathcal{L}}{\delta\phi_{i,\mu}}\ \phi^\nu_{i,} \tag{9}$$

is the stress tensor. If the field changes are restricted to those satisfying

$$\delta\phi_i = 0 \tag{10}$$

Then the invariance of the action $\delta I\big|_{\delta\phi=0} = \delta_x I = 0$ means

$$\partial_\mu(T^{\mu\nu}\delta x_\nu) = 0 \tag{11}$$

Under coordinate translations

$$\delta x^{\mu} = a^{\mu} = \text{constants}$$

so invariance under translation simplifies to

$$\partial_{\mu} T^{\mu\nu} = 0 \tag{12}$$

Consider now a transformation

$$x^{\mu} \to \xi^{\mu}(x) = x^{\mu} + \varepsilon^{\mu}(x) \tag{13}$$

A conformal transformation is one such that

$$g_{\alpha\beta}(x) \to g'_{\alpha\beta}(\xi) = \frac{dx^{\mu}}{d\xi^{\alpha}} \frac{dx^{\nu}}{d\xi^{\beta}} g_{\mu\nu}(x) = \rho(x) g_{\alpha\beta}(x) \tag{14}$$

where $\rho(x)$ is an arbitrary scalar function. If (13) is conformal and infinitesimal, then

$$g'_{\alpha\beta}(\xi) = (\delta^{\mu}_{\alpha} - \varepsilon^{\mu}_{\ ,\alpha})(\delta^{\nu}_{\beta} - \varepsilon^{\nu}_{\ ,\beta}) g_{\mu\nu} = \rho g_{\alpha\beta} \tag{15}$$

That is, ε^{μ} must satisfy

$$(1-\rho) g_{\alpha\beta} = g_{\alpha\nu} \varepsilon^{\nu}_{\ ,\beta} + g_{\beta\nu} \varepsilon^{\nu}_{\ ,\alpha} \tag{16}$$

From (16) and (12) it can be derived that

$$\partial_{\mu}(T^{\mu\nu} \varepsilon_{\nu}) = \frac{1}{2} (1-\rho) \mathrm{Tr}(T) \tag{17}$$

Therefore, the vanishing of the trace of the stress tensor

$$\mathrm{Tr}(T) = 0 \tag{18}$$

is a necessary condition for the action to be invariant under conformal coordinate transformations.

3. CONFORMAL WARD IDENTITIES

Here we show the stress tensor is a generator of variations arising from coordinate transformation in expectation values such as correlation functions. Consider an operator

$$\mathcal{O}(\phi(\xi)) \equiv \mathcal{O}(\phi_1(\xi_1), \phi_2(\xi_2), \cdots) \tag{19}$$

and the expectation values

$$\langle \mathcal{O}(\phi(\xi)) \rangle = \int [\mathcal{D}\phi] \, \mathcal{O}(\phi(\xi)) e^{-I[\phi]} \tag{20}$$

$$\langle \mathcal{O}(\phi(\xi+\varepsilon)) \rangle = \int [\mathcal{D}\phi] \, \mathcal{O}(\phi(\xi+\varepsilon)) e^{-I[\phi]} \tag{21}$$

where

$$\xi^\mu \to \xi^\mu + \varepsilon^\mu(z) \tag{22}$$

is a coordinate transformation. By changing the path-integration variable from $\phi(\xi+\varepsilon)$ to $\phi(\xi)$, the right-hand side of (21) becomes

$$\int [\mathcal{D}\phi] \mathcal{O}(\phi(\xi)) e^{-(I-\delta_\varepsilon I)[\phi]} \tag{23}$$

where $\delta_\varepsilon I$ represents the change in I under a coordinate transforamtion corresponding to (22). Comparing (20) and (21), we obtain the change in the expectation value

$$\langle \delta_\varepsilon \mathcal{O}(\phi(\xi)) \rangle = \langle \mathcal{O}(\phi(\xi)) \delta_\varepsilon I[\phi] \rangle \tag{24}$$

If I is invariant under (22), such as in the case of translations and rotations, then there is no change in the expectation value. In the case of conformal transformation, it is generally not possible for $\varepsilon(z)$ to be analytic and small everywhere. However, one can always devise a transformation that is conformal and small inside a domain M_1 enclosing all the points ξ_i. In the complement of M_1, which we call M_2, the transforation need not be conformal, so it can be made small, analytic and approach identity at infinity. In M_2, $\delta_\varepsilon I \neq 0$, so from (8), (9) and (24)

$$\begin{aligned} \langle \delta_\varepsilon \mathcal{O}(\phi(\xi)) \rangle &= -\langle \int_{M_2} d^D x \, \partial_\mu \xi_\nu(x) \, T^{\mu\nu}(x) \mathcal{O}(\phi(\xi)) \rangle \\ &= -\int_{M_2} \partial_\mu \varepsilon_\nu(x) \, \langle T^{\mu\nu}(x) \mathcal{O}(\phi(\xi)) \rangle \end{aligned} \tag{25}$$

The interchange of integration and computation of expectation value is permitted because M_2 excludes all the points ξ_i, so the expectation values are analytic functions. We can now use Stoke's theorem to convert the integration over M_2 to that over the boundary ∂M_2, which has two pieces, one at infinity and the other is the boundary between M_2 and M_1. Since $\varepsilon \to 0$ at infinity, only the latter, with $\partial M_2 = -\partial M_1$ contributes, so

$$\langle \delta_\varepsilon \mathcal{O}(\phi(\xi)) \rangle = \int_{\partial M_1} n_\mu \varepsilon_\nu(x) \, \langle T^{\mu\nu}(x) \mathcal{O}(\phi(\xi)) \rangle$$

$$= \int_{M_1} \partial_\mu \varepsilon_\nu(x) \, \langle T^{\mu\nu}(x) \mathcal{O}(\phi(\xi)) \rangle \tag{26}$$

where n is the outward normal of ∂M_1. Note that even though $x \to x + \varepsilon(x)$ is a conformal transformation in M_1, so that $\partial_\mu \varepsilon_\nu T^{\mu\nu} = 0$, the last expression in (26) does not vanish because here taking expectation value and integrating do not commute, since the expectation value can be singular at points ξ_i, which are enclosed in M_1. Indeed, since the integrand is analytic everywhere except at $x = \xi_i$, M_1 can be deformed to a set of isolated domains M_i each surrounding a point ξ_i. Then (26) becomes the conformal Ward identity[1]

$$\langle \delta_\varepsilon \mathcal{O}(\phi(\xi_1, \xi_2 \cdots)) \rangle = \sum_i \int_{\partial M_i} n_\mu \varepsilon_\nu(x) \, \langle T^{\mu\nu}(x) \mathcal{O}(\phi(\xi)) \rangle \tag{27}$$

We now specialize to two dimensions. In Euclidean space, the line element is

$$d^2 s = (dx^1)^2 + (dx^2)^2 \tag{28}$$

Define light-cone coordinates $z = z^+$, $\bar{z} = z^-$

$$z^{\pm} = x^1 \mp i x^2 \equiv z_{\mp} \tag{29}$$

then

$$d^2 s = dz \, d\bar{z} \tag{30}$$

Under any analytic transformation

$$z \to \zeta(z) = z + \varepsilon(z), \quad \bar{z} \to \bar{\zeta}(\bar{z}) = \bar{z} + \bar{\varepsilon}(\bar{z}) \tag{31}$$

$$d^2 s \to d\zeta d\bar{\zeta} = \left(\frac{d\zeta}{dz}\right)\left(\frac{d\bar{\zeta}}{d\bar{z}}\right) d^2 s \tag{32}$$

so that, according to (14), (31) is always conformal. In terms of light-cone components

$$\mathrm{Tr}(T) = 2T^{+-} = 0 \tag{33}$$

$$T^{++} = \frac{1}{4}(T^{11} - T^{22} - 2iT^{12}) \tag{34a}$$

$$T^{--} = \frac{1}{4}(T^{11} - T^{22} + 2iT^{12}) \tag{34b}$$

$$\partial_\mu T^{\mu +} = \partial_+ T^{++} = \partial T^{++}/\partial z = 0 \tag{35a}$$

$$\partial_\mu T^{\mu -} = \partial_- T^{--} = \partial T^{--}/\partial \bar{z} = 0 \tag{35b}$$

So T^{++} (T^{--}) is a function of $\bar{z}(z)$ but not of $z(\bar{z})$. This enables us to write

$$T(z) \equiv \pi T^{--}(z), \quad \bar{T}(\bar{z}) \equiv \pi T^{++}(\bar{z}) \tag{36}$$

Note that $\bar{T}$ is not the complex conjugate of T. Often $T(\bar{T})$ is called the (anti)analytic component of $T^{\mu\nu}$.

In light-cone components, the right-hand side of (27) reads

$$\frac{1}{2\pi}\sum_i \int_{\partial M_i} ds\{n_{\bar{z}}\, \varepsilon(z) \langle T(z)\mathcal{O}(\phi(\xi))\rangle + n_z \bar{\varepsilon}(\bar{z})\langle \bar{T}(\bar{z})\mathcal{O}(\phi(\xi))\rangle\} \tag{37}$$

where the line integration along the contours ∂M_i is explicitly expressed and $n_{\bar{z}} = n_1 + in_2$ ($n_z = n_1 - in_2$) is the outward normal of ∂M_i along $\bar{z}(z)$. Note that each ∂M_i surrounds a point ξ_i in the x^1x^2-plane where at least one of the integrands is singular. Suppose the counter-clockwise contour $C_i \in \{\partial M_i\}$ is around a point $\zeta_i = \xi_i$ where $\langle T(z)\mathcal{O}\rangle$ is singular. Then around this point

$$ds\, n_{\bar{z}} = |z-\zeta_i|e^{i\sigma}d\sigma = -i\, d(z-\zeta_i) = -i\, dz \tag{38}$$

where σ is the angle relative to ζ_i on C_i. Thus for this point the integral can be converted to an integral in the complex z-plane

$$\frac{1}{2\pi i}\int_{C_1} dz\, \varepsilon(z) \langle T(z)\mathcal{O}(\phi(\xi))\rangle$$

This conversion can be carried out for all the integrals in (37) and we have

$$\langle\delta_\varepsilon \mathcal{O}(\phi(\zeta,\bar{\zeta}))\rangle = \frac{1}{2\pi i}\{\sum_i \oint_{C_i} dz\varepsilon(z)\ \langle T(z)\mathcal{O}(\phi(\zeta,\bar{\zeta}))\rangle$$

$$- \sum_i \oint_{\bar{C}_i} dz\ \bar{\varepsilon}(z)\ \langle\bar{T}(z)\mathcal{O}(\phi(\zeta,\bar{\zeta}))\rangle\} \tag{39}$$

where $\mathcal{O}$ is written as a function of ζ_i and $\bar{\zeta}_i$. Note that z in (39) is just a dummy integration variable, and because T and $\bar{T}$ are mutually independent, the two sets of contours $\{C_i\}$ and $\{\bar{C}_i\}$ are <u>different</u>. $C_i(\bar{C}_i)$ surround points $z = \zeta_i(\bar{\zeta}_i)$ where $\langle T\mathcal{O}\rangle$ ($\langle\bar{T}\mathcal{O}\rangle$) is singular. In the following, the $\bar{T}$ term, which generally is just a copy of the T term, will often be suppressed, as will the factor $1/2\pi i$ accompanying the contour integrals.

The conformal Ward identity (39) expresses concretely the assertion that the stress tensor T(z) (and $\bar{T}(z)$) is a generator of conformal transformations. One therefore expects an algebra associated with T(z). To see this, define

$$T_\varepsilon \mathcal{O}(\zeta) \equiv \oint_C dz\ \varepsilon(z)\ T(z)\mathcal{O}(\zeta) \tag{40}$$

where C is a circle centred at z = 0 enclosing all the points in the operators to the right of T(z). The interpretation of this definition is as follows. If we express $z = e^{\tau+i\sigma}$ then the contour is defined by holding the "time" variable τ constant. The convention (40) assures that the time associated with T_ε is later than the time associated with the operators on which T_ε acts. In other words, the operators on both sides of (40) are time ordered. It is then clear that

$$[T_\varepsilon, \mathcal{O}(\zeta)] = \oint_{C_\zeta} dz\ \varepsilon(z)\ T(z)\mathcal{O}(\zeta) = \delta_\varepsilon\mathcal{O}(\zeta) \tag{41}$$

where C_ζ is a contour around ζ and <u>not</u> centred at z = 0. Then (39) can be written as

$$\langle\delta_\varepsilon\mathcal{O}(\phi(\zeta))\rangle = \langle[T_\varepsilon, \mathcal{O}(\phi(\zeta))]\rangle \tag{42}$$

The fact that the integral in the Ward identity picks out singularities at $z = \zeta_i$ reveals the decisive dependence of the conformal change to $\mathcal{O}(\phi(\zeta))$ on the short distance behavior of the operator product $T(z)\mathcal{O}(\phi(\zeta))$. A particularly important case is when $\mathcal{O}$ is $T(\zeta)$ itself. In the following we first show by examples how $\delta_\varepsilon T(\zeta)$ is computed from (39) and then discuss the algebra content of (42).

4. EXAMPLES

We are interested in examining the conformal transformation of $T(\zeta)$. This involves, from the right-hand side of (39), evaluating the operator product $T(z)T(\zeta)$. On the other hand, (40) and the left-hand side of (41) show that the transformation is related to the commutator between the two T's. So we expect the change to be linear in ε, and at most linear in T. In two dimensions, in powers of momentum, T and ε have dimensions 2 and -1 respectively, so we expect $\delta_\varepsilon T$ to have the form

$$\delta_\varepsilon T \sim a\varepsilon\partial T + b\partial\varepsilon T + c\partial^3\varepsilon$$

where a, b and c are dimensionless constants. We will in fact show that for z near ζ, $T(z)T(\zeta)$ has the universal expression

$$T(z)T(\zeta) = \frac{c/2}{(z-\zeta)^4} + \frac{2}{(z-\zeta)^2} T(\zeta) + \frac{1}{z-\zeta} \partial_\zeta T(\zeta)$$
$$+ \text{ nonsingular terms} \tag{43}$$

4.1 Free Scalar Field

Consider the lagrangian for the scalar field X

$$\mathcal{L} = -\frac{1}{2} :\partial_\mu X \partial^\mu X: = -\frac{1}{2}:\left(\partial_+ X \partial^+ X + \partial_- X \partial^- X\right): \tag{44}$$

where operators bracketed by the :: are normal ordered. For normal ordered operator products, (39) holds without the need to take expectation values. Then (within an unimportant normalisation)

$$T(z) = -\frac{1}{2}:(\partial_z X)^2: , \qquad \bar{T}(\bar{z}) = -\frac{1}{2}:(\partial_{\bar{z}} X)^2: \tag{45}$$

and the relation (35) is satisfied by virtue of the equation of motion

$$\partial_z \partial_{\bar{z}} X = 0$$

We now use Wick's expansion to express $T(z)T(\zeta)$ in normal ordered products and then expanded around $z = \zeta$. For this purpose we need the propagator

$$\langle X(x)X(y)\rangle = -2\, \ell n|x-y| = -\ell n(z-\zeta) - \ell n(\bar{z}-\bar{\zeta}) \tag{46}$$

where z, $\bar{z}$ $(\zeta,\bar{\zeta})$ are the light-cone coordinates for $x(y)$. Only the first term on the right-hand side is relevant to $\delta_\varepsilon T(\zeta)$. In the following we use the simplifying notation $W \equiv X(x)$, $X \equiv X(y)$, $\bar{\partial} \equiv \partial_z$, $\partial = \partial_\zeta$, $\eta \equiv z-\zeta$, then

$$T(z)T(\zeta) = \frac{1}{4}:(\bar{\partial}W)^2:\ :(\partial X)^2:$$

$$= \frac{1}{4}\left(4:\bar{\partial}W\partial X:\ \langle\bar{\partial}W\partial X\rangle + \langle\bar{\partial}W\partial X\rangle^2\right) \tag{47}$$

Substituting

$$\langle\bar{\partial}W\ X\rangle = -\frac{1}{\eta}$$

and other derivatives computed from (45) and expanding W around ζ, we obtain

$$T(z)T(\zeta) = \frac{2}{\eta^2}\, T(\zeta) + \frac{1}{\eta}\, \partial T(\zeta) + \frac{1}{2\eta^4} + \mathcal{O}(1) \tag{48}$$

Similarly,

$$T(z)X(\zeta) = -\frac{1}{2}:(\bar{\partial}W)^2:X = -\bar{\partial}W\langle\bar{\partial}W\ X\rangle$$

$$= \frac{1}{\eta}\, \partial X + \partial^2 X \tag{49}$$

$$T(z)\partial X(z) = \frac{1}{\eta^2}\, \partial X + \frac{1}{\eta}\, \partial^2 X + \mathcal{O}(1) \tag{50}$$

$$T(z)e^{i\alpha X(z)} = \frac{\alpha^2}{2\eta^2}\, e^{i\alpha X} + \frac{1}{\eta}\, \partial e^{i\alpha X} + \mathcal{O}(1) \tag{51}$$

We apply (39), and remember that the contour integral picks out only pole terms, to obtain

$$\delta_\varepsilon T = \varepsilon\partial T + 2\partial_\varepsilon T + \frac{1}{12}\,\partial^3\varepsilon \tag{52}$$

$$\delta_\varepsilon X = \varepsilon\partial X \tag{53}$$

$$\delta_\varepsilon \partial X = \varepsilon\partial^2 X + \partial\varepsilon\cdot\partial X \tag{54}$$

$$\delta_\varepsilon e^{i\alpha X} = \varepsilon\partial e^{i\alpha X} + \frac{\alpha^2}{2}\,\partial\varepsilon\; e^{i\alpha X} \tag{55}$$

Note that $\delta_\varepsilon T$ indeed has the expected form.

4.2 Free Majorana Fermions

The lagrangian is

$$\mathcal{L} = -\frac{1}{2}{:}\psi\partial_{\bar{z}}\psi{:} - \frac{1}{2}{:}\bar{\psi}\partial_z\bar{\psi}{:} \tag{56}$$

with the stress tensors

$$T(z) = -\frac{1}{2}{:}\psi\partial_z\psi{:}$$

$$T(\bar{z}) = -\frac{1}{2}{:}\bar{\psi}\partial_{\bar{z}}\bar{\psi}{:} \tag{57}$$

Use notation $\phi \equiv \psi(z)$, $\psi \equiv \psi(\zeta)$ and ∂, $\bar{\partial}$, η as before, then the propagator is

$$\langle\phi\psi\rangle = \frac{1}{\eta} \tag{58}$$

and

$$T(z)T(\zeta) = \frac{1}{4}{:}\phi\bar{\partial}\phi{:}\;{:}\psi\partial\psi{:} = \frac{1}{4}{:}\{\phi\partial\psi\langle\bar{\partial}\phi\psi\rangle + \langle\phi\partial\psi\rangle\partial\phi\psi - \phi\psi\langle\bar{\partial}\phi\partial\psi\rangle$$

$$- \langle\phi\psi\rangle\bar{\partial}\phi\partial\psi + \langle\phi\partial\psi\rangle\langle\bar{\partial}\phi\psi\rangle - \langle\phi\partial\psi\rangle\langle\bar{\partial}\psi\psi\rangle\}{:} \tag{59}$$

Expand

$$\phi = \psi + \eta\partial\psi + \frac{1}{2}\,\eta^2\partial^2\psi + \cdots$$

and note that

$${:}\psi\psi{:} = 0$$

Then

$$T(z)T(\zeta) = -\frac{1}{2}\eta:\psi\partial^2\psi: - \frac{1}{\eta^2}:\psi\partial\psi: + \frac{1}{4\eta^4} \tag{60}$$

so that

$$\delta_\varepsilon T = \varepsilon\partial T + 2\partial\varepsilon T + \frac{1}{24}\partial^3\varepsilon \tag{61}$$

Similarly, one finds

$$\delta_\varepsilon\psi = \varepsilon\partial\psi + \frac{1}{2}\partial\varepsilon\psi \tag{62}$$

$$\delta_\varepsilon\partial\psi = \varepsilon\partial^2\psi + \frac{3}{2}\partial\varepsilon\ \partial\psi + \frac{1}{2}\partial^2\varepsilon\psi \tag{63}$$

5. SCALING DIMENSION AND CONFORMAL ANOMALY

The examples (52) - (55) and (61) - (63) all have the form

$$\delta_\varepsilon\mathcal{O} = \varepsilon\partial\mathcal{O} + h\mathcal{O}\partial\varepsilon \quad + \text{higher derivatives of } \varepsilon \tag{64}$$

where h is a constant. What is the interpretation of this result? Consider the translation

$$\zeta \to \zeta + \varepsilon(\zeta) = \zeta + a$$

under which

$$\partial_\varepsilon\mathcal{O}(\zeta) = \mathcal{O}(\zeta+a) - \mathcal{O}(\zeta) = a\partial\mathcal{O} = \varepsilon\partial\mathcal{O}$$

This shows that the first term on the rhs of (64) is associated with translation.

For the second term, suppose the scaling dimension of $\mathcal{O}$ is d, then under a scale transformation

$$\zeta \to \lambda\zeta = \zeta + (\lambda-1)\zeta$$

$$\mathcal{O} \to \lambda^d\mathcal{O}$$

so that

$$\partial\varepsilon = \lambda - 1$$

$$\delta_\varepsilon\mathcal{O} = (\lambda^d - 1)\mathcal{O} = [(\partial\varepsilon+1)^d - 1]\mathcal{O} = d\mathcal{O}\partial\varepsilon\ .$$

This shows the constant h in (64) is just the scaling dimension of the operator. The higher derivative terms are changes in $\mathcal{O}$ induced by other types of conformal transformation.

The changes in the stress tensor has the form

$$\delta_\varepsilon T = \varepsilon\partial T + 2T\partial\varepsilon + \frac{c}{12}\partial^3\varepsilon \tag{65}$$

Here h = 2, which is just the naive dimension of T. The constant c is known as the conformal anomaly. We have seen that for scalars, c = 1 and for Majorana or two-component fermions, c = 1/2. Although the c-term in (65) arises perfectly naturally, in quantum field theory c quantifies the breaking of conformal invariance leading to the tracelessness of T (see (18)) by quantum fluctuation at the one-loop level, hence its name.

6. VIRASORA ALGEBRA

To expose the algebraic content of (65), or equivalently, (43), consider the Lourant expansion of T(z)

$$T(z) = \sum_{n=-\infty}^{\infty} \frac{1}{z^{n+2}} L_n; \qquad L_n = \oint dz\, z^{n+1}\, T(z) \tag{66}$$

Then

$$[L_n, L_m] = \oint dz \oint d\zeta \Big|^{\ell n z < \ell n\zeta}_{\ell n z > \ell n\zeta}\, z^{n+1}\zeta^{m+1} T(z)T(\zeta)$$

$$= \oint d\zeta \oint_{C_\zeta} dz\, z^{n+1}\zeta^{m+1} T(z)T(\zeta) \tag{67}$$

The only effect of the ordering of the two L's is on the relative magnitude of the radii of the circular contours for the z and ζ integrations. As explained in §3, the result is a small contour C_ζ around $z = \zeta$ in the z-plane and unrestricted integration in the ζ-plane. Now substitute (43) to obtain for the rhs of (67)

$$\begin{aligned} &\oint d\zeta\, \zeta^{m+1}\left[\frac{c}{12}\partial^3 + 2T(\zeta)\partial + (\partial T(\zeta))\right]\zeta^{n+1} \\ &= \oint d\zeta\, \left[\frac{c}{12} n(n^2-1)\, \zeta^{m+n-1} + (n-m)\zeta^{m+n+1} T(\zeta)\right] \end{aligned} \tag{68}$$

Therefore

$$[L_n, L_m] = (n-m)L_{n+m} + \frac{c}{12}(n^3-n)\delta_{m+n,o} \qquad (69)$$

which identifies the L_n's as the generators of the infinite dimensional Virasora algebra[11] $\mathcal{L}_c$, whose commutation relation is characterized by the existence of a central extension - the last term in (69) - with central charge c.

In parallel with $\mathcal{L}_c$, there is a Virasoro algebra $\bar{\mathcal{L}}_c$ corresponding to $\bar{T}$ and spanned by the generators $\bar{L}_n$. The groups of transformations generated by $\{L_n\}$ and $\{\bar{L}_n\}$ are respectively Γ_c and $\bar{\Gamma}_c$, and the combined group $\Gamma_c \times \bar{\Gamma}_c$ contains all conformal transformations.

The subset $\{L_s; s = 0, \pm 1\}$ closes by (69) and forms a subalgebra $\mathcal{L}_o$ without central extension, which is the largest subalgebra of $_c$. The group corresponding to $\mathcal{L}_o$ is the group of Möbius (or projective) transformations

$$z \to \zeta = \frac{az+b}{cz+d}, \qquad ad-bc = 1 \qquad (70)$$

which are bijective (one-to-one and onto) mappings of $\mathcal{C}^1$ to $\mathcal{C}^1$. To see the relation between $\mathcal{L}_o$ and (70), consider the infinitesimal transformation around the identity, a = d = 1, b = c = 0, then

$$\delta z = \alpha + \beta z + \gamma z^2 \qquad (71)$$

with the α, β, γ terms generated by

$$\delta_{-1} = \partial_z, \qquad \delta_o = z\partial_z, \qquad \delta_{+1} = z^2\partial_z \qquad (72)$$

respectively. It is then see that for any function admitting a Laurant series, δ_{-1}, δ_o and δ_{+1} form an algebra isomorphic to $\mathcal{L}_o$ with the identification $L_s \simeq -\delta_s$. One then sees that L_{-1} generates translation and L_o scale transformation.

In order to utilize the algebraic properties of (69), it is desirable to construct a Hilbert space on which L_n acts. For this purpose, we once again express z as

$$z = e^{\tau + i\sigma} \qquad (73)$$

and identify τ as the time variable. Then the infinite past (future) with $\tau \to -\infty(+\infty)$ corresponds to $z = 0(\infty)$, and $L_o + \bar{L}_o$ generates time shift. The ground state $|0\rangle$ must be such that $T(z)|0\rangle$ and $\langle 0|T(z)$ are regular when $z \to 0$ and $z \to \infty$, respectively. This requires

$$L_n|0\rangle = \langle 0|L_{-n} = 0, \qquad n \geq -1 \tag{74}$$

and similarly for the $\bar{L}_n$ operators. It can be shown that the adjoint of an operator $\mathcal{O}(z)$ with scaling dimension h is

$$(\mathcal{O}(z))^\dagger = z^{-2h}\mathcal{O}(-\frac{1}{z}) \tag{75}$$

In particular, this gives

$$L_n^\dagger = L_{-n} \tag{76}$$

We thus define

$$H = L_o + \bar{L}_o \tag{77}$$

as the Hamiltonian and, from (75), identify

$$L_{-n} = L_n^\dagger , \qquad n \geq 1 \tag{78}$$

and

$$L_n = L_{-n}^\dagger , \qquad n \geq 1 \tag{79}$$

as the lowering and raising operators, respectively.

7. PRIMARY FIELDS AND CONFORMAL FAMILIES

Our examples in section 3 showed that field operators with the simplest conformal property transform as

$$\delta_\varepsilon \phi = \varepsilon \partial \phi + h \phi \partial \varepsilon \tag{80}$$

which implies that, under $z \to z'$

$$\phi(z) \to (\frac{dz'}{dz})^h \phi(z') \tag{81}$$

BPZ called such fields primary fields, which are the only fields that satisfy the simple conformal relation (0). For every formal theory specified by its conformal anomaly c, there may be one or more primary fields ϕ_h, each having a scaling dimension h; it is assumed that the primary field with h = 0 is the identity operator, which must be a field in every theory. BPZ showed that all fields appearing in a theory can be grouped into "conformal families", one associated with each primary field, and every field within a family $[\phi_h]$ is related to the primary field ϕ_h by conformal transformation, and can therefore be obtained from ϕ_h by repeated operation of the stress tensor on the latter.

To see how this comes about consider (80). From it and (41) we have ($\eta \equiv z-\zeta$, and we suppress the label h on ϕ_h)

$$T(z)\phi(\zeta) = \sum_{k=o}^{\infty} \eta^{k-2}\phi^{(-k)}(\zeta) \tag{82}$$

with the first two terms in the rhs, which are singular, known

$$\phi^{(0)} = h\phi, \quad \phi^{(-1)} = \partial\phi \tag{83}$$

and the regular terms ($k \geq 2$) given by

$$\phi^{(-k)}(\zeta) = \oint dz \frac{T(z)}{(z-\zeta)^{k-1}} \phi(\zeta) \tag{84}$$

where the integration contour encloses ζ. The conformal properties of the "secondary" fields are obtained by applying δ_ε on both sides of (82)

$$(\delta_\varepsilon T(z))\phi(\zeta) + T(z)\delta_\varepsilon \phi(\zeta) = \sum_{k=o}^{\infty} \eta^{k-2}\delta_\varepsilon \phi^{(-k)}(\zeta) \tag{85}$$

The ℓhs is

$$\left[\varepsilon(z)T'(z) + 2\varepsilon'(z)T(z) + \frac{c}{12}\varepsilon'''(z)\right]\phi(\zeta) + T(z)\left[\varepsilon(\zeta)\phi'(\zeta) + h\varepsilon'(\zeta)\phi(\zeta)\right] \tag{86}$$

Substitute (82), expand around $z = \zeta$ and compare with the rhs of (86) and we get, for $k \geq 2$

$$\delta_\varepsilon \phi^{(-k)} = \varepsilon \partial \phi^{(-k)} + (h+k)\phi^{(-k)}\partial\varepsilon + \frac{c}{12}\phi\frac{1}{(k-z)!}\partial^{k+1}\varepsilon$$

$$+ \sum_{\ell=1}^{k} \frac{k+\ell}{(\ell+1)!}\phi^{(\ell-k)}\partial^{\ell+1}\varepsilon \quad (87)$$

This in turn implies

$$T(z)\phi^{(-k)}(\zeta) = \sum_{\ell=1}^{k} \frac{k+\ell}{\eta^{\ell+2}}\phi^{(\ell-k)}(\zeta) + \frac{c}{12}\frac{k(k^2-1)}{\eta^{k+2}}\phi(\zeta)$$

$$+ \sum_{\ell=0}^{\infty} \eta^{\ell-2}\phi^{(-\ell,-k)}(\zeta) \quad (88)$$

where

$$\phi^{(0,-k)} = (h+k)\phi , \qquad \phi^{(-1,-k)} = \partial\phi^{(-k)} \quad (89)$$

and the regular terms are

$$\phi^{(-\ell,-k)}(\zeta) = \oint dz \frac{T(z)}{(z-\zeta)^{\ell-1}}\phi^{(-k)}(\zeta) \equiv L_{-\ell}(\zeta)\phi^{(-k)}(\zeta) \quad (90)$$

Knowledge of the conformal properties of $\phi^{(-\ell,-k)}$ can again be obtained by operating δ_ε on both sides of (88) and repeating the steps from (85) to (90). Repetition of this cycle then yields all fields belonging to the conformal block [ϕ]. The conformal property of $\phi^{(-\{k\})}$, $\{k\} = \{k_1,k_2,\cdots,k_n\}$ has the form

$$T(z)\phi^{(-\{k\})}(\zeta) = \sum_{\ell=1}^{N} \frac{1}{\eta^{\ell+2}}\psi_\ell(\zeta) + \sum_{\ell=0}^{\infty} \eta^{\ell-2}\phi^{(-\ell,-\{k\})}(\zeta) \quad (91)$$

where $N = k_1 + k_2 + \cdots + k_n$ is the level of the field, $\psi_\ell(\zeta)$ are known fields,

$$\phi^{(0,-\{k\})} = (h + N)\phi, \qquad \phi^{(-1,-\{k\})} = \partial\phi^{(-\{k\})} \quad (92)$$

$$\phi^{(-k_o,-\{k\})}(\zeta) = L_{-k_o}(\zeta)\phi^{(-\{k\})}(\zeta) = [\prod_{i=o}^{n} L_{-k_i}(\zeta)]\phi(\zeta) \quad (93)$$

where $L_{-k}(\zeta)$ is defined in (90), i increases from left to right and the integration contour for ζ_i encloses ζ and all ζ_j, $j > i$.

8. KAC'S FORMULA, THE FQS CLASSIFICATION AND NULL STATES

States in the Hilbert space associated with the fields discussed in the last section form a basis for representations of the Virasoro algebra. Here the usual rules for constructing representations are followed: first find a highest weight state, then use the lowering operators of the algebra to generate all other states. In this instance, each representation is labelled by the scaling dimension of the primary field, and every representation is infinite dimensional.

We begin by observing that, from (41), (66) and (80), $\phi(\equiv \phi_h)$ satisfies

$$[L_m, \phi(z)] = z^{m+1}\partial\phi + (m+1)z^m h\phi \tag{94}$$

so that the state

$$|\phi\rangle \equiv \phi(0)|0\rangle \tag{95}$$

satisfies

$$L_o|\phi\rangle = h|\phi\rangle \tag{96}$$

$$L_n|\phi\rangle = 0 \qquad\qquad n > 0 \tag{97}$$

This identifies $|\phi\rangle$ as the highest weight state of the representation labelled by h, from which lower states are generated by repeated applications of L_{-n}, $n > 0$. Let

$$\{k\} = \{k_1 \leq k_2 \leq \cdots \leq k_n\}, \qquad \Sigma_{\{k\}} = N \tag{98}$$

then

$$|\phi;\{k\}\rangle = \prod_{\{k\}} L_{-k_i}|\phi\rangle \tag{99}$$

is a level-N state with eigenvalue h+N for L_o, which can be viewed as the Hamiltonian for such states. It is obvious that L_o has spectrum h, h+1, h+2, $\cdots$, and that the degeneracy P(N) for states with eigenvalue h+N is equal to the number of partitions of k_i's satisfying (98).

The infinite number of states defined by (98) and (99) form a basis for an infinite dimensional representation of $\mathcal{L}_c$, and span a space known as the Verma modulus V_h. Note that from (98), states in V_h have ordered $\{k\}$ sets. The Virasoro algebra assures that a level N state with an unordered {k} set is a linear combination, with integer coefficients, of ordered-{k} level N states in V_h. If we view the fields in $[\phi_h]$ as maps from $\mathbb{C}^1$ to $\mathbb{C}^1$, then comparison of (93) to (99) reveals an isomorphism between V_h and the subset of ordered-{k} maps. The discussion above also shows that an unordered-{k} map is a linear combination of ordered-{k} maps.

Let us denote the level N state in V_h by $|\phi_i\rangle$, i = 1,2,•••, P(N), and consider the overlap matarix M_N, whose matrix elements

$$(M_N)_{ij} = \langle\phi_i|\phi_j\rangle \tag{100}$$

are implicit functions of h and the central charge c. Some useful relations for computing M_N are given below; the notation is $\langle\mathcal{O}\rangle \equiv \langle\phi|\mathcal{O}|\phi\rangle$, and it is understood that $n \geq 1$.

$$A_n \equiv \langle(L_{-1})^n(L_{-1})^n\rangle = n(2h + n - 1)A_{n-1} \tag{101a}$$

$$B_n \equiv \langle(L_{-n})^\dagger(L_{-1})^n\rangle = (n+1)!h \tag{101b}$$

$$C_n \equiv \langle(L_{-n})^\dagger L_{-n}\rangle = 2nh + cn(n^2-1)/12 \tag{101c}$$

$$D_n \equiv \langle(L_{-1}L_{-n+1})^\dagger(L_{-1})^n\rangle = 2(h+n-1)B_{n-1} + nD_{n-1} \tag{101d}$$

$$E_n \equiv \langle(L_{-1}L_{-n+1})^\dagger L_{-n}\rangle = (n+1)C_{n-1} \tag{101e}$$

$$F_n \equiv \langle(L_{-1}L_{-n+1})^\dagger L_{-1}L_{-n+1}\rangle = 2(h+n-1)C_{n-1} + n^2C_{n-2} \tag{101f}$$

$$A_0 = 1, \qquad C_0 = h^2, \qquad D_1 = 2h^2 \tag{101g}$$

The two lowest level overlap matrices are

$$M_2 = \begin{pmatrix} 4h+c/2 & 6h \\ 6h & 4h(2h+1) \end{pmatrix} \tag{102}$$

$$M_3 = \begin{pmatrix} 6h+2c & 16h+2c & 24h \\ 16h+2c & 8h^2+34h+hc+2c & 36h(h+1) \\ 24h & 36h(h+1) & 24h(h+1)(2h+1) \end{pmatrix} \tag{103}$$

The relation $P(N)=N$ holds true only for $N=2$ and 3. For $N \geq 4$, $P(N) > N$.

The determinant of M_N is given by the Kac formula[12]

$$\det M_N = \prod_{pq \leq N} \left(h - h(c)_{p,q}\right)^{P(N-pq)} \tag{104}$$

where c is parametrized by the real number m

$$c = c(m) = 1 - 6/m(m+1) \tag{105}$$

and

$$h_{p,q}(c) = h_{p,q}(c(m)) = \frac{[(m+1)p - mq]^2 - 1}{4m(m+1)} \tag{106}$$

with p,q being nonnegative integers.

If the states $|\psi_i\rangle$ are to have physical meaning, then they should have nonnegative norms. Friedan, Qiu and Shenker[2] (FQS) used this criterion, which they called the unitarity condition, to constrain possible values for c and h. They found that whereas the condition gives no constraint when $c \geq 1$, for $c < 1$ the necessary condition for a unitary representation is

$$c_{\text{unitary}} = c(m), \qquad m = 2,3,\cdots \tag{107}$$

$$h_{\text{unitary}} = h_{p,q}(c(m)), \qquad 1 \leq q \leq p \leq m-1 \tag{108}$$

Goddard, Kent and Olive[13] subsequently proved that representations indeed exist for all the values of $c < 1$ and h given by (107,8). Cardy[14] showed that for a given c, if the representation is unitary and the number of h's is finite, then $c < 1$ follows as a consequence of modular invariance.

There is a subtlety to the application of the restriction (108), which arises from the symmetry relation $h_{p,q} = h_{m-p,m-q+1}$. If $p < q$, then $m-p \geq m-q+1$. So the unitary scaling dimensions of FQS actually include all the distinct $h_{p,q}$'s with $1 \leq q,p \leq m-1$. In particular, for any m, $h_{1,m-1}$ is a unitary scaling dimension in spite of (108), and appears together with $h_{m-1,1}$ as roots of $\det M_N$ for N at the lowest possible level, that is, at $N = m-1$.

Since there is a one-to-one correspondence between states in the Verma modulus and conformal fields, the FQS restriction implies the following classification of unitary models for $c < 1$: Each model is labelled by a positive integer $m \geq 3$, with conformal anomaly c given by Kac's formula (105), and within each model there is a finite number of primary fields $\phi_{(p,q)}$, $1 \leq q \leq p \leq m-1$, with scaling dimension $h_{p,q}$ given by the Kac formula (106). This classification solved a long standing puzzle of why, at critical point, should every known statistical model in two dimensions possess only a finite number of scaling fields, each having a scaling dimension equal to a rational number. Each of the unitary classification of FQS has been identified with at least one specific statistical model. For example the $m = 3$ classification with $c = 1/2$, and $h = 0$, $1/2$ and $1/16$, and the $m = 4$ classification, with $c = 7/10$ and $h = 0$, $7/16$, $3/2$, $3/80$, $3/5$ and $1/10$ correspond respectively to the Ising and tricritical Ising models, and the $m = 5$ and $m = 6$ classifications correspond respectively to the Q-state and tricritical Q-state Potts models with $Q = 3$. It is typical of the universal nature of the BPZ theory that each classification can be realized in many apparently different ways[4,15].

The fact that each allowed scaling dimension h is a root of the determinant of M_N for some N implies that not all the P(N) states $|\psi_i\rangle$ associated with h are linearly independent. Let X be the $P(N) \times P(N)$ matrix that diagonalizes M_N (X is orthogonal, since M_N is real and symmetric), X_α be the column vectors of X and λ be the diagonalized M_N, then

$$M_N \cdot X = X \cdot \lambda \tag{109}$$

and λ is the overlap matrix of the orthogonal states

$$|\chi_\alpha\rangle = X^i_\alpha |\psi_i\rangle \tag{110}$$

$$\langle\chi_\alpha|\chi_\beta\rangle = \delta_{\alpha\beta}\lambda_\alpha \tag{111}$$

Since $\det M_N = 0$, at least one of the λ_α's, say λ_h, is zero. It then follows from (111) that the corresponding $|\chi_h\rangle$, given by (110) where X_h is the solution to

$$M_N \cdot X_h = 0 \tag{112}$$

is a null state, since it has zero norm and vanishing overlaps with all the other level N states, and therefore has vanishing overlaps with all states in V_h. This means V_h is degenerate and owing to its isomorphism with the conformal family $[\phi_h]$, the latter is also degenerate. This has a significant consequence in the computability of correlation functions, a subject to which we will return in the next section.

9. CORRELATION FUNCTIONS

For the discussion in this section, we shall use the simplified notations $\phi_i \equiv \phi_{h_i}$ for primary field, and $\langle \ \rangle = \langle 0| \quad |0\rangle$ for vacuum expectation values. Consider the correlation function of primary fields

$$\langle\phi_1(z_1)\phi_2(z_2)\cdots\phi_n(z_n)\rangle \equiv \langle\Phi(z_i)\rangle \tag{113}$$

Then from (94) and that L_s, $s = 0,\pm1$ annihilate the vacuum in the infinite past and infinite future it follows that

$$\Lambda_s\langle\Phi\rangle = 0 \qquad s = 0,\ \pm 1 \tag{114a}$$

$$\Lambda_{-1} = \sum_i^n \partial_i \tag{114b}$$

$$\Lambda_0 = \sum_i^n (z_i\partial_i + h_i) \tag{114c}$$

$$\Lambda_{+1} = \sum_i^n (z_i^2\partial_i + 2z_ih_i) \tag{114d}$$

Recall L_s are the generators of the infinitesimal projective transformation (see (70)), which is the only conformal transformation from $\mathbb{C}^1$ to $\mathbb{C}^1$. The set of equations (114) is a reflexion of this uniqueness; $\langle\Phi\rangle$ does not satisfy any other differential equation not restricted to specific values of c and h. On the other hand, since the derivation of (114) relies on the commutation relations between ϕ_i and only L_s but not the other generators, it is clear that (114) will still hold if the ϕ_i's are replaced by fields that are not primary but satisfy (94) at least for $m = 0, \pm1$. BPZ called such fields quasiprimary.

It is readily recognized that (114b) and (114c) are consequences of translation and scale invariances, respectively. The three relations are sufficient to determine all two-point and three-point correlation functions to within a proportional constant[10]:

$$\langle\phi_1(z_1)\phi_2(z_2)\rangle \sim \delta_{h_1,h_2}(z_1-z_2)^{-2h_1} \tag{115}$$

$$\langle\phi_1(z_1)\phi_2(z_2)\phi_3(z_3)\rangle \sim \prod_{i,j,k\ \text{cyclic}} (z_i-z_j)^{-h+2h_k}; \qquad h = h_1+h_2+h_3 \tag{116}$$

In the above, as has been our practice, it is understood that ϕ_i also depends on $\bar{z}_i$ and that factors depending on $\bar{z}_i$, $\bar{h}_i$, etc. similar to those shown on the rhs have been suppressed.

The four-point function is restricted by (114) to have the form

$$\langle \prod_{i=1}^{4} \phi_i(z_i)\rangle = \Big[\prod_{i<j} z_{ij}^{h_i+h_j}\Big](z_{13}z_{24})^a(z_{14}z_{23})^b\, f\Big(\frac{z_{12}z_{34}}{z_{13}z_{24}}\Big) \tag{117}$$

where $z_{ij} = z_i - z_j$; a,b are arbitrary constants; f is an undetermined function.

Although conformal invariance does not lead to further general differential equations for the primary correlation function (113), by (93) it reduces all correlation functions to the form of partial differential operators operating on primary correlation functions that satisfy (0). For example, one can show (with the help of (114b)) that

$$\langle\phi_o^{(-k_1,\cdots,-k_m)}(\zeta)\Phi(z_i)\rangle = \Big(\prod_{\ell=1}^{n} \mathcal{L}_{-k_\ell}(\zeta)\Big)\langle\phi_o(\zeta)\Phi(z_i)\rangle \tag{118}$$

where

$$\mathcal{L}_{-k}(\zeta) = \sum_{i=1}^{n}\Big[\frac{(k-1)h_i}{(z_i-\zeta)^k} - \frac{1}{(z_i-\zeta)^{k-1}}\partial_i\Big] \tag{119}$$

and $\mathcal{L}_{-k_i}$ is to the right of $\mathcal{L}_{-k_j}$ if $j > i$. The general form of the reduction is

$$\langle \prod_{i=1}^{n} \phi_i^{(-\{k\}_i)} \rangle = \Delta(z_1,\cdots,z_n) \langle \Phi(z_i) \rangle \tag{120}$$

where Δ is a known differential operator.

To make further progress, recall it was pointed out in the last section that there exists null states $|\chi_h\rangle$ in all the degenerate models of FQS. Owing to the isomorphism between the Verma modulus and space of conformal fields, the corresponding field $\chi_h(z)$ must be a null field, which means that

$$\langle \chi_h(z)\ \Phi(z_i) \rangle = 0 \tag{121}$$

for any Φ. Since the null field is a linear combination of secondary fields (with coefficients X_h^i given by (112)), (120) in turn implies

$$\Delta(z,z_1\cdots)\langle \phi_h(z)\Phi(z_i)\rangle = 0 \tag{122}$$

where the differential operator is known. BPZ has shown that, expressed in the FQS classification, all unitary models with $m \geq 3$ have at least two level-2 null fields

$$\chi_{2,h} = \phi_h^{(-2)} - \frac{3}{2(2h+1)}\,\phi_h^{(-1,-1)}, \qquad h = h_{(2,1)} \text{ or } h_{(1,2)} \tag{123}$$

and all models with $m \geq 4$ have at least two level-3 null fields

$$\chi_{3,h} = \phi^{(-3)} - \frac{2}{h+2}\,\phi_h^{(-1,-2)} + \frac{1}{(h+1)(h+2)}\,\phi_h^{(-1,-1,-1)}, \quad h=h_{(3,1)} \text{ or } h_{(1,3)} \tag{124}$$

These are obtained from solving (112) for M_2 and M_3 given respectively in (102) and (103). The corresponding differential equations for the correlation functions are, respectively,

$$\left[\mathcal{L}_{-2}(\zeta) - \frac{3}{2(2h+1)}\,\partial_\zeta^2\right] \langle \phi_k(\zeta)\Phi(z_i)\rangle = 0 \tag{125}$$

$$\left[\mathcal{L}_{-3}(\zeta) - \frac{2}{h+2}\,\mathcal{L}_{-2}(\zeta)\partial_\zeta + \frac{1}{(h+1)(h+2)}\,\partial_\zeta^3\right]\langle \phi_h(\zeta)\Phi(z_i)\rangle = 0 \tag{126}$$

Being functions of ∂_i, the $\mathcal{L}_{-k}$'s are partial differential operators. However, when Φ is a product of three primary functions, the three ∂_i's can be solved by (114) in terms of ∂_ζ. The result is

$$\partial_i = -[(z_i-z_j)(z_i-z_k)]^{-1}\{(\zeta-z_j)(\zeta-z_k)\partial_\zeta$$
$$+ h_i(2z_i-z_j-z_k) + (h_j-h_k)(z_j-z_k) + h(2\zeta-z_j-z_k)\} \tag{127}$$

with i,j,k running over 1,2,3 cyclicly. Substituting this into (125) (or (126)) then turns the latter into a simple second (third) order ordinary differential equation for a four-point correlation function with ϕ_h being the primary field for $h = h_{(2,1)}$ or $h_{(1,2)}$ and $m \geq 3$ ($h = h_{(3,1)}$ or $h_{(3,1)}$ and $m \geq 4$) and Φ a product of any three primary functions.

BPZ has also shown that (125) and (126) lead to selection rules for scaling dimensions in three-point correlation functions. For a more recent development in this topic see Dotsenko and Fateev[16].

10. SOME PHYSICS

Here we give two examples to demonstrate the universal properties of two dimensional systems at critical point resulting from conformal invariance. In the first example[3] it is shown the width of an infinite strip in units of the correlation length is equal to the scaling dimension times a universal constant. In the second example[17,18] it is shown that the free energy density of an infinite strip has a universal edge effect proportional to the conformal anomaly.

For the first case we start with the two-point correlation function (115) and consider the transformation

$$z \to w = \frac{L}{2\pi} \ln z = \tau + i\sigma \tag{128}$$

which transforms the infinite z-plane to a semi-infinite strip of width L in the w-plane, with the two edges $\sigma = 0$ and $\sigma = L$ identified. The strip can be thought of as the surface of a cylinder. It follows from (115) and (0) (it is assumed that the fields are primary) that on the strip

$$\langle\phi(w_1)\phi(w_2)\rangle_{strip} \quad [e^{2\pi r/L} + e^{-2\pi r/L} - 2\cos(\frac{2\pi}{L}(\sigma_1-\sigma_2))]^{-h} \tag{129}$$

where $r = |\tau_1 - \tau_2|$. Average across the strip and let $r >> L$, then

$$g(r) \equiv \lim_{r/L\to\infty} \frac{1}{L} \int_0^L d\sigma_1 \langle\phi(w_1)\phi(w_2)\rangle_{strip} \quad e^{-2\pi hr/L} \tag{130}$$

This shows that, unlike the original correlation function, the one-dimensional $g(r)$ is no longer singular, but decays with a correlation length

$$\xi_{strip} = L/2\pi h \qquad \text{(periodic edges)} \tag{131}$$

Thus, for a strip with periodic edges, the ratio L/ξ is the scaling dimension times the universal constant 2π.

As a variant of this result, consider the transformation

$$z \to w = \frac{L}{\pi} \ell nz = \tau + i\sigma \tag{132}$$

which transforms the half z-plane to a semi-infinite strip with free edges. The result is then

$$L/\xi_{strip} = \pi h \qquad \text{(free edges)} . \tag{133}$$

Next consider the free energy of a system with partition function Z

$$F = -\ell nZ \tag{134}$$

Owing to the conformal anomaly c, the free energy is not invariant under a coordinate change

$$z \to w(z) \tag{135}$$

even as the theory is conformally invariant. In two dimensions, the change in F induced by (135) is given by the formula due to Polyakov[19]

$$\delta F = -\frac{c}{48\pi} \int d^2z \left[\frac{1}{2} (\partial_a \ell n\rho)^2 + \mu^2(\rho-1)\right] \tag{136}$$

where $\rho \equiv |dw/dz|^2$ and μ^2 is a constant depending on the system. Note that $\delta F = 0$ in the absence of coordinate change, when $\rho = 1$. Based on (136), Blöte et al.[17] and Affleck[18] independently derived a remarkable result, showing that c can in principle be measured, as follows.

Let the coordinate change be

$$z \to w = e^{-2\pi z/\ell} \tag{137}$$

which transforms a strip of width ℓ and length L to a unit disc with a hole of radius $e^{-2\pi L/\ell}$. Then from (136)

$$\delta F = F_{strip} - F_{disc} = -\frac{c}{48\pi}\left[\frac{8\pi^2 L}{\ell} - \mu^2 \ell L + \ (\ell)\right] \tag{138}$$

Now keep ℓ fixed and send L to ∞. The disc then becomes a unit disc without a hole. Clearly F_{dics} must remain finite but F_{strip} need not. Therefore

$$\lim_{L\to\infty} \frac{F_{strip}}{\ell L} = \text{constant} - \frac{c\pi}{6\ell^2} \tag{139}$$

The constant term depends on the system, but the second term, which gives the finite size effect, is universal and is proportional to the conformal anomaly.

Conformal field theory has given us astounding new insights in the critical behavior of two dimensional systems, but in ways that often seem mysterious. For example, why are the allowed scaling dimensions for $c < 1$ limited to roots of the Kac determinant? What is the meaning of the demarcation between the $c \geq 1$ continuum and the $c < 1$ discrete spectrum? Are there ways to detect the degeneracy of the Verma modulus other than by the unitarity requirement? No doubt these and other issues will eventually be clarified.

Finally, conformal field theory in two dimensions has been extended to include supersymmetry[20] and some of the extended models have been identified with supersymmetric critical systems[21]. There is also a very extensive literature on constructing concrete two dimensional theories[22] and on using the theory as a basis for the theory of strings[23]. More recently efforts have been initiated to derive fundamental principles of quantum field theories from the study of all conformally invariant theories in two dimensions[24].

References

1. A.A. Belavin, A.M. Polyakov and A.B. Zamolodchikov, J. Stat. Phys. 34(1984)763; Nucl. Phys. B241(1984)333.
2. D. Friedan, Z. Qiu and S. Shenker, in "Vertex Operators in Mathematics and Physics", J. Lepowsky et al., eds. (Springer, Berlin, 1985) p.419; Phys. Rev. Lett. 52(1984)1574.
3. J.L. Cardy, J. Phys. A17(1984)L385.
4. J.L. Cardy, in "Phase Transitions and Critical Phenomena", Vol. II, C. Domb and J.L. Leibowitz, eds., (Academic, New York, 1987) p.55.
5. D. Friedan, E. Martinec and S. Shenker, Nucl. B271(1986)93.
6. Y. Saint-Aubin, Phenomènes critiques en deux dimensions et invariance conforme, Centre Res. Math. report 1472 (Montreal 1987).
7. L.P. Kadanoff, in "Phase Transitions and Critical Phenomena", Vol. 5a, C. C. Domb & M.S. Green, eds., (Academic, New York, 1976) p.1.
8. L.P. Kadanoff, Physics 2(1966)263.
9. K.G. Wilson, Phys. Rev. B4(1971)3174;3184.
10. A.M. Polyakov, Sov. Phys. JETP Lett. 12(1970)381.
11. M.A. Virasoro, Phys. Rev. D1(1970)2933.
12. V.G. Kac, in "Group Theoretical Methods in Physics", W. Beiglbock and A. Bohm, eds. (Springer, New York, 1979) p.441; B.L. Feigin and D.B. Fuchs, Funct. Anal. Appl. 16(1982)114.
13. P. Goddard and D. Olive, Nucl. Phys. B257(1985)226; P. Goddard, A. Kent and D. Olive, Phys. Lett. 152(1985)88.
14. J.L. Cardy, Nucl. Phys. B270(1986)186.
15. D.A. Huse, Phys. Rev. B30(1984)3908.
16. Vl. S. Dotsenko and V.A. Fateev, Nucl. Phys. B240(1984)312; Nucl. Phys. B251(1985)691.
17. H.W.J. Blöte, J.L. Cardy and M.P. Nightingale, Phys. Rev. Lett. 56(1986)742.
18. I. Affleck, Phys. Rev. Lett. 56(1986)746.
19. A.M. Polyakov, Phys. Lett. 103B(1981)207.
20. Refs. 2 and 5, and M.A. Bershadsky, V.G. Kniznik and M.G. Teitelman, Phys. Lett. 151B(1985)31.
21. Z. Qiu, Nucl. Phys. B270(1986)205.
22. C. Itzykson and J.-B. Zuber, Nucl. Phys. B275(1986)580; I. Affleck and F.D.M. Haldone, preprint PUPT-1052 (1987).
23. Ref. 5, E. Martinec, Nucl. Phys. B281(1987)157; B. Sakita, contribution to these Proceedings.
24. D. Friedan and S. Shenker, Nucl. Phys. B281(1987)509.

CAP/NSERC Summer Institute in Theoretical Physics
July 10-24, 1987

List of participants

I. Affleck	Univ. of British Columbia
N. Arimitsu	Yokahama National University
T. Arimitsu	University of Tsukuba
P. Asthana	University of Alberta
C. Attanasio	University of Alberta
S. Bell	University of Manitoba
L. Benoit	Université de Montréal
W. Bentz	University of Tokyo
J. Biel	University of Calgary
M. Boyce	University of Alberta
W.J.L. Buyers	Chalk River Nuclear Lab.
W. Brouwer	University of Alberta
E. Calzetta	University of Maryland
B.A. Campbell	University of Alberta
A.Z. Capri	University of Alberta
J.P. Carbotte	McMaster University
S. Chugh	University of Alberta
P. Coleman	Rutgers University
M. Couture	Chalk River Nuclear Lab.
G. Crabtree	Argonne National Lab.
E.A. de Kerf	I.T.P., Amsterdam
R. de Luca	U. of Southern California

J. Dixon	University of Calgary
S. Durand	Université de Montréal
R. Dymarz	University of Alberta
D. Evens	University of Winnipeg
Hong-Yi Fan	Univ. of New Brunswick
R. Floreanini	M.I.T.
S. Fujiki	Dalhousie University
R. Fukuda	Keio University
V. Galina	Simon Fraser University
S.J. Gates, Jr.	University of Maryland
H.F. Gautrin	Université de Montréal
J.D. Gegenberg	Univ. of New Brunswick
Z.W. Gortel	University of Alberta
G. Gruebl	University of Innsbruck
S. Hama	University of Alberta
I. Hardman	University of Alberta
A.P. Harrington	University of Oxford
G. Hayward	University of Alberta
M. Harvey	Chalk River Nuclear Lab.
J. Hebron	University of Alberta
K. Heiderich	Univ. of British Columbia
B.L. Hu	University of Maryland
C.J. Isham	Imperial College
K. Isler	ETH, Switzerland
M. Jacques	Université de Montréal
F. Jetzer	Université de Montréal

J. Johansson	University of Alberta
A.N. Kamal	University of Alberta
R. Kantowski	University of Oklahoma
T.-D. Kieu	University of Edinburgh
F.C. Khanna	University of Alberta
D.Y. Kim	University of Regina
J.R. Klauder	AT&T Bell Laboratories
R. Kobes	Memorial U. of Newfoundland
G. Kunstatter	University of Winnipeg
J.M. Laget	CEN-Saclay
N.P. Landsman	I.T.P., Amsterdam
I.D. Lawrie	Univ. of British Columbia
Y. Leblanc	M.I.T.
A. Leclair	University of Alberta
H.C. Lee	Chalk River Nuclear Lab.
M. Legaré	Université de Montréal
J. Le Tourneux	Université de Montréal
H.-P. Leivo	University of Toronto
J.M. Lina	Université de Montréal
J. Lopuszanski	University of Wroclaw
A.J. MacFarlane	University of Cambridge
R. MacKenzie	University of Cambridge
K. Maki	U. of Southern California
R.B. Mann	University of Waterloo
L. Marleau	Université Laval
T. Matsuki	Simon Fraser University

M. Mayrand	Université de Montréal
D. McManus	University of Alberta
H. Minn	Seoul National University
M. Mitchard	University of Cambridge
N. Mobed	University of Alberta
J. W. Moffat	University of Toronto
D.C. Morgan	Univ. of British Columbia
A. Munoz-Sudupe	Lawrence Berkeley Lab.
K. Nakamura	Meiji University
Y. Nakawaki	University of Alberta
W. Newman	U.C.L.A.
K. Ng	University of Alberta
I. Ojima	Kyoto University
M. Otwinowski	University of Calgary
N. Papastamatiou	U. of Wisconsin, Milwaukee
B.V. Paranjape	University of Alberta
M. Paranjape	Université de Montréal
J. Pasupathy	Indian Institute of Science
D. Pattarini	University of Sussex
K. Peterson	University of Alberta
A. Pimpale	University of Alberta
A. Rebhan	Tech. Univ. Wien
M. Razavy	University of Alberta
M. Rho	CEN-Saclay
W. Rozmus	University of Alberta
P. Sahay	University of Alberta

Y. Saint-Aubin	Université de Montréal
B. Sakita	City Univ. of New York
M.N. Sanielevici	M.I.T.
P.U. Sauer	Univ. Hannover
H. Schiff	University of Alberta
G. Semenoff	Univ. of British Columbia
J.J.A. Shaw	University of Cambridge
H.S. Sherif	University of Alberta
N. Sinha	University of Alberta
R. Sinha	University of Alberta
R. Skinner	University of Saskatchewan
S.J. Smith	University of Calgary
K. Stelle	Imperial College
J. Stephenson	University of Alberta
Y. Takahashi	University of Alberta
J.G. Taylor	King's College
P.K. Townsend	University of Cambridge
G. Tsoupros	University of Alberta
P.A. Tuckey	University of Cambridge
H. Umezawa	University of Alberta
M. Vassanji	University of Toronto
D. Vincent	University of Winnipeg
L. Vinet	Université de Montréal
K.S. Viswanathan	Simon Fraser University
M. Walton	Stanford University
M.E. Wehlau	Brandeis University

G.L. Wiersma	I.T.P., Amsterdam
J.G. Williams	Brandon University
O. Wong	University of Alberta
Y.S. Wu	University of Utah
Y. Yamanaka	University of Alberta
Z. Ye	University of Alberta
X. Zhu	University of Toronto
Z. Zhu	Chalk River Nuclear Lab. and Academia Sinica, Beijing, China.